ublished by
mit Pal Jain *for*
lew India Publishing Agency
)1, Vikas Surya Plaza, CU Block, L.S.C. Mkt.,
tam Pura, New Delhi- 110 088, (India)
one: 011-27341717, Fax: 011-27341616
mail: newindiapublishingagency@gmail.com
eb: www.bookfactoryindia.com

N : 978-93-80235-25-7

posed and Designed by NIPA

MOLECULAR MARKERS AND PLANT BIOTECHNOLOGY

MOLECULAR MARKER
&
PLANT BIOTECHNOLO

by :

Rukam S. Tomar
Manoj V. Parakhia
Sunil V. Patel
B.A. Golakiya

NIPA

New India Publishing Agen
Pitam Pura, New Delhi-110 088

Preface

Scientific dogma stands on practical work pertaining its theme. In nut shell science means more of laboratory or field work supporting or not supporting the hypothesis. The book entitled "Molecular Markers and Plant Biotechnology" is for those who work in the laboratories with their own hands. It is the integration of the advancement of organic chemistry, biology and instrumentation which inspired us to put this sort of title.

This collection covers wide range of aspects right from laboratory safety rules to genotypic barcoding. It's a very good text for biological laboratories.

Conventional breeding was dependent on phenotypic markers. Evolution of molecular markers changed the principles and practices of breeding crops and animals. It also opened up new avenues of utilizing molecular markers. It can also be used not only for varietal identification but also for selection (MAS) and to identify the agroclimatic conditions in which it has grown. A branch of agricultural forensic is emerging on the foundation of molecular markers. Beside this many laboratory techniques and instrumentation described in this text will improve the skill of manpower in the biological laboratories.

Everything about the laboratory techniques is written in the respective heads. What is impossible to put into black and white is the skill in working. The laboratory skill perhaps develops through liking of on hand working. Interest and affection towards work evolves skill and dexterity in working hands and it is the ultimate aim behind this text.

Authors

Contents

Laboratory Safety Rules

The Laboratory Worker is Surrounded by many Dangers such as from......

- Handling of infectious materials.
- Handling of broken glassware.
- Accidental spill of corrosive reagents.
- Swallowing of corrosive reagents such as concentrated Sulphuric acid,
- Hydrochloric acid, sodium hydroxide, trichloroacetic acid etc.
- Swallowing of infectious specimen.
- Inhalation of poisonous fumes.
- Infection in the conduct of necropsies.
- Potential hazards in the form of inflammable chemicals and gas leakages.

Important Practical Precautions

- The use of rubber gloves while handling corrosive substances such as strong acids or alkalies and also while handling poisonous chemicals such as potassium cyanide is essential.

- The use of laboratory coats is meant to protect the wearer from chemical splashes and infectious materials. Cotton is a better material for a laboratory coat since it has a greater absorptive capacity and is generally more resistant to chemical splashes.
- Safety spectacles or goggles should be used while carrying out any procedures where there is a risk to the eyes from reagent splashes.
- All chemicals should be considered as potentially dangerous, Contact with skin and clothing should be avoided. The solutions such as concentrated Sulphuric acid, concentrated sodium hydroxide and solutions containing sodium or potassium cyanide should never be pipette out by mouth.
- A gas cylinder and gas taps should be handled carefully, Accidents can happen through ignorance and incorrect use.
- All body fluids such as blood, serum, plasma, urine, CSF etc. should be handled with great care since they may be potential source of infections.
- Laboratory workers must know the meaning of safety signs for the following types of harmful substance.

THINGS TO BE REMEMBERED WHILE WORKING IN THE LABORATORY

A) Personal Mesures

1. Behave safely and ethically at all times

Think ahead for each action what could go wrong and determine in advance how you will respond. Keep in mind the safety and health of your colleagues and the community, as well as your own, and the wildlife and environment that are downstream.

2. Do no harm—don't damage or delay your colleagues' work

Before you take anything from someone's bench, ask and put it back where you found it. Put completed PCRs in the enzyme freezer. Turn the shaker back on after checking your cultures. Turn back on power supplies when somebody else is using the same power supply. If you think you've messed something up, tell someone – they may be able to salvage things. Shut (and lock if necessary) freezers and refrigerators firmly. Treat lab strains with the utmost caution, minimizing the time they are out of the freezer. Periodically add to the collection (and database) strains from your own work that could be useful or important for later use.

3. Take responsibility for your actions and take initiative in replacing things

If you make a mess or spill something, clean it up. If you break it, fix it or get it repaired. If you use it up, replace it, refill it, order it, purchase it or make more of it. DATE everything new chemical, antibody, restriction enzyme etc, when you open it. For common solutions, new solution has to be made when the level is on or under the line drawn on the bottle. If you generate waste, dispose of it. Don't keep jamming waste into containers once they are filled within 3" of the top—fill out a Request for Disposal and call EHS for disposal. If you are using a large number of autoclaved flasks, test tubes, or tips, you should prepare these materials by yourself. Tips racked by Alice are only for everyday-type use.

4. Keep common areas immaculate and ready for the next person to use

Clean up all spills. Remove tape and marker from glassware and rinse. If you've used something up or below the line drawn on the bottle, replace it, refill it, order it, purchase it or make more of it. Don't abandon items (razor blades, gel combs, glassware, etc.) on counters. Keep bleaching cell or phage items at your own bench.

5. Label every sample or solution with contents, initials and date

This is essential in case of an emergency or if you forget to dispose of the solution when you leave the group. You may know what's inside but others need that information too. Also everything put in the cold room or warm room should be labeled. If not labeled, it can be discarded.

6. When you join the group, get the training you need

Know how to use equipment properly before you try it; see the person in charge of the equipment for training. Common early mistakes are spilling in centrifuges or autoclaves due to overfilled tubes or glassware, overdialing centrifuges or pipettors past their ranges, forgetting to turn shakers back on after checking cultures, leaving someone's PCR at room temperature, sloppy pipetting that contaminates common stocks, and using up the last of something and failing to replace it. The university also requires that you get the hazard communication site-specific training to get information about safety aspects (see Kitty) and radiation aspects (see radiation safety person).

7. When someone joins the group, help them get trained

Train new people on your permanent equipment as soon as they need to use it. Assist people in finding supplies, reagents, etc.

8. When you leave the group, don't leave behind any messes

Dispose of unneeded tubes, plates, chemicals, etc. Be sure to check all areas for your old items: hoods, freezers, environmental rooms. Anything left behind should be useful to others and clearly labeled. Preferably, inform the person who might benefit most from your material and turn your material over. Clean your bench thoroughly.

B) Communal Enzyme Stock

In the –20°C freezer, located near the gel area, a communal stock of (restriction) enzymes is present. In this freezer at least the following restriction enzymes from New England Biolabs can be found (when purchased from 1 vendor will make it easier to do double/triple digestion in one buffer) as well as the other listed enzymes. Zephyr will maintain this collection; so let her know in time when we are running out of an enzyme. High fidelity enzymes are not in this collection. When you open a new vial, label with DATE. When you use an enzyme KEEP IT COLD all the time. If you feel a specific enzyme should be added to this collection, let Zephyr know.

SAP, Klenow, Taq polymerase, T4 ligase, *BamHI, BsaI, EcoRI, EcoRV, HindIII, KpnI, NcoI, NdeI, NotI, PstI, SfiI, XbaI, XhoI.*

C) Others

1. Sinks, gels and water baths

Keep surfaces clutter-free (i.e., no razors, gel combs or abandoned glassware). Clean up counters after yourself. If you use up the supply of paper towels, soap, ethanol or bleach, replace or refill from the backup supply under the sink; if you notice that we are running out, put it on the board or get more yourself. (Towels and soap are available in ESB, ethanol and bleach in chemistry storeroom.) If you empty a squirt bottle, refill it. Store icebuckets and autoclaving bins under sinks. Help maintain the supply of sterile tips, sufficient 10X and 1X TAE, 10 mg/mL ethidium bromide solution, bromophenol blue loading dye solution, and molecular weight standards for common use. Empty biohazard bag when you fill it. Keep gel casting supplies and electrophoresis apparatus clean, organized and in good repair.

Note : Keep an eye on the aera's. Clean-up once a week and check whether there is enough EtBr, loading dye, molecular weight standard and TAE. Maintain supply of ethanol, bleach, soap and towels. Make new 70% ethanol when necessary. Refill water baths every 2-3 days.

2. Balance and pH meter

Clean balance and surrounding area after each use. Don't abandon items. Return chemicals to the proper storage place. Clean pH meter area after use. Put pH meter tip back in storage buffer. Close NaOH and HCl bottles tight (otherwise the pH will change). Empty waste beaker and refill wash bottle as needed.

Note: Clean balance area every 1-2 days, keeping area decluttered and surface clean. Maintain a supply of large and small weighing boats and paper and make sure there are clean spatulas. Keep a supply of pH 4, 7 and 10 buffers for calibration, concentrated and 1 M NaOH and HCl for adjusting pH, and water wash bottle. Dispose of unneeded buffers or waste acids or bases (adjust to pH 7 first)(ie, if you want to neutralize a strong base like 10M NaOH, use about a 10x less strong acid and start with adding drop wise to prevent extreme heat development). Keep area around pH meter clean and clutter-free. Calibrate pH meter weekly.

3. UV/VIS spectrophotometer and ELISA plate reader

Keep area clean and clutter-free. After use of the spectrophotometer, dispose of the biohazard liquid in the bleach-containing flask. Put the cuvette in the biohazard waste bag (replace bag when full). Other hazardous material should be discarded according to the regulations. All non-hazardous liquid waste should be discarded in the sink. Clean up any spills. Turn lamps off when not in use. Leave note if you're planning to return later. After use of one of the ELISA-plate readers, take you plate(s) and turn the reader off.

Note: Wash counters weekly. Provide a box of 1000 uL tips, dedicated P1000 pipettor, box of Kimwipes, wash bottle of ddH2O, a small biohazard bag with stand and flask with bottom of bleach (minimal 10% of total volume). Empty biohazard bag as needed. Clean quartz cuvettes and store nearby to the spec. Make sure sufficient supplies of plastic cuvettes are on hand.

4. Distributing

All delivered items will be placed onto the metal shelf together with the invoice slips. All cold items will be dealt with immediately by people. (Please be on the lookout for any cold items you may have ordered). Open incoming boxes, distribute to the appropriate person or shelf, store at 4°C or –20°C if necessary. Any dry ice in the packaging should be placed into the dry ice freezer next to the main –80°C freezer. Check receiving reports against package contents and notify Central Receiving of any shorted or damaged items (if there are shortages or damage, work with the person that ordered the item(s) to correct it). Cut open cold packs and drain to sink. Any packaging (peanuts, bubble wrap etc) should be discarded and the empty box should be placed in the hallway. If the vendor re-accepts used packaging, tape up (usually Styrofoam) packing cases and place these packages on the shelf. These packages should be taken down to the MBB mailroom at least once a week. Check MBB and CPE mailboxes when you are in the mailroom, and distribute mail to lab members. Return invoices weekly to Eddie Ibarra (or Ellington lab for MURI/ Beckman grant items only).

5. Warm/cold/dark/autoclave rooms

Check these rooms every 1-2 days for working equipment. Discard any unlabeled flasks or plates (name/initials and date are required) at the end of each day. Handle usage problems such as abandoned biohazard bags, sloppy treatment of equipment or space, etc. Maintain a supply of razor blades, paper towels, gloves and ethanol in a wash bottle in the darkroom, and treated waste labels in the autoclave room. Make sure that orange gloves are available in the autoclave room drawer.

6. Pipettes and glassware

Wash pipettes as needed, place in canisters and sterilize. Sterilize washed flasks with closures and tubes with caps. Maintain a sufficient supply of these items at all times.

Dirty glass pipettes are kept in a basket inside a pipette jar containing diluted bleach. When the basket is full, pull it up, drain the liquid and move it into the washer (usually in the sink by Barrett's bench). Add 4 Alcotabs detergent tablets (available in Biology stock room, stored under the sink) to the basket. Connect the clear tubing to the hot water faucet and place the other end into the washer.

IMPORTANT: Secure the tubing with a clip to the top of the basket (otherwise the tubing can pull loose and flood the bench). Turn on hot water, adjust the flow rate to low (~2L/min) and run water at least 3 hr. Turn off hot water, switch the tubing to deionised water (prop the white faucet knob open with a bottle cap) and rinse the pipettes at least 3 hr. Drain pipettes in basket at least 24 hr. Sort pipettes into 25, 10 and 5 mL, place in canisters, autoclave and store on shelf.

7. Electrophoresis/buffers

The following buffers are for common use, use them properly. Running buffer (SDS-Glycine), destain, stain, anode-buffer I, anode-buffer II, and cathode-buffer at all times. Make new buffer/stock when after your usage the level dropped on or below the line drawn on the bottle.

8. Waste disposal

Autoclave biohazard waste in your aisle and common areas when necessary (don't overfill the waste bin), making sure not to leave full cans over the weekends. Call EH&S for disposal when solid/liquid ethidium bromide, coomassie, solid chemical waste, non/halogenated organic solvents, radioactive waste, or sharps containers are filled. Check fume hoods and biological safety cabinets for waste chemicals and dispose as necessary. Arrange for recycling pickups or dropoffs of styrofoam, plastics and aluminum cans.

9. Microbiological waste

Collect solid microbiological waste in the red bags. Liquid waste should NEVER go into the red bags, therefore empty cuvettes, pipet tips, ELISA plates (only in red bag when biohazard material is used) etc. Close the bag and autoclave it with program GRAV125. Make an entry in the log, noting temperature, pressure and time autoclaved. Attach sticker, discard in trash and rinse out spills from the waste bin. (Stickers can be found in the left drawer in the autoclave room. Check with Ron or EH&S for more.)

Each person is responsible for treating the liquid microbiological waste they generate. Add household chlorine bleach (don't dilute further then 10 times) and let it stand for at least 4 hours. Keep it on your own bench, not in the sink.

Ethidium Bromide Waste

SOLID – A specific trashcan for ethidium bromide (EtBr) waste is in the darkroom. EtBr containing gels can be discarded here.

LIQUID – There are 5 gallon plastic containers located under the gel benches. These containers are labeled "Liquid EtBr Waste". When changing the TAE buffer in the gel rigs, the old buffer should be discarded in these containers. Any other liquid EtBr waste you may generate may also be placed here. The containers should NOT be filled more than 3 inches from the top.

Coomassie Stain/destain

Use the can for organic chemicals coupled to the phast-gel system. Do NOT fill this container more than 3 inches from the top.

Solid Chemical Waste

Request a container for solid chemical waste. Record the chemical name and mass (g) on an attached sheet of paper or EH&S tag.

Non-halogenated Organic Solvents

Dispose in the 4L bottle marked "non-halogen organic waste" in hood. Organic solvents not containing a halogen should be places in this bottle. REMEMBER to record the volume of each solvent (mL) on the sheet taped to the hood.

Halogenated Organic Solvents

Dispose in the 4L bottle marked "halogen organic waste" in the hood. ANY solvent mixture containing even a single halogenated solvent should be disposed of in this container. REMEMBER to record the volume of each solvent (mL) on the sheet taped to the hood.

Radioactive Waste

Before you can start working with radioactive compounds, mark the area with radiation tape. Discuss planed experiments with radiation safety person before initiating experiments. Make sure nobody can be exposed to radiation. Make sure that you do not contaminate the surrounding area, but if you do, clean and screen. If you are not sure how, seek help! Solid waste goes in the special trashcan. Some liquid waste can go down the sink (make sure you know what and how much). Other liquid waste should be discarded in a 1 to 4 L bottle that is clearly labeled as radioactive. Percentages of each and every component in the liquid should be written on a sheet of paper and attached to the bottle. Wipe test must be done and documented before, during and after experiment. Once after receipt, store radioactive material in locked –80°C freezer. Document exactly every use & how much get in (forms come with order) (i.e. on 5/20/06, used 30uCi, 15uCi in solid waste, 15uCi in liquid waste). All documentation needs to be kept in orange radioactive folder until further notice.

Glass

Glass is disposed in the special "glass only" boxes. However, do NOT put the following items into the boxes:

- Intact, empty chemical containers (rinse 3X, write "EMPTY" across label, place in hall)
- Glassware contaminated w/ potentially infectious material (place in sharps container)
- Empty chemical containers that held an acutely hazardous chemical (dispose of as chemical waste, following EH&S procedures)
- Razor blades

Containers should not be filled over the top of the box. Place the lid on full boxes and fill out the label on the lid (request a new box). Place the full box in the hallway for pick-up.

Sharps

All sharps must be placed into sharp containers (located on one of the sinks in each lab). Sharps include but are not limited to the following, **regardless of contamination**:

- hypodermic needles
- hypodermic syringes with attached needles
- razor blades, disposable razors, and disposable scissors used in surgery or other medical procedures

In addition different kinds of glass material contaminated with infectious material should be considered sharps.

Do NOT include broken glass. Do not fill containers more than 3/4 full. Keep contents of sharps containers as free of chemicals as possible. To avoid accidental sticks, place needles directly into containers and **do NOT resheath**, bend, break, clip, or remove needles from disposable syringes. Do not dispose of sharps containers with the regular trash. Submit a request for disposal form for biological wastes and sharps to EH&S. Indicate when new containers are needed.

Empty bottles Intact, empty chemical containers that did not hold an acutely hazardous material can be placed in the hallway after rinsing the bottle 3X and writing "EMPTY" across the label.

Empty boxes can be placed in the hallway for pick-up.

The laboratory worker should take following precautions regarding fire and explosions

- Open flames should not be left unattended. Leakages of gas from gas cylinder or from the gas taps should be promptly reported to the superior.
- Smoking should be strictly prohibited in the laboratory.
- Burning matchsticks should not be thrown in the wastebasket.
- The laboratory workers should know the location of following requirement:

- A charged fire Extinguisher.
- Bucket of sand with scoop and
- Fire Blankets.

Precautionary Measures Against Fire

- For a small blaze, water, sand and fire blanket should be used to put out the fire.
- A fire extinguisher is used for a longer blaze.
- Water should not be used on electrical fire. After putting off the main electrical switch, carbon dioxide fire extinguisher should be used.
- Water should not be used on a fire caused by organic solvents such as ether, alcohol, petrol, oil or grease. Sand should be used to smother the, fire..
- During the fire escape, it is safe to stay close to the floor and crawl by covering the mouth with a damp cloth. (This may help to filter out some of the flames and also may reduce the danger of inhaling the flames).

Laboratory First Aid

- The laboratory first aid kit should contain (1) Cotton Wool and Guaze (2) Spirit (3) Roller Bandage (4) Medicinal Adhesive Tape (5) A pair of scissors (6) Tincture iodine (1g iodine in 95% alcohol) (7) A disinfectant solution (diluted dettol) (8)Sterile saline (9) 2% (W/V) Sodium carbonate (10) 5% (W/V) Acetic acid (11) 8% (W/V) Magnesium hydroxide (Milk of Magnesium) (12) 5% (W/V) Soap Solution.

The First aid Measures given below must be Applied with Common Sense

1) Contact with corrosive chemicals and reagents

Acid splashes on the skin

- Wash the affected skin with plenty of tap water.
- Bathe the affected skin with cotton wool soaked in 5% solution of sodium carbonate.
- Rinse the affected skin in mild detergent and then contact a physician.

Alkali splashes on the skin

- Wash the affected skin with plenty of tap water.
- Bathe the affected skin with cotton wool soaked in 5% acetic acid.
- Seek medical help.

Contact with phenol

- Wash the affected skin with large volumes of the tap water.
- Use polyethylene glycol (PEG) mixed with water for further irrigation of affected skin.
- Consult a physician.

2) Eye Accidents

Burns of the eye by alkali or acid are among the most urgent ocular emergencies. Alkali burns are more disastrous than those of acid burns.

Acid or Alkali splashes in the eyes

- Immediately wash the eyes with plenty of tap water, by holding part of the eyelids manually.
- Rinse the eyes in sterile saline
- Consult an ophthalmologist immediately.

3) Accidental swallowing of poisonous reagent

- Spit it out immediately.
- Rinse the mouth promptly with tap water and then induce vomiting by drinking warm salt water. (One tablespoon of common salt in glassful of water).

4) Accidental swallowing of infectious specimen

- Spit it out immediately.
- Wash the mouth with dilute antiseptic lotion.
- Rinse the mouth thoroughly with tap water.

5) Accident swallowing of corrosive reagents

Acids

- Promptly rinse the mouth with tap water.
- Antidote such as 5% soap solution, 8% magnesium hydroxide, or white of egg mixed with about 500 ml of water can be used orally to neutralize the acid.
- Seek medical help immediately.

Alkalies

- After rinsing the mouth with tap water, antedate such as lemon juice or 5% acetic acid can be taken orally to neutralize the alkalies.
- Seek medical help immediately.

6) Contact of lip and tongue with corrosive reagents

Acids

- Immediately rinse in tap water.
- Bathe the affected part in 2% aqueous sodium Carbonate.

Alkalies

- Immediately rinse in tap water.
- Bathe the affected part in 5% acetic acid.

7) Injuries caused by broken glass

- Wash the wound immediately with a disinfectant solution (diluted dettol).
- Cover with gauze and adhesive tape.

8) Bleeding

- Make the patient to lie down.
- Try to stop bleeding by applying direct pressure to the wound with a sterilized pad and by means of a firm bandage.
- Clean the affected area by using an antiseptic such as tincture iodine and apply sterile gauze and bandage.

9) Burns and Management

Minor Burns

- Immediately bathe the affected area in cold water.
- Cover the burnt area with sterile dressing as early as possible.
- Seek medical help.

Severe Burns

- If the victim is on fire, put out the fire by using any material such as sheets, towels or coats.
- Pour cold water on the affected area.
- Remove smoldering clothes as quickly as possible.
- Call a physician, immediately.

Other Precautions to be taken while working in a Laboratory

(a) Chemicals. A number of chemicals used in the laboratory are hazardous. All manufacturers of hazardous materials are required by law to supply the user with pertinent information on any hazards associated with their chemicals. This information is supplied in the form of Material Safety Data Sheets, or MSDS. This information contains the chemical name, CAS#, health hazard data, including first aid treatment, physical data, fire and explosion hazard data, reactivity data, spill or leak procedures, and any special precautions needed when handling this chemical. In addition, MSDS information can be accessed on the Web on the Biological Sciences Home Page. You are strongly urged to make use of this information prior to using a new chemical, and certainly in the case of any accidental exposure or spill.

The following chemicals are particularly noteworthy:

Phenol—can cause severe burns

Acrylamide—potential neurotoxin

Ethidium bromide—carcinogen.

These chemicals are not harmful if used properly: always wear gloves when using potentially hazardous chemicals, and never mouth-pipette them. If you accidentally splash any of these chemicals on your skin, immediately rinse the area thoroughly with water and inform the instructor. Discard waste in appropriate containers.

(b) Ultraviolet Light. Exposure to ultraviolet (UV) light can cause acute eye irritation. Since the retina cannot detect UV light, you can have serious eye damage and not realize it until 30 minutes to 24 hours after exposure. Therefore, always wear appropriate eye protection when using UV lamps.

(c) Electricity. The voltages used for electrophoresis are sufficient to cause electrocution. Cover the buffer reservoirs during electrophoresis. Always turn off the power supply and unplug the leads before removing a gel.

(d) General Housekeeping. All common areas should be kept free of clutter and all dirty dishes, electrophoresis equipment, etc., should be dealt with appropriately. Since you have only a limited amount of space of your own, it is to your advantage to keep that area clean. Since you will use common facilities, all solutions and everything stored in an incubator, refrigerator, etc., must be labeled. In order to limit confusion, each person should use his initials or another unique designation for labeling plates, etc. Unlabeled material found in the refrigerators, incubators, or freezers may be discarded. Always mark the backs of the plates with your initials, the date, and relevant experimental data, e.g., strain numbers.

Disposal of Buffers and Chemicals

a. Any uncontaminated, solidified agar or agarose should be discarded in the trash, not in the sink, and the bottles rinsed well.

b. Any media that becomes contaminated should be promptly autoclaved before discarding it. Petri dishes and other biological waste should be discarded in biohazard containers, which will be autoclaved prior to disposal.

c. Organic reagents, e.g., phenol, should be used in a fume hood and all organic waste should be disposed of in a labeled container, not in the trash or the sink.

d. Ethidium bromide is a mutagenic substance that should be treated before disposal and handled only with gloves. Ethidium bromide should be disposed off in a labeled container.

Glossary

5' add-on mutagenesis a form of PCR mutagenesis in which the 5' end of a primer is designed to introduce into the amplification product a DNA sequence not present in the target DNA.

Allele one of several alternative forms of a gene or DNA sequence at a specific chromosomal location (*locus*). At each autosomal locus an individual possesses two alleles, one inherited from the father and one from the mother.

Allele-specific oligonucleotide (ASO) a synthetic oligonucleotide, often about 20 bases long, which hybridizes to a specific target sequence and whose hybridization can be disrupted by a single base pair mismatch under carefully controlled conditions. ASOs are often labeled and used as allele-specific hybridization probes or as allele specific primers in PCR.

Allelic association particular alleles at two or more neighboring loci show allelic association if they occur together with frequencies significantly different from those predicted from the individual allele frequencies. Often called *linkage disequilibrium.*

Allelic exclusion the mechanism whereby only one of the two immunoglobulin alleles in B lymphocytes, or T-cell receptor alleles in T lymphocytes, is expressed. More generally, any naturally occurring mechanism that causes only one allele to be expressed.

Alphoid DNA (or α-satellite DNA) a class of satellite DNA with an average repeat length of about 170 bp; found at centromeres.

Alternative splicing the natural use of different sets of splice junction sequences to produce more than one product from a single gene.

***Alu*-PCR** PCR using a primer that anneals to *Alu* repeats to amplify DNA located between two oppositely oriented *Alu* sequences. Used as a method of obtaining a fingerprint of bands from an uncharacterized human DNA.

Anchored PCR PCR where one primer anneals to an oligonucleotide linker which has been ligated onto the end of an uncharacterized DNA fragment.

Aneuploidy a chromosome constitution with one or more chromosomes extra or missing from a full euploid set.

Annealing the association of complementary DNA (or RNA) strands to form a double helix.

Anonymous DNA DNA not known to have a function.

Anticipation a phenomenon in which the age of onset of a disorder is reduced and/or the severity of the phenotype is increased in successive generations. Characteristic of *dynamic mutations*.

Anticodon the 3-base sequence in a tRNA molecule that base-pairs with the codon in mRNA.

Antisense oligonucleotide a synthetic oligonucleotide designed to be complementary to a naturally occurring mRNA molecule. Formation of a double helix can prevent expression of the mRNA

Antisense RNA a transcript from the antisense strand of a gene. Naturally occurring antisense RNAs may negatively regulate gene expression.

Antisense strand (template strand) the DNA strand of a gene which, during transcription, is used as a template by RNA polymerase for synthesis of mRNA.

Apoptosis programmed cell death.

Archaeon (pl. archaea) a single-celled prokaryote superficially resembling a bacterium, but with molecular features indicative of a third kingdom of life.

ARMS (Amplification refractory mutation system) allele-specific PCR.

Assortative mating marriage between people of similar phenotype or genotype (e.g. tall people tend to marry tall people, deaf people tend to marry deaf people; some people prefer to marry relatives). Assortative mating can produce a non-Hardy–Weinberg distribution of genotypes in a population.

Autonomously replicating sequence (ARS) in yeast, a DNA sequence having an origin of replication.

Autosome any chromosome other than the sex chromosomes, X and Y.

Autozygosity in an inbred person, homozygosity for alleles identical by descent.

Autozygosity mapping a form of genetic mapping for autosomal recessive disorders in which affected individuals are expected to have two identical disease alleles by descent.

Bacterial artificial chromosome (BAC) a recombinant plasmid in which inserts up to 300 kb long can be propagated in bacterial cells.

Barr body (sex chromatin) a blob of condensed chromatin visible at the edge of the nucleus in some interphase cells of females that represents the inactive X chromosome.

Bias of ascertainment distortions in a set of data caused by the way cases are collected – for example, severely affected people are more likely to be ascertained than mildly affected people.

Biotin–streptavidin the very high affinity of streptavidin for biotin allows biotin-labeled molecules to be isolated efficiently from a mixture.

Bivalent the four-stranded structure seen in prophase I of meiosis, comprising two synapsed homologous chromosomes.

BLAST a family of programs that search sequence databases for matches to a query sequence.

Blunt-ended a DNA fragment having no single-stranded extensions.

Boundary elements sequences that define the boundaries of coordinately regulated chromatin domains in chromosomes.

Branch site in mRNA processing, a rather poorly defined sequence (consensus CTRAY; R = purine, Y = pyrimidine) located 10–50 bases upstream of the splice acceptor, containing the adenosine at which the lariatsplicing intermediate is formed.

Bubble-linker PCR (vectorette-PCR) a form of *anchored PCR* that uses a linker containing an unpaired region (vectorette) to amplify adjacent uncharacterized sequences.

Cap a specialized chemical group that cells add to block the 5' end of mRNA

cDNA (complementary DNA) DNA synthesized by the enzyme reverse transcriptase using mRNA as a template, both experimentally or *in vivo*.

cDNA selection a hybridization-based method for retrieving genomic clones that have counterparts in a cDNA library.

CentiMorgan (cM) a unit of genetic distance equivalent to a 1% probability of recombination during meiosis.

CentiRay (cR) a mapping unit when using radiation hybrids, which is dependent on the intensity of the irradiation. A distance of $1cR_{8000}$ represents 1% frequency of breakage between two markers after exposure to a dose of 8000 rad.

Central dogma a summary of the usual information flow in organisms, from DNA to RNA to protein.

Centric fusion strictly, formation of an abnormal chromosome by fusion of two chromosome arms at the centromeres. Loosely applied to the common human Robertsonian translocations, which are in fact produced by exchange between the proximal short arms of acrocentric chromosomes.

Centromere the primary constriction of a chromosome, separating the short arm from the long arm, and the point at which spindle fibers attach to pull chromosomes apart during cell division.

Chiasma (pl. chiasmata) a visible crossover between paired homologous chromosomes in prophase I of meiosis.

Chimera an organism derived from more than one zygote.

Chimeric insert a clone insert consisting of two or more originally noncontiguous DNA sequences.

Chromatid from the end of S phase of the cell cycleuntil anaphase of cell division, chromosomes consist of two sister chromatids. Each contains a complete double helix and the two are exact copies of each other.

Chromatin fiber the 30nm coiled coil of DNA and histones that is believed to be the basic conformation of chromatin

Chromosome painting fluorescence labeling of a whole chromosome by a *FISH* procedure in which the probe is a cocktail of many different DNA sequences from a single chromosome.

Chromosome walking a method of isolating sequences adjacent on the chromosome to a characterized clone.

***Cis*-acting** *cis*-acting regulatory sequences control the activity of a gene only when it is part of the same DNA molecule or chromosome.

Clone fingerprinting involves generating a pattern of fragments from a clone by restriction digestion or *Alu-PCR*.

Coding DNA DNA that encodes the amino acid sequence of a polypeptide, or a functional mature RNA which does not specify a polypeptide.

Codon a nucleotide triplet (strictly in mRNA, but by extension, in genomic coding DNA) that specifies an amino acid or a translation stop signal.

Coefficient of selection (s) s = 1 - f, where f is the biological *fitness* of a genotype.

Comparative genome hybridization (CGH) use of competitive fluorescence *in situ* hybridization to detect chromosomal regions that are amplified or deleted, especially in tumors.

Competent bacterial cells in a competent state are able to take up high molecular weight DNA.

Complementary strands two nucleic acid strands are said to be complementary in sequence if they can form sufficient base pairs so as to generate a stable double-stranded structure.

Complementation two alleles complement if in combination they restore the wild-type phenotype. Normally, alleles complement only if they are at different loci, although some cases of interallelic complementation occur.

Concatemer multiple copies of the same sequence joined tandemly end to end.

Concerted evolution the process whereby individual members of a DNA family within one species are more closely related to each other than to members of the same type of DNA family in other species.

Conditional knock-out an engineered mutant that causes loss of gene function under some circumstances (for example, raised temperature), or in some cells, but not others

Congenic strains strains of a laboratory animal that are identical except at one specific locus.

Conservation of synteny when genetic maps of two organisms are compared, synteny is conserved if loci that are located on the same chromosome in one organism are also located together on a single chromosome in the other.

Conservative substitution a mutation causing a codon to be replaced by another codon that specifies a different amino acid, but one which is related in chemical properties to the original amino acid.

Constitutional an abnormality or a mutation of a genotype that was present in the fertilized egg, and is therefore present in all cells of a person.

Constitutive expression a state where a gene is permanently active. Mutations that result in inappropriate constitutive expression are often pathogenic.

Constitutive heterochromatin chromatin that is always heterochromatic (principally at centromeres of chromosomes).

Contig a list or diagram showing an ordered arrangement of cloned overlapping fragments that collectively contain the sequence of an originally continuous DNA strand.

Contiguous gene (segmental aneusomy) syndrome a syndrome caused by deletion of a contiguous set of genes, several or all of which contribute to the phenotype.

Continuous character a character like height, which everybody has, but to a differing degree – as compared to a *dichotomous character* like polydactyly, which some people have and others do not have.

Copy number the number of different copies of a particular DNA sequence in a genome.

Cosmid a vector constructed by inserting the *cos* sequences of lambda phage into a plasmid. Permits cloning of inserts 30–46kb long.

Cot-1 DNA a fraction of DNA consisting largely of highly repetitive sequences. Obtained from total genomic DNA by selecting for rapidly reassociating DNA sequences during renaturation of DNA.

CpG dinucleotide the sequence 5' CG 3' within a longer DNA molecule. *CpG dinucleotides* are targets of a specific DNA methylation system in mammals that is important in control of gene expression.

CpG island short stretch of DNA, often <1kb, containing frequent unmethylated *CpG dinucleotides*. CpG islands tend to mark the 5' ends of genes.

Cre-lox system A technique for generating predefined chromosomal deletions. *Cre* is a bacteriophage P1 gene whose product facilitates recombination between *lox*P sequences. So called because it creates recombination.

Cryptic splice site a sequence in pre-mRNA with some homology to a splice site. Cryptic splice sites may be used as splice sites when splicing is disturbed or after a base substitution mutation that increases the resemblance to a normal splice site.

Degenerate oligonucleotides (or primers) a panel of synthetic oligonucleotides designed so that collectively they correspond to various codon permutations for a given sequence of amino acids.

Denaturation dissociation of complementary strands to give single-stranded DNA and/ or RNA.

Dichotomous character a character like polydactyly, which some people have and others do not have – as opposed to a *continuous character* like height, which everybody has, but to differing degree.

Dideoxynucleotides (ddNTPs) synthetic nucleotides lacking 2' and 3' hydroxyl groups. They act as chain terminators during DNA replication.

Diploid having two copies of each type of chromosome; the normal constitution of most human somatic cells.

Displacement (D) loop region in mitochondrial DNA, a short triple-stranded region.

DNA chip a microarray of oligonucleotides or cDNA clones fixed on a glass surface. They are commonly used in a form of reverse hybridization assay to test for sequence variation in a known gene, or to profile gene expression in an mRNA preparation.

DNA fingerprinting a method which produces a pattern of hybridizing bands in a Southern blot that can be used to identify a person for legal or forensic purposes.

DNA ligase an enzyme able to form a phosphodiester bond between adjacent but unlinked nucleotides in a double helix.

DNA profiling using genotypes at a series of polymorphic loci to recognize a person, usually for legal or forensic purposes.

DNAse I-hypersensitive sites regions of chromatin that are rapidly digested by DNAse I. They are believed to be important long-range control sequences.

Dominant (in human genetics) describes any trait that is expressed in a heterozygote. See also *Semi-dominant.*

Dominant negative mutation (antimorph) a mutation which results in a mutant gene product that can inhibit the function of the wild-type gene product in heterozygotes.

DOP-PCR (degenerate oligonucleotide primer PCR) PCR using a mixture of closely related oligonucleotides rather than a single species as primer.

Dosage compensation any system that equalizes the amount of product produced by genes present in different numbers. In mammals, describes the X-inactivation mechanism that ensures equal amounts of X-encoded gene products in XX and XY cells.

Dot-matrix analysis a technique for comparing two DNA sequences.

Downstream in the 3' direction when a gene sequence is written in the conventional way (showing the sense strand, in the 5' to 3' direction).

Duplicative (copy or replicative) transposition transposition of a copy of a DNA sequence while the original remains in place.

Dynamic mutation an unstable expanded repeat that changes size between parent and child.

Embryonic stem cells (ES cells) undifferentiated, pluripotent cells derived from an embryo. Mouse ES cell lines are commonly used as a vehicle for transferring foreign DNA into the germline in order to generate transgenic mice.

Empiric risks risks calculated from survey data rather than from genetic theory. Genetic counseling in most nonmendelian conditions is based on empiric risks.

Enhancer a set of short sequence elements which stimulate transcription of a gene and whose function is not critically dependent on their precise position or orientation.

Epigenetic heritable (from mother cell to daughter cell, or sometimes from parent to child), but not produced by a change in DNA sequence. DNA methylation is the best understood mechanism.

Episome any DNA sequence that can exist in an autonomous extra-chromosomal form or can be integrated into the chromosomal DNA of the cell. Often used to describe self-replicating and extra-chromosomal forms of DNA.

Epitope a part of an antigen with which a particular antibody reacts.

ES cells see *Embryonic stem cells.*

EST see *Expressed sequence tag.*

Euchromatin the fraction of the nuclear genome which contains transcriptionally active DNA and which, unlike *heterochromatin*, adopts a relatively extended conformation.

Eugenics 'improving' a population by selective breeding from the 'best' types (positive eugenics) or preventing 'undesirable' types from breeding (negative eugenics).

Eukaryote an animal or plant, having cells with a membrane-bound nucleus and organelles. May be multicellular (e.g. *metazoa*) or unicellular (*protist*).

Euploidy the state of having one or more complete sets of chromosomes with none extra or missing; the opposite of aneuploidy.

Exclusion mapping genetic mapping with negative results, showing that the locus in question does not map to a particular location. Particularly useful for excluding a possible *candidate gene* without the labor of mutation screening.

Exon a segment of a gene that is represented in the mature RNA product. Individual exons may contain coding DNA and/or noncoding DNA (untranslated sequences).

Exon shuffling evolution of a gene by combining exons from other pre-existing genes.

Exon skipping alternative splicing in which splice junction sites that are normally used in RNA splicing are by-passed.

Exon trapping a technique for detecting sequences within a cloned genomic DNA that are capable of splicing to exons within a specialized vector.

Expressed pseudogene a copy of a gene that is transcribed but not functional.

Expressed sequence tag (EST) a short (typically 100–300) bp partial cDNA sequence.

Expression cloning cloning of cDNAs in specialized vectors permitting expression of a gene product from the insert.

FACS (fluorescence activated cell sorter) a machine that can separate cells, or individual chromosomes in suspensions, according to their ability to bind fluorescent dyes. Used, among other purposes, for *flow cytometry*.

Facultative heterochromatin chromatin that may exist as euchromatin or heterochromatin, depending on the state of the cell.

FISH see *Fluorescence* in situ *hybridization*.

Fitness (f) in population genetics, a measure of the success in transmitting genotypes to the next generation. Also called biological fitness or reproductive fitness. f always lies between 0 and 1.

Flow cytometry (flow sorting) fractionation of individual chromosomes according to size and base composition in a fluorescence-activated chromosome (or cell) sorter.

Fluorescence *in situ* hybridization (FISH) *in situ* hybridization using a fluorescently labeled DNA or RNA probe. A key technique in modern molecular genetics.

luorophore a fluorescent chemical group, used for labeling probes etc.

Footprinting a method of identifying sequences within a cloned DNA molecule that can specifically bind protein molecules, such as transcription factors.

Founder effect high frequency of a particular allele in a population because the population is derived from a small number of founders, one or more of whom carried that allele.

Frameshift mutation a mutation that alters the normal translational *reading frame* of a mRNA by adding or deleting a number of bases that is not a multiple of three.

Functional genomics analysis of gene function on a large scale, by conducting parallel analyses of gene expression/function for large numbers of genes, even all genes in a genome.

Fusion (hybrid) gene a gene containing coding sequence from two different genes, usually created by unequal crossover or chromosomal translocations.

Fusion protein the product of a natural or engineered *fusion gene* a single polypeptide chain containing amino acid sequences that are normally part of two or more separate polypeptides.

Gain-of-function mutation a mutation that causes inappropriate expression or function of the gene product, rather than simply loss of function.

Gene conversion a naturally occurring nonreciprocal genetic exchange in which a sequence of one DNA strand (acceptor sequence) is altered to become identical to the sequence of another DNA strand (donor sequence).

Gene dosage the number of copies of a gene. Abnormal dosage of some genes (dosage-sensitive genes) can cause developmental abnormalities.

Gene frequency the proportion of all alleles at a locus that are the allele in question. Strictly the term should be allele frequency, but the use of gene frequency is too well established now to change.

Gene pool the totality of genes, either alleles at a given locus or over all loci, in a population.

Gene targeting a form of *in vivo* mutagenesis whereby the sequence of a predetermined gene is selectively modified within an intact cell.

Gene trap a method of selecting *transgene* insertions that have occurred into a gene.

Genetic distance distance on a genetic map, defined by recombination fractions and the *mapping function*, and measured in centiMorgans.

Genetic enhancement the possible application of molecular genetic technologies to alter the human *germ line* and as a result alter the normal phenotype in some way that is considered to be beneficial.

Genetic load the relative decrease in the average fitness, compared to what it would be if all individuals had the fittest genotype, i.e. $(w_{max} - w_{av})/w_{max}$. Genetic load is made up of *mutational load* and segregational load.

Genetic redundancy performance of the same function in parallel by genes at more than one locus, so that loss of function mutations at one locus do not cause overall loss of function.

Genome the total genetic complement of an organism or virus. For cellular organisms, the set of different DNA molecules, e.g. the human genome comprises 25 different DNA molecules.

Genome-wide *p* value in testing for linkage or association, the probability on the null hypothesis of observing the statistic in question anywhere in a screen of the whole genome – see also *pointwise p value*.

Genotype the genetic constitution of an individual, either overall or at a specific locus.

Germ cells (or gametes) sperm cells and egg cells.

Germ line the germ cells and those cells which give rise to them; other cells of the body constitute the soma.

Germinal (gonadal or gonosomal) mosaic an individual who has a subset of germ line cells carrying a mutation that is not found in other germ line cells.

Haploid describing a cell (typically a gamete) which has only a single copy of each chromosome (i.e. 23 in man).

Haploinsufficiency a locus shows haploinsufficiency if producing a normal phenotype requires more gene product than the amount produced by a single copy.

Haplotype a series of alleles found at linked loci on a single chromosome.

Hardy–Weinberg distribution the simple relationship between gene frequencies and genotype frequencies that is found in a population under certain conditions.

Hemimethylation DNA methylation on only one strand of the double helix. A transient state occurring after DNA replication. Hemimethylation allows proof-reading enzymes to work out which strand is the original when they find a replication error.

Hemizygous having only one copy of a gene or DNA sequence in diploid cells. Males are hemizygous for most genes on the sex chromosomes. Deletions occurring on one autosome produce hemizygosity in males and in females.

Heritability the proportion of the causation of a character that is due to genetic causes.

Heterochromatin a chromosomal region that remains highly condensed throughout the cell cycle and shows little or no evidence of active gene expression.

Heteroduplex double-stranded DNA in which there is some mismatch between the two strands.

Heteroplasmy mosaicism, usually within a single cell, for mitochondrial DNA variants.

Heterozygous an individual is heterozygous at a locus if (s)he has two different alleles at that locus.

Heterozygote advantage the situation when somebody heterozygous for a mutation has a reproductive advantage over the normal homozygote. Sometimes called overdominance. Heterozygote advantage is the reason why several severe recessive diseases remain common.

Homologous chromosomes (homologs) the two copies of a chromosome in a diploid cell. Unlike sister chromatids, homologous chromosomes are not copies of each other; one was inherited from the father and the other from the mother.

Homologous genes (homologs) two or more genes whose sequences are significantly related because of a close evolutionary relationship, either between species (*orthologs*) or within a species (*paralogs*).

Homoplasmy of a cell or organism, having all copies of the mitochondrial DNA identical.

Homozygous an individual is homozygous at a locus if (s)he has two identical alleles at that locus. The exact meaning depends on the stringency with which identity is established. For clinical purposes a person is often described as homozygous AA or aa if they have two normally functioning or two pathogenic alleles, regardless of whether they are in fact completely identical. Homozygosity for alleles *identical by descent* is called *autozygosity*.

Hotspot a sequence associated with an abnormally high frequency of recombination or mutation.

Housekeeping gene a gene whose expression is essential for the function of most or all types of cell.

Hybrid cell panel a collection of *somatic cell hybrids* or *radiation hybrids* used for physical mapping.

Hybridization assay testing for the presence of a given sequence in a DNA or RNA sample by mixing single DNA (or RNA) strands from a known probe with those of the poorly characterized target sample, then allowing complementary strands to anneal.

Hybridization stringency the degree to which mismatches are tolerated in a *hybridization assay*. High stringency is achieved by using a high temperature and low salt concentration.

Identity by descent (IBD) alleles in an individual or in two people that are identical because they have both been inherited from the same common ancestor, as opposed to *identity by state*.

Identity by state (IBS) coincidental possession of alleles that appear identical. The alleles may or may not be truly identical.

Illegitimate (ectopic) transcription low-level transcription of a gene in a cell in which the gene is not normally expressed. Probably a near-universal phenomenon.

Imprinting determination of the expression of a gene by its parental origin.

***In situ* hybridization** hybridization of a labeled nucleic acid to a target DNA or RNA which is typically immobilized on a microscopic slide.

***In vitro* mutagenesis** introduction of a predefined mutation into a cloned sequence.

Inbreeding marrying a blood relative. The term is comparative, since ultimately everybody is related. The coefficient of inbreeding is the proportion of a person's genes that are *identical by descent*.

Informative meiosis in linkage analysis, a meiosis is informative if the genotypes in the pedigree allow us to decide whether it is recombinant or not (for a given pair of loci).

Insertional inactivation inactivation of a gene by insertion of a foreign DNA sequence within it – a transposon or a transgene,

Interference in meiosis, the tendency of one crossover to inhibit further crossing-over within the same region of the chromosomes.

Interphase all the time in the cell cycle when a cell is not dividing.

Intron noncoding DNA which separates neighboring exons in a gene. During gene expression introns are transcribed into RNA but then the intron sequences are removed from the pre-mRNA by splicing. Can be classified according to the mechanism of splicing or, if separating coding DNA sequences, their precise location within codons

Isochromosome an abnormal symmetrical chromosome, consisting of two identical arms, which are normally either the short arm or the long arm of a normal chromosome.

Isoforms / isozymes alternative forms of a protein/ enzyme.

Karyogram an image showing the chromosomes of a cell sorted in order and arranged in pairs.

Karyotype a summary of the chromosome constitution of a cell or person, such as 46, XY

Knock-in mutation a targeted mutation that replaces activity of one gene by that of an introduced gene (usually an allele).

Knock-out mutation the targeted inactivation of a gene within an intact cell.

Lagging strand in DNA replication, the strand that is synthesized as Okazaki fragments

Leading strand in DNA replication, the strand that is synthesized continuously.

Library a collection of clones.

Ligase see DNA ligase.

Ligation formation of a 3'–5' phosphodiester bond between nucleotides at the ends of two molecules (intermolecular ligation) or the two ends of the same molecule (intramolecular ligation, cyclization).

LINE (long interspersed nuclear element) a class of repetitive DNA sequences that make up about 10% of the human genome. Some are active transposable elements.

Linkage the tendency of genes or other DNA sequences at specific loci to be inherited together as a consequence of their physical proximity on a single chromosome.

Linkage disequilibrium see *allelic association*.

Linker (or adaptor oligonucleotide) a double-stranded oligonucleotide which is ligated to a DNA fragment of interest to assist cloning or PCR amplification.

Liposome a synthetic lipid membrane designed to transport a molecule of interest into a cell.

Locus a unique chromosomal location defining the position of an individual gene or DNA sequence.

Locus control region (LCR) a stretch of DNA containing regulatory elements which control the expression of genes in a gene cluster that may be located tens of kilobases away.

Locus heterogeneity determination of the same disease or phenotype by mutations at different loci. A major problem for linkage analysis.

Lod score (z) a measure of the likelihood of genetic linkage between loci. The log (base 10) of the odds that the loci are linked (with recombination fraction q) rather than unlinked. For a mendelian character, a lod score greater than +3 is evidence of linkage; one that is less than -2 is evidence against linkage.

Manifesting heterozygote a female carrier of an X-linked recessive condition who shows some clinical symptoms, presumably because of skewed X-inactivation.

Mapping function a mathematical equation describing the relation between recombination fraction and *genetic distance*. The mapping function depends on the extent to which *interference* prevents close double recombinants.

Marker a **genetic marker** is a polymorphic DNA or protein sequence deriving from a single chromosomal location, which is used in genetic mapping. A **marker chromosome** is an extra chromosome of unidentified origin.

Matrilineal inheritance transmission from just the mother, but to children of either sex; the pattern of mitochondrial inheritance.

Mean heterozygosity of a marker, the likelihood that a randomly selected person will be heterozygous. A measure of the usefulness of the marker for linkage analysis.

Metaphase the stage of cell division (mitosis or meiosis) when chromosomes are maximally contracted and lined up on the equatorial plane (metaphase plate) of a cell.

Metazoon (pl. metazoa) a multicellular animal as opposed to a unicellular protozoon.

Microsatellite small run (usually less than 0.1kb) of tandem repeats of a very simple DNA sequence, usually 1–4bp, for example $(CA)_n$. The primary tool for genetic mapping during the 1990s.

Microsatellite instability a phenomenon characteristic of certain tumor cells, where during DNA replication the repeat copy number of microsatellites is subject to random changes. Abbreviated to MIN, MSI or RER (replication error).

MIM number the catalog number for a gene or mendelian character, as listed in Victor McKusick's Mendelian Inheritance in Man, available as a book and electronically (*OMIM*).

Minisatellite DNA an intermediate size array (often 0.1–20kb long) of short tandemly repeated DNA sequences. See Table 7.11. Hypervariable minisatellite DNA is the basis of DNA fingerprinting and many *VNTR* markers.

Mismatch repair a natural enzymic process that replaces a mis-paired nucleotide in a DNA duplex (most likely present because of an error in DNA replication) to obtain perfect Watson–Crick base-pairing.

Missense mutation a nucleotide substitution that results in an amino acid change.

Modifier gene a gene whose expression can influence a phenotype resulting from mutation at another locus.

Monoallelic expression expression of only one of the two copies of a gene in a cell, because of X-inactivation, imprinting or other epigenetic change, or because of the gene rearrangements that take place with immunoglobulin and T-cell receptor genes.

Monoclonal antibody (mAb) pure antibodies with a single specificity, produced by hybridoma technology, as distinct from polyclonal antibodies that are raised by immunization.

Mosaic an individual who has two or more genetically different cell lines derived from a single zygote. The differences may be point mutations, chromosomal changes, etc.

mRNA differential display a PCR-based technique for comparing the mRNA species that are expressed in two related sources of cells to pick out differentially expressed genes.

Multigene family a set of evolutionarily related loci within a genome, at least one of which can encode a functional product.

Mutational load the loss of fitness in a population due to deleterious recessive mutations. The mutational load and the segregational load (due to loss of homozygotes where there is heterozygote advantage) are the two components of the *genetic load*. **Mutator gene** a gene with an error-checking function. When such a gene is disabled by mutation, there is a general increase in the mutation rate in a cell.

Nondisjunction failure of chromosomes (sister chromatids in mitosis or meiosis II; paired homologs in meiosis I) to separate (disjoin) at anaphase. The major cause of numerical chromosome abnormalities.

Nonhomologous recombination recombination between sequences that either have no homology or have limited local homology. A major cause of insertions and deletions, at the genetic or chromosomal level.

Nonpenetrance the situation when somebody carrying an allele that normally causes a dominant phenotype does not show that phenotype. Due to the effect of other genetic loci or of the environment. A pitfall in genetic counseling.

Nonsense mutation a mutation that occurs within a codon and changes it to a stop codon.

Northern blot a membrane bearing RNA molecules that have been size-fractionated by gel electrophoresis, used as a target for a hybridization assay.

Nucleolar organizer region (NOR) the satellite stalks of human chromosomes 13, 14, 15, 21 and 22. NORs contain arrays of ribosomal DNA genes and can be selectively stained with silver. Each NOR forms a nucleolus in telophase of cell division; the nucleoli fuse in interphase.

Nucleosome a structural unit of chromatin.

Oligogenic determined by a small number of genes acting together.

Oncogene a gene involved in control of cell proliferation which, when overactive can help to transform a normal cell into a tumor cell. Originally the word was used only for the activated forms of the gene, and the normal cellular gene was called a *proto-oncogene*, but this distinction is now widely ignored.

Open reading frame (ORF) a significantly long sequence of DNA in which there are no termination codons in at least one of the possible reading frames. Six reading frames are possible for a DNA duplex because each strand can have three reading frames.

Ortholog one of a set of homologous genes in different species (e.g. *PAX3* in humans and *Pax3* in mice).

Overdominant phenotypes showing *heterozygote advantage*. A term used in population genetics.

Palindrome a DNA sequence such as ATCGAT that reads the same when read in the 5' 3' direction on each strand. DNA–protein recognition, for example by restriction enzymes, often relies on palindromic sequences.

Paracentric inversion inversion of a chromosomal segment that does not include the centromere.

Paralog one of a set of homologous genes within a single species.

Partial digestion digestion, usually of DNA by a restriction enzyme, that is stopped before all target sequences have been cut. The object is to produce overlapping fragments.

Penetrance the frequency with which a genotype manifests itself in a given phenotype.

Pericentric inversion inversion of a chromosomal segment that includes the centromere.

Phage display an expression cloning method in which foreign genes are inserted into a phage vector and are expressed to give polypeptides that are displayed on the surface (protein coat) of the phage.

Phase of linked markers – the relation (coupling or repulsion) between alleles at two linked loci. If allele A1 is on the same physical chromosome as allele B1, they are in coupling; if they are on different parental homologs they are in repulsion. **Of the**

cell cycle – G_1, S, G_2, M and G_0 phases. **Of an intron** – a term used to classify introns in coding sequences according to the position at which they interrupt the message.

Phenotype the observable characteristics of a cell or organism, including the result of any test that is not a direct test of the *genotype*.

Physical distance distance between genes or sequences measured in kilobases, megabases or (in radiation hybrid mapping) centiRays.

Physical mapping a method of mapping genes or DNA sequences on a chromosome which does not rely on meiotic segregation (*genetic mapping*).

Point mutation a mutation causing a small alteration in the DNA sequence at a locus. The meaning is a little imprecise when being compared to chromosomal mutations, the term 'point mutation' might be used to cover quite large (but submicroscopic) changes within a single gene, whereas when mutations at a single locus are being discussed, 'point mutations' would mean the substitution, insertion or deletion of just a single nucleotide.

Pointwise *p* value in linkage analysis, the probability on the null hypothesis of exceeding the observed value of the statistic at one given position in the genome. Compare with the *genome-wide* p *value*.

Polyadenylation addition of typically 200 A residues to the 3' end of a mRNA. The poly(A) tail is important for stabilizing mRNA.

Polygenic character a character determined by the combined action of a number of genetic loci. Mathematical polygenic theory assumes there are very many loci, each with a small effect.

Polymorphism (1) strictly, the existence of two or more variants (alleles, phenotypes, sequence variants, chromosomal structure variants) at significant frequencies in the population. Looser usages among molecular geneticists include (2) any sequence variant present at a frequency >1% in a population, (3) any nonpathogenic sequence variant, regardless of frequency.

Polyploid having multiple chromosome sets as a result of a genetic event that is abnormal (e.g., constitutional or mosaic triploidy, tetraploidy, etc.) or programmed (e.g. some plants and certain human body cells are naturally polyploid).

Primary transcript the initial RNA product of transcription, before intronic sequences are spliced out.

Primer a short oligonucleotide, often 15–25 bases long, which base-pairs specifically to a target sequence to allow a polymerase to initiate synthesis of a complementary strand. Suitable primers are crucial for PCR, RT-PCR and DNA sequencing.

Primer extension a method of identifying the transcription initiation site.

Primordial germ cells cells in the fetus which give rise eventually to the sperm or egg cells.

Prior probability in Bayesian statistics, the initially estimated probability of an outcome before all relevant information has been taken into account.

Probe a known DNA or RNA fragment (or a collection of different known fragments) which is used in a *hybridization assay* to identify closely related DNA or RNA sequences within a complex, *poorly understood* mixture of nucleic acids (the *target*). In standard hybridization assays, the probe is labeled, but in reverse hybridization assays the target is labeled.

Processed pseudogene (retropseudogene) a pseudogene which lacks intronic sequences and flanking sequences of the related functional gene. Originates by reverse transcription.

Prokaryote a unicellular organism, either a bacterium or an *archaeon*, where the cell has a simple internal structure.

Promoter a combination of short sequence elements to which RNA polymerase binds in order to initiate transcription of a gene.

Proofreading an enzymic mechanism by which DNA replication errors are identified and corrected.

Protein truncation test a method of screening for chain-terminating mutations by artificially expressing a mutant allele in a coupled transcription–translation system.

Proteome the complete protein repertoire of an organism (named by analogy to the *genome*, the complete gene repertoire).

Protist any unicellular eukaryote. Can resemble an animal (protozoon), or a plant.

Proto-oncogene a cellular gene which can be converted by activating mutations into an *oncogene*. The term *oncogene* is now widely used for both the normal and activated forms of such genes.

Provirus a phage or viral genome that has integrated into the chromosomal DNA of a cell.

Pseudoautosomal region a region at the tips of the sex chromosomes marked by homology between the X and Y chromosomes which may be involved in recombination during male meiosis.

Pseudogene a DNA sequence which shows a high degree of sequence homology to a nonallelic functional gene but which is itself nonfunctional.

Quantitative trait locus (QTL) a locus important in determining the phenotype of a continuous character.

Radiation hybrid in human physical mapping, a rodent cell that contains numerous small fragments of human chromosomes. Produced by fusion with a lethally irradiated human cell. Radiation hybrid panels allow very rapid mapping of *STS*s.

Rare cutter a restriction nuclease which cuts DNA infrequently because the sequence it recognizes is large and/or contains one or more CpGs. Examples are *Not*I, *Sac*II and *BssH*II.

Reading frame during translation, the way the continuous sequence of the mRNA is read as a series of triplet codons. There are three possible reading frames for any mRNA, and the correct reading frame is set by correct recognition of the AUG initiation codon.

Reassociation kinetics The rates at which complementary DNA strands reassociate. Highly repetitive sequences reassociate rapidly; single copy sequences reassociate slowly.

Recessive a character is recessive if it is manifest only in the homozygote.

Recombinant In linkage analysis, a person who inherits from a parent a combination of alleles that is the result of a crossover during meiosis.

Recombinant DNA is DNA containing covalently linked sequences with different origins, for example a vector with an insert.

Recombination fraction for a given pair of loci, the proportion of meioses in which they are separated by recombination. Usually signified as q. q values vary between 0 and 0.5.

Renaturation (reannealing, reassociation) reformation of double helices from complementary single strands. The basis of all *hybridization assays*.

Repetitive DNA a DNA sequence that is present in many identical or similar copies in the genome. The copies can be tandemly repeated or dispersed.

Replication fork the point of bifurcation when a DNA double helix is being replicated. Two replication forks proceeding outwards from a single starting point create a replication bubble.

Replication slippage a mistake in replication of a tandemly repeated DNA sequence, that results in the newly synthesized strand having extra or missing repeat units compared to the template.

Reporter gene a gene used to test the ability of an upstream sequence joined on to it to cause its expression. Putative *cis*-acting regulatory sequences can be coupled to a reporter gene and transfected into suitable cells to study their function. Alternatively, transgenic mice or other organisms can be made with a promotorless reporter gene integrated at random into its chromosomes, so that expression of the reporter indicates the presence of an efficient promoter.

Reporter molecule a molecule whose presence is readily detected (for example, a fluorescent molecule) that is attached to a DNA sequence we wish to monitor.

Response elements sequence usually located a short distance upstream of promoters that makes gene expression responsive to some chemical in the cellular environment.

Restriction fragment length polymorphism (RFLP) a polymorphic difference in the size of allelic restriction fragments as a result of the polymorphic presence or absence of a particular restriction site. RFLPs can be assayed by *Southern blotting*.

Restriction map a diagram showing the positions of restriction sites within a DNA sequence. In short sequences these can be used as landmarks.

Restriction site a short DNA sequence, often 4–8bp long and usually *palindromic*, which is recognized by a restriction endonuclease.

Retrotransposon (retroposon) a transposable DNA element that transposes by means of an RNA intermediate. Retroposons encode a *reverse transcriptase* that acts on the RNA transcript to make a cDNA copy, which then integrates into chromosomal DNA at a different location.

Retrovirus an RNA virus with a reverse transcriptase function, enabling the RNA genome to be copied into cDNA prior to integration into the chromosomes of a host cell.

Reverse transcriptase an enzyme that can make a DNA strand using an RNA template. Used to make cDNA libraries and for RT-PCR. Reverse transcription is an essential part of the retroviral life cycle and is also encoded by some LINE-1 elements.

Riboprobe a labeled RNA probe prepared by *in vitro* transcription from a cloned DNA sequence.

Ribozyme a natural or synthetic catalytic RNA molecule.

RNA editing a natural process in which specific changes occur post-transcriptionally in the base sequence of an RNA molecule. Occurs rarely in human genes.

Robertsonian fusion a chromosomal rearrangement that converts two acrocentric chromosomes into one metacentric. Sometimes called *centric fusion*, although the point of exchange is actually in the proximal short arm, and not at the centromere.

RT-PCR (reverse transcriptase-PCR) a PCR reaction that amplifies a cDNA made by reverse transcription of a mRNA.

Satellite (1) this word has two different meanings in genetics (2) **satellites on chromosomes** are stalked projections variably present on the short arms of human acrocentric chromosomes (13, 14, 15, 21, 22). (3) **satellite DNA** originally described a DNA fraction that forms separate minor bands on density gradient centrifugation because of its unusual base composition. The DNA is composed of arrays of tandemly repeated DNA sequences.

Secondary structure regions of a single-stranded nucleic acid or protein/polypeptide molecule where chemical bonding occurs between distantly spaced nucleotides or amino acids resulting in complex structures. Secondary structure is often due to intra-strand hydrogen bonding.

Segregation ratio the proportion of offspring who inherit a given gene or character from a parent.

Semiconservative DNA replication is semiconservative because each daughter duplex contains one old and one newly synthesized strand.

Semidominant describes a mutation which, in the heterozygote, produces a phenotype intermediate (but not necessarily halfway) between the wild type and the homozygote. A term widely used in mouse genetics, but better avoided, at least in human genetics, since dominance is a property of a character and not of an allele.

Sense strand the DNA strand of a gene that is complementary in sequence to the template (antisense) strand, and identical to the transcribed RNA sequence (except that DNA contains T where RNA has U). Quoted gene sequences are always of the sense strand, in the 5'–3' direction.

Sequence homology a measure of the similarity in the sequences of two nucleic acids or two polypeptides.

Sequence tagged site (STS) any unique piece of DNA for which a specific PCR assay has been designed, so that any DNA sample can be easily tested for its presence or absence.

Sib-pair analysis a form of linkage analysis in which markers are tested for linkage to a disease or phenotypic trait by measuring the extent to which affected sib pairs share marker haplotypes.

Signal sequence (leader sequence) a sequence of about 20 aminoacids at the N-terminus of a polypeptide that controls its destination within or outside the cell.

Silencer combination of short DNA sequence elements which suppress transcription of a gene.

Silent (synonymous) mutation a mutation that changes a codon but does not alter the amino acid encoded. Such mutations may still have effects on mRNA splicing or stability.

SINE (short interspersed repetitive element) a class of moderate to highly repetitive DNA sequence families, of which the best known in humans is the *Alu* repeat family.

Sister chromatid two chromatids present within a single chromosome and joined by a centromere. Nonsister chromatids are present on different but homologous chromosomes.

Sister chromatid exchange (SCE) a recombination event involving sister chromatids. Since sister chromatids are duplicates of each other, such exchanges should have no effect unless they are *unequal*. However, an increased frequency of SCEs is evidence of DNA damage.

Site-directed mutagenesis production of a specific predetermined change in a DNA sequence. Can be done *in vitro* on cloned DNA (section 6.4) or *in vivo* by homologous recombination.

Slipped strand mispairing (replication slippage) a process in which the complementary strands of a double helix pair out of register at a tandemly repeated sequence. The resulting daughter DNA strands will have extra or missing repeat units.

SNP (single nucleotide polymorphism) any polymorphic variation at a single nucleotide. SNPs include RFLPs, but also other polymorphisms that do not alter any restriction site. Although less informative than microsatellites, SNPs are more amenable to large-scale automated scoring.

snRNA small nuclear RNAs, often part of the RNA splicing mechanism.

Somatic cell any cell in the body except the gametes.

Somatic cell hybrid an artificially constructed cell in which chromosomes have been stably introduced from cells of different species.

Southern blot transfer of DNA fragments from an electrophoretic gel to a nylon or nitrocellulose membrane (filter), in preparation for a hybridization assay.

Splice acceptor site the junction between the 3' end of an intron and the start of the next exon. Consensus sequence y_{11}nyagR (Y = pyrimidine, R = purine, upper case = exon).

Splice donor site the junction between the end of an exon and the start (5' end) of the downstream intron. Consensus sequence AGgtragt (R = purine, upper case = exon).

Spliceosome a ribonucleoprotein complex used in RNA splicing.

Splicing normally RNA splicing, in which RNA sequences transcribed from introns are excised from a primary transcript and those transcribed from exons are spliced together in the same linear order as the exons. Natural DNA splicing is a much rarer event but occurs in B and T lymphocytes.

Splicing enhancer sequences sequences that increase the probability that a nearby potential splice site will actually be used.

SSCP or SSCA single stranded conformation polymorphism or analysis, a commonly used method for point mutation screening.

Stem cell a cell which can act as a precursor to differentiated cells but which retains the capacity for self-renewal.

Sticky ends (cohesive termini) short single-stranded projections from a double-stranded DNA molecule, typically formed by digestion with certain restriction enzymes. Molecules with complementary sticky ends can associate, and can then be covalently joined using *DNA ligase* to form recombinant DNA molecules.

STS see *Sequence tagged site.*

STS content mapping checking a series of clones for the presence or absence of particular sequence tagged sites. Used as a tool in assembling clone contigs.

Suppressor tRNA a mutant transfer RNA molecule with a nucleotide substitution in the anticocon. Shows an altered coding specificity and is able to translate a nonsense (or missense) codon.

Synonymous (silent) substitution a substitution that replaces one codon with another that encodes the same amino acid.

Synteny loci are syntenic if on the same chromosome. Syntenic loci are not necessarily linked if they are sufficiently far apart on the chromosome they will not cosegregate more than by random chance.

Target DNA *i*template DNA used in a PCR reaction; (ii) DNA to which a known probe DNA is designed to hybridize in a *hybridization assay*.

Telomere a specialized structure at the tips of chromosomes. It consists of an array of short tandem repeats, $(TTAGGG)_n$ in humans, which form a closed loop and protect the chromosome end.

Template strand in transcription, the DNA strand that base-pairs with the nascent RNA transcript.

Termination codon a UAG (amber), UAA (ochre), or UGA (opal) codon in a mRNA (and by extension, in a gene) that signals the end of a polypeptide.

Tissue *in situ* hybridization a form of molecular hybridization in which the target nucleic acid is RNA within cells of tissue sections immobilized on a microscope slide.

Tissue-specific transcription factors proteins that cause a gene to be expressed only in a certain tissue.

***Trans*-acting** *trans*-acting factors (protein or maybe RNA) affect expression of all copies of a gene, whereas *cis*acting factors (normally DNA sequences) regulate only the DNA molecule of which they are part.

Transcription the synthesis of RNA on a DNA template by RNA polymerase.

Transcription unit a stretch of DNA that is naturally transcribed in a single operation to produce a single primary transcript. In straightforward cases, a transcription unit is the same thing as a gene.

Transfection uptake of DNA by eukaryotic cells (includes the exact equivalent of *transformation* in bacteria, but that word has a different meaning for eukaryotic cells). Also uptake of plasmid DNA by bacterial cells.

Transformation of a cell (1) uptake by a *competent* cell of naked high molecular weight DNA from the environment. (2) Alteration of the growth properties of a normal cell as a step towards evolving into a tumor cell.

Transgenic animal an animal in which artificially introduced foreign DNA becomes stably incorporated into the germ line. This can be accomplished by pronuclear microinjection or, in the case of transgenic mice, by the use of embryonic stem cells.

Transition G-A (purine for purine) or C-T (pyrimidine for pyrimidine) nucleotide substitution.

Translocation transfer of chromosomal regions between nonhomologous chromosomes.

Transmission disequilibrium test (TDT) a statistical test of allelic association.

Transposon a mobile genetic element

Transversion a nucleotide substitution of purine for pyrimidine or vice versa.

Triploid of a cell, having 3 copies of the genome; of an organism, being made of triploid cells.

Trisomy having three copies of a particular chromosome, e.g. trisomy

Tumor-suppressor (TS) genes genes whose normal function is to inhibit or control cell division. Tumors always have inactivating mutations in TS genes.

Two-hit hypothesis Knudson's theory that hereditary cancers require two successive mutations to affect a single cell.

Two-hybrid system see Yeast two-hybrid system.

Unequal crossover (UEC) recombination between nonallelic sequences on nonsister chromatids of homologous chromosomes.

Unequal sister chromatid exchange (UESCE) recombination between nonallelic sequences on sister chromatids of a single chromosome.

Uniparental disomy a cell or organism with normal chromosome numbers, but abnormal parental origin, in that both copies of one particular chromosome pair are derived from one parent. Depending on the chromosome involved, this may or may not be pathogenic.

Untranslated regions (5′UTR, 3′UTR) regions at the 5' end of mRNA before the AUG translation start codon, or at the 3' end after the UAG, UAA or UGA stop codon. See **Upstream** in the direction of the 5' end when a gene sequence is written in the conventional way (showing the sense strand, in the 5' to 3' direction).

Variable expression Variable extent and intensity of phenotypic signs among people with a given genotype.

Variable number of tandem repeat (VNTR) polymorphism *microsatellites, minisatellites* and megasatellites are arrays of tandemly repeated sequences that often vary between people in the number of repeat units. The term VNTR is often used to mean specifically minisatellites

Vector a nucleic acid that is able to replicate and maintain itself within a host cell and that can be used to confer similar properties on any sequence covalently linked to it.

Vectorette PCR see *Bubble-linker PCR*

VNTR see *Variable number of tandem repeat polymorphisms.*

Western blotting a process in which proteins are size-fractionated in a polyacrylamide gel, then transferred to a nitrocellulose membrane for probing with an antibody.

Whole-genome amplification a PCR method using highly degenerate primers that can amplify a very large number of random sequences spread across the genome. Can be used to allow repeated testing of DNA from a single cell, for example in typing single sperm.

X-inactivation (lyonization) the inactivation of one of the two X chromosomes in the cells of female mammals by a specialized form of genetic *imprinting*.

YAC transgenic a transgenic mouse in which the transgene is a complete *YAC*, allowing studies of the regulatory effects of sequences surrounding the gene involved.

Yeast artificial chromosome (YAC) a vector able to propagate inserts of a megabase or more in yeast cells.

Yeast two-hybrid system a yeast-based system for identifying and purifying proteins that bind to a protein of interest.

Zinc finger a polypeptide motif which is stabilized by binding a zinc atom and confers on proteins an ability to bind specifically to DNA sequences. Commonly found in transcription factors.

Zoo blot a Southern blot containing DNA samples from different species.

Zygote the fertilized egg cell.

1

Milestones in DNA History and Biotechnology

BC:

1750 The Sumerians brew beer.

500 The Chinese use moldy soybean curds as an antibiotic to treat boils.

250 The Greeks practice crop rotation to maximize soil fertility.

100 Powdered chrysanthemum is used in China as an insecticide.

AD: BEFORE THE 20TH CENTURY

1590 The microscope is invented by *Janssen*.

1663 Cells are first described by *Robert Hooke*.

1675 *Leeuwenhoek* discovers protozoa and bacteria.

1745 Maupertuis proposes an adaptationist account of organic design

1797 *Jenner* inoculates a child with a viral vaccine to protect him from smallpox.

1802 The word "biology" first appears.

1824 *Dutrochet* discovers that tissue is composed of living cells.

1830 Proteins are discovered.

1833 The cell nucleus is discovered

1839 The cell theory, or cell doctrine, states that all organisms are composed of similar units of organization, called cells. The concept was formally articulated in 1839 by *Schleiden & Schwann* and has remained as the foundation of modern biology.

1855 The *Escherichia coli* bacterium is discovered. It later becomes a major research, development and production tool for biotechnology. *Pasteur* begins working with yeast, eventually proving they are living organisms.

1859 Concepts of evolution was given by *Charles Darwin*

1863 *Mendel*, in his study of peas, discovers that traits were transmitted from parents to progeny by discrete, independent units, later called genes. His observations lay the groundwork for the field of genetics.

1865 *Gregor Mendel* laid down the rules of inheritance by distinct "factors" acting dominantly or recessively.

1866 *Mendel's* paper is published: units of inheritance in pairs; dominance and recessiveness; equal segregation; independent assortment. These ideas are not recognized for 34 years.

1869 *Miescher* discovers "nuclein" (DNA) in the cells from pus in open wounds- cells composed mostly of nuclear material. It became known as nucleic acid after 1874, when Miescher separated it into a protein and an acid molecule. The significance of DNA is not appreciated for over 70 years.

1877 A technique for staining and identifying bacteria is developed by *Koch*.

1878 The first centrifuge is developed by *Laval*.

The term "microbe" is first used.

1879 *Flemming* discovers chromatin, the rod-like structures inside the cell nucleus that later come to be called "chromosomes."

1883 *F. Galton* gave Quantitative aspects of heredity.

The first rabies vaccine is developed.

1888 The chromosome is discovered by *Waldyer*.

1889 *R. Altmann* first used the term "nucleic acid"

1892 *R. Ivanowski* first used the term "virus" introduced

1897 *Eduard Buchner*, discovered by accident that fermentation actually does not require the presence of living yeast cells. Buchner made an extract of yeast cells by grinding them and filtering off the remaining cell debris. Then he added a preservative – sugar – to the resulting cell-free solution to preserve it for future study. He observed that fermentation, the formation of alcohol from sugar, occurred. Buchner then realized that living cells were not required for carrying out metabolic processes such as fermentation. Instead, there must be some small entities capable of converting sugar to alcohol. These entities were enzymes. Buchner's accidental discovery won him the 1907 Nobel Prize in chemistry.

DURING EARLY 20TH CENTURY

1900 *H. de Vries, E.Tschermak and K. Correns,* independently recognized Mendel's discovery

Landsteiner identified AB0 blood group system

1902 Some diseases in man inherited according to Mendelian rules *(W. Bateson, A. Garrod)*

T. Boveri proved the individuality of chromosomes

W. Sutton stated Chromosomes and Mendel's factors are related

McClung identified Sex chromosomes

The term "immunology" first appears.

1906 *W. Bateson* proposed the term "genetics"

1907 The first *in vivo* culture of animal cells is reported.

1908 *Hardy, Weinberg* gave the principles of Population genetics

1909 Genes are linked with hereditary disorders.

British physician *Archibald Garrod* first proposes the relationship between genes and proteins. He hypothesizes that genes might be involved in creating the proteins that carry out the chemical reactions of metabolism.

W. Johannsen coined the terms "gene," "genotype," phenotype"

Janssens identified Chiasma formation during meiosis.

C. Little first inbred mouse strain DBA.

1910 *T. H. Morgan* begin work on *Drosophila* genetics

First *Drosophila* mutation *(white-eyed)*

1911 The first cancer-causing Sarcoma virus is discovered by *Peyton Rous*

1912 Crossing-over *(Morgan and Cattell)*
Genetic linkage *(Morgan and Lynch)*
First genetic map *(A. H. Sturtevant)*

1913 *A. Carrel* produced first cell culture

1914 *C. B. Bridges* identified Nondisjunction

Bacteria are used to treat sewage for the first time in Manchester, England.

1915 Genes located on chromosomes (chromosomal theory of inheritance) *(Morgan, Sturtevant, Muller, Bridges)*

Phages, or bacterial viruses, are discovered.

1919 The word "biotechnology" is first used by a Hungarian agricultural engineer.

1920 The human growth hormone is discovered by *Evans and Long*.

1922 *F. Blakeslee* characterized phenotypes of different trisomies in the plant *Datura stramonium.*

1924 Blood group genetics *(Bernstein)*

Fisher gave statistical analysis of genetic traits

1926 Enzymes are proteins *(J. Sumner)*

1927 *Muller* formulates the chief principles of spontaneous gene mutation as point effects of ultramicroscopic physico-chemical accidents; he induces such changes using X-rays

S.Wright gave the term Genetic drift

1928 Euchromatin/heterochromatin *(E. Heitz)*

Genetic transformation in pneumococci *(F. Griffith)*

Fleming discovers penicillin, the first antibiotic.

1933 Pedigree analysis *(Haldane, Hogben, Fisher, Lenz, Bernstein)*

Polytene chromosomes *(Heitz and Bauer, Painter)*

1935 *C. B. Bridges* gave first cytogenetic map in *Drosophila*

1937 *P. Gorer* identified mouse H2 gene locus

1938 The term "molecular biology" is coined.

1940 *E. B. Ford* explained the term Polymorphism.

Rhesus blood groups *(Landsteiner and Wiener).*

1941 *E. B. Lewis* gave the theory evolution through gene duplication.

Beadle and Tatum proved the genetic control of enzymatic biochemical reactions.

Auerbach identified mustard gas as Mutagen agent.

The term "genetic engineering" is first used by a Danish microbiologist.

1942 The electron microscope is used to identify and characterize a bacteriophage- a virus that infects bacteria.

1943 Avery demonstrates that DNA is the "transforming factor" and is the material of genes.

1944 *Oswald Avery* identifies nucleic acids as the active principle in bacterial transformation

DNA is shown to be the material substance of the gene.

1946 Genetic material can be transferred laterally between bacterial cells, as shown by *Lederberg and Tatum.* (Genetic recombination)

1947 Genetic recombination in viruses*(Delbrück and Bailey, Hershey)*

1949 Sickle cell anemia, a genetically determined molecular disease *(Neel, Pauling)*

Hemoglobin disorders prevalent in areas of malaria *(J. B. S. Haldane)*

X chromatin *(Barr and Bertram)*

1950 In DNA, there are equal amounts of A and T, and equal amounts of C and G, as shown by *Erwin Chargaff*. However, the A+T to C+G ratio can differ between organisms.

DURING 50'S

1951 *B. McClintock* discovers transposable elements, or "jumping genes," in corn.

1952 *Alfred Hershey and Martha Chase* show that on infection of the host bacterium by a virus, at least 80% of the viral DNA enters the cell and at least 80% of the viral protein remains outside.

Plasmids *(Lederberg)*

Transduction by phages *(Zinder and Lederberg)*

Cori and Cori first identified enzyme defect in man.

Mohr identified first linkage group in man.

Colchicine and hypotonic treatment in chromosomal analysis *(Hsu and Pomerat).*

J. Warkany identified exogenous factors as a cause of congenital malformations.

1953 *Watson and Crick* determine that deoxyribonucleic acid (DNA) is a double-strand helix of nucleotides. Each nucleotide consists of a deoxyribose sugar molecule to which is attached a phosphate group and one of four nitrogenous bases: two purines (adenine and guanine) and two pyrimidines (cytosine and thymine). The nucleotides are joined together by covalent bonds between the phosphate of one nucleotide and the sugar of the next, forming a phosphate-sugar backbone from which the nitrogenous bases protrude. The two strands are linked by selective hydrogen bonds: the purine adenine bonds only with the pyrimidine thymine, and the purine cytosine only with the pyrimidine guanine.

Nonmendelian inheritance *(Ephrussi).*

Howard and Pelc gave cell cycle.

Dietary treatment of phenylketonuria *(Bickel)*

1954 *Muller* gave the mechanism of DNA repair.

Leukocyte drumsticks *(Davidson and Smith)*

Polani identified cells in Turner syndrome as X-chromatin negative.

Cell-culturing techniques are developed.

1955 *F. Sanger* gave amino acid sequence of insulin.

Lysosomes *(C. de Duve)*

Buccal smear *(Moore, Barr, Marberger)*

A. Pardee and R. Litman identified 5-Bromouracil, an analogue of thymine, to be an inducer of mutations in phages.

An enzyme involved in the synthesis of a nucleic acid is isolated for the first time.

1956 46 Chromosomes in man *(Tijo and Levan, Ford and Hamerton)*

The fermentation process is perfected in Japan.

Kornberg discovers the enzyme DNA polymerase I, leading to an understanding of how DNA is replicated.

Genetic heterogeneity *(Harris, Fraser)*

1957 During a dysentery epidemic in Japan, biologists discover that some strains of bacterium are resistant to antibiotics. Later scientists will find that this resistance is transferred by olasmids.

Amino acid sequence of hemoglobin molecule *(Ingram)*

Benzer identified cistron, as the smallest nonrecombinant unit of a gene

Genetic complementation *(Fincham)*

DNA replication is semi-conservative, as shown by *Meselson and Stahl* using equilibrium density gradient centrifugation.

Genetic analysis of radiation effects in man *(Neel and Schull)*

Sickle cell anemia is shown to occur due to a change of a single amino acid.

1958 Somatic cell genetics *(Pontocorvo)*

Ribosomes *(Roberts, Dintzis)*

Human HLA antigens *(Dausset)*

Cloning of single cells *(Sanford, Puck)*

Synaptonemal complex, the area of synapse in meiosis *(Moses)*

1959 Messenger RNA is the intermediate between DNA and protein.

First chromosomal aberrations described in man: trisomy 21 *(Lejeune, Gautier, Turpin)*,

Turner syndrome: 45,XO *(Jacobs)*

Klinefelter syndrome: 47 XXY *(Ford)*

Isoenzymes *(Vesell, Markert)*

Pharmacogenetics *(Motulsky, Vogel)*

1960 Phytohemagglutinin-stimulated lymphocytecultures *(Nowell, Moorehead,Hungerford)*

Exploiting base pairing, hybrid DNA-RNA molecules are created.

Messenger RNA is discovered.

DURING 60's

1961 *Sidney Brenner and Francis Crick* establish that groups of three nucleotide bases, or codons, are used to specify individual amino acids.

Nirenberg, Mathaei, and Ochoa determined the genetic code.

X-chromosome inactivation *(M. F. Lyon)*, confirmed by *Beutler, Russell, Ohno)*

Jacob and Monod gave the gene regulation, concept of operon

Galactosemia in cell culture *(Krooth)*

Cell hybridization *(Barski, Ephrussi)*

Thalidomide embryopathy *(Lenz, McBride)*

1962 *Nowell and Hungerford* identified Philadelphia chromosome.

Xg, the first X-linked human blood group *(Mann, Race, Sanger).*

Screening for phenylketonuria *(Guthrie, Bickel).*

Edelman and Franklin made the molecular characterization of immunoglobulins.

Identification of individual human chromosomes by 3H-autoradiography *(German, Miller).*

Jacob and Brenner gave the concept of Replicon.

S. Brenner gave the term "codon" for a triplet of (sequential) bases.

1963 Lysosomal storage diseases *(C. de Duve)*

First autosomal deletion syndrome (cridu-chat syndrome) *(J. Lejeune)*

1964 Excision repair *(Setlow).*

Bach and Hirschhorn, Bain and Lowenstein gave the concept of MLC test.

Terasaki and McClelland gave the concept of Microlymphotoxicity test.

Selective cell culture medium HAT *(Littlefield)*

German, Schroeder gave the concept of Spontaneous chromosomal instability.

Cell fusion with Sendai virus *(Harris and Watkins)*

Cell culture from amniotic fluid cells *(H. P. Klinger)*

Hereditary diseases studied in cell cultures *(Danes, Bearn, Krooth, Mellman)*

Population cytogenetics *(Court Brown)*

Fetal chromosomal aberrations in spontaneous abortions *(Carr, Benirschke)*

The existence of reverse transcriptase (RT) is predicted.

1965 Limited life span of cultured fibroblasts *(Hayflick, Moorehead)*

1966 The genetic code is cracked by a number of researchers (including *Nirenberg, Matthaei, Leder, and Khorana*) using RNA homopolymer and heteropolymer experiments as well as tRNA labeling experiments.

The genetic code is deciphered when biochemical analysis reveals which codons determine which amino acids.

Catalog of Mendelian phenotypes in man *(McKusick)*

1967 The first automatic protein sequencer is perfected.

1968 HLA-D the strongest histocompatibility system *(Ceppellini, Amos)*

Repetitive DNA *(Britten and Kohne)*

Biochemical basis of the AB0 blood group substances *(Watkins)*

DNA excision repair defect in xeroderma pigmentosum *(Cleaver)*

Restriction endonucleases *(H. O. Smith, Linn and Arber, Meselson and Yuan)*

First assignment of an autosomal gene locus in man *(Donahue, McKusick)*

Khorana synthesied gene in vitro.

1969 An enzyme is synthesized in vitro for the first time.

1970 *Hamilton Smith,* at Johns Hopkins Medical School, isolates the first restriction enzyme, an enzyme that cuts DNA at a very specific nucleotide sequence. Over the next few years, several more restriction enzymes will be isolated.

D. Baltimore, H. Temin gave the concept of Reverse transcriptase, independently.

Renwick gave the new term Synteny, that refers to all gene loci on the same chromosome.

Enzyme defects in lysosomal storage diseases *(Neufeld, Dorfman)*

Individual chromosomal identification by specific banding stains *(Zech, Casperson, Lubs, Drets and Shaw, Schnedl, Evans)*

Y-chromatin *(Pearson, Bobrow, Vosa)*

Thymus transplantation for immune deficiency *(van Bekkum)*

Specific restriction nucleases are identified, opening the way for gene cloning.

RT is discovered independently in murine and avian retroviruses.

DURING 70's

1971 Two-hit theory in retinoblastoma *(A. G. Knudson)*

RT is shown to have ribonuclease H (Rnase H) activity.

1972 *Stanley Cohen and Herbert Boyer* combine their efforts to create recombinant DNA. This technology will be the beginning of the biotechnology industry.

High average heterozygosity *(Harris and Hopkinson, Lewontin)*

Association of HLA antigens and diseases

The DNA composition of humans is discovered to be 99% similar to that of chimpanzees and gorillas.

Purified RT is first used to synthesize cDNA from purified mRNA in vitro.

1973 Receptor defects in the etiology of genetic defects, genetic hyperlipidemia *(Brown, Goldstein, Motulsky)*

S. A. Latt gave demonstration of sister chromatid exchanges with BrdU.

Philadelphia chromosome as translocation *(J. D. Rowley)*

Cohen and Boyer perform the first successful recombinant DNA experiment, using bacterial genes.

1974 Chromatin structure, nucleosome *(Kornberg, Olins and Olins)*

P. C. Doherty and R. M. Zinkernagel gave the concept of dual recognition of foreign antigen and HLA antigen by T lymphocytes.

Clone of a eukaryotic DNA segment mapped to a specific chromosomal location *(D. S. Hogness)*

The National Institute of Health forms a Recombinant DNA Advisory Committee to oversee recombinant genetic research.

1975 Asilomar conference First protein-signal sequence identified *(G. Blobel)*

Southern blot hybridization *(E. Southern).*

Monoclonal antibodies *(Köhler and Milstein).*

Colony hybridization and Southern blotting are developed for detecting specific DNA sequences.

The first monoclonal antibodies are produced.

1976 *Herbert Boyer* cofounds Genentech, the first firm founded in the United States to apply recombinant DNA technology

Overlapping genes in phage X174 *(Barell, Air, Hutchinson)*

R. Jaenisch made first transgenic mouse.

Loci for structural genes on each human chromosome known at Baltimore Conference on Human Gene Mapping.

The tools of recombinant DNA are first applied to a human inherited disorder.

Molecular hybridization is used for the prenatal diagnosis of alpha thalassemia.

Yeast genes are expressed in *E. coli* bacteria.

1977 *R. J. Roberts, P. A. Sharp,* independently gave the concept that genes contain coding and noncoding DNA segments ("split genes") (exon/ intron structure).

First recombinant DNA molecule that contains mammalian DNA *F. Sanger; Maxam and Gilbert* gave the methods to sequence DNA.

Finch et al. gave X-ray diffraction analysis of nucleosomes.

Genetically engineered bacteria are used to synthesize human growth protein.

1978 Somatostatin, which regulates human growth hormones, is the first human protein made using recombinant technology.

Leder, Weissmann, Tilghman and others gave Globulin gene structure.

North Carolina scientists *Hutchinson and Edgell* show it is possible to introduce specific mutations at specific sites in a DNA molecule.

1979 The first monoclonal antibodies are produced.

1980 The U.S. Supreme Court, in the landmark case *Diamond v. Chakrabarty,* approves the principle of patenting genetically engineered life forms.

The U.S. patent for gene cloning is awarded to *Cohen and Boyer.*

Genes for embryonic development in *Drosophila* studied by mutational screen *(C. Nüsslein-Volhard and others).*

DURING 80's

1981 The North Carolina Biotechnology Center is created by the state's General Assembly as the nation's first state-sponsored initiative to develop biotechnology. Thirty-five other states follow with biotechnology centers of various kinds.

The first gene-synthesizing machines are developed.

The first genetically engineered plant is reported.

Mice are successfully cloned.

S. Anderson, S. G. Barrell, A. T. Bankier did the sequencing of a mitochondrial genome.

1982 Humulin, Genentech's human insulin drug produced by genetically engineered bacteria for the treatment of diabetes, is the first biotech drug to be approved by the Food and Drug Administration.

Tumor suppressor genes *(H. P. Klinger)*

Prions (proteinaceous infectious particles) proposed as cause of some chronic progressive central nervous system diseases (kuru, scrapie, Creutzfeldt Jakob disease) *(S. B. Prusiner)*

1983 The Polymerase Chain Reaction (PCR) technique is conceived. PCR, which uses heat and enzymes to make unlimited copies of genes and gene fragments, later becomes a major tool in biotech research and product development worldwide.

The first genetic transformation of plant cells by TI plasmids is performed.

The first artificial chromosome is synthesized.

The first genetic markers for specific inherited diseases are found.

Efficient methods are developed to synthesize double-stranded DNA from first-strand cDNA involving minimal loss of sequence information.

L. Montagnier, R. Gallo reported HIV virus.

1984 The DNA fingerprinting technique is developed.

The first genetically engineered vaccine is developed.

Chiron clones and sequences the entire genome of the HIV virus.

Gusella Localizes the gene for Huntington disease.

Tonegawa Identifies the T-cell receptor.

Variable DNA sequences as "genetic fingerprints" *(A. Jeffreys)*

McGinnis discovers homeotic (Hox) regulatory genes, responsible for the basic body plan of most animals. In subsequent work, his team demonstrates that a single mutation in a Hox gene suffices to suppress all limb development in the thoracic region of fruit flies.

1985 Fully active murine RT is cloned and overexpressed in *E. coli.*

*Gitschie*Characterizes the gene for clotting factor VIII *(r).*

Sequencing of the AIDS virus Localization of the gene for cystic fibrosis Hypervariable DNA segments Genomic imprinting in the mouse *(B. Cattanach)*

1986 PCR is developed by *Kary Mullis*. The first field tests of genetically engineered plants (tobacco) are conducted.

The first biotech-derived interferon drugs for the treatment of cancer, Biogen's Intron A and Genentech's Roferon A, are approved by the FDA. In 1988, the drugs are used to treat Kaposi's sarcoma, a complication of AIDS.

The first genetically engineered human vaccine, Chiron's Recombivax HB, is approved for the prevention of hepatitis B.

First cloning of human genes. First identification of a human gene based on its chromosomal location (positional cloning) *(Royer-Pokora et al.)*

RNA as catalytic enzyme *(T. Cech)*

1987 Humatrope is developed for treating human growth hormone deficiency.

Advanced Genetic Sciences' Frostban, a genetically altered bacterium that inhibits frost formation on crop plants, is field tested on strawberry and potato plants in California, the first authorized outdoor tests of an engineered bacterium.

Genentech's tissue plasminogen activator (tPA), sold as Activase, is approved as a treatment for heart attacks.

Reverse transcription and PCR are combined to amplify mRNA sequences.

Cloned murine RT is engineered to maintain polymerase and eliminate Rnase H activity.

Fine structure of an HLA molecule *(Björkman, Strominger et al.)*

Cloning of the gene for Duchenne muscular dystrophy *(Kunkel)*

Knockout mouse *(M. Capecchi)*

A genetic map of the human genome *(H. Donis-Keller et al.)*

Mitochondrial DNA and human evolution *(R. L. Cann, M. Stoneking, A. C. Wilson)*

1988 Congress funds the Human Genome Project, a massive effort to map and sequence the human genetic code as well as the genomes of other species.

Molecular structure of telomeres at the ends of chromosomes *(E. Blackburn and others)*

1989 Amgen's Epogen is approved for the treatment of renal disease anemia.

Microorganisms are used to clean up the Exxon Valdez oil spill. The gene responsible for cystic fibrosis is discovered.

Cloning of a defined region of a human chromosome obtained by microdissection *(Lüdecke, Senger, Claussen, Horsthemke)*

1990 The first federally approved gene therapy treatment is performed successfully on a 4-yearold girl suffering from an immune disorder.

Evidence for a defective gene causing inherited breast cancer *(Mary-Claire King)*

DURING 90's

1991 *Amgen* develops Neupogen, the first of a new class of drugs called colony stimulating factors, for the treatment of low white blood cells in chemotherapy patients.

Complete sequence of a yeast chromosome.

Increasing use of microsatellites as polymorphic DNA markers

1992 The three-dimensional structure of HIV RT is elucidated. Recombinate, developed by Genetics Institute and used in the treatment of hemophilia A, becomes the first genetically engineered blood clotting factor approved in the U.S.

Trinucleotide repeat expansion as a new class of human pathogenic mutations.

High density map of DNA markers on human chromosomes

O. Smithies identifies X chromosome inactivation center

p53 knockout mouse *(O. Smithies)*

1993 Chiron's Betaseron is approved as the first treatment for multiple sclerosis in 20 years.

The FDA declares that genetically engineered foods are "not inherently dangerous" and do not require special regulation.

The Biotechnology Industry Organization (BIO) is created by merging two smaller trade associations.

Gene for Huntington disease cloned

1994 Genentech's Nutropin is approved for the treatment of growth hormone deficiency.

The first breast cancer gene is discovered.

Calgene's Flavr Savr tomato, engineered to resist rotting, is approved for sale.

Physical map of the human genome in high resolution.

Mutations in fibroblast growth factor receptor genes as cause of achondroplasia and other human diseases.

1995 The first baboon to human bone marrow transplant is performed on an AIDS patient.

The first full gene sequence of a living organism other than a virus is completed for the bacterium *Hemophilus influenzae*.

The three-dimensional structure of a catalytically active fragment of murine RT is elucidated.

Master gene of the vertebrate eye, *sey* (small-eye) *(W. J. Gehring)*

STS-band map of the human genome *(T. J. Hudson et al.)*

1996 Scottish scientists clone identical lambs from early embryonic sheep.

Yeast genome sequenced.

Mouse genome map with more than 7000 markers *(E. S. Lander)*

1997 A group of Oregon researchers claims to have cloned two Rhesus monkeys.

A new DNA technique combines PCR, DNA chips, and a computer program, providing a new tool in the search for disease-causing genes.

F. R. Blattner et al. Sequences *E. coli.*

Wilmut cloned Mammal by transfer of an adult cell nucleus into an enucleated oocyte.

Embryonic stem cells

1998 University of Hawaii scientists clone three generations of mice from nuclei of adult ovarian cumulus cells.

Human skin is produced in vitro.

Embryonic stem cells are used to regenerate tissue and create disorders mimicking diseases.

The first complete animal genome for the elegans worm is sequenced. A rough draft of the human genome map is produced, showing the locations of more than 30,000 genes.

Cloned vain RT with fully active polymerase and minimized Rnase H activity is engineered.

The Biotechnology Institute is founded by BIO as an independent national, 501(c)(3) education organization with an independent Board of Trustees.

Nematode *C. elegans* genome sequenced

1999 The complete genetic code of the human chromosome is first deciphered.

The rising tide of public opinion in Europe brings biotech food into the spotlight.

First human chromosome (22) sequenced Ribosome crystal structure

2000 A rough draft of the human genome is completed by Celera Genomics and the Human Genome Project.

Pigs are the next animal cloned by researchers, hopefully to help produce organs for human transplant.

"Golden Rice," modified to make vitamin A, promises to help third-world countries alleviate blindness.

The 2.18 million base pairs of the commonest cause of bacterial meningitis, Neisseria meningitidis, are identified.

Drosophila genome sequenced First draft of the complete sequence of the human genome First complete plant pathogen *(Xylella fastidiosa)* genome sequence *Arabidopsis thaliana*, the first plant genome sequenced.

DURING 21's CENTURY

2001 The sequence of the human genome is published in Science and Nature, making it possible for researchers all over the world to begin developing treatments.

2002 Scientists complete the draft sequence of the most important pathogen of rice, a fungus that destroys enough rice to feed 60 million people annually. By combining an understanding of the genomes of the fungus and rice, scientists will elucidate the molecular basis of the interactions between the plant and pathogen.

2003 Dolly, the cloned sheep that made headlines in 1997, is euthanized after developing progressive lung disease. Dolly was the first successful clone of a mammal.

2007 Controversies continue over human and animal cloning, research on stem cells, and genetic modification of crops.

2

Equipments Required in a Molecular Marker Laboratory

1. pH meter

pH meter/electrode systems are the most common method of measuring pH in the biology laboratory. Although more expensive and more complicated to use than chemical indicators, pH meters provide greater accuracy, sensitivity, and flexibility. When used properly, pH meters can measure the pH of a solution to the nearest 0.1 pH unit or better, and they can be used with a variety of samples.

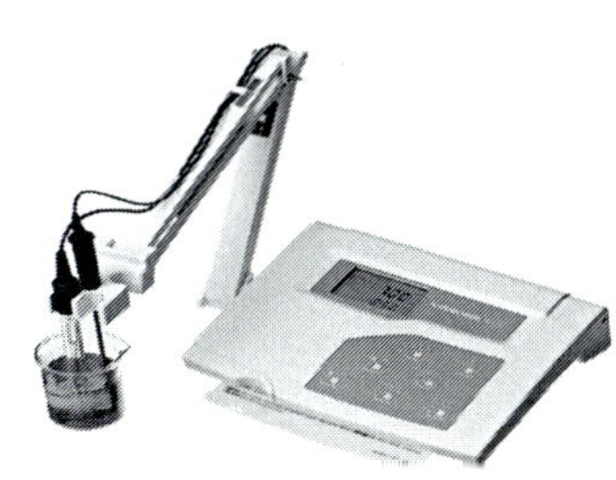

2. UV Spectrophotometer

Perhaps there is no other technique except UV-visible spectroscopy, which has been so extensively employed and has such a profound impact on progress in biochemical investigations. UV-visible spectroscopy has undoubtedly become an indispensable tool and is routinely used for identification and quantitative estimation of most of the biomolecules. Besides other applications, most of the enzyme assays and the screening of biomolecules recovered in different fractions by separation processes such as column chromatography, density gradient centrifugation etc, rely heavily on spectrophotometer or colorimetric technique.

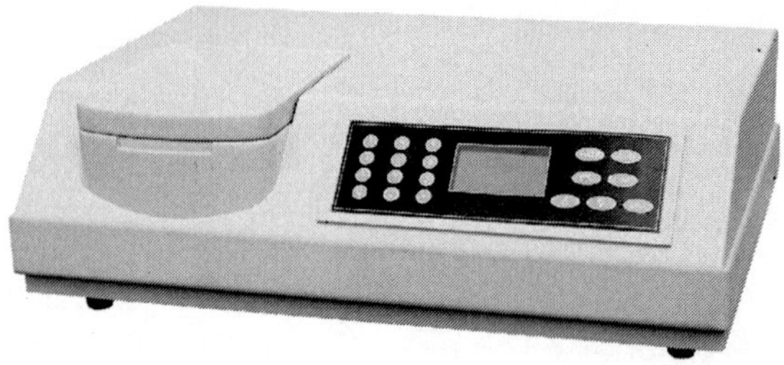

Points to Remember While Using UV Spectrophotometer

The detailed operation of a particular instrument must, of course, be obtained by carefully reading the instruction manual, but a few general points concerning the use and care of calorimeters and spectrophotometers are given below.

1. *Cleaning cuvettes:* Cuvettes are cleaned by soaking in 50 per cent v/v nitric acid and then thoroughly rinsed in distilled water.

2. *Using the cuvettes:* First of all, fill the cuvettes with distilled water and check them against each other to correct for any small differences in optical properties. Always wipe the outside of the cuvettes with soft tissue paper before placing in the cell holder and do not handle them by the optical faces. When all the measurements have been taken, wash them with distilled water and leave in the inverted position to drain.

3. *Absorption of radiation by cuvettes:* All cuvettes absorb radiation and the wavelengths at which significant absorption occurs depend on the material from which the cuvette is made. Silica cuvettes are the most transparent to UV. Glass cuvettes are expensive so than silica and so they are used whenever possible and invariably in the visible region of the spectrum. However, they do absorb UV and cannot be used below 360 nm. Plastic disposable cuvettes also absorb in the UV but much less than glass and they can be conveniently used at 340 nm for the assay of dehydrogenease enzymes that require NAD or NADP as coenzymes.

4. *Light source:* A tungsten lamp produces a broad range of radiant energy down to about 360 nm. To obtain the ultraviolet reaction of the spectrum a deuterium is used as the light source. If the tungsten lamp is used in the range 360-400 nm, then a blue filter is placed in the light beam.

5. *Blanks:* The extinction of a solution is read against a reagent blank, which contains everything, except the compound to be measured. The blank is first placed in the instrument and the scale adjusted to zero extinction (100 per cent transmittance) before reading any test solutions. Alternatively, the extinction can be read against distilled water and the absorbance of the blank subtracted from that of the test solution.

6. *Replicates:* It is essential to prepare all blanks and standard solutions in duplicate so that an accurate standard curve can be constructed. In addition, the test solutions should also be prepared in duplicate wherever possible.

3. Centrifuge

Biologists seldom need to separate gravel and sand from one another; rather, they are interested in such material as cells, organelles, bacteria, and viruses. These biological materials, called particles in the context of centrifugation, are much smaller than sand and would take a long time to sediment due to the force of gravity alone.

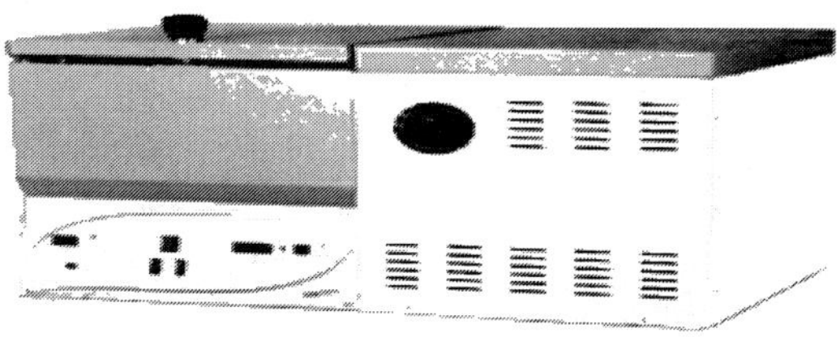

In a simple centrifuge there is a central drive shaft that rotates during centrifugation. A rotor sits on top of the drive shaft and holds tubes, bottles, or other sample containers. As the drive shaft rotates, the sample containers spin rapidly, creating the force that facilitates the sedimentation of particles.

$$RCF = 11.2 \times r\left(\frac{RPM}{1,000}\right)^2$$

Where,

RCF = the relative centrifugal field in units of X g

RPM = the speed of rotation in revolutions per minute

r = the radius of rotation in centimeters

You may also see this equation rearranged as:

RCF = 1.12×10^{-5} (RPM)^2 (r)

Examples of Applications of Preparative Centrifugation

1. Separation of intact, single cell suspension such as bacteria, viruses, and blood cells from a liquid medium.
2. Separation of organelles, such as mitochondria, and, ribosomes from disrupted cells.
3. Separation of biological macromolecules, including DNA, RNA, and protein from a solution.
4. Separation of plasma (the portion of blood) from blood cells.
5. Separation of immiscible liquids from one another.
6. Separation of cells from broth after fermentation.

Analytical centrifugation is used in specialized research settings to determine the molecular weight, purity, shape, and other physical characteristics of proteins and other macromolecules Analytical centrifugation involves specially designed, computer controlled centrifuges that allow the operator to monitor the movement of particles.

4. Autoclave

The laboratory apparatus designed to use steam under regulated pressure to achieve sterilization is called autoclave. It used for sterilization of nutrient media and glass wares.

Principle

Water boils at 100°C depending on the vapour pressure of the atmosphere. If the atmospheric pressure is raised the boiling temperature for water will also rise. So if the steam pressure inside a close vessel is increased to 15 lb/ inch2. The temperature can go up to 121.6°C. An autoclave is equipped with device that can maintain saturated steam at a designated temperature and pressure for any period of time.

Description

It is essentially a double jacketed vertical cylinder made up of a strong metal and is separated by a case made up of iron sheet. The lid is very heavy and made up of gun metal and is screwed by buffer files screws. It is rendered air tight with interposition of an asbestos washer. The cylinder contains water up to certain level. The articles to be sterilized are placed on a perforated diaphragm serving as a platform situated above the water level. Heating can be done either by gas burner or by electricity. The chamber is furnished with a steam tap at the top and a pressure gauge with a safety valve. The pressure employed is 15 lb/inch for 15-20 minutes under this pressure H_2O boils at 121°C under operation it is necessary that air in the chamber is replaced by saturated steam otherwise required by temperature will not be obtained. It is not the pressure that kills the organism but the temperature of steam. The time of operation to achieve sterility depends upon the nature of the material being sterilized, the type of the containers and the volumes.

Sterilization by autoclaving all organisms including spores. Death of cells and spores is due to coagulation of proteins.

Uses : Sterilization of most type of solid or liquid media, distilled water, normal saline, discarded cultures, contaminated media, aprons, rubber tubes, gloves etc.

5. Hot Air Oven

Dry heat or hot air sterilization is recommended where it is either undesirable or unlikely that steam under pressure will make direct and complete contact. With the materials to be sterilized. The apparatus employed for this type of sterilization may be special electric or gas oven or even the kitchen stove oven.

Principle

Sterilization is carried out by dry heat at high temperature. Bacterial cells and spore die due to dehydration.

Description

It consists of a chamber having triple wall. The inner two walls are made up of copper sheet and outer of asbestos. The chamber is filled with several adjustable shelves and a thermometer is interested to record temperature. There are adjustable holes which are kept open during rising the temperature and are again open partially when sterilization completes clean glassware, like petridishes, test tube, flasks, pipettes etc. are place inside and the door is closed. Electric heated is put ON and the temperature of chamber is opened only after the temperature comes down to normal.

Uses : It is useful device for sterilization of laboratory glassware such as petridishes and pipettes. It is also used to sterilize substances which can not b e sterilized by moist heat like oils, fats and powders.

6. Deep Freeze

It is observed that in case of most of the bacteria, their preservation needs low temperature. This is because their metabolic activities are retarded. It is used for routine preservation of laboratory cultures. It is also used for storage of vaccines, sera, blood, plasma, antibiotics, media and several heat sensitive substances and many chemical. In DNA isolation it used for given a low (-20°C) temperature during precipitation.

It is a cupboard like chamber with storage facility for wide variety of material. The cooling in done by pheron gas and it is circulated in the chamber by a compressor.

7. BOD Incubator

Specially designed to meet the requirements for incubation of bacteria to decompose organic matter in waste water. It is used in Biology, Botany Virology, and Oceanography. Water pollution, sewage, agriculture, food and research departments also use it. In addition to this, it helps in preservation of vaccines. It provides a wide range of conditions for the study of synthesizing organisms. Temperature range, 5°C to 50°C, is attainable within the cabinet to the study the growth of organisms under tropical or temperate and summer or winter conditions. Temperature is controlled with an Electronic Digital Temperature Controller cum Indicator with an accuracy of ± 0.5°C or better.

8. Micro Oven

It is use for maintaining the specific temperature during DNA isolation as the lysis of the cell require high temperature.

9. Gel-Documentation

n gel-documentation system uses UV lights to make ethidium bromide or other dye fluorescence. With fluorescent staining of nucleic acids, a fluorescent substance that has bound to nucleic acids is excited by ultraviolet irradiation and emits fluorescent light. The fluorescent substance EtBr binds specifically to nucleic acid and the amount

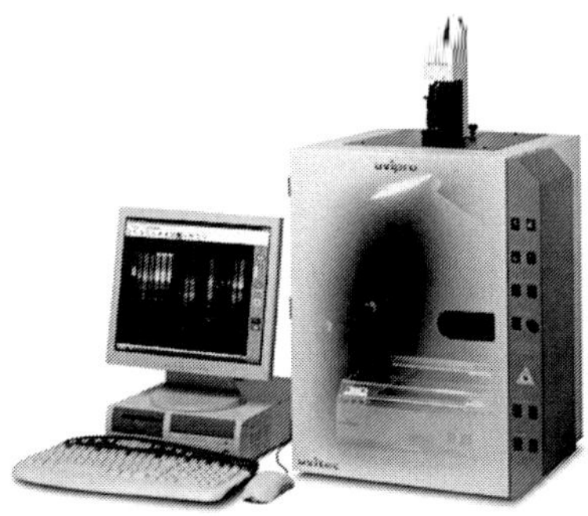

of bonding depends on the molecular weight and concentration of the nucleic acid. In other words, a band for a large molecular weight or large amount will shine brighter; conversely, fluorescence will be weaker for a band for a small molecular weight or small amount. This instrument is very useful in analyzing the band in gel.

10. Deep Freeze Centrifuge

A **centrifuge** is a piece of equipment that accelerates the rate of sedimentation by rapidly spinning the samples, thus, creating a centrifugal force many times that of gravity.

It is use for the separation of cells, organelles, bacteria, and viruses. In separation of DNA &RNA from rest of the material by using high speed centrifugation.

11. Micropipette

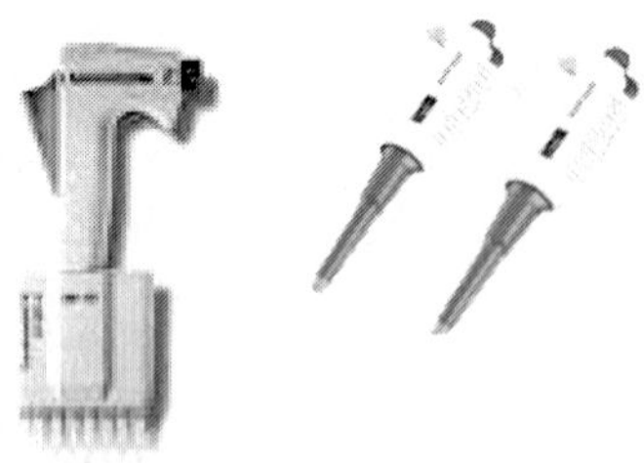

In molecular biology practical the most important thing is that to take a reagent / chemical a specific volume by using automated pipette we can take a micro liter volume of any chemical which can not take sing conventional pipette.

12. PCR

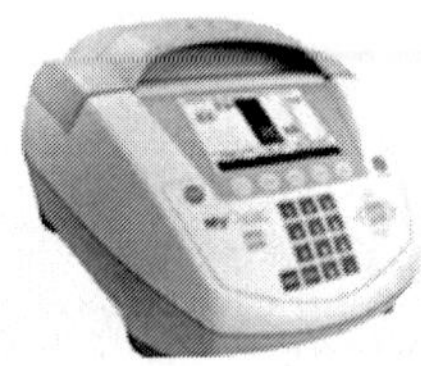

The purpose of a PCR (Polymerase Chain Reaction) is to make a huge number of copies of a gene. This is necessary to have enough starting template for sequencing.

13. Mortar & Pestle

Mortar and pestle is used to crush/grind the leaf tissues / plant tissues to isolate DNA/RNA/Isoenzyme etc. for molecular marker study.

14. Digital Balance

Digital balance is used to precisely weigh chemicals, plant material and other things used in molecular marker study and plant biotechnology.

15. Liquid Nitrogen Cylinder

Isolation of DNA and storage of plant tissues. Helps in grinding of tissues to the powder form and avoid enzymatic activities that harm/degrade DNA/isoenzyme etc. while working with them.

16. Water Bath

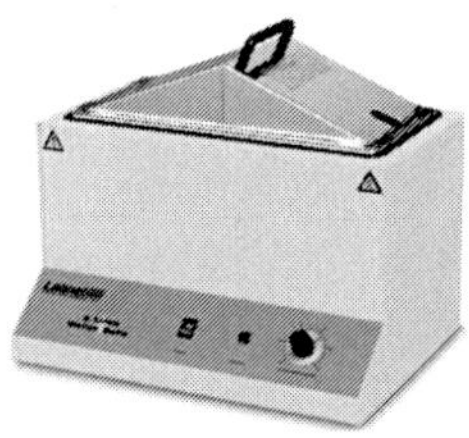

Maintains water temperature to incubate DNA (or any other) reagent / samples/ restriction digestion mixture etc. at a constant temperature.

17. Ice Bucket

Ice bucket is used to keep the working samples of DNA / isoenzymes etc. while working with any molecular biology protocol.

18. Ice Flaker

Ice flaker is used to prepare ice flakes, which is further used in different place while working with DNA/RNA/Protein, like it is put in the ice buckets for storage of working DNA / Isoenzyme samples.

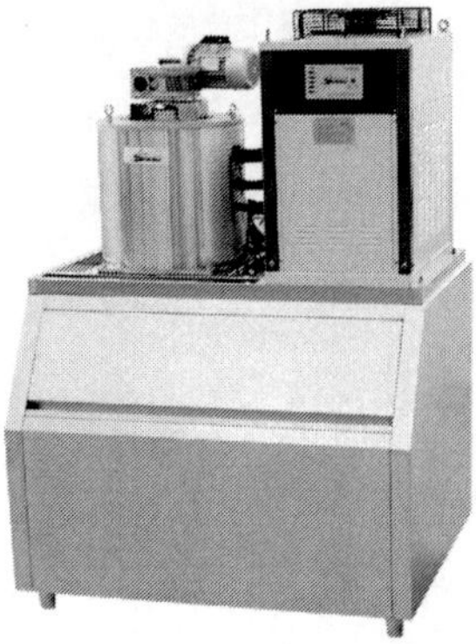

19. Electrophoresis Apparatus

Electrophoresis is a method of separating out large molecule (DNA or protein). An electric current is passed through a medium containing the molecules, and each molecule travels at a different rate depending on its electrical charge, size and shape. Separation by electrophoresis is based on these differences. In electrophoresis, agarose and acrylamide gels are used for electrophoresis of nucleic acids and proteins.

20. Sequencing Gel Apparatus

Sequence gel apparatus is to run PAGE (Polyacrylamide gel electrophoresis). SDS-PAGE is the most widely used method for qualitatively analyzing any protein mixtures. It is particularly useful for monitoring protein purity and determines their relative mass.

21. UV transilluminator

UV transilluminator is used for gel documentation. In gel-documentation system uses UV lights to make ethidium bromide or other dye fluorescence.

22. Genome Sequencer

The purpose of genome sequencer is to determine the order of the nucleotides of a gene.

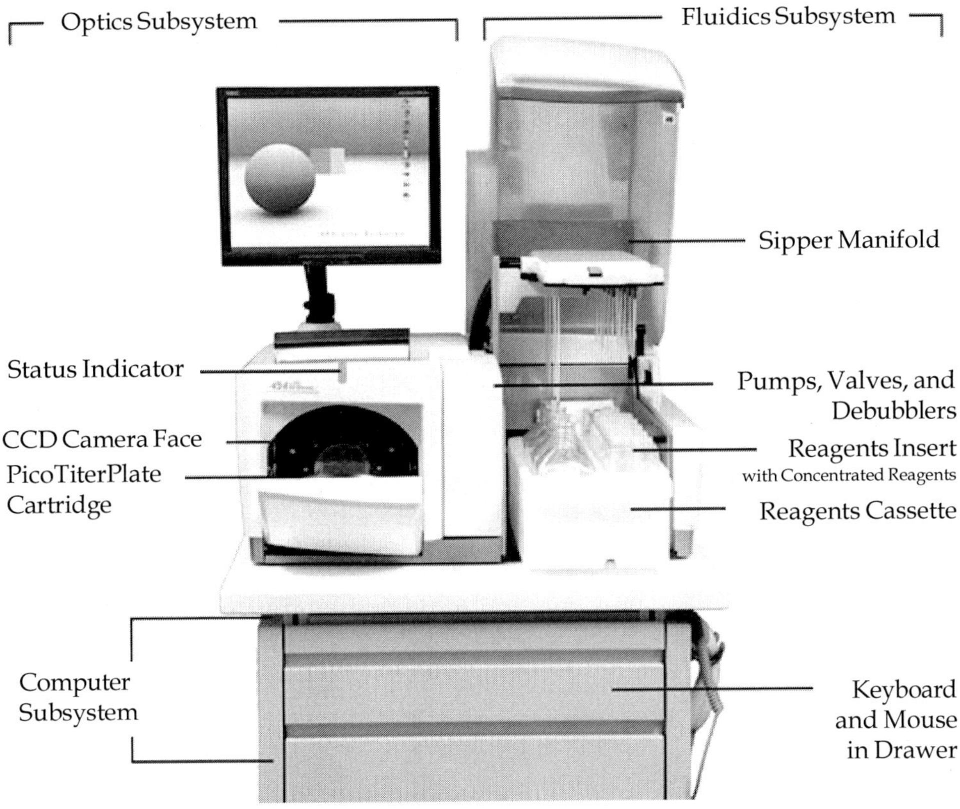

23. Aminoacid Sequencer

The purpose of amino acid sequencer is to determine the order of the nucleotides of proteins. Sequence information can be used to identify a protein or homologous proteins through searches in databases; the minimal number of residues for a succesful search is ten, but often more residues are required. The N-terminal sequence is often used for confirmation of the identity of a protein; in that case a few residues are enough. The sequence also gives information about post translational cleavage points. In addition, the sequence results give informations about the purity of a preparation; limits of detectable contamination depend on the sequences of the analysed proteins.

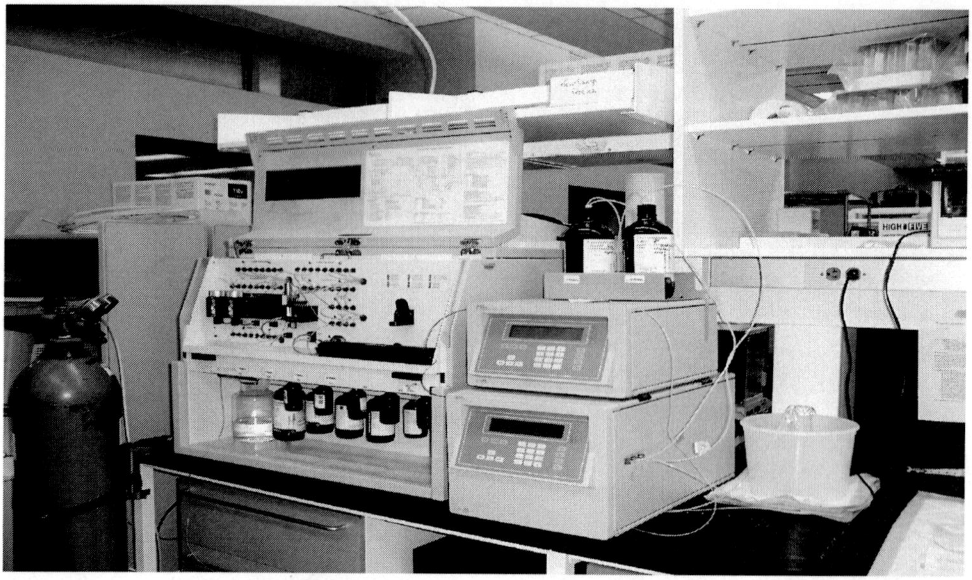

24. Western Blotting Apparatus

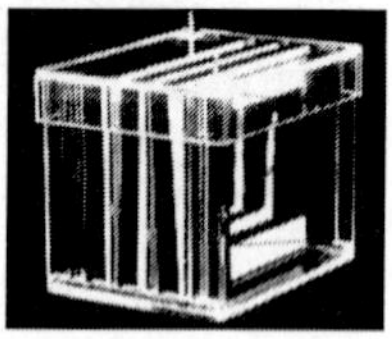

The Western Blotting Apparatus is made with a view to blotting a maximum of three gels at one time. It is used to transfer DNA to the nylon membrane from the gel.

25. Picodrop

Picodrop uses UV/Vis measurements to be taken directly through the pipette tip for finding out purity and concentration of DNA. It requires only 2μl of sample for measurement. It can be further used for RNA and Protein estimation.

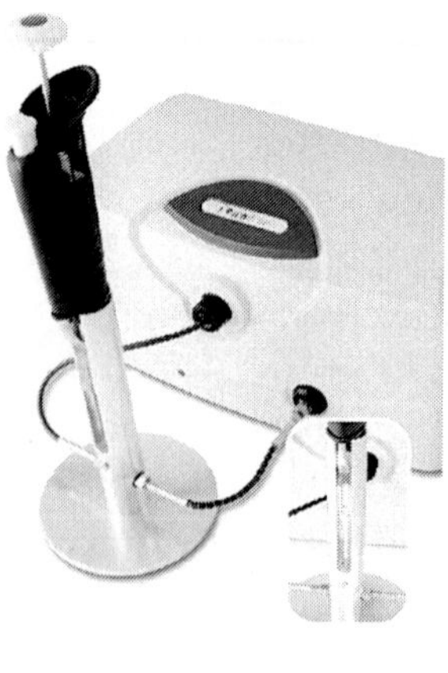

26. Laminar Air Flow

Laminar air flow work on basis of filtration in this HEPA filter are used for filter the contaminates air flow prevent the enter the other bacteria inside the chamber.

In microbiology one major problem is contamination and in molecular biology practical /plant tissue culture practical to need sterile condition. Laminar air flow provide the sterile condition for specific region in this chamber is totally sterile.

3

Isolation, Purification and Quantification of Genomic DNA from Plants

What is DNA

1. DNA is a simple molecule
2. DNA is negatively charged
3. DNA has lots of ring structures; ring structures are generally quite hydrophobic.
4. DNA is a very robust molecule.
5. DNA work has lots of toys.

Nucleotides are the Building Stones of DNA

There are 4 different nucleotides :

- dATP : deoxyadenosine triphosphate
- dGTP : deoxyguanosine triphosphate
- dTTP : deoxythymidine triphosphate
- dCTP : deoxycytidine triphosphate

For convenience, these 4 nucleotides are called dNTP's (deoxynucleoside triphosphates). A nucleotide is made of three major parts : a **nitrogen base**, a **sugar** molecule and a **triphosphate**. Only the nitrogen base is different in the 4 nucleotides.

Nucleotide = base + sugar + phosphate

4 different dNTP's (deoxynucleoside triphosphate) :

deoxyadenosine triphophate = dATP

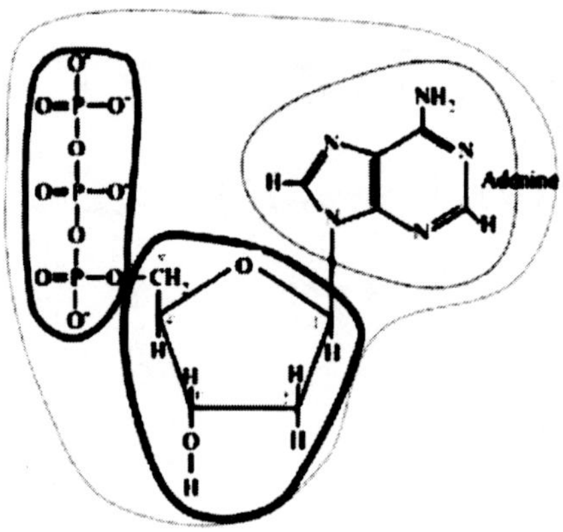

deoxyguanosine triphosphate = dGTP

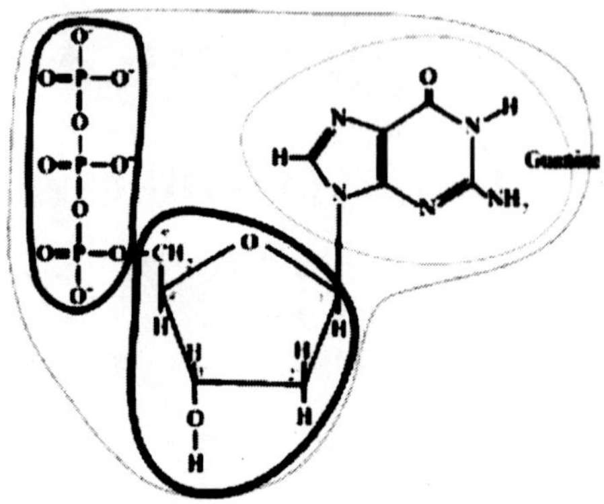

deoxycytidine triphosphate = dCTP

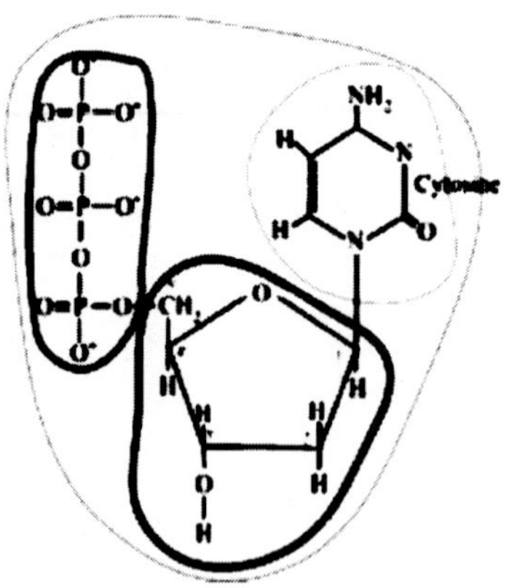

deoxythymidine triphosphate = dTTP

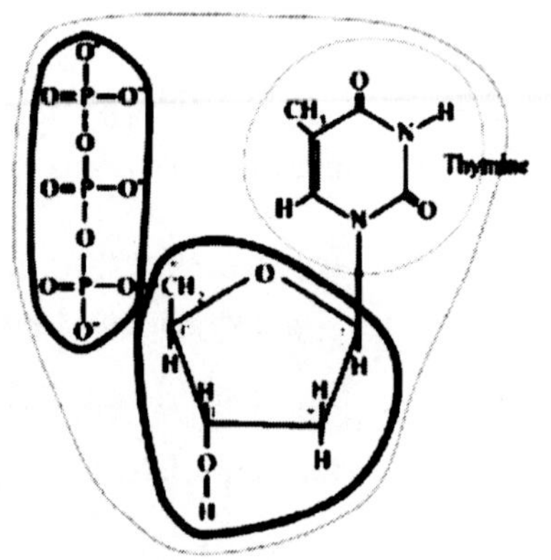

Figure : The components of nucleotides.

How do the Nucleotides form a DNA Chain ?

DNA is formed by coupling the nucleotides between the phosphate group from a nucleotide (which is positioned on the **5th C-atom** of the sugar molecule) with the hydroxyl on the **3rd C-atom** on the sugar molecule of the previous nucleotide. To accomplish this, a diphosphate molecule is split off (and releases energy). This means that new nucleotides are always added on the 3' side of the chain.

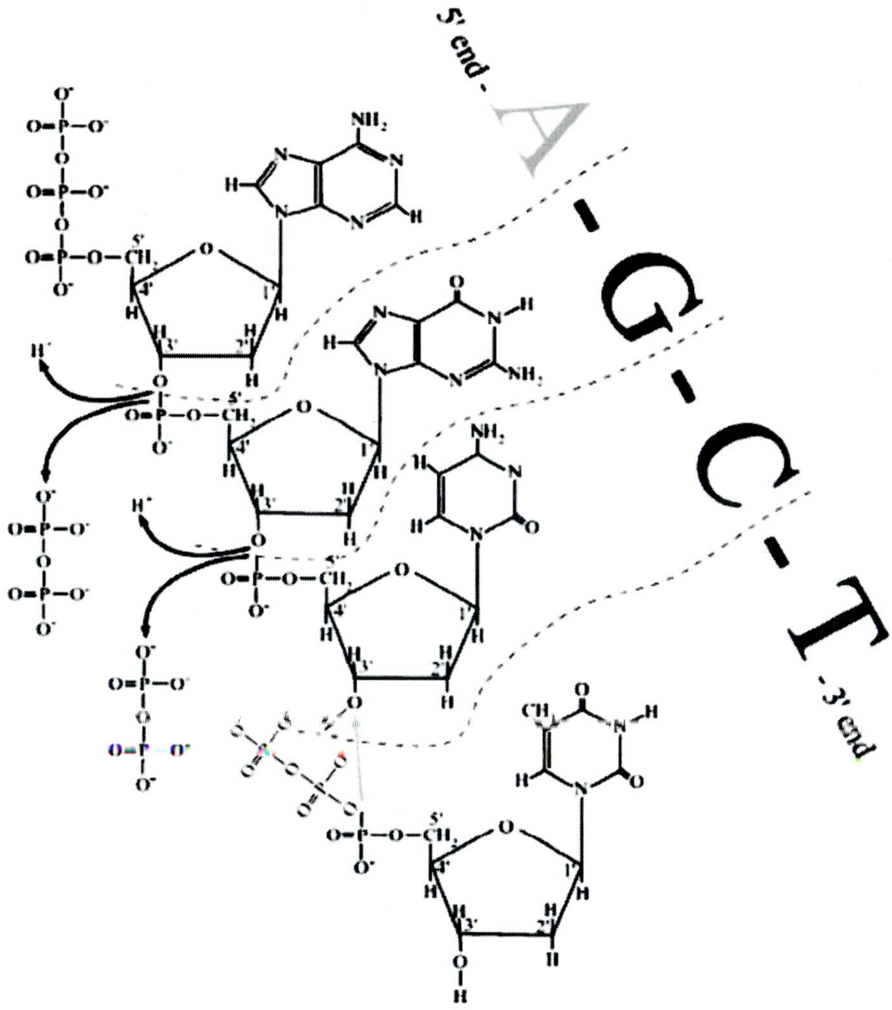

Figure : From nucleotide to DNA.

DNA in a Cell

- Which organelles contain DNA ?

Eucariotic cells contain several organelles. The **nucleus** contains most of the DNA in a cell and this DNA is called the **chromosomal DNA**. It is separated from the rest of the cell (cytoplasm) by a double layer of membrane. The **mitochondria**, which have a role in the oxidative degradation of nutrient molecules, also contain DNA, called the **mitochondrial DNA**. Eucariotic cells that are capable of photosynthesis contain **chloroplasts** with **chloroplast DNA**.

To have an idea of the size of this : every million bases take up a linear space of 0,34 mm. So when you take one human cell, uncoil all the chromosomal DNA and put it on a line, you would have 204 cm of DNA (a human cell contains in total 6×10^9 nucleotide pairs). To store this information on paper, you need a few pages of A4 paper : one page stores 3100 letters on one side (font courier 12), so 6×10^9 nucleotides, printed on both sides of a sheet, would need 967.742 pages (you'll have a pile of 120,96 m paper). The table below shows the size of genetic material in different organisms.

Table : Size of genetic material

Type of DNA	Organism	Size in base pairs
chromosomal DNA	mammals	6×10^9
	plants	$2 \times 10^8 - 2 \times 10^{11}$
	fungi	$2 \times 10^7 - 2 \times 10^8$
mitochondrial DNA	animals	$16 \times 10^3 - 19 \times 10^3$
	higher plants	$150 \times 10^3 - 250 \times 10^4$
	fungi	$17 \times 10^3 - 78 \times 10^3$
	green alga	16×10^3
	protozoa	$22 \times 10^3 - 40 \times 10^3$
chloroplast DNA	higher plants	$120 \times 10^3 - 200 \times 10^3$
	green alga	180×10^3

There are three types of genes :

1. Protein-coding genes : these are transcribed into RNA and then translated into proteins.
2. RNA-specifying genes : these are only transcribed into RNA.
3. Regulatory genes : according to a narrow definition, these include only untranscribed sequences.

The first two types are also called 'structural genes'.

GENOMIC DNA: What and Why.

Genomic DNA? What is it? Well, it's actually a reasonably vague term, but consider it to be the DNA sample that contains the entire set of genes in a cell, organelle and/or virus. In other words think of it as the whole shabang and assume that (most of the time), you are trying to isolate all DNA from your sample. By the way, genomic DNA is big, complicated and often ends up looking like snot.

Why are you Getting Genomic DNA? (A Couple of Reasons)

1. Profiling/Fingerprint: This is the stuff you use to compare blood/semen samples. Essentially, your genomic DNA can lead to some form of reproducible data that provides a unique representation of the sample (i.e. band profile on a PCR DNA fingerprint for example)
2. First attempt at cloning a new gene. If you plan on looking for something completely new and novel. Therefore, you will need to go to the original source of that code which is the genome. And although, you may be able to get protein or mRNA information before this step, you will ultimately need to define it within a genomic context (i.e. what chromosome is it on, promoter regions, intron/exon organization, etc.)
3. Characterization/Diagnotics. Lots of examples in basic research *i.e.* you've made a transgenic/knockout organism, and you need to check if indeed, the modification has occurred – this starts with a genomic prep. You have an interesting phenotype and you want to look deeper. i.e. a disease state. This hunt may often start from a genomic prep.
4. Studying genome for clues into gene expression and regulation: Although, usually the cDNA for a gene gets a lot of attention (this is a DNA copy of the messenger RNA directly responsible for coding the protein), this edited form of the gene misses out a lot of stuff that may be interesting or relevant to your research *i.e.* study gene expression regulation, euchromatin nuances, tissue/developmental specificity, spliced isoforms, etc.

GENERAL METHOD FOR EXTRATION, PURIFICATION AND PRECIPITATION OF GENOMIC DNA

• Cell Lysis

First things first: Large pieces of DNA can shear. (not actually a big deal for most procedures but why take the risk). In this case, dealing with a piece of DNA that is ~100kb + in size. No vortexing, no vigorous pipetting/snip the end of your tip before use.

Lysis Procedure: (SDS/Proteinase K)

1. Get the cells ready. Keep it cold, Do something physical if need be. In our case, we are working with notoriously tough tissue – i.e. that from a corn plant. We will use a mortar and pestle in the presence of liquid nitrogen to work at our sample. NOTE that in our case, this step is more about breaking open the plant cell wall so that cells can get out, rather than "lysing cells" per se.

2. Then Lyse the cells open -> to get at DNA since it is inside your cell. Make DNA easily assessible for subsequent steps. Lyse cells with **Tris**(buffering agent -> good @pH6-8)

EDTA chelates divalent cations which are necessary cofactors for DNase activity (way of shutting down nucleases)

NaCl at physiological concentration (generally considered to be 100 – 150mM) Keeps all molecules happy (particularily proteinase k)/prevents unwanted aggregation.

SDS, nasty ionic detergent/ good at breaking membrane, general denaturant (inhibit enzyme activity). Since DNA is so robust, not really adversely affected by SDS treatment.

Note: When dealing with plant material, a very common detergent in place of SDS is **Sarkosyl**. This essentially behaves in a similar manner to SDS.

Proteinase K->serine protease (works well at 55°C), used because it is very effective and not particularily susceptible to SDS, and other denaturants such as urea. Proteinase K will chew up protein, which helps lysis in general and frees up the DNA from any protein gunk associated with it (euchromatin structures/histones, etc). Best used FRESH (*quite an important step).

Incubation step generally a minimum of an hour. Although most will allow the procedure to go longer (this may be largely dependant on material being used – i.e. many procedures outline an overnight or >16hrs incubation time).

It's important to realize that there are many variations of the lysis procedure. Some are quicker, some are more efficient, some are more expensive, some only work in certain situations.

The SDS/proteinase K is a very standard procedure.

- **DNA Purification**

Purification Procedure (phenol/chloroform).

Note that Phenol is pretty nasty! It will burn. (Apparently, someone once told me that if you cover 5% of your body, you will die). Since we are using reasonable amounts, do phenol addition in fumehood, but all manipulations can occur at the nech. If you get it on you -> not panic mode. Quickly rinse off with cold water. Phenol, by the way, is usually buffer saturated. (i.e. when you buy it or make it, the solution comes in two layers -> top layer is excess buffer, and bottom layer is buffer saturated phenol. This is because the pH is important - In order for our purification to work, we need the phenol at a neutral pH.(acidic pH is not good as DNA becomes soluble in phenol). The fact that it is saturated is also important because this means that any additional aqueous (i.e. water) solution you add will create it's own fluid layer.

This purification procedure works on the principle of **"differential solubility".**

1. To your lysate, you will add an equivalent volume **Phenol/Chloroform/ Isoamyl** alcohol. (usually at a volume ratio of 24:23:1). Phenol – organic solvent/ nucleic acids not soluble at all. Therefore, DNA/RNA will stay dissolved in aq phase. Lipids and polysaccharides preferentially go into the phenol phase. Proteins will also selectively go into phenol solution.

2. Furthermore phenol also acts as a denaturant, proteins denature form aggregates and will collect at the interface. You will see GUNK @ the interface.

3. **Choloroform,** also has same general attributes as phenol (as far as solvent properties) but also stabilizes the rather unstable boundary between aq and organic layers. Isoamyl alcohol also contributes to interface stability and also helps prevent frothing.

Generally you do this step ~2 or 3 times. The more times you do it, the cleaner your sample (you may even note that the interphase gets cleaner and cleaner with each step). Note this procedure is very reliable and does not lose much DNA yield. This is probably why a lot of labs still like to use it. (BACK EXTRACT: adding extra aqueous to your organic samples).

Sometimes do a final **Choloform** step. Here, the interface is a little trickier to handle. But this step is a good safeguard to prevent any organic carrying over to your final aqueous solution. Likely finish with a chloroform step because it evaporates easily. In other words, this give you the option of leaving the lid open to really make sure all your organics are gone from your prep.

• DNA Precipitation.

Precipitation of DNA (alcohol and salt procedure):

You will add **Ammonium Acetate**. 0.5volume. Why? Helps in the precipitation of the DNA in **EtOH**. Can use **NaCl**, can omit entirely (dependant on concentration of DNA). Salt will help neutralize negative charge of DNA (will also sequester the solvent molecules - in this case water) Salt will also interact with water, thereby weakening it's solvating prowess. This is commonly known as "**salting out**" Use 100% EtOH. Time frame (show graph) EtOH generally helps because it is a much crappier solvent than something like water (which is very polar). DNA will tend to stay precipitated in **>65% Ethanol**. NOTE that efficiency of EtOH precipitation is dependant on a number of things. Temp, time, amount of DNA. Can also use isopropanol for precipitation steps. RNA tends to stay soluble in this solvent (selective precipitation). which is why some people use it for this purpose. DNA will tend to stay precipitated in **>50% isopropanol**.

Use glass pasteur pipette to spool DNA out carefully. Dip in **70% ethanol**, and resuspend in TE. The glass pipette technique seems to be favoured solely for speed. It is a very quick way to retrieve your DNA and also wash it.

Note: the 70% ethanol wash is included to get rid of excess salts.

Alternative. Some labs like to centrifuge the DNA pellet down, rather than pool it out. This may be more effective if you have very low yield. However, you will still need to wash the pellet in 70% ethanol and respin your sample.

• Quantitation

Spectrophotometry readings: Using **UV absorbance** to ascertain DNA/RNA amounts and purity. Ring structures can absorb UV wavelengths. BUT Lots of things have ring-structures (including nucleic acids, proteins, organics, detergents, lipids, the list goes on and on...) i.e. Both DNA and RNA -> ring structure absorbs strongly at ~ 260nm and ~280nm.

Protein => some amino acids like W and P and Y also flouresce strongly at ~ 260nm and ~280nm.

Bottom line is that you have to take numbers with a grain of salt. Very rough estimate, since lots of things absorb at UV wavelengths. Also quite

sensitive to pH which is why dilutions are often done with TE buffer at a particular pH, that was also used to calibrate the spectrophotometer.

You can also assess purity by looking at ratios of 260 over 280.

DNA A260/A280 ~1.8

RNA A260/A280 ~2.0

These are good ratios for purified product. If you get good numbers here, then maybe you can also get good quantitation numbers.

Note: that the conversion for O.D. numbers to DNA amounts is the constant **50ug/ml per O.D260nm** value, likewise, the constant for RNA is **40ug/ml per O.D.260nm** value).

Quantitation Assays : Usually rely on stoichiometry binding between reagents like Ethidium Bromide, SYBR dyes to correlate with a DNA amount. Generally a more robust, more expensive manner of determining amounts. ALONG the same lines, DNA fragment quantitation can be done by comparing band intensities in an ethidium bromide stained gel - i.e. use a marker as a gauge.

PLANT DNA

Plants contain genomic and organelle DNA. This includes nuclear, mitochondrial (mt) and chloroplast (cp) DNA. Quite elaborate methods exist for the isolation of each type of DNA. All DNA preparation methods involve the removal of the envelopes (cell wall and nuclear membrane) around the DNA, the separation of the DNA from all other cell components such as cell wall debris, proteins, lipids and RNA; and avoiding damage to the DNA during the procedure, i.e. the protection from nucleuses and mechanical shearing. In the most common method, which is applicable to a whole range of plant material, grinding in liquid nitrogen opens cells. Generally freezing of the harvested tissue is done in liquid nitrogen followed by lyophilizing it and grinding it to ensure a long term supply of tissue. The low temperature prevents nuclease activity that degrade the DNA.

Problems and their Solutions in Isolating DNA

The problems in isolating DNA, especially from plants, are (1) the presence of DNAase activities that degrade the DNA and (2) the presence of other macromolecules which co-purify with, or polymerize to, the DNA during the isolation procedure. The nuclease problem is reduced by removing cations such as Mg^{++} which are required for nuclease activity. Agents such as EDTA, EGTA and phenanthroline have been used at a range of concentrations, in different protocols, depending oh the plant or animal species being analyzed. In addition detergent agents such as sodium dodecylsulphate (SDS) are often used to inhibit enzyme activities. Plant researchers often encounter undesirable macromolecules, other than DNA, which create problems in the DNA isolation procedure.

The problems created by the presence of phenolic compounds can be reduced by the addition of 1% (or higher) polyvinylpyrollidine (PVP) in the initial isolation buffer. Detergents such as SDS dissolve membranes and dissociate proteins from DNA and make them more accessible to degradation by proteinases used in the DNA isolation. A reagent, which is used in several procedures, is cetyltrimethylammonium bromide (CTAB) which binds strongly to DNA displacing protein and preventing degradation.

The CTAB itself is removed by chloroform extractions leaving DNA in the aqueous phase to be ethanol precipitated. The aqueous phase contains DNA, RNA, polysaccharides and some protein, while lipids are found in the organic phase and cell debris and most of the protein aggregate in the interphase. Ethanol precipitation concentrates the DNA.

Method 1

CTAB Method

Principle of CTAB method

This is the most commonly used method for isolation of total nucleic acids from a wide range of plants. Plant tissue is ground to a fine powder in presence of liquid nitrogen to break cell walls and prevent action of nucleases (nucleases are most active when the griding is not done in cold environment). Cetyl trimethylammonium bromide (CTAB) is a nonionc detergent which complexes with carbohydrates and other contaminating macrmolecules such as protein and cell wall debris and causes their precipitation on phenol extraction. Nucleic acids upon centrifugation comein the aqueous phase and are pericipitated with isopropanol orethanol.

The CTAB method of DNA isolation is widely used on plants because of its versatility. Total genomic DNA has been isolated from many genera of monocotyledons and dicotyledons. Various types of tissues can be used, including whole seedlings, leaves, cotyledons, seeds/grains, endosperm, embryos, tissue culture callus, and pollen. In addition, milligram amounts of tissue can be used when sample size is limiting. The starting material can be lyophilized, dehydrated (even mummified), frozen, or fresh.

The nonionic detergent cetyltrimethylammonium bromide (CTAB) is used to liberate and complex with total cellular nucleic acids. This general procedure has been used on a wide array of plant genera and tissue types. Many modifications have been published to optimize yields from particular species. The protocol is relatively simple, fast, and easily scaled from milligram to grams of tissue; it requires no cesium chloride density gradient centrifugation.

- Reagents required to perform CTAB method

A. CTAB buffer
   1.4 M NaCl
   2.0 % CTAB
   100 mM Tris-CI, pH 8.0
   20 mM EDTA
   0.2 % β-mercaptoethanol (to be added just before use)

B. Chloroform: Isomyl alcohol mixture (24 : 1 v/v)

C. 3 M Sodium acetate, pH 5.2

D. TE Buffer
   10 mM Tris-CI, pH 8.0
   1 mM EDTA

E. Isopropanol

F. RNase A 10 mg/ml

   Dissolve RNase A in 10 mM Tris-CI (pH 7.5), 15 mM NaCl. Heat in boiling water bath for 10-15 min. Cool to room temperature and store at -20°C in small aliquots.

G. Phenol saturated with TE.

Procedure

1. Add 2-ME to the required amount of CTAB extraction solution to give a final concentration of 2% (v/v). Heat this solution to 65°C.
2. Chill a mortar and pestle with liquid nitrogen (-196°C). Grind approximately 5g of fresh leaf tissue to a fine powder and transfer the frozen tissue into a 50ml polypropylene centrifuge tube. Use young tissue and avoid larger stems and veins to achieve the highest DNA yield with minimal polysaccharide contamination.

3. Add 15ml of preheated 2- ME / CTAB extraction solution to the pulverized tissue and mix to wet thoroughly.

4. Incubate for 60-90min. with occasional mixing. A 60min. incubation results in larger DNA yields. If maximum yield is not important, 15min. should be adequate. If the tissue contains large amounts of phenolic compounds, 1% (v/v) polyvinylpyrrolidone may be added to absorb them.

5. Extract the homogenate with an equal volume of 24 : 1 chloroform / octonol or chloroform / isoamyl alcohol. Mix well by inversions. Octonol, rather than isoamyl alcohol, is used because it may enhance isolation of nuclei.

6. Centrifuge for 5min. at 7500 x g (8000 rpm) at room temperature. Slower centrifugation speeds are possible if centrifugation time is increased accordingly. After centrifugation, two phases should be evident with tissue debris at the interface.

7. Transfer the aqueous phase to another 50ml polypropylene centrifuge tube using a large bore pipette. Add heat-treated RNase A to a final concentration of 100 mg / ml. Mix, and incubate at room temperature for 30min.

8. Add 0.6 volume of ice-cold isopropanol. Mix by very gentle inversions. At this stage, DNA-CTAB complex may precipitate as a whitish network.

9. Spool the precipitate out of the solution using a glass hook (e.g. a bent pasteur pipette). Continue with OPTIONS A, B and C.

   OPTION A : DNA purification : phenol extraction to obtain DNA of higher purity.

10. Place hook with DNA in eppendorf tube containing 1ml of TE; gently twirl hook until DNA slides off the hook. Dissolve DNA at 4° C overnight. If the DNA is needed immediately, it may be incubated at 37° C for one hour.

11. Add equilibrated phenol or 1 : 1 phenol : chloroform equal to the total amount of TE added in step 10 above. Mix thoroughly but gently for 5-10min. and spin at room temperature in a microcentrifuge for 10min. Upon centrifugation, the phenol layer, which contains some proteins and lipids, moves toward the bottom of the tube, while the aqueous layer containing the DNA remains on top. The denatured protein forms an insoluble layer at the interphase.

Diagramatically Representation of CTAB Method

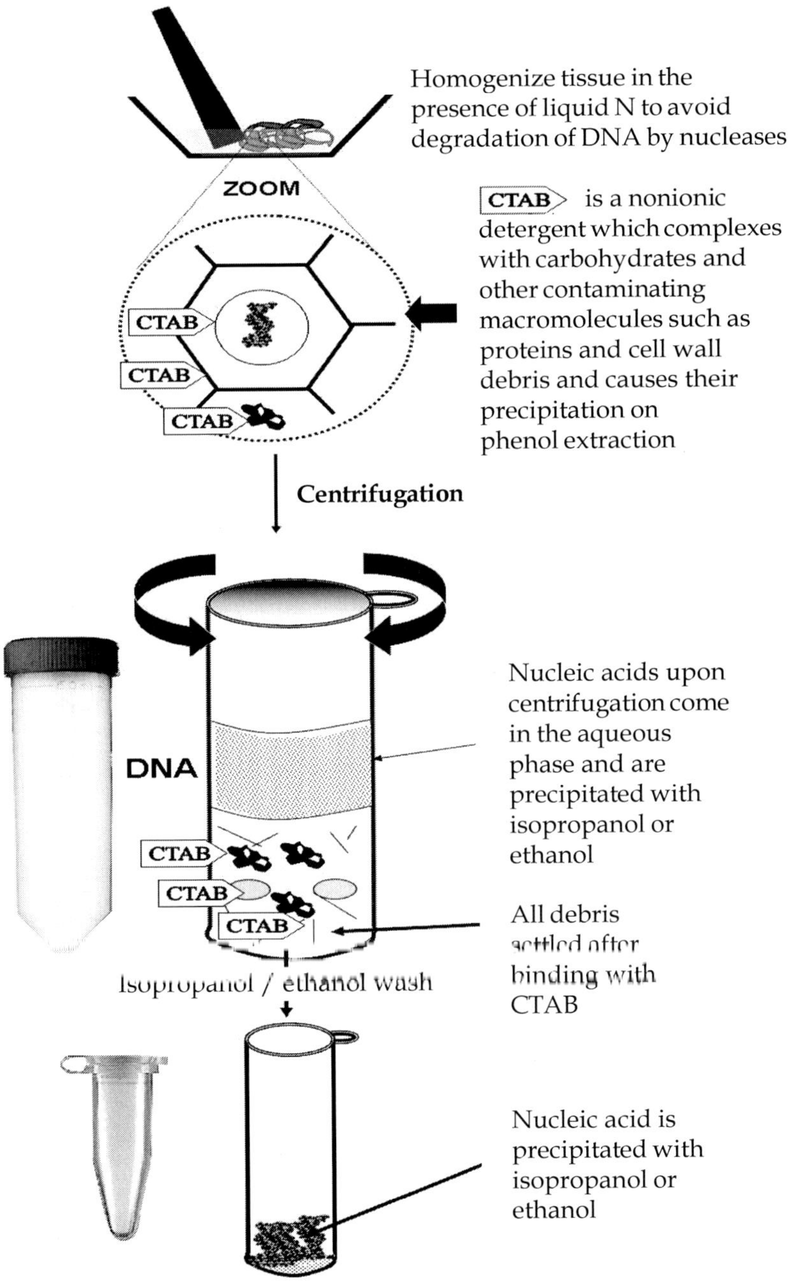

12. Transfer the top (aqueous) layer to a new tube and extract with an equal volume of chloroform. Mix well and spin for 5min. Transfer the supernatant to a new tube.

OPTION B : Ethanol precipitation

13. Place hook with DNA in an eppendorf tube containing 1ml of TE; gently twirl hook until DNA slides off the hook. Keep it for overnight to dissolve DNA.
14. Precipitate DNA by adding 1/10th volume of 3 M NaAc, pH 5.5 and 2 volumes of 100% ethanol and mix well.
15. Spool precipitated DNA with glass hook.

OPTION C : DNA washes

16. Place hook with DNA in eppendorf tube containing 1-2ml of wash solution 1. Leave DNA on hook in tube for about 20min.
17. Rinse DNA on hook briefly in 1-2ml of wash solution 2 and dry the DNA. DNA should neither contain residual ethanol nor should they be too dry. In both case, redissolving may be difficult.
18. Transfer DNA to new eppendorf tube containing 0.3-1.0ml of TE; gently twirl hook until DNA slides off the hook and dissolve the DNA. Store samples at 4°C. Intactness of DNA and the presence of contaminating polysaccharides are the main determinants of the duration of this step. High molecular weight DNA may take several hours to dissolve.

b) Removal of RNA

1. To the DNA solution add 20 μg/ml DNase free RNase and incubate at 37°C for 1 h.
2. Add 0.1 volume of 3 M sodium acetate, pH 5.2.
3. Add an equal volume of Phenol: Chloroform mixture (1:1) and mix well by swirling. Centrifuge and collect the aqueous phase.
4. Repeat the above extraction step till no white interface is seen
5. Precipitate the DNA by adding 2 volumes of absolute ethanol. Recover the DNA by centrifugation, wash it with 70% ethanol, air dry and dissolve in TE buffer.

c) Estimation of DNA Concentration, Purity and Quality

Assessment by Agarose Gel Electrophoresis

Quantification of the DNA in the extract may be achieved by analysis on an agarose electrophoresis gel with ethidium bromide staining. Comparison to a set of standards e.g. of lambda DNA, gives an estimate of the amount of DNA in an unknown sample. In contrast to the spectrophotometric method this technique allows at the same time, 1) DNA quantification, 2) estimation of the extent of contamination by RNA (which usually runs ahead) and 3) evaluation of DNA quality (i.e. extent of degradation).

Solutions

- **Ethidium Bromide : 10 mg / ml in dH_2O**
- 0.8% Agarose in electrophoresis buffer
- Electrophoresis buffer : TAE, TBE or TPE
- 10X Loading buffer : 40% bromophenol blue, 40% Xylene Cyanol FF, 50% Glycerol.

Method

1. To check for molecular weight, load about 2 mg of undigested DNA on the gel along with 0.5 mg of molecular weight standard such as lambda DNA cut with HindIII.
2. Compare the migration of the samples on the gel with that of the standard. The largest fragment in a lambda HindIII measures 23.1 Kb.
3. The bands for the sample must appear as a distinct solid fluorescence. Diffused bands reaching toward lower molecular weight indicates degradation of DNA fragments.

The concentration of total DNA may be estimated on the gel by comparing the fluorescence of digested samples of each preparation with those of digested concentration standards (made also of total DNA). This method also gives information on the digestibility of the sample DNA and can be routinely carried out with the checking of molecular weight.

1. Assume that all samples to be checked have a uniform concentration of, say, 1.0 mg / ml. Pipette out 5.0 ml of DNA from each sample.
2. Digest the DNA with a restriction enzyme by following the steps described in the section on restriction digest. The enzyme EcoR1 is relatively inexpensive and is a common choice for this purpose.

3. Also digest 2.0 ml of DNA from each stock of the known concentration standards. At least three (3) standards will do (i.e., Standards with concentrations of 1.5, 1.0 and 0.5 mg/ml).

4. Visualize the DNA on a gel following electrophoresis and after staining with ethidium bromide. With the photograph of the gel, compare the intensity of fluorescence of each sample with those of the standards. Since equal volumes of the unknown samples and the standards have been pipette out from their stocks, the concentration of an unknown can be directed equated to that of the standard based on intensity of fluorescence. Estimate concentrations for unknowns that fall between two standards.

Assessment by Spectrophotometer

Outlined below is the stepwise procedure for using the Beckman DnR-64 Spectrophotometer for DNA sample quantitation.

1. Plug the instrument and wait for the self-checkup to be completed.
2. Turn the uv lamp by pressing the "uv" button.
3. Warm up the machine for at least 30 mins. When ready, "uv" becomes "UV". To calibrate
4. Press 280 and then the lambda button. The wavelength will be automatically set at 280 nm.
5. Wash the cell with distilled H_2O. Dry with kimwipes.
6. Using a pipetman, clean the insides of the cell by pipetting TE buffer in and out of the cell.
7. Insert the cell containing 100 ml of TE into its chamber.
8. Press "CALB" button. "0.000" will appear on the display after a few seconds.
9. Press "260" and then the lambda button.
10. Press "CALB". To take measurements of samples
11. Remove the cell from its compartment and discard the TE.
12. Put 5.0 ml of the DNA sample in the cell . Add 95 ml of TE. Mix solution thoroughly by press-release method.

13. Introduce the cell into the compartment and close the cover tightly to avoid any leak of light.

14. Read the O.D. value directly from the screen. This gives the DNA concentration when multiplied by 1000. That is, if O.D. equals 0.200 then DNA concentration equals 0.200 x 50 x 20 = 200 mg / ml (50 because 1 unit O.D. corresponds to a concentration of 50 g / ml and 20 because the DNA solution was diluted 20X).

    To assess purity of the samples

15. Press "280" then lambda button.

16. Read the O.D. at 280 nm.

17. Compute for O.D. 260 / O.D. 280. A ratio value of 1.8 suggests a highly pure preparation of DNA.

    If this ratio is 1.8-2.0, the absorption is probably due to nucleic acids. A ratio less than 1.8 indicates that there may be proteins and / or other UV absorbers in the sample, in which cases it is advisable to reprecipitate the DNA. A ratio higher than 2.0 indicate the samples may be contaminated with chloroform or phenol and should be reprecipitated with ethanol (OPTION B)

18. Read the O.D. values for the other samples at 260 and 280 nm by repeating the steps described above. Make sure to wash the cell thoroughly with TE between DNA samples.

19. Unplug the instrument.

Safety Measures

1. Frequent freezing and thawing of plant material should be avoided
2. Carefully handle liquid nitrogen as it causes bums.
3. Carefully handle phenol as it also causes severe burns. It must be extra pure and buffered with TE.
4. Over drying of DNA pellet should be avoided as the over dried DNA pellet is difficult to dissolve in TE.
5. No harsh treatments should be given to DNA, i.e. handle DNA GENTLY to avoid its shearing.

Method 2

PEX Method of DNA Isolation

This is also a quick method of DNA isolation from small amount of fresh plant tissues (eg., single leaf of young wheat seedlings). The quantity of DNA isolated from this tissue is sufficient to carry out 50 PCRs approximately.

Steps involved

1. Make set of four 1.5ml eppendorf tubes.
2. Loosely 1/2 fill the eppendorf tube with freeze-dried leaf tissue.
3. Add 800 ml PEX buffer.
4. Incubate at 65°C (vortex after every 15 minutes) for 1 hr.
5. Centrifuge at 14,000 rpm for 5′ and collect all the supernatant in fresh tube (squeeze the sample).
6. Centrifuge again for 5′ at 14,000 rpm and collect all the supernatant in fresh tube. (400-600 ml supernatant).
7. Add 1/10th vol (60 ul for 600 ul supernatant) of 3M Sod Acetate pH 5.2 & 1 vol (660 ml) of isopropanol (100%).
8. Vortex and incubate at -20°C O/N.

Next Day

1. Centrifuge at top speed (14,000 rpm) for 30′
2. Discard supernatant
3. Add 1/2 ml of 70% ETOH (cold) to the pellet. Mix by inverting to wash the pellet.
4. Centrifuge at top speed (14,000 rpm) for 5′
5. Repeat ETOH wash.
6. Discard supernatant.
7. Air dry for 30′
8. Re-suspend pellet in 100 ml 1xTE, vortex
9. Incubate samples at 65°C for 10-15′ to help dissolve the pellet - Spin for 1′
10. Transfer the supernatant in fresh tube. The supernatant has DNA.
11. Quantitate DNA (as already explained in CTAB method).

Composition of PEX Buffer

12.5 mM PEX (potassium ethyl xanthogenate; Fluka #60045)

100 mM Tris:HCL, pH 7.5 (Trisma base adjusted to pH 7.5 with HCl)

700 mM NaCl

10 mM EDTA

Method 3

A Miniprep Method

(Source: Manual on Molecular Markers: Theory and Practices)

DNA isolation for PCR analysis can be a limiting step in MAS because generally the population size in a breeding program is quite large. A technically simple, rapid, and reproducible DNA isolation procedure is needed for marker-assisted breeding. A simple protocol to isolate DNA for PCR analysis which does not require liquid nitrogen, needs only a very small amount of tissue sample, and can be done very quickly (Zeng *et al.* 1995). The quality of the DNA isolated with the protocol is sufficient for PCR analysis.

Procedure 1

Principle: The procedure for DNA isolation must be able to lyse the cell walls and cell membranes and release the DNA in the soluble media. With the use of SDS (detergent) plant cells are lysed and nucleo-protein complex is dissociated. EDTA is a chelating agent which inhibits nuclease activity on DNA. The large number of phenolic compounds present in plant cells are removed by PVP and PVPP.

Reagents

1. Extraction buffer

   50 mM Tris Cl pH 8.0

   100 mM EDTA pH 8.0

   150 mM NaCI

   1.8% SDS

   1 .5% mixture of PVP and PVPP

2. Phenol: Chloroform: Isoamyl mix (25 : 24 : 1)

3. Ethanol

4. TE Buffer

Procedure

1. To 20 mg of tissue add 100µl extraction buffer. Macerate using sterile pipette tip.
2. Incubate at 65°C for 30′ with frequent shaking and inversions.
3. Add equal volume of phenol: chloroform: isoamyl alcohol mix. Mix gently and thoroughly for 2′ to form a blend.
4. Centrifuge at 10,000 rpm for 5′ at 4°C.
5. Remove top aqueous phase (supernatant) to a fresh eppendorf tube and add 2x volume of chilled ethanol. Mix by inversion. Incubate at -20°C for 5′ and centrifuge at 10,000 rpm for 5′ at 4°C. Discard the supernatant, air dry pellet and dissolve it in 50µl 1x TE buffer for storage.
6. Check quantity and quality of DNA as already explained in CTAB method.

Procedure 2

This method is used in order to get larger quantity and good quality DNA in short duration of time. Seeds are grown in ptriplates in dark, from germinated shoot tips (in which chlorophyll contents is less) DNA is extracted. It is sufficient for about 100 PCR reactions.

Protocol

1. Grow the plants in dark 2-3 days prior to tissue harvest to ensure low carbohydrate content.
2. Harvest tissues into 2ml, microfuge tube. Grind the tissue in liquid nitrogen.
3. Add 750ml of extraction buffer to the ground tissue, vortex to mix thoroughly and incubate at 65°C for 10 min.
4. Add about 150 ml of 5M K-acetate/Sodium Acetate, vortex, incubate on ice for 20 min. Spin @ 12k x g in microfuge for 10 min.
5. Transfer 750-800ml of supernatant to a new microfuge tube and add equal volume of isopropanol. Immediately spin down nucleic acid by centrifugation @ 12k x g for 4min. (Some small pieces of tissue may pellet at this stage, but they will be removed in the next step).
6. Wash pellet with 75% ethanol and resuspend in 200ml sterilized double distilled water. Spin @12k x g for 8 min.

7. Remove 180ml of supernatant to a new tube and add 20ml 3M Sodium -acetate (or Potassium acetate). Mix and add 500ml of ethanol, the incubate at -20°C for 10 min. Pellet the nucleic acid by centrifugation at 12 k x g for 5 min. This step will remove chlorophyll as well as any debris carried over from step 5.
8. Wash the pellet with 75% ethanol, dry and resuspend in 10-50 ml of TE.

Extraction Buffer

50 mM Tris pH 8.0

10 mM EDTA pH 8.0

100 mM NaCl

1% (w/v) SDS (sodium dodecyl sulphate)

10 mM β mercaptoethanol (add just before use)

Method 4

DNA Extraction Method for High Throughput Marker Analysis

(Source: Manual on Molecular Markers: Theory and Practices)

Another recent flexible method for plant DNA extraction is presented. It allows a high-throughput of samples in a short time without the need for freezing or lyophilizing the plant material. The method allows the isolation of genomic DNA with a yield of ≈ 100 mg from a minimal amount of 200 mg of leaf material. This is sufficient for work with large-genome plant species such as hexaploid wheat, where 20 mg of genomic DNA are required for a single RFLP analysis. DNA extraction in amounts suitable for RFLP analysis represents the most time-consuming and labour-intensive step in molecular marker analysis of plant populations.

Protocol

1. Take two leaves of two week old wheat seedlings or (200-300 mg) of leaf tissue of wheat plant.
2. Transfer harvested leaves to plastics bags by cutting them into small pieces.
3. Add 1.2 ml of extraction buffer and remove air carefully of the bag. Ensure double sealing to avoid any leakage.
4. Place bags on a piece of cardboard covered with a layer of PE attached to the bench and a second layer of PE foil over the bags.
5. Homogenize leaf material using a hand homogenizer.

6. Incubate for 1 hr at 65°C.
7. Separate bags from each other & transfer solution to a 2.2 ml micro centrifuge tube.
8. Extract solution with 800ml of dichloromethane: isoamylalcohol (24:1) by vertically rotating in a shaker for 15 min at 4°C.
9. Centrifuge at 10,000g for 15 min.
10. Pippet out 800 ml of supernatant in to the slots of a 2.0 ml polypropylene deep well plate and treat with 5ml of RNA seA for 15 min at 37°C.
11. Seal the deep well plate for incubation with a cap mat.
12. Add 0.7 vol. of isopropanol (560 ml) to each sample.
13. Seal the wells with cap mat and mix the solution by inverting the block five times.
14. Pellet down the DNA by centrifugation (20 min at 1850 g at room temp.).
15. Remove the supernatant by inverting the block.
16. To eliminate traces of supernatant tap the block upside down on stack of paper towel.
17. Wash the DNA by 0.7 ml of washing solution. (76% ethanol, 200mM sodium acetate).
18. Seal the block with the cap mat tilling it to each side to wash the walls of the wells.
19. Incubate for 5 min at room temp.
20. Centrifuge 10 min at 1850 g.
21. Remove the supernatant & give washing once again
22. Dry the pellet for at least 10 min & finally resuspend in TE buffer.

Extraction Buffer

2%(w/v) CTAB

200 mM Tris/HCl pH-8.0

20 mM EDTA pH 8.0

1.4M NaCl 1%(w/v)

PVP 1%

1% (v/v) BME

Method 5

Midi-prep CTAB DNA Extraction from Wheat Leaves

1. Collect young leaf samples from each plant (one leaf/plant) in a small polythene bag (10x10cm) and place in ice (these leaves can be used immediately for DNA extraction or these can be stored at -80°C deep freezer until needed)
2. Grind the leaf in liquid nitrogen in a 3″ mortar and transfer the ground material in 2-ml tubes (place tubes on ice or at -70°C if the number of samples is more).
3. Add 900ml of extraction buffer. Vortex to suspend ground samples.
4. Incubate for 40-50 minutes at 65°C. Stir after regular intervals (10min. interval) to homogenize the mixture.
5. Add 800 ml chloroform, mix well for 15-20 minutes and spin for 7 min at 13,000 rpm in a microcentrifuge.
6. Transfer the supernatant into a clean tube and add 600ml of isopropanol. Mix well by slowly inverting the tubes until DNA precipitates.
7. Fish out precipitated DNA or spin down by centrifugation.
8. Rinse the DNA with 70% ethanol and air dry by inverting tubes on a clean filter paper at room temperature.
9. Add about 50 - 100ml of 1 X TE
   Note: This procedure yields up to 50mg DNA.

Extraction Buffer

2 X CTAB Extraction Buffer (100ml)
lM Tris, pH 8.0
5M NaCl
0.5M EDTA
10% CTAB
Sod. bisulphite
β merceptoethanol
dds H_2O

Method 6

DNA Extraction from Seeds

The DNA extraction procedures requiring leaf tissues have a limitation of necessity of the presence of proper seed germinators / growth chambers, particularly in the non-crop growing season. This sometimes makes difficult to extract the DNA. However, a DNA extraction procedure is available using which it can be extracted from non germinative portion of the seed. This method also does not necessarily require Liquid Nitrogen, which is must in other methods explained earlier.

Protocol

1. Grind 10 seeds to fine powder using sterile, dry pestle and mortar.
2. Transfer the powder in 2 ml centrifuge tube.
3. Add 900 ml of extraction buffer to each tube. Invert the tubes several times with gentle tapping to mix the contents thoroughly.
4. Incubate the samples at 65°C in water bath for 1 hour- mix the contents after every 15 min by inverting the tubes several times.
5. Add 900 ml of chloroform: isoamylalcohol solution (24:1) in each tube and mix the contents by inverting the tubes gently for 5 min.
6. Centrifuge the tubes for 10 minutes at 5000 rpm.
7. Carefully remove the supernatant (upper layer) and transfer it in fresh tubes.
8. Repeat the steps 5, 6, 7.
9. Centrifuge the tubes at 6500 rpm for 10 minutes to pellet the protein contamination left during pipetting of supernatant.
10. Remove the supernatant in fresh tube.
11. Add 2 volumes of ice-cold isoproparal and mix the contents by inverting the tubes several times and keep the tubes at 4°C for 30 to 45 min.
12. Centrifuge for 10 min at 6000 rpm. Pipette off solution without disturbing the DNA pellet/settled at the bottom of tubes.
13. Add 500 ml 70% ethanol, centrifuge and remove the supernatant.
14. Air dry the pellet till smell of ethanol goes off.
15. Add 100 ml of TE buffer or sterile double distilled water.

Solutions

Extraction buffer: 100 mM TrisCl

500 mM NaCl

50 mM Sodium EDTA

1.2% SDS

3.8 g/l sodium bisulfite

Chloroform : isoamylalalcohol (24:1)

Isopropanol

Ethanol

TE buffer (10 mM Tris Cl pH8.0 and 1mM EDTA pH8.0)

Method 7

Isolation of Genomic Plant DNA Using BN QuickPick™ Protocol

For the rapid purification of genomic plant DNA

Sourcs: http://www.metachem.co.uk/PickPen.htm

The QuickPick genomic Plant DNA purification kit provides a rapid and simple means of isolating genomic DNA from plant tissue. The method does not require any organic solvents and eliminates the need for centrifugation, vacuum filtration or column separation.

Principle

DNA in the sample is released using Plant DNA Proteinase K Solution and Plant DNA Lysis Buffer. Released DNA is bound to Plant DNA MagaZorb™ Magnetic Particles in the presence of Plant DNA Binding Buffer. Plant DNA MagaZorb™ Magnetic Particles with the bound DNA are captured with the PickPen™ and the contaminants are removed by washing with Plant DNA Wash Buffer. The DNA is then eluted from the magnetic particles with the Plant DNA Elution Buffer. The procedure starting from homogenised plant sample and ending with purified DNA lasts less than 40 minutes.

BN QuickPick™ Genomic Plant DNA Isolation Protocol

1. Number tubes 1 to 5 and pipette QuickPick genomic Plant DNA kit reagents into tubes as follows :

   Tube 1 : 5µL Plant DNA Proteinase K Soultion is added to the homogenised leaf tissue suspended in 30µL of the Plant DNA Lysis Buffer. Mix by gentle pulse vortexing and lyse at 65°C for 15-30 minutes.

   1

   Lyse at 65°C for 15-30 minutes

   Tube 2 : 125µL Plant DNA Binding Buffer, 5µL Plant DNA Magnetic Particles.

Tube 3 : 200µL Plant DNA Wash Buffer.

Tube 4 : 200µL Plant DNA Wash Buffer.

Tube 5 : 100µL Plant DNA Elution Buffer

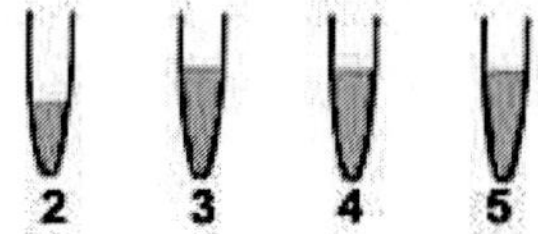

2. Remove **Tube 1** from 65°C. Centrifuge tube at 16,000 x g for 5 minutes. Gently transfer supernatant into **Tube 2** (Plant DNA Binding Buffer. Plant DNA MagaZorb™ Magnetic Particles). Mix tube gently and incubate at room temperature for 2-10 minutes.

   Bind at room temperature with continuous mixing for 2-10 minutes

3. Collect the magnetic particles from **Tube 2** with the PickPen and release them into **Tube 3** (Plant DNA Wash Buffer). Repeat washing step in **Tube 4.**

4. Collect the magnetic particles from **Tube 4** with the PickPen and release them into **Tube 5** (Plant DNA Elution Buffer). Mix **Tube 5** gently and continuously and elute at room temperature for 5 - 10 minutes.

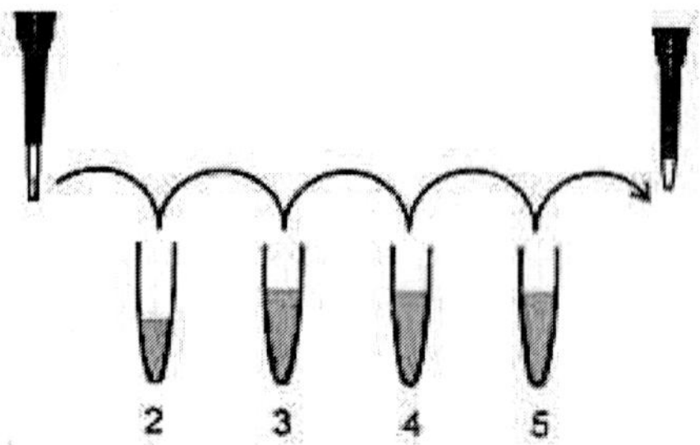

5. Collect the magnetic particles from **Tube 5** and discard them. The eluate containing genomic DNA is ready for further analysis.

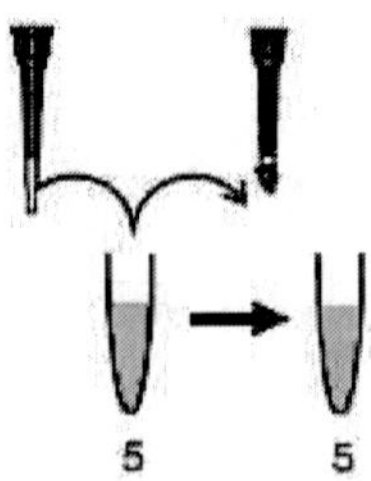

Collect the magnetic particles and discard

Genomic Plant DNA from *Hordeum vulgare* isolated with QuickPick Plant DNA kit

Kits are available for 48 and 96 preparations of plant DNA.

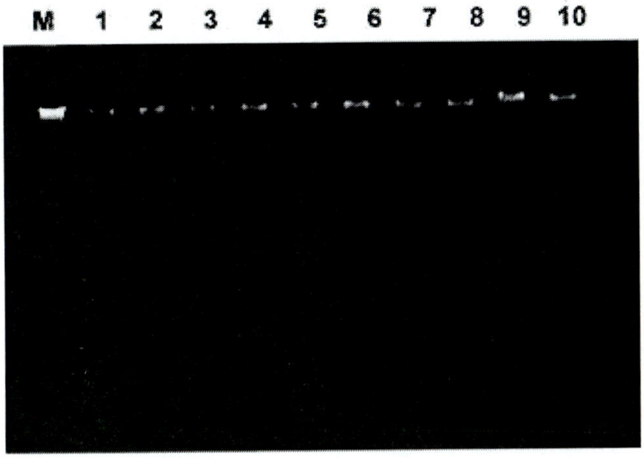

M = Marker

1-10 = DNA isolated from 25 mg of barley

Method 8

Protocol for Total DNA Isolation: Dellaporta *et al.* (1983)

Miniprep Procedure

1. Weigh 5 g of 5-day old etiolated seedlings, quick freeze in liquid nitrogen and grind to a fine powder in a prechilled mortar and pestle. Transfer powder with liquid nitrogen into a 30 ml Oak Ridge tube.

   Note: Avoid thawing of tissue once frozen until buffer is added, and do not cap the tubes while nitrogen is evaporating.

1. Add 20 ml of preheated (65°C) Extraction Buffer (Dellaporta Buffer), mix gently to avoid shearing of DNA {Dellaporta Buffer: 100 mM Tris-HCI ph 8.0, 50 mM EDT A pH 8.0, 100 mM NaCl, 2% of 10% SDS}.
2. Add 100 f.11 of 10 mg/ml proteinase K (final concentration of 0.05 mg/ml), mix gently and incubate at 65°C in hot water both for 1 h with occasional gentle inversion to mix tube contents.
3. Add 10 ml phenol + 10 ml chloroform (24 chloroform:1 isoamyl alcohol) in each tube, mix gently by inversion to form an emulsion. Make sure that the samples are completely mixed.

4. Centrifuge at 2000 rpm for 20 min in bench-top centrifuge to separate phases, take supernatant and pour into fresh tubes. The phenol/ chloroform extraction may be repeated.
5. Add equal volume (20 ml) of chloroform in each tube, mix gently and centrifuge at 2000 rpm for 5-10 min, and transfer the supernatant to the fresh tubes.
6. To the supernatant, add 0.6 volume (15 ml) of isopropanol and mix by inversion. DNA precipitate appears at interphase. Spool out the precipitate with a glass hook, and rinse in an excess of 70% ethanol. Place precipitate into a 1.5 ml polypropylene centrifuge tube, rinse 2-3 times with 70% ethanol and air dry/vacuum dry briefly (avoid over drying).
7. Dissolve the pellet in 700 μl of $T_{50}E_{10.}$ Warming (50°C) helps, but keep for longat 4°C.
8. Add 7 μl RNase (final concentration of 50 μg /ml) incubate at 37°C for 1 h.
9. Add 350 μl phenol + 350 μl chloroform, mix gently by inversion and centrifuge at 12000 rpm for 5 min. Save supernatant, and repeat phenol/ chloroform extraction.
10. To the supernatant, add equal volume (700 μl) chloroform, mix gently by inversion and centrifuge at 12000 rpm for 5 min. Repeat chloroform extraction if interphase is excessive.
11. Add equal volume of chilled isopropanol, mix by inversion. DNA precipitate appears. Spool out DNA on glass hook, rinse in 70% ethanol, place in a sterile 2 ml screw cap tube and drain excess 70% ethanol after quick centrifugation. Air dry O/N or vacuum dry briefly.
12. Add 0.5 to 1.5 ml of $T_{10}E_1$ (depending upon the size of DNA pellet), and allow DNA to dissolve.
13. Samples are stored at 4°C for use.

Method 9

Large-Scale DNA Extraction

Source: Laboratory protocols; CIMMYT Applied Molecular Genetics Laboratory

Lyophilization

1. Harvest leaves from greenhouse or field grown plants. It is preferable to use young leaves without necrotic areas or lesions, although older leaves which are not senescent may be used.

2. If the midrib is thick and tough, remove it. Cut or fold leaves into 10-15 cm sections and place in a plastic screen mesh bag along with the tag identifying the sample. (Aluminum foil or paper bags may be substituted if holes are punched to allow good air flow.) Place bags in an ice chest or other container with ice to keep samples cool (but do not allow them to freeze). Make sure samples do not get wet.

3. Place leaf samples in a Styrofoam container or another type container that will to hold liquid nitrogen. Add liquid nitrogen to quick-freeze samples. *Once frozen, do not allow samples to thaw until freeze-dried!*

   Note: Leaf samples may be frozen and stored at -80°C until ready to be lyophilized.

4. Transfer frozen leaf samples to lyophilizer. Make sure the lyophilizer is down to temperature (the chamber is = -50°C) and pulling a good vacuum (= 10 microns Hg) before loading samples. Do not overload lyophilizer: make sure the vacuum is always = 100 microns and condenser temperature is = -50°C. Samples should be dry in 72 hours. Typically, fresh weight ~ 10X dry weight.

5. Dried leaf samples may be stored in sealed plastic bags at room temperature for a few days or, preferably, at -20°C for several years.

6. Fill out a harvesting record sheet.

Grinding

1. Grind to a fine powder with a mechanical mill (Tecator Cyclotec Sample Mill, Model 1093), into a plastic scintillation vial or any other appropriate plastic container that can be closed airtight.

   Note: If the plant material weighed less than 4 g fresh weight, grind to a powder in a coffee mill or a mortar and pestle with liquid nitrogen. The finer the grind, the greater the yield of DNA from a given amount of material.

2. Store ground samples tightly capped at -20°C. Samples are stable for several years.

Genomic DNA Isolation

(Based on Method of Saghai-Maroof *et al.*, 1984[1])

1. Weigh 300-400 mg of ground, lyophilized tissue, into a 15 ml polypropylene centrifuge tube. DNA yields range from 50 to more than 100 μg DNA/100 mg dry tissue. If higher amounts are needed, start

with 1 g lyophilized tissue into a 50 ml polypropylene centrifuge tube, and triple all the amounts given below. If lower amounts are needed, then weigh 100-150 mg lyophilized tissue into a 5 ml polypropylene centrifuge tube, and use 1/3 of the amounts given below.

2. Add 9.0 ml of warm (65°C) CTAB extraction buffer to the 300-400 mg ground, lyophilized tissue. It is best to distribute tissue along the sides of the tube before adding buffer, to avoid clumping of dry tissue in the bottom. Mix several times by gentle inversion.
3. Incubate for 60-90 min, with continuous gentle rocking in a 65°C oven.
4. Remove tubes from oven, wait 4-5 min for tubes to cool down, and then add 4.5 ml chloroform/octanol (24:1). Rock gently to mix for 5-10 min.
5. Spin in a table-top centrifuge for 10 min at 1300-1500 x g[2] at RT.

   Note: Below 15°C the CTAB/nucleic acid complex may precipitate. This could ruin the preparation and damage the centrifuge.
6. Pour off top aqueous layer into new 15 ml tubes. Add 4.5 ml chloroform/octanol and rock gently for 5-10 min.
7. Spin in a table-top centrifuge for 10 min at 1300-1500 x g[2] at RT.
8. Pipette top aqueous layer into new 15 ml tubes containing 30 μl of 10 mg/ml RNase A (pre-boiled). Mix by gentle inversion and incubate for 30 min at RT.
9. Add 6.0 ml of isopropanol (2-propanol). Mix by very gentle inversion.
10. Remove precipitated DNA with glass hook.[3] Continue with **OPTION A, B, or C**.

OPTION A: Phenol Extraction to Obtain DNA of Higher Purity

This option is usually not necessary for RFLP analyses, unless DNA does not digest properly. In fact, it is better to perform phenol extraction only after restriction digestion; this improves DNA band separation and resolution after electrophoresis (see later sections for details).

1. Place hook with DNA in 5 ml plastic tube containing 1 ml of TE; gently twirl hook until DNA slides off the hook. Cap tubes and rock gently overnight at room temperature to dissolve DNA.
2. Phenol extract each sample with 1 ml (1x original TE volume) of equilibrated phenol or 1:1 phenol:chloroform. Centrifuge the sample 10 min at 1300 x g1 in swinging bucket rotor.

3. Transfer top (aqueous) layer to new 5 ml tube. Extract DNA with 1 ml (1x original TE volume) of chloroform/octanol. Centrifuge the sample 10 min at 1300 x g[1] in swinging bucket rotor. Transfer top (aqueous) layer to new 5 ml tube.

OPTION B: Ethanol Precipitation

4. Place hook with DNA in 5 ml plastic tube containing 1 ml of TE; gently twirl hook until DNA slides off the hook. Cap tubes and rock gently overnight at room temperature to dissolve DNA.
5. Precipitate DNA by adding 50 µl of 5 M NaCl and then 2.5 ml absolute EtOH (2.5 original TE volume); mix by gentle inversion.
6. Remove precipitated DNA with glass hook.

OPTION C: DNA Washes

7. Place hook with DNA in 5 ml plastic tube containing 3-4 ml of WASH 1. Leave DNA on hook in tube for about 20 min.
8. Rinse DNA on hook briefly in 1-2 ml of WASH 2 and transfer DNA to 2 ml microfuge tube (preferably Sarsted with screw-on lids to avoid possible evaporation of the TE) containing 0.3-1.0 ml TE (based on experience, we use 0.3-0.5 ml for maize and 0.5-1.0 ml for wheat); gently twirl hook until DNA slides off the hook. Cap tube and rock gently overnight at room temperature to dissolve DNA. Store samples at 4°C.

CTAB Extraction Buffer[1]

STOCK	[FINAL]	1 RXN 10 ml	5 RXN 50 ml	10 RXN 100 ml	20 RXN 200 ml	50 RXN 500 ml	60 RXN 600 ml
dH2O		6.5 ml	32.5 ml	65.0 ml	130.0 ml	325.0 ml	390.0 ml
1 M Tris-7.5	100 mM	1.0 ml	5.0 ml	10.0 ml	20.0 ml	50.0 ml	60.0 ml
5 M NaCl	700 mM	1.4 ml	7.0 ml	14.0 ml	28.0 ml	70.0 ml	84.0 ml
0.5 M EDTA-8.0	50 mM	1.0 ml	5.0 ml	10.0 ml	20.0 ml	50.0 ml	60.0 ml
CTAB[2]	1 %	0.1 g	0.5 g	1.0 g	2.0 g	5.0 g	6.0 g
14 M BME[3]	140 mM	0.1 ml	0.5 ml	1.0 ml	2.0 ml	5.0 ml	6.0 ml

[1] Use freshly made; warm buffer to 60-65°C before adding the CTAB and BME.

[2] CTAB = Mixed alkyltrimethyl-ammonium bromide (Sigma M-7635).

[3] Add BME (β-mercaptoethanol) just prior to use, under a fume hood

WASH 1: 76% EtOH, 0.2 M NaOAc

STOCK	100 ml	200 ml	300 ml	400 ml	500 ml
Absolute EtOH	76 ml	152 ml	228 ml	304 ml	380 ml
2.5 M NaOAc	8 ml	16 ml	24 ml	32 ml	40 ml
dH2O	16 ml	32 ml	48 ml	64 ml	80 ml

WASH 2: 76% EtOH, 10 mM NH4OAc

STOCK	100 ml	200 ml	300 ml	400 ml	500 ml
Absolute EtOH	76 ml	152 ml	228 ml	304 ml	380 ml
1 M NH4OAc	1 ml	2 ml	3 ml	4 ml	5 ml
dH2O	23 ml	46 ml	69 ml	92 ml	115 ml

CHLOROFORM:OCTANOL: 24:1

STOCK	100 ml	200 ml	300 ml	400 ml	500 ml
Chloroform	96 ml	192 ml	288 ml	384 ml	480 ml
Octanol	4 ml	8 ml	12 ml	16 ml	20 ml

DNA Extraction from Small Amounts of lyophilized Tissue

1. To extract DNA from small amounts of lyophilized tissue (~ 50 mg), use 2 ml tubes and proceed as follows:
2. Add 1 ml of CTAB buffer.
3. Incubate for 60 min with continuous movement?
4. Remove tubes from incubator, let them cool, and add 1 ml of chloroform:octanol. Mix for 10 min.
5. Centrifuge for 10 min.
6. Remove 700 μl of the top aqueous layer.
7. Add 10 μl of 10 mg/ml RNase A. Mix and incubate for 30 min.
8. Add 1 ml of isopropanol and mix.
9. Centrifuge tubes for 15 min at 12000 rpm to precipitate DNA.
10. Remove the supernatant and dry the DNA at RT.
11. Re-suspend in 200 μl TE.

Method 10

DNA Extraction Using the Sap Extractor

(Based on Method of Clarke et al., 1989)

1. Setting up and using the sap extractor : Make sure that the rollers are completely clean and that the flushing system for cleaning the rollers between samples is connected to a high pressure source of de-ionized water. If you can only use tap water to flush the rollers, make sure that you finally rinse them thoroughly with de-ionized or dH_2O between samples. Always wipe the rollers dry using clean, soft tissue paper before initiating the following sample extraction. Position the buffer feeding tip over the upper half of the rollers to ensure that the buffer will mix effectively with the pressed tissue sample. Feed the tissue sample between the rotating rollers at a slight angle to ensure even pressure is applied to a single layer of the tissue (the tissue will wrap around one roller in a spiral).

2. Use 150-250 mg of freshly harvested leaf tissue kept in ice (within a tube) or frozen at -80°C (within a tube). It is critical that as you feed the tissue into the extractor, between the rollers, the buffer should already be at that position in the rollers. So make sure that you synchronize this operation well with the pumping of the buffer; otherwise, the DNA will be degraded.

   Pump 1.0 ml of extraction buffer and collect the extract in 2 ml tubes at the tips of the rollers.

3. Incubate the extracts in a water bath or an oven at 65°C for 20-40 min; mix gently twice or continuously during this incubation. Remove the tubes from the heat and let cool for 5-10 min.

4. Extract the samples with 1 ml of octanol-chloroform (1:24). Mix by inversion for 5 min; then spin in a table-top centrifuge at 3200 rpm for 10 min.

5. Transfer the aqueous supernatant containing the DNA to 2.0 ml Eppendorf tubes.

   If the DNA has to be quantified precisely at the end of the extraction, add 10-20 µl of RNAse A + T1 (see other protocols) in the tube and incubate for 30 min at 37°C, or for one hour at RT.

6. Add 75 µl of 5M NaCl and precipitate DNA with 1 ml of cold absolute ethanol.

7. Spin DNA down, decant ethanol, and dry under a weak vacuum for 30 min.
8. Re-suspend overnight in the cold room in 200-500 µl TE, pH 8.0.
9. Quantify using a gel method or a TKO fluorometer. With this method, a minimum of 15 µg of DNA can be obtained.

Extraction Buffer[1]

STOCK	[FINAL]	10 ml	50 ml	100 ml	200 ml
dH_2O		1.7 ml	8.5 ml	17.0 ml	34.0 ml
1 M Tris-8.0	100 mM	1.0 ml	5.0 ml	10.0 ml	20.0 ml
5 M NaCl	2100 mM	4.2 ml	21.0 ml	42.0 ml	84.0 ml
0.5 M EDTA-8.0	150 mM	3.0 ml	15.0 ml	30.0 ml	60.0 ml
PVP[2]	0.5%	0.05 g	0.25 g	0.5 g	1.0 g
CTAB[3]	2.0%	0.2 g	1.0 g	2.0 g	4.0 g
14 M BME[4]	140 mM	0.1 ml	0.5 ml	1.0 ml	2.0 ml

[1] Use freshly made; warm buffer to 60-65°C before adding the CTAB and BME.

[2] We recommend using Sigma PVP, catalog PVP-40 (polyvinyl pyrrolidone with 40,000 average molecular weight).

[3] CTAB = Mixed alkyltrimethyl-ammonium bromide (Sigma M-7635).

[4] Add BME (β-mercaptoethanol) just prior to use, under a fume hood.

Method 11

Small-Scale Extraction of High Quality DNA

The grinding of fully lyophilized leaf tissue before extraction can give very high quality DNA in quantities that depend on the methods used. The large-scale grinding and extraction process used on Method 9 for RFLPs can be conveniently scaled down to grinding in a coffee grinder or by using small metal beads in a 1.5 ml tube. These methods provide a cheap, fast, and easy way to obtain small-to-medium amounts of very high quality DNA.

Lyophilization

1. Harvest the youngest fully mature leaf from plants grown in the greenhouse or field. It is best to use young plants without necrotic or damaged areas, but mature plants may be used if they are not yet beginning to senesce.

2. The final amount of DNA needed will determine which of the two procedures (stainless steel balls or coffee grinder) you will use. Each uses a different amount of leaf tissue. When material is scarce or only very low quantities of DNA are needed from each individual plant, the stainless steel ball procedure is recommended as more samples can be processed at a time. If more DNA is needed or if DNA will be extracted from many plants and bulked for population analysis, the coffee grinder procedure should be used.

Stainless Steel Balls

1. Small portions of the leaf (0.5-1.5 cm) are cut from each plant and placed in a 1.5 or 2.0 ml tube. Leaves from more than one plant can be placed in the same tube, which will accommodate a maximum of 6 leaves.
2. Keep tubes of leaves cool until they can be frozen, but freeze as soon as possible. Freeze in a -80°C freezer overnight or using liquid nitrogen. Samples must not thaw before lyophilization.
3. Place trays of the open tubes containing frozen leaf materials into the lyophilizer. Lids of the tubes must be OPEN.
4. Be sure that the lyophilizer chamber is at -60°C at all times. Verify that it has reached the proper vacuum level after loading the samples, and that it maintains a vacuum level of - 100 microns. Fortunately, the small leaf sizes in each tube makes it hard to overload the machine. Typically, samples must dry for 72 h, but may take less time using this method.
5. Dried tissue may be stored in the tubes (with the lids CLOSED) at room temperature for a few days, or can be stored for longer periods at -20°C. DNA extraction can be started in the same tubes.

Coffee grinder

6. Cut one leaf per plant (8-10 cm or so) and place the leaves in a glassine bag. 15 – 20 leaves can be placed in the same bag. Keep the samples cool until they can be frozen, but freeze as soon as possible. Freeze in the -80°C freezer overnight or using liquid nitrogen.
7. For the analysis of populations via bulks, we recommend the use of 15 plants, which must be the same age. Cut the youngest fully mature leaf of each one, with a size at 10 cm in longitude. The size and maturity of the leaves must be exactly the same, as the quantity of DNA depends on both factors, and equal quantities of DNA must be extracted from each plant.

8. Glassine bags with samples can be stored in a sealed plastic bag at -80°C until lyophilized. Keep samples in the freezer for at least 12 h, unless liquid nitrogen is used to accelerate the procedure; samples can be placed in the lyophilizer directly from the liquid nitrogen. Samples must not thaw before lyophilization.
9. Transfer samples to the lyophilizer. Be sure the lyophilizer chamber is at -60°C at all times. Verify the proper vacuum level has been reached after loading the samples, and that a vacuum level of -100 microns is maintained. Do not overload the chamber. Samples typically dry in 72 hours, but may take longer if many satellite chambers are placed in the lyophilizer.
10. Dried leaf samples may be stored at room temperature for a few days in a sealed plastic bag. They may be stored for longer periods at -20°C.

Grinding

Stainless Steel Balls

1. The stainless steel balls used in this procedure are 5/32" (4 mm) and may be purchased by the thousand at "Baleros y Bandas Sánchez," Juárez Sur No. 340, Texcoco, Mex., Tel. 9540687.
2. If leaves were dried in glassine bags before grinding, they may still be cut and placed into 1.5 ml tubes; however, once the leaves are dry, cutting them is difficult as they tend to disintegrate.
3. Place 2-3 stainless steel balls (4 mm in diameter) into each tube and close securely. Place the tubes in a Styrofoam holder and secure the lid of the holder with several strong rubber bands.
4. Place the Styrofoam holder with tubes on an orbital shaker and secure to the shaker with
5. Rubber bands or laboratory tape. Keep the tubes in constant vibration on the shaker at 400 rpm for 2 h or until leaf tissue is ground to a fine powder.
6. Alternatively, the Styrofoam holder can be taped or secured to a vortexer, which should be left on for 1-2 h until samples are finely ground.
7. Use a magnet to remove the stainless steel balls from the powder before beginning extraction.
8. Leaf powder can be stored in the closed tubes, or DNA extraction can begin immediately in the same tubes.

9. If samples are not fully dried before grinding, grinding will be inefficient and DNA yield will be poor. The finer the powder, the higher the yield of DNA will be.

Coffee grinder

10. Coffee grinders can be any brand, but we buy Braun grinders in Texcoco at Carrillo Alonzo, Art. 123 No. 7, Col. Centro, Tel. 55123046. Coffee grinders are modified by taping clear plastic over the lids; otherwise, leaves will become trapped in the lids during grinding and will not be ground.
11. Place the dried leaf tissue in the coffee mill and grind to a fine powder (from 30 sec to 2 min). The finer the powder, the higher the yield of DNA will be.
12. Collect leaf powder into a 15 ml tube using a paintbrush and a paper funnel.
13. Place the cap on the tube and keep sealed until ready to extract. DNA extraction can begin in the same tubes.
14. Using a damp cloth, fine brush, or compressed air, clean the coffee grinder after each sample is ground to avoid contamination.

Genomic DNA Isolation

With this method, from 50 to 100 ug of DNA per each 100 mg leaf tissue may be obtained. When extracting DNA from larger amounts of tissue, increase the amounts given below (up to 1000 mg).

1. Preheat the CTAB isolation buffer to 65°C.
2. Place 50 mg of lyophilized ground leaf tissue in a 2.0 ml tube (if using a 1.5 ml tube, all volumes may be scaled down by 25%).
3. Add 1 ml of CTAB isolation buffer. Mix by gentle swirling to homogenize the tissue with the buffer.
4. Incubate the samples at 65°C for 90 min with continuous gentle rocking.
5. Remove tubes from the oven and allow them to cool for 5-10 min.
6. Add 500 µl of chloroform:octanol (24:1). Mix gently with continuous rocking for 10 min at room temperature.
7. Centrifuge at 3500 rpm at room temperature for 10 min to generate a yellow aqueous phase and a green organic phase.

8. Remove approximately 750 µl of the yellow aqueous phase and place in a new 1.5 or 2.0 ml tube containing 5 µl RNAse. Optional step: Repeat the chloroform treatment on the aqueous phase. This produces cleaner *DNA,* but a lower yield.
9. Mix with gentle inversion and incubate at 37°C for 30 min.
10. Add ½ volume ice-cold 100% isopropanol (2-propanol). Mix very gently to precipitate the nucleic acid. *Optional step: Incubate samples at -20°C overnight, especially if precipitated DNA is not visible.*
11. Centrifuge at 3500 rpm at room temperature for 30 min to form a pellet at the bottom of the tube. Discard the supernatant. *Optional step: Phenol extract each sample with 0.5 ml 1:1 phenol:chloroform according to phenol extraction procedures on page 3. This is rarely necessary when using lyophilized tissue.*
12. Add 1 ml of 75% ethanol. Wash the DNA pellet gently. Discard ethanol by decantation. Wash once again. Allow pellet to air-dry until ethanol evaporates completely. Any remaining alcohol smell indicates pellet is not completely dry. Re-suspend the DNA pellet in 1 ml of TE or double-distilled water. Store samples at 4°C until use; if DNA will not be used for a long time, store at -20°C.

    Note: DNA that is re-frozen after being thawed begins to break after each freezing session, so freeze DNA only for long-term storage and preferably after all testing is finished. If DNA will be used for multiple projects with long periods of time between projects, it can be aliquoted into several tubes and frozen, so that each aliquot is thawed only once at the start of each project.

4

Isolation, Purification and Quantification of RNA from Plants

Although it can be argued that due to its overall trendiness, they have been a lot of recent developments in making it more fun to work with. As a molecule, it is very similar to DNA (the most notable chemical difference is its flexibility), but suffers because the nucleases (RNases) that can destroy it are notoriously very stable and very difficult to get rid of.

• Care to be Taken While Working with RNA

1. Use disposable plasticware: Essentially, there are different levels of care. There is sterile plastic ware that you take good care of, there are RNase free grade plastic ware that you take care of, and it's also a good idea to use filtered tips if possible.
2. Decontaminate other equipment from RNAses: There's a couple different ways of doing this. For instance, with glassware, you can thoroughly clean with detergent, rinse, and then ovenbake at >25°C for >4hrs. Things that are not oven friendly (please don't put your gel box in the oven) often use DEPC. DEPC is a very useful reagent for making buffers and equipment RNase free. What does it do? DEPC results in a covalent modification of nucleases rendering them inactive. Histidine modifier – imine -> carbonate >NH >N-CO3.

3. In general, Work in fumehood when treating water/buffers or equipment. Usually let your buffer + DEPC or equipment/glassware incubate at ~0.1% DEPC concentration for >12hrs. Then leave it in fumehood O/N with venting (lid slightly ajar) SO THAT FUMES ESCAPE. Or autoclave for faster evaporation (if possible). NOTE that technically, things are "DEPC treated" meaning that the DEPC is no longer present. Some nuances regarding DEPC: comes as a liquid/ can buy at different percentages.

• TRICKS

1. Lots of quickie protocols out there in the web especially. For example, a gel box can be cleaned with 0.5% SDS. Rinsed with water, dried with ethanol. Filled with 3% hydrogen peroxide (10min). rinse thoroughly with Rnase free dH2O.
2. Chloroform also denatures RNAses, (rinse with chloroform). Can be quite harmful to plasticware.
3. Fancy reagents like "Rnase AWAY." "RNase ZAP" Can wipe apparatus clean. Not harmful to plastics.
4. RNase inihibitors: so many classes of RNases to worry about (A, B, C, I, T1, T2...) can get expensive.

RNA ISOLATION

Method 1

Reagents

a) Liquid nitrogen

b) Extraction buffer: 0.25 M NaCl, 0.05 M Tris-HCl (pH 7.5), 20 mM EDTA, 1% (w/v) sodium dodecyl sulphate (SDS) and 4% (w/v) PVP (M.W. 360,000)

c) Chloroform : Isoamyl alcohol (CI, 24:1 v/v)

d) Phenol: Chloroform isoamyl alcohol (PCI, 1:1 v/v)

e) 3 M sodium acetate (adjusted to pH 5.2 with acetic acid)

f) Cold 70 % (v/v) ethanol

g) Cold absolute ethanol

h) 0.1 % (v/v) diethyl pyrocarbonate (DEPC)-treated-autoclaved water

i) 10 M LiCl

Equipment

- Pre-cooled pestle and mortar
- High speed centrifuge
- Spectrophotometer
- Agorose gel electrophoresis equipment
- Power supply
- Vortex mixer

Procedure

1. Add 7.5 ml of extraction buffer and 7.5 ml of CI to a 30 ml-round bottom microfuge tube
2. Grind 1 g frozen sample to fine powder with a mortar and pestle in liquid nitrogen.
3. Transfer 1 g of ground sample to a tube containing the extraction buffer and CI and vortex vigorously.
4. Centrifuge at 13,000 *g* for 2 min at room temperature.
5. Transfer the supernatant to a new 30 ml-round bottom microfuge tube. and purify with an equal volume of PCI. Centrifuge at 13,000g for 2 min at room temperature. Repeat this step until there is a clean interface observed.
6. Transfer the supernatant to a new 30 ml-round bottom microfuge tube and add one tenth volume of 3 M sodium acetate pH 5.2 and 2.5 volume of cold absolute ethanol, mix well, and incubate at 4°C for 30 min.
7. Recover the nucleic acids by centrifugation at 13000g for 20 min at 4°C.
8. Wash the pellet with 70% (v/v) ethanol, air-dry, and add 200µl DEPC-treated water to dissolve the pellet.
9. Transfer the supernatant to a 1.5 ml tube and add 10 M LiCl to a final concentration of 2 M and keep on ice for 1 h. Centrifuge at 19000g for 20 min at 4°C.
10. Wash the pellet with 70% (v/v) ethanol, air-dry, and add 20 µl DEPC-treated water to dissolve the pellet. Centrifuge at 19000g for 10 min at 4°C.
11. Add 0.1 volume of 3 M sodium acetate pH 5.2 and 2.5 volume of cold absolute ethanol, mix well, and store at -80°C.

    RNA can be stored in ethanol at -80°C for at least 2 months.

Analysis of RNA quality

12. The total RNA was quantified with a spectrophotometer at 230, 260, 280 nm. The integrity of total RNA was verified by analyzing approximately 1μg RNA sample on 1% (w/v) formaldehyde denaturing agarose gel (Sambrook *et al.*, 1989).

The protocol developed in this study also gave the best quality of RNA the $A_{260/280}$ and $A_{260/230}$ obtained were 1.79 and 2.75.

Method 2

Extraction of RNA with CTAB and Macaloid

Reagents needed :

Washing buffer :

76 % ethanol and 10 mM ammoniumacetate

CTAB : (store at room temperature)

2 % (w/v) CTAB (hexadecyltrimethylammoniumbromide)

1,4 M NaCl

0,2 % (v/v) 2-mercaptoethanol

20 mM EDTA

100 mM Tris/HCl pH 8,75

100 μg/ml proteinase K : (add just before use)

Macaloid :

Procedure :

Everything has to be Done on Ice

1. Put in a 2 ml tube with screwcap 500 μl CTAB and ± 0,1 g glass beads and 6 μl proteinase K (1 mg/100 μl)
2. Add a nematode drop (from a culture dropped into liquid nitrogen)
3. Beadbeate during 30 seconds at 5000 cycles/minute (if the ice is almost completely defrosted)
4. Incubate 30 minutes at 60°C and shake regularly
5. Add 500 μl chloroform/isoamylalcohol (24/1) and shake gently
6. Centrifuge 10 minutes at 8.000 rpm (4°C)
7. Transfer the aqueous fase into an other tube and add an equal volume cold isopropanol

8. Put 1 hour at -20°C
9. Centrifuge 15 minutes at 14.000 rpm (4°C)
10. Remove supernatans
11. Dissolve pellet in 500 µl DEPC destilled water, add 170 µl macaloid and 500 µl phenol/chlorophorm/isoamylalcohol pH 8 and shake for 1 minute
12. Centrifuge 15 minutes at 10.000 rpm (4°C)
13. Transfer the aqueous fase into an other tube
14. Add 500 µl DEPC destilled water, add 170 µl macaloid and 500 µl phenol/chlorophorm/isoamylalcohol pH 8 and shake for 1 minute
15. Centrifuge 15 minutes at 10.000 rpm (4°C)
16. Transfer the aqueous fase into an other tube
17. Add 0,1 volumes sodiumacetate pH 5 and an equal volume cold isopropanol
18. Put 1 hour at -20°C
19. Centrifuge 15 minutes at 14.000 rpm (4°C)
20. Remove supernatans and wash the pellet with 500 µl 75% ethanol
21. Add 50 µl DEPC destilled water to the pellet (to store the RNA, is best to leave it in 75% ethanol) use 1 to 5 µl for RT-PCR.

Method 3

Some of kit also available for isolation of RNA

Agilent Plant RNA Isolation Mini Kit Protocol

An important advantage of this rapid and convenient method is the removal of contaminating genomic DNA (gDNA) from the isolated total cellular RNA. Virtually all gDNA is removed in the initial stage of the protocol by centrifuging the cell or tissue homogenate through a unique mini prefiltration column. Typically, this method results in a 1000- to 10,000-fold lower level of gDNA, relative to typical silica or glass fiber isolation methods for plant material. For the majority of downstream applications, including microarray gene expression analysis, no supplemental DNase treatment step is necessary when using total RNA purified by this method. An additional advantage of the kit, for applications requiring high RNA concentrations, is the relatively small elution volume for the isolation column - as little as 10 µL of water. Total RNA is recovered with high yield, and without the addition of buffers or solvents that may interfere with subsequent enzymatic, chemical, or physical manipulations.

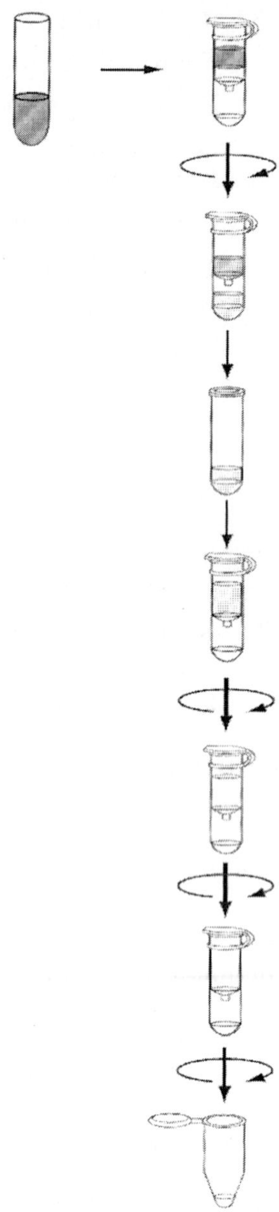

The intact and highly-purified RNA prepared by this method is ideal for cDNA synthesis and labeling, RT-PCR and quantitative RT-PCR, Microarray Gene Expression experiments, Northern Blotting, RNase protection assays and other downstream applications.

Homogenize Tissue or Cells

1. Prepare tissue or seed homogenate in appropriate volume of Extraction Solution/β-ME mixture.
2. Centrifuge up to 600μL of homogenate through the mini prefiltration column (natural) for 3 minutes at 16,000 × g. Capture RNA on mini isolation column
3. Add an equal volume of isopropanol to the filtrate and mix until homogeneous. Incubate 5 minutes.
4. Apply to mini isolation column (blue) and centrifuge 30 seconds at 16,000 × g. Discard flow-through and replace RNA-loaded column in same collection tube.

   Wash two times and spin dry

5. Add 500μL of prepared Wash Solution to the mini isolation column and centrifue 30 seconds at 16,000 × g. Discard flow-through and replace
6. isolation column in the same collection tube. Add 500μL wash solution and centrifuge 2 minutes at 16,000 × g.

Elute pure intact RNA

Elute the purified RNA by adding 10–50 μL of nuclease-free water, waiting 1 minute, and centrifuging 1 minute at 16,000 × g.

5

Electrophoresis

Electrophoresis is an analytical method frequently used in molecular biology and medicine. It is applied for the separation and characterization of proteins, nucleic acids and subcellular-sized particles like viruses and small organelles. Its principle is that the charged particles of a sample migrate in an applied electrical field. An electric current is passed through a medium containing the molecules, and each molecule travels at a different rate depending on its electrical charge, size and shape. Separation by electrophoresis is based on these differences. In electrophoresis, agarose and acrylamide gels are used for electrophoresis of nucleic acids and proteins.

What is Gel Electrophoresis?

A gel is a solid form. The term electrophoresis describes the migration of charged particle under the influence of an electric field. *Electro* refers to the energy of electricity. *Phoresis,* from the Greek verb *phoros,* means "to carry across." Thus, gel electrophoresis refers to the technique in which molecules are forced across a span of gel, motivated by an electrical current. Activated electrodes at either end of the gel provide the driving force. A molecule's properties determine how rapidly an electric field can move the molecule through a gelatinous medium.

Principles of Electrophoresis

Proteins and nucleic acids - that differ in size, charge or conformation. As such, it is one of the most widely-used techniques in biochemistry and molecular biology.

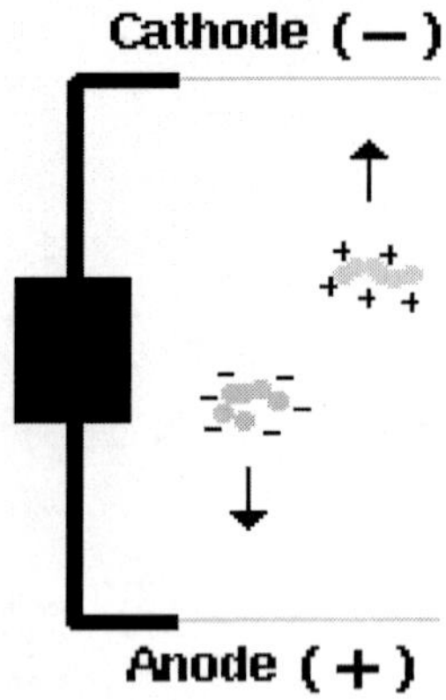

When charged molecules are placed in an electric field, they migrate toward either the positive (anode) or negative (cathode) pole according to their charge. In contrast to proteins, which can have either a net positive or net negative charge, nucleic acids have a consistent negative charge imparted by their phosphate backbone, and migrate toward the anode. Those charged molecules will therefore migrate in an electrical field in a velocity and direction depending on their electrophoretic mobility.

Electrophoretic mobility = migration velocity (cm/s) / field strength (V/cm).

Support Matrices

Generally the sample is run in a support matrix such as paper, cellulose acetate, starch gel, agarose or polyacrylamide gel. The matrix inhibits convective mixing caused by heating and provides a record of the electrophoretic run: at the end of the run, the matrix can be stained and used for scanning, autoradiography or storage. In addition, the most commonly used support matrices - agarose and polyacrylamide - provide a means of separating molecules by size, in that they are porous gels. A porous gel may act as a sieve by retarding, or in some cases completely obstructing, the movement of large macromolecules while allowing smaller molecules to migrate freely. Because dilute agarose gels are generally more rigid and easy to handle than polyacrylamide of the same concentration, agarose is used to separate larger macromolecules such as nucleic acids, large proteins and protein complexes. Polyacrylamide, which is easy to handle and to make at higher concentrations, is used to separate most proteins and small oligonucleotides that require a small gel pore size for retardation.

Electrophoresis technique/apparatus may vary depending upon, methods can be classified based on

i) biological molecule separated,
ii) principle of separation
iii) size of the molecule,
iv) type of support matrix used.

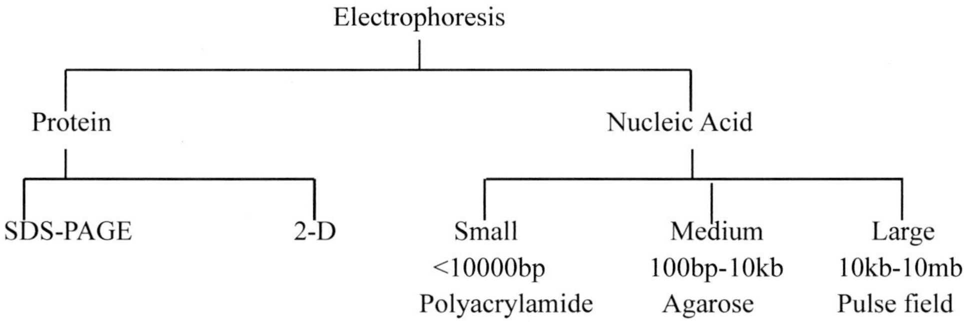

Electrophoresis can be classified based on:

1. Molecules to be separated
   a. Nucleic acids (DNA, RNA)
   b. Protein
      i. 1D
      ii. 2D
2. Kind of gel matrix used
   a. Agarose gel electrophoresis
   b. Polyacrylamide gel electrophoresis (PAGE)
      i. Native PAGE
      ii. Denaturing PAGE
   c. SDS-PAGE
3. Size of the molecules: Pulse Field Gel Electrophoresis

Agarose Gel Electrophoresis

Electrophoresis through agarose or polyacrilamide gels is the standard method to separate, identify, quantify and purify DNA fragments. Although agarose gels have a lower resolving power than polyacrilamide gels, they have a greater range of separation and are much simpler and easier to handle than the polyacrilamide gels, and have been more widely used in DNA electrophoresis.

Agarose

Agarose is a polysaccharide extracted from seaweed. It is typically used at concentrations of 0.5 to 2%. The higher the agarose concentration the "stiffer" will be the gel. Agarose gels are extremely easy to prepare: you simply mix agarose powder with buffer solution, melt it by heating, and pour the gel. It is also non-toxic.

Agarose gels have a large range of separation, but relatively low resolving power. By varying the concentration of agarose, fragments of DNA from about 200 to 50,000 bp can be separated using standard electrophoretic techniques. Agarose gels have lower resolving power than polyacrylamide gels but have a greater range of separation. DNA from 200 bp to approximately 50 kb in length can be separated on agarose gels of various concentrations.

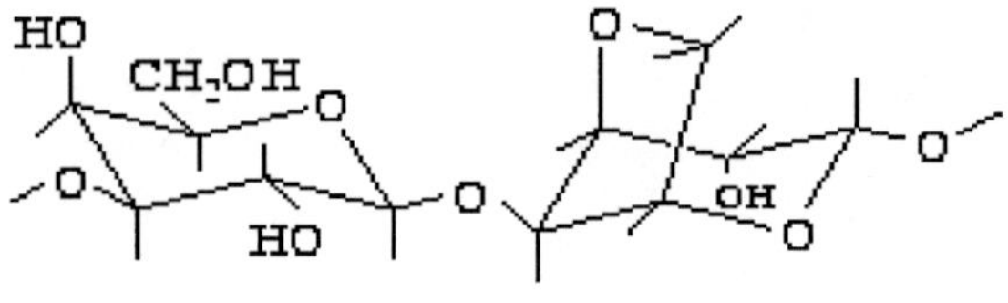

-galactose 3,6-anhydro L-galactose

Agarose is a linear polysaccharide (average molecular mas about 12,000) made up of the basic repeat unit agarobiose, which comprises alternating units of galactose and 3,6-anhydrogalactose. Agarose is usually used at concentrations between 1% and 3%. Agarose gels are formed by suspending dry agarose in aqueous buffer, then boiling the mixture until a clear solution forms. This is poured and allowed to cool to room temperature to form a rigid gel.

Electrophoresis of Proteins

Proteins can be separated and purified. Methods for separating proteins take advantage of properties such as charge, size, and solubility, which vary from one protein to the next. Because many proteins bind to other biomolecules, proteins can also be separated on the basis of their binding properties. The source of a protein is generally tissue or microbial cells. The cell must be broken open and the protein must be released into a solution called a crude extract. If necessary, differential centrifugation can be used to prepare subcellular fractions or to isolate organelles. Once the extract or organelle preparation is ready, a variety are available for separation of proteins. Ion-exchange chromatography can be used to separate proteins with different charges (similar to the way amino acids are separated). Other chromatographic methods take advantage of differences in size, binding affinity, and solubility. Nonchromatographic methods include the selective precipitation of proteins with salt, acid, or high temperatures. In addition to chromatography, another important set of methods is available for the separation of proteins, based on the migration of charged proteins in an electric field, a process called (gel) electrophoresis. Gel electrophoresis is especially useful as an analytical method. Its advantage is that proteins can be visualized as well as separated, permitting a researcher to estimate quickly the number of proteins in a mixture or the degree of purity of a particular protein preparation. Also, gel electrophoresis allows determination of crucial properties of a protein such as its isoelectric point and approximate molecular weight.

Amino acids differ not only in R-group characteristics but also in molecular weight. Different amino acids are linked together in a linear chain by peptide bonds in various combinations and sequences to form specific proteins. A protien may be comprised of amino acids from all of the categories. The net charge of a protein will depend on its amino acid composition. If it has more positively charged amino acids such that the sum of the positive charges exceeds the sum of the negative charges, the protein will have an overall positive charge and migrate to the cathode (negatively charged electrode) in an electrical field. Proteins even with a variation of one amino acids will have a different overall charge, and thus are electrophoretically distinguishable.

If a protein is electrophoresed before the disulfide bonds are broken, it will yield only one band of 125,000 Daltons molecular weight on the gel. That is, the protein will move as one entity and the peptide chains cannot be distinguished. However, after treatment with a reducing agent, the protein will yield four distinct bands with a combined molecular weight of 125,000 Daltons—the same as the original protein.

The gel results will show that some of the high molecular weight bands from samples not treated with the disulfide reducing agent are missing in the samples treated with the disulfide reducing agent. However, two or more bands are imaged on the gel lane containing the treated samples, thereby replacing each of the missing high molecular weight bands from the samples that were not treated. The additional bands that appear in the treated samples represent the individual polypeptides that make up complex proteins. The break up of complex proteins into their respective polypeptides allows us to study the structure of proteins that result from the interaction of several genes. A gene is a discrete unit of hereditary information that usually specifies a protein. A single gene provides the genetic code for only one polypeptide. Thus, a protein consisting of four polypeptides requires the interaction of four genes to synthesize that specific protein.

A molecular weight protein marker is used to prepare a standard separation curve with which various unknown proteins or polypeptide fractions can be identified.

Electrophoresis of Nucleic Acids

Gel electrophoresis is the process by which scientists can sort pieces of DNA cut with restriction enzymes by size. An agarose or polyacrylamide gel is loaded with the DNA fragments and current is passed through the gel. Since DNA is negatively charged, it will migrate towards the positive pole. The DNA will not migrate at the same rate, however, larger pieces of DNA collide with the gel matrix more often and are slowed down, while smaller pieces of DNA move through more quickly.

Since different genes have different nucleotide sequences, restriction enzymes will cut them at different places, generating different size DNA fragments. By using gel electrorphoresis, biologists can tell which gene is which based upon the sizes of the fragments generated when a gene is treated with a restriction enzyme.

Factors Affecting the Rate of DNA Migration in Agarose Gels

1. Composition of the Biomolecules

Different forms of DNA (super helical circular, nicked circular and linear form of DNA of the same molecular weight migrate through agarose gel at different rate. Molecules of linear double-stranded DNA, which tend to become oriented in an electric field in an end-on position, migrate through gel matrices at rates that are inversely proportional to the log10 of the number of base pairs. Larger molecules migrate more slowly because of greater frictional drag and because they form their way through the pores of the gel less efficiently than smaller molecules.

2. Agarose Concentration

By using gels with different concentrations of agarose, one can resolve different sizes of DNA fragments. Higher concentrations of agarose facilitate separation of small DNAs, while low agarose concentrations allow resolution of larger DNAs.

3. Conformation of DNA

Amount of agarose in gel(% [w/v])	**Efficient range of separation of linear DNA molecules (kb)**
0.3	5-60
0.6	1-20
0.7	0.8-20
0.9	0.5-7
1.2	0.4-6
1.5	0.2-3
2.0	0.1-2

Superhelical circular (form I), nicked circular (form II), and linear (form III) DNAs of the same molecular weight migrate through agarose gels at different rates. The relative mobilities of the three forms depend primarily on the agarose concentration in the gel, but they ionic strength of the buffer, and the density of superhelical twists in the form I DNA. Under some conditions, the order is reversed.

Identification of Conformational forms of DNA

An unambiguous method for identifying the different conformational forms of DNA is to carry out electrophoresis in the presence of increasing quantities of ethidium bromide. As the concentration of ethidium bromide increases, more of the dye becomes bound to the DNA. The negative superhelical turns in form I molecules are progressively removed, the radii of the molecules increase, and their rate of migration decreases. At the critical free-dye concentration, where no superhelical turns remain, the rate of migration of form I DNA reaches its minimum value. As still more ethidium bromide is added, positive superhelical turns are generated, the DNA molecules become more compact, and their mobility increases rapidly. Simultaneously, the mobilities of form II and form III DNA decrease differentially due to charge neutralization and the greater stiffness imparted to the DNA by the ethidium bromide. For most preparations of form I DNA, the critical concentration of free ethidium bromide is in the range of 0.1mg/ml to 0.5mg/ml.

4. Applied Voltage

At low voltages, the rate of migration of linear DNA fragments is proportional to the voltage applied. However, as the electric field strength is raised, the mobility of high-molecular-weight fragments of DNA increases differentially. *Thus the effective range of separation in agarose gels decreases as the voltage is increased.* To obtain maximum resolution of DNA fragments greater than 2 kb in size, agarose gels should be run at no more than 5 V/cm.. Distance is measured as the shortest path between the electrodes and is not merely the length of the gel itself.

5. Direction of the Electric Field

DNA molecules larger than 50-100 kb in length migrate through agarose gels at the same rate if the direction of the electric field remains constant. However, if the direction of the electric field is altered periodically, the DNA molecules are forced to change course. Because larger molecules of DNA take longer to realign themselves to the new direction of the field, pulsed-field gel electrophoresis can be used to fractionate populations of extremely large molecules of DNA (up to 10,000 kb).

6. Base Composition and Temperature

The electrophoretic behavior of DNA in agarose gels (in contrast to polyacrylamide gels) is not significantly affected by either the base composition of the DNA or the temperature at which the gel is run. Thus, in agarose gels the relative electrophoretic mobilities of DNA fragments of different sizes do not change between 4°C and 30°C. In general, agarose gels are run at room temperature. However, gels containing less than 0.5% agarose and low-melting-temperature agarose gels are rather flimsy, and it is best to run them at 4°C, where they are stronger.

7. Presence of Intercalating Dyes

Ethidium bromide, a fluorescent dye that is used to detect DNA in agarose and polyacrylamide gels, reduces the electrophoretic mobility of linear DNA by about 15%. The dye intercalates between stacked base pairs, extending the length of linear and nicked circular DNA molecules and making them more rigid

Ethidium bromide is a carcinogen and should be handled with care. All solutions containing ethidium bromide should be decontaminated before disposal.

8. Effects of Ethidium Bromide

Ethidium bromide is a fluorescent dye that intercalates between bases of nucleic acids and allows very convenient detection of DNA fragments in gels, as shown by all the images on this page. As described above, it can be incorporated into agarose gels, or added to samples of DNA before loading to enable visualization of the fragments within the gel. As might be expected, binding of ethidium bromide to DNA alters its mass and rigidity, and therefore its mobility.

9. Composition of Electrophoresis Buffer

Several different buffers have been recommended for electrophoresis of DNA. The most commonly used for duplex DNA are TAE (Tris-acetate-EDTA) and TBE (Tris-borate-EDTA). DNA fragments will migrate at somewhat different rates in these two buffers due to differences in ionic strength. Buffers not only establish a pH, but provide ions to support conductivity. If you mistakenly use water instead of buffer, there will be essentially no migration of DNA in the gel Similarly, if you use concentrated buffer (e.g. a 10 x stock solution), enough heat may be generated in the gel to melt it.

Types of Electrophoresis Buffers

Several different buffers are available for electrophoresis of native double-stranded DNA such as Tris-acetate EDTA (TAE), Tris-borate EDTA (TBE), or Tris-phosphate EDTA (TPE). TAE is the most commonly used buffer. However, its buffering capacity is rather low, and it tends to become exhausted during extended electrophoresis (the anode becomes alkaline, the cathode acidic). Replacement of the buffer or recirculation between the two reservoirs is therefore advisable when carrying out electrophoresis for long periods of time at high current. Both TPE and TBE are slightly more expensive than TAE, but they have significantly higher buffering capacity. Double-stranded linear DNA fragments migrate approximately 10% faster through TAE than through TBE or TPE, but the resolving powers of these systems are almost identical, with the exception that the resolution of supercoiled DNAs is better in TAE than in TBE. The most commonly used buffer for electrophoresis of denatured single-stranded DNA is 50 mM NaOH, 1 mM EDTA (alkaline electrophoresis buffer).

Agarose Gel Electrophoresis Apparatus

Several manufactures have introduced variety of models for various purposes, based on the number of samples and the level of resolution required one has to choose the apparatus. In this class few models of gel apparatus are dealt with.

Several models of both vertical and horizontal gel apparatuses are available in the market. But most commonly used are the horizontal slab gel units developed by Walter Schaffner.

Advantages of these horizontal gel units are:

1. Low agarose concentrations can be used
2. Gels can be casted in a variety of sizes
3. simple to load, pour and handle
4. Less expensive long durable.

Some of the models available are (based on the number of samples can be run):

1. Submarine Midi
2. Submarine-Midi
3. Submarine XL & XXL systems

Submersible Gel Electrophoresis Apparatus- Agarose

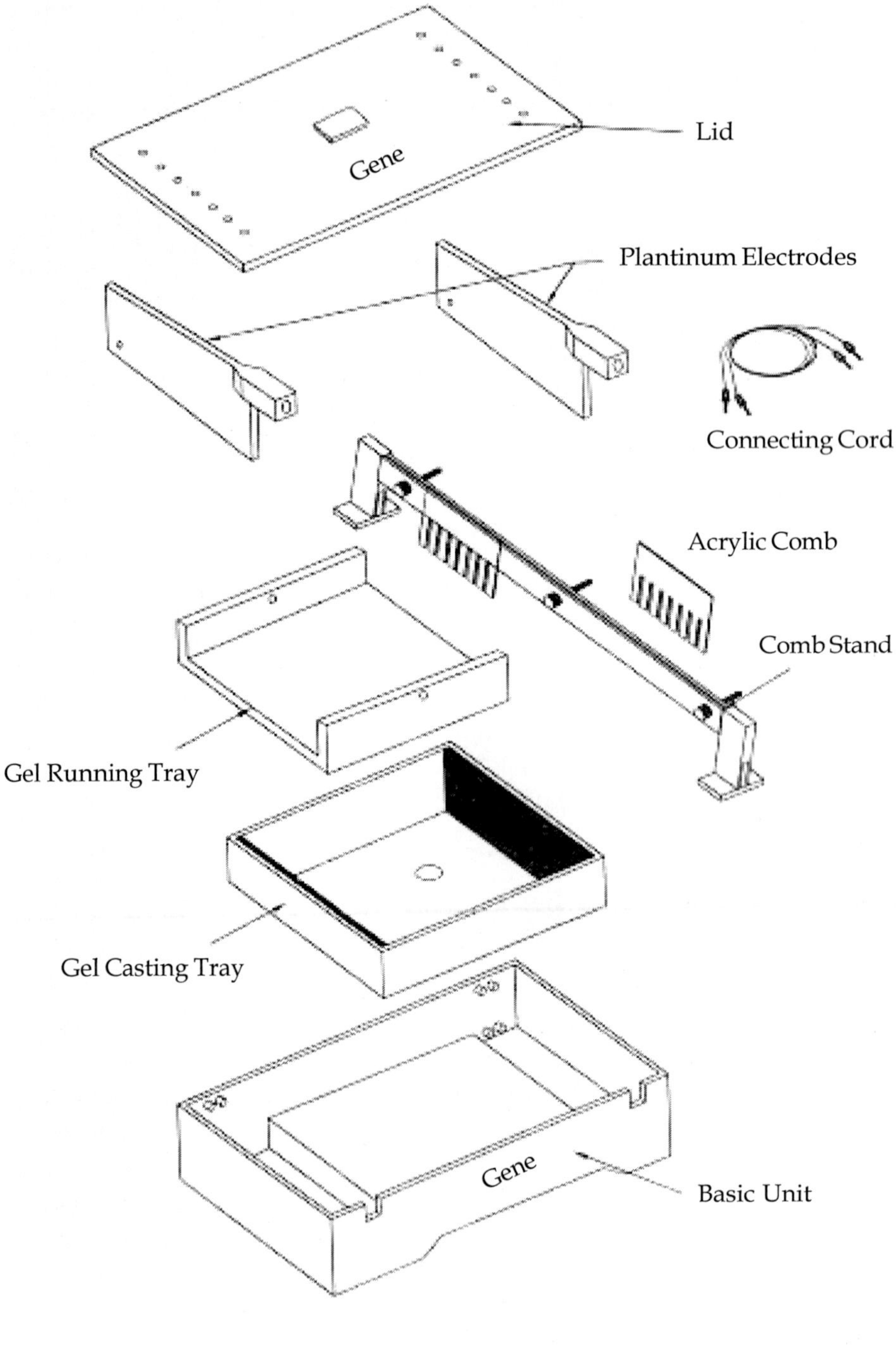

Steps Involved in Agarose Gel Setup

1. Complete assembly

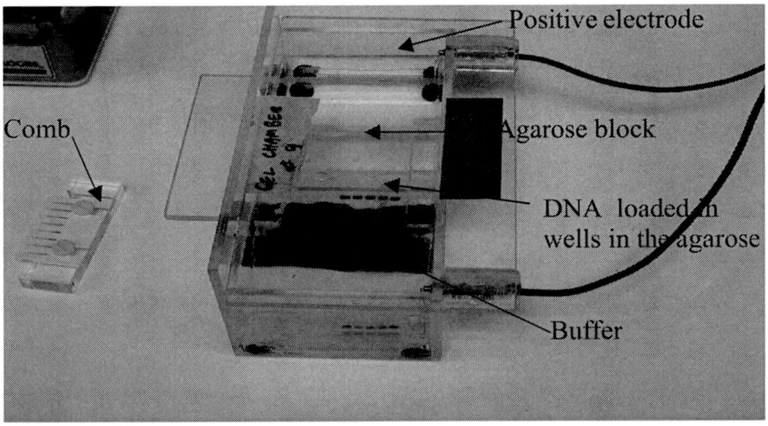

2. Weight the agarose powder as required.

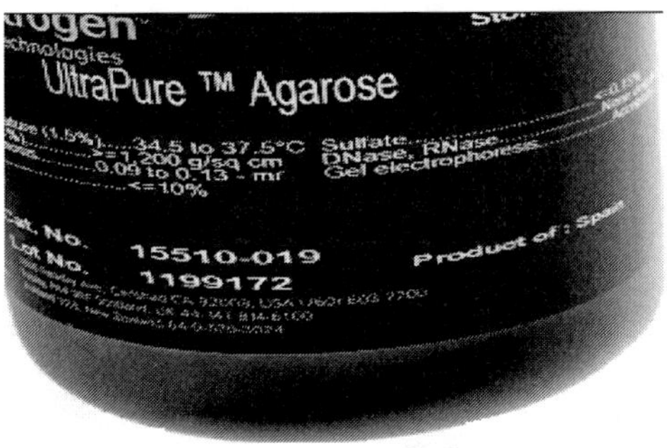

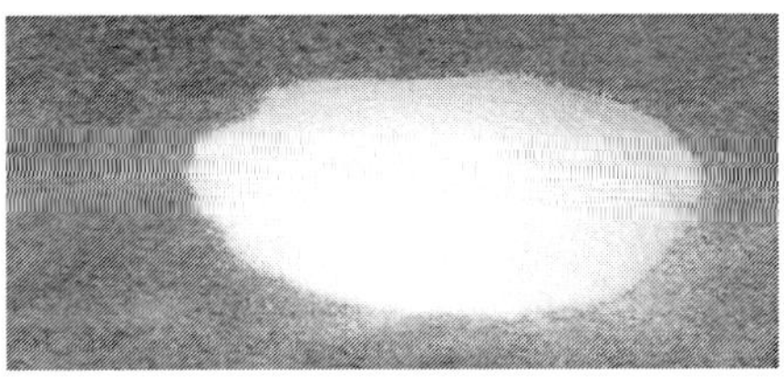

3. Prepare a required volume of electrophoresis buffer) to fill the electrophoresis tank and to prepare the gel.

4. Add the needed amount of powdered agarose (typically 8-15%) to the electrophoresis buffer in an Erlenmeyer flask. The buffer should not occupy more than 50% of the volume of the flask.

5. Heat the agarose solution in a oven or in water bath to allow all of the grains of agarose to dissolve. If part of the buffer evaporated during the heating, bring the solution back to the original volume through the addition of buffer.

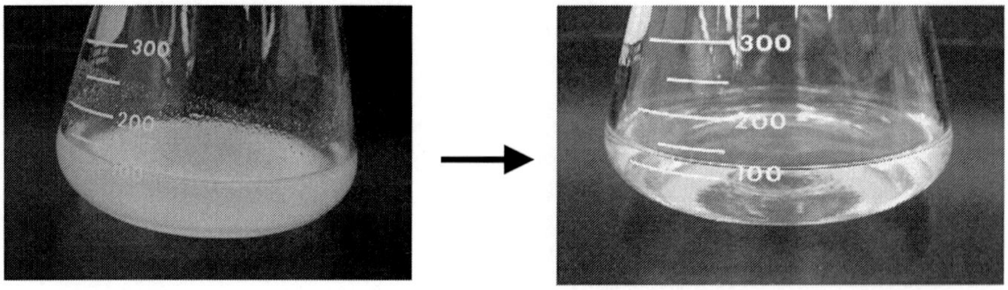

6. Cool down the solution to 60ºC. You may add ethidium bromide (Et Br) (10 mg/l in water to a final concentration of 0.5 um/ml) in this step or you can submerge the gel in an Et Br solution once it solidifies.

7. Take the Electrophoresis Equipment

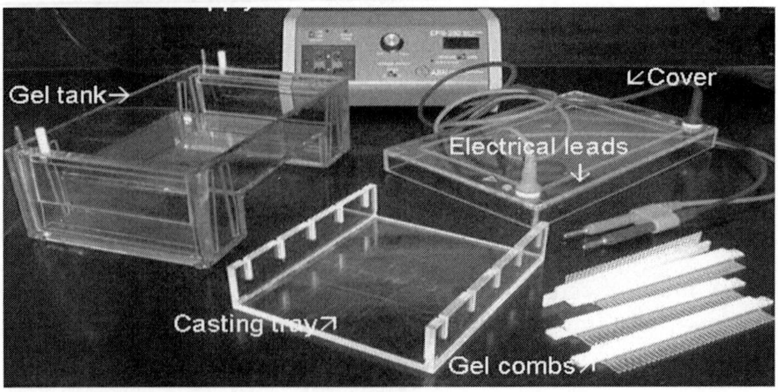

8. Position the comb 0.5-1.0 mm above the plate, thus permitting that a complete well is formed when the agarose solidifies. Air bubbles under or between the teeth of the comb should be avoided. Using a Pasteur pipette seal the glass plate with small amounts of agarose solution. Once the seals are set, pour the gel in the glass plate.

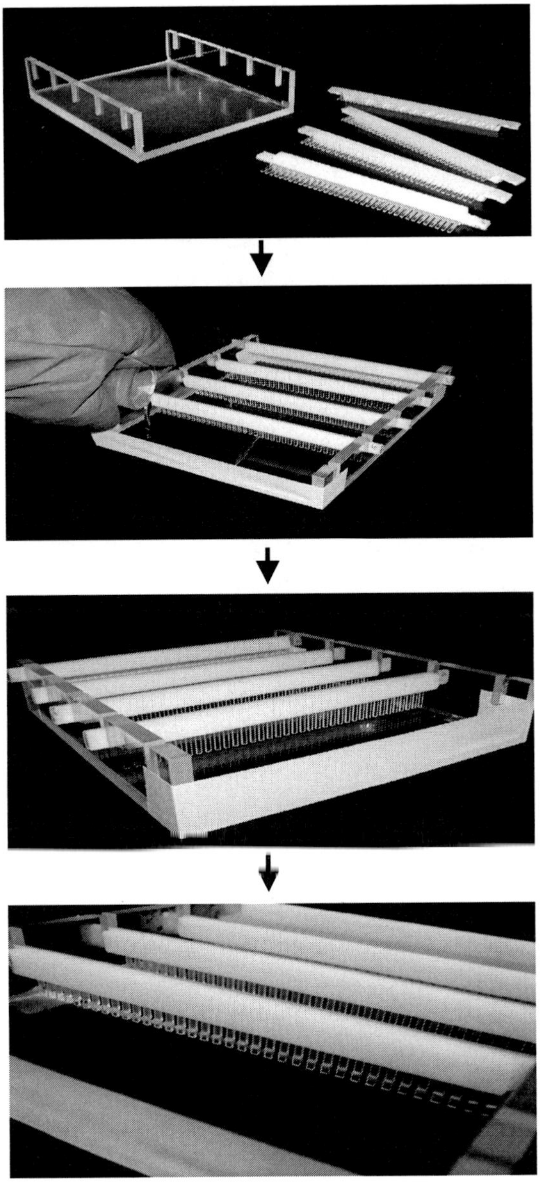

9. Allow the agarose solution to cool slightly (~60°C) and then carefully pour the melted agarose solution into the casting tray. Avoid air

When cooled, the agarose polymerizes, forming a flexible gel. It should appear lighter in color when completely cooled (30-45 minutes). Carefully remove the combs and tape.

10. Remove the comb and the tape when the gel has completely hardened (30 to 40 minutes at room temperature) and place the gel in the electrophoresis tank. Add enough electrophoresis buffers to the tank to cover the gel (about 1 mm of depth). The top of the wells should be submerged.
11. Place the gel in the electrophoresis chamber

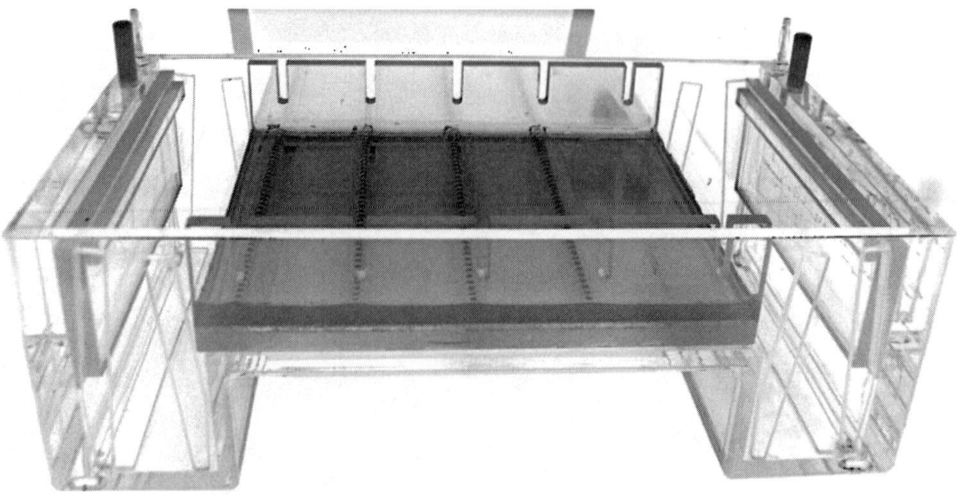

Add enough electrophoresis buffer to cover the gel to a depth of at least 1 mm. Make sure each well is filled with buffer

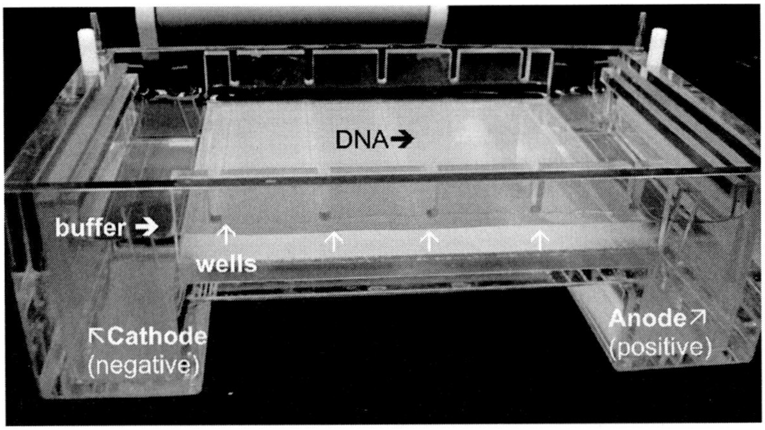

12. Mix the samples of DNA with the 10X sample loading buffer (w/ tracking dye). This allows the samples to be seen when loading onto the gel, and increases the density of the samples, causing them to sink into the gel wells.

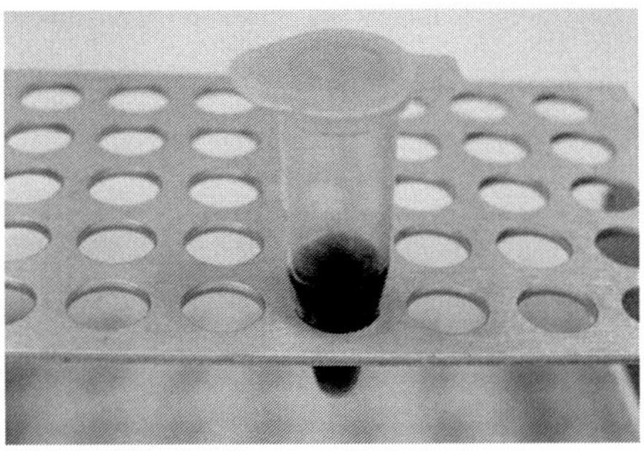

10 x Loading Buffer:

- Bromophenol Blue (for color)
- Glycerol (for weight)

13. Carefully place the pipette tip over a well and gently expel the sample. The sample should sink into the well. Be careful not to puncture the

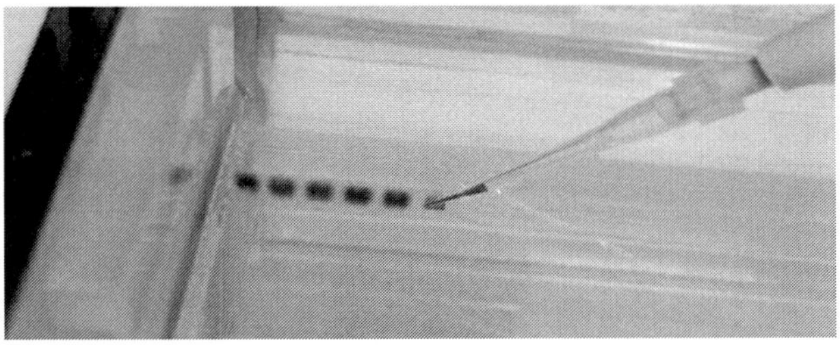

14. Close the lid of the tank. Connect the electrical leads to the power supply. Be sure the leads are attached correctly - DNA migrates toward the anode (red). When the power is turned on, bubbles should form on the electrodes in the electrophoresis chamber.

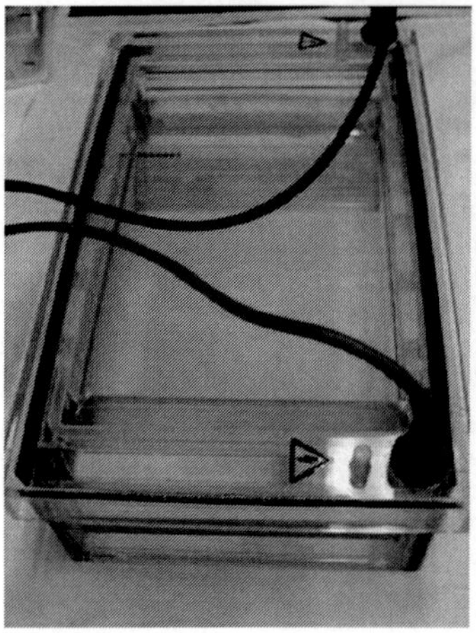

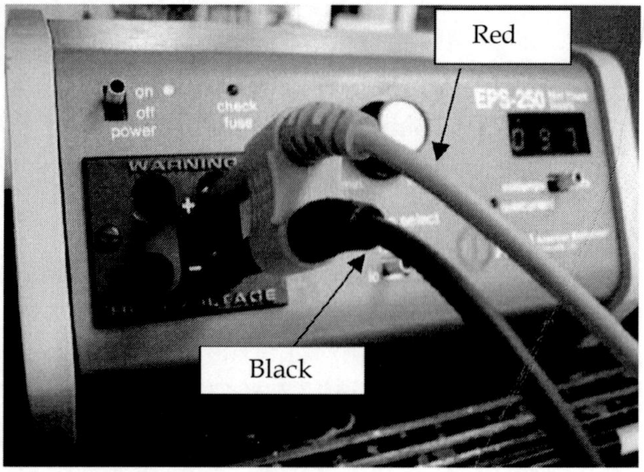

15. If the unit is working you should see bubbles that are formed in the buffer due to the electrical field that has been created; and later you should see the dyes running in the gel. After the current is applied, make sure the Gel is running in the correct direction. Bromophenol blue will run in the same direction as the DNA.

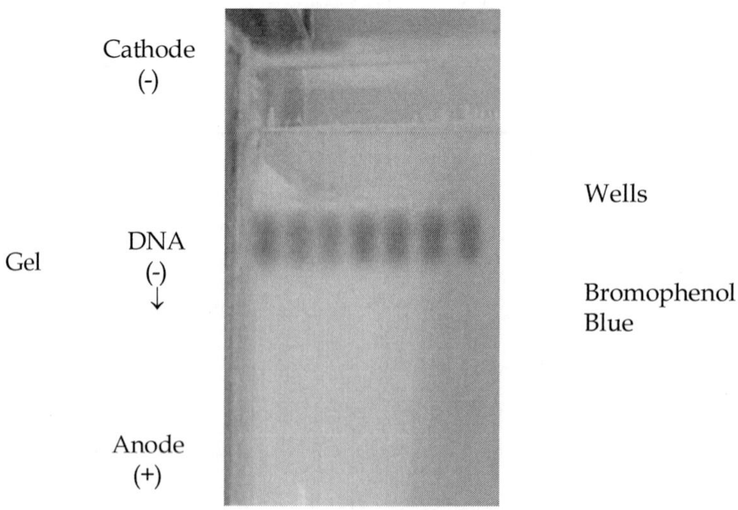

16. If ethidium bromide was present in the agarose gel, DNA may be photographed by with UV light (>500uW/cm2). The location of DNA within the gel may be determined by a film image of light transmitted by a fluorescing DNA. Thus, a wide range of bands or fragments of DNA can be detected on the film (from 200 bp to ~ 50 kb in length). Size and amount of DNA can be calculated by running standards of known size and concentration. DNA bands may be recovered from the gel and utilized for a diversity of cloning designs.

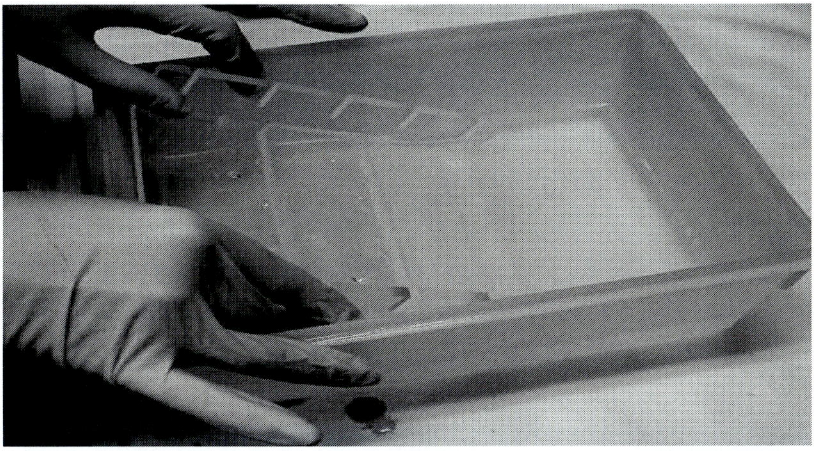

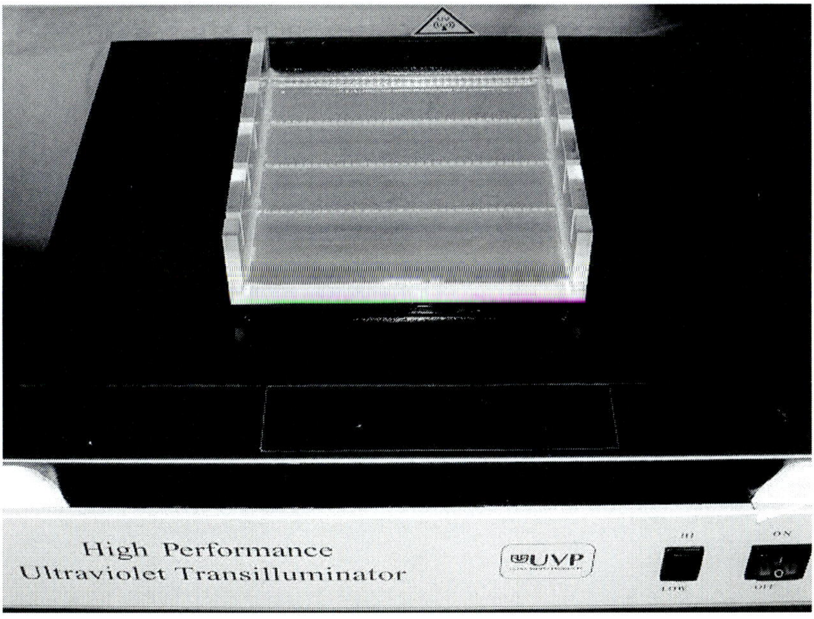

17. Visualization of DNA bands with UV light
DNA Ladder Standard

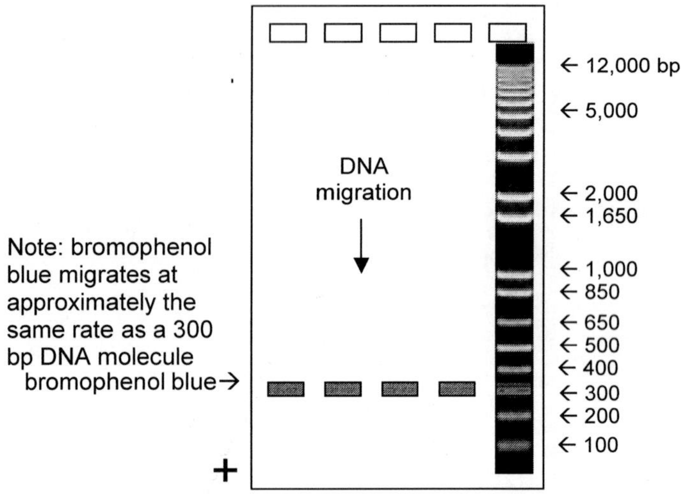

Inclusion of a DNA ladder (DNAs of know sizes) on the gel makes it easy to determine the sizes of unknown DNAs.

Visualizing the DNA (ethidium bromide)

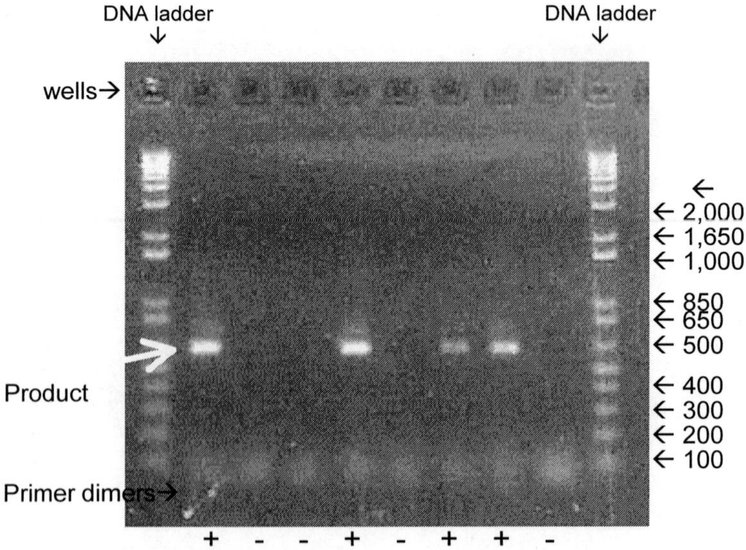

Samples 1, 4, 6 & 7 were positive for Wolbachia DNA

Protocol to be used for Agarose Gel Electrophorosis

1. Weigh out 1gm of agarose and transfer it to a conical flask.
2. Add 100 ml of 1 × Tris-Acetate- EDTA (TAE) buffer.
3. Melt the agarose completely into a clear solution, using a microwave oven.
4. Cool the molten agarose to~ 50°C and add Ethidium bromide to a final concentration of 0.5mg/ ml (Add 5ml of 10 mg/ml solution of Ethidium bromide to 100 ml gel mixture) and mix well.
5. Pour the molten agarose in a pre-set template with well-forming comb (Set in advance, an agarose gel template by sealing its ends with adhesive tapes. Place it horizontally on a leveling table and put an appropriate well-forming comb.
6. Allow gelation for 30 minutes to 45 minutes.
7. Remove the comb and sealing tapes, carefully.
8. Mount the template (carrying the gel) in an appropriate electrophoresis tank and fill the tank with 1 × TAE buffer, just to immerse the gel.
9. Load the DNA samples into the wells (The DNA sample should be made electrophoresis-ready by mixing the sample with required volume of sample/loading buffer).
10. Connect the electrophoresis apparatus to a power pack and perform electrophoresis at 50 volts, using 1 × TAE as the running buffer.
11. Stop electrophoresis, when the dye, Bromophenol blue crosses more than 2/3 of the length of the gel (in no case it should move out of the gel).
12. Visualize DNA bands(s) on an UV trans-illuminator.
13. If required, this profile can be photographed through a red filter for documentation purpose.

Buffers and reagents required

1. Tris-acetate EDTA buffer (50 ×):
   (TAE)
   242 g Tris base
   57.1 ml glacial acetic acid
   100 ml 0.5 M EDTA (pH 8.0)
2. 10mg/ ml Ethidium bromide solution.

NOTE

1. The above procedure is for making 100ml of 1 % agarose gel. Depending upon the nature of DNA sample to be analyzed and the volume of the template, the experimenter should decide the percentage and the volume of the gel. For instance, plasmid DNA could be analyzed on 0.6- 0.8 % gels. For analyzing chromosomal DNA, 0.4 % gel is preferable. However, as the handling of a 0.4 % gel is difficult, chromosomal DNA can be analysed on a higher percentage gel. For analyzing PCR products, 1.5-2.0% gel is required.
2. The gel can be stained with Ethidium bromide after completion of electrophoresis. In this case a brief destaining either with water or 1 mM Magnesium sulphate solution is necessary.

   Caution : Ethidium bromide is a powerful carcinogen. Wear gloves, when handling.

Care and Handling of Electrophoresis Apparatus

Electrophoresis apparatus may represent a significant electrical hazard in laboratories if used incorrectly. The potential for injury arises principally from the high levels of electrical energy involved, and from the supporting gel medium being immersed in aqueous solutions. Many workers mistakenly believe that there is little hazard associated with electrophoresis apparatus except when operating at higher voltages, such as might be required for procedures like DNA sequencing. However, even routine agarose gel electrophoresis operating at 100 volts (V) can cause a lethal shock at a current of 25 milliamps (mA) and minor leaks in gel tank devices can result in electric shock. 10-100mA direct (DC) current can cause muscular paralysis and cardiac problems, and such currents are within the output range of most electrophoresis power packs and typical of contacts made through wet skin. Minor shocks may also cause secondary injury due to involuntary muscle contraction, while handling chemicals or flammable reagents, for example. Other techniques such as electroporation and electrotransformation, which also apply electricity to biological materials, may present a similar electrical hazard.

The hazards associated with the use of electrophoresis apparatus may be controlled by a combination of design features on the equipment, strict adherence to a safe system of work, adequate training, and ongoing supervision.

Power Packs

(a) All power supplies should be clearly labelled with their maximum output voltage.

(b) Electrical supply to the power pack should be protected with a Residual Current Device (RCD). This will protect against faults with the main 230V AC electrical supply to the power pack.

(c) All power packs must incorporate a safety device that disables the high voltage supply to the electrophoresis equipment when the cover of the gel tank is opened or removed.

(d) It is recommended that power packs also have the following additional safety features. They should:

   (i) be isolated or 'floating', meaning that terminals are not referenced to the case (earth).

   (ii) have a load detection system, preventing power being supplied if there is no load on the output terminals.

   (iii) have a 'reset on / turn on' feature, meaning that if the electrical supply is turned off and then on, the apparatus must be reset to recommence the electrophoresis run.

(e) Power packs capable of operating at 1000 volts and above must utilise 2mm shrouded connectors on the high voltage outlets so that they cannot be connected to apparatus which is not designed for this voltage range. However, a given supplier's power pack, used exclusively with the same supplier's dual moulded 4mm plug connectors is regarded as satisfactory.

(f) Multi-outlet powerpacks may be used with multiple electrophoresis devices, providing a safe system of work is followed.

(g) Multi-purpose power packs are frequently purchased to fulfil a number of functions in the laboratory but their use must be strictly monitored by supervisors. Where, for example, a maximum voltage of 200 volts is required for a process it is inappropriate to use a power pack supplying 2000 volts. In such cases a low voltage power pack should be purchased for low voltage work.

Gel Equipment

(a) Gel equipment must be used with a cover in place such that neither electrodes nor buffer chambers are accessible.

(b) Gel equipment should be interlocked with the power supply. Removable covers must be fitted with deep insulated sockets that engage with contact pins in the electrophoresis tanks and be so arranged that the current is cut off to the tank immediately the cover is removed. The live parts of the sockets must not be inadvertently accessible. An integrity check of the cover must be performed before use.

(c) Apparatus that is not designed for use above 1000 volts must not be fitted with leads that can be plugged into a power pack capable of delivering more than 1000 volts. The use of adaptors which convert 4mm plugs to 2mm plugs is prohibited.

(d) During operation all electrophoresis apparatus should have prominently displayed safety warning signs to indicate that the equipment is 'live'.

(e) The use of customized or locally made electrophoresis devices is discouraged.

High Voltage Cables

(a) All connecting cables between power packs and gel equipment must be fitted with shrouded connectors so that live parts are inaccessible. Stackable plugs must not be used. Both the cable and connectors must be correctly rated for the maximum voltage that the power pack can deliver in use.

(b) Electrophoresis sets fitted with their own set of permanently attached connector leads that eliminate jacks and plugs entirely are acceptable.

(c) Leads must be regularly inspected to ensure that there is no damage to insulation and that all parts of the conductor are covered.

(d) Leads must never be left connected to power packs when the electrophoresis run is complete.

Training

Only trained individuals should be allowed to operate electrophoresis apparatus.

Additional Cautions

- Does not autoclave or dry-heat sterilize the apparatus or components.
- Do not expose the apparatus or components to phenol, acetone, benzene, halogenated hydrocarbon solvents, or undiluted laboratory alcohols.
- Avoid prolonged exposure of the apparatus or components to UV light.

Maintenance of Electrophoresis Units

The following simple, routine inspection and maintenance procedures will help ensure both the safety and the performance of the apparatus.

Electrical Connections

- Inspect electrical connections and power cords regularly. If the banana plug connectors to either the unit or the power supply do not exhibit reasonable friction or can be rocked easily, replace the plugs or power cords.
- If power cords show any signs of wear or damage (*e.g.*, cracks nicks, abrasions, or melted insulation), replace them immediately.
- Examine the electrode mounting nuts to ensure that they are tight and free of corrosion. If mounting nuts are corroded, they should be replaced.

Aluminum Plates

After each use, rinse the aluminum plates thoroughly with distilled water to remove salts and urea. Failure to keep plates properly clean may affect performance and introduce a potential electrical hazard.

6

Molecular Weight Markers for Gel Electrophoresis

Two types of molecular weight (MW) standards are routinely used. The Lambda/*Hin*dIII and PhiX174/*Hae*III MW standards provide a useful reference for calculating molecular weights of large and small DNA fragments, respectively, after electrophoretic separation; the "internal MW standards" provide a means for normalizing fragment migration distances within each lane to facilitate comparisons between lanes on the same or different luminographs in fingerprinting studies.

End-labeled Lambda (λ) DNA as a Molecular Weight Standard for Luminographs

Digestion of λ DNA with *Hin*dIII

Stock	**Final or amount**	**50 µl RXN**
ddH2O	--	31.8 µl
10X Buffer	1X	5.0 µl
0.1 M Spermidine	2.5 mM	1.2 µl
ë DNA (0.45 µg/µl)*	5 µg	11.0 µl
*Hin*dIII (10 U/µl)	2 U/µg DNA	1.0 µl

* Check the concentration of commercial λ and adjust quantities accordingly.

1. Allow to digest at 37°C for 2-3 h.
2. Check that digestion is complete by running about 50 ng on a 0.7% agarose gel. When it is complete, move to step 3 or 4.
3. If you are going to use the digested λDNA as a MW marker without end-labeling it, inactivate the enzyme by incubating at 65°C for 10 min. Then add 110µl TE and 40µl 5X SGB to bring to a concentration of 25 ng/µl. Aliquot and keep at 4°C or in the freezer.
4. For end-labeling, precipitate by adding 5µl of 2.5 M NaOAc and 125µl of absolute EtOH, mix well by inversion, and place at -80°C for 30 min.
5. Centrifuge in a microfuge for 10-15 min at full-speed. Pour off supernatant and invert tubes to drain. It is very important to allow the pellet to dry.
6. Re-suspend the pellet in 15µl ddH_2O. Assuming little or no DNA was lost during precipitation, the concentration should be about 5 µg/15 µl or 0.33 µg/µl.

End-labeling of ë/*Hin*dIII DNA with digoxigenin-dUTP (dig-dUTP)

Stock	Final or amount	50 µl RXN
ddH2O	--	25.0µl
10X Klenow Buffer	1X	5.0µl
10 mM dATP	100µM	0.5µl
10 mM dCTP	100µM	0.5µl
10 mM dGTP	100µM	0.5µl
1 mM dig-dUTP	40µM	2.0µl
λ/*Hin*dIII DNA[1]	5µg	15.0µl
2U/µl Klenow[2]	3 U	1.5µl

[1] Check the concentration of commercial λ and adjust accordingly.

[2] Purchase from Fisher Scientific (cat. # PR-M2201 Promega-Biotec) or BRL (cat. # 80125B).

7. Incubate at 37°C for 1.5 h.
8. Stop the reaction by placing at 65°C for 15 min.
9. EtOH precipitate as in (2) above.
10. Re-suspend in 250µl TE to bring to a final concentration of 20 ng/µl. This stock can then be diluted to 10 or 1 ng/µl with TE.

11. Verify incorporation of dig-dUTP following the protocol "Checking the activity of incorporated digoxigenin-dUTP".

    Use 5 ng/lane of λ DNA digested with HindIII and end-labeled with digoxigenin-dUTP.

12. Prepare working solutions from the stocks based on the following proportions:

1 ng/µl	10 ng/µl	20 ng/µl	
Stock	**Stock**	**Stock**	**Stock**
λ DNA end labeled	5µl	0.50µl	0.25µl
TE	11µl	15.50µl	15.75µl
5X SGB	4µl	4.00µl	4.00µl

Digestion of ØX174 DNA with *Hae* III

Stock	Final or amount	150 µl RXN
ddH$_2$O	--	68.25µl
10X Buffer	1X	15.00µl
0.1 M Spermidine	2.5 mM	3.75µl
öX174 DNA (0.25 µg/µl)[1]	15 µg	60.00µl
*Hae*III (10 U/µl)	2 U/µg DNA	3.00µl

1 Check the concentration of commercial ϕ X174RF plasmid DNA and adjust quantities accordingly.

1. Allow to digest at 37°C for 2-3 h.
2. Check that digestion is complete by running about 50 ng on a 0.7% agarose gel.
3. Inactivate enzyme by incubating at 65°C for 10 min. Then add 300 µl TE and 150 µl 5X SGB to bring to a concentration of 25 ng/µl. Aliquot (200 µl per 0.5 ml tubes) and keep at 4°C or in the freezer.

Internal Molecular Weight Markers for Fingerprinting with RFLPs

Two markers, a "top" and a "bottom" λ DNA fragments, are used routinely as internal MW standards in each and every lane of a fingerprinting gel, including the MW marker lane(s). They were chosen because of their easy preparation and detection, as well as their convenient size for normalization purposes in most fingerprinting experiments using RFLPs.

Preparation of a "Top MW Standard"

1. Digest λ DNA with *Xba*I to generate 2 large fragments (24.5 and 24 kb) that will co-migrate after the short migrations used in these protocols (see, for example, the following protocol).

Stock	Final or amount	50 µl RXN
ddH2O		30.3µl
10X Buffer	1X	5.0µl
0.1 M Spermidine	2.5 mM	1.2µl
λ DNA (0.4µg/µl)[1]	5µg	12.5µl
*Xba*I (10 U/µl)	2 U/µg DNA	1.0µl

[1] Check the concentration of commercial λ and adjust quantities accordingly.

2. Allow to digest at 37°C for 1-2 h. Verify the digestion by running a small sample (say, 0.5 µl) in a 0.7% agarose microgel. Add more enzyme to digestion reaction and incubate for another hour if necessary.

3. Precipitate by adding 5 µl of 2.5 M NaOAc and 125 µl of absolute EtOH, mix well by inversion, and place at -80°C for 30 min.

4. Centrifuge in a microfuge for 10-15 min at full-speed. Pour off supernatant and invert tubes to drain. It is very important to allow the pellet to dry.

5. Re-suspend the pellet in 500 µl ddH$_2$O. Assuming little or no DNA is lost during precipitation, the concentration should be about 10 ng/µl. This amount will be enough for at least 150 gels with 120 wells each.

   Since it is important to obtain a very "clean" fragment, treat the resulting DNA with proteinase K at 37°C for 30 min, then perform a phenol/chloroform extraction followed by a back extraction to minimize losses of DNA, and finally EtOH precipitate before re-suspending in TE.

6. Digest 10 µg of the plasmid-containing DNA in a 30 µl reaction with 2 units each of *Eco*RI and *Bam*HI (same buffer).

7. Check digestion by loading 1 µl (i.e., about 300 ng) on a minigel.

8. If digestion is complete, add 6 µl of 5X SGB and load on a 1.2% low melting point (LMP) agarose gel. You can load up to 5 µg/lane (load in 2 to 4 wells). Include EtBr in the gel and running buffer.

9. Run the gel in the cold room at 40 mA. Check separation with portable UV lamp after 30 min (if running in a minigel).
10. When plasmid and insert are well separated, take out the insert either by cutting it out or by electroelution of the λ fragment onto DEAE-cellulose membrane (e.g., S&S NA-45).
11. Adjust to a final concentration of 10 ng/μl. If you have cut the fragment out, melt the gel at 65°C before adding TE to adjust the concentration.

    Remember that 10 μg plasmid DNA will yield 3.5 μg insert DNA.
12. Check on a minigel (50-100 ng are enough for this purpose).

    Note: Use 0.25 ng/lane of the 24.5 kb lambda fragment, and 0.50 ng/lane of the 1.5 kb one, and detect by using 500 ng of labeled λ DNA per large hybridization bottle. Label λ by random-priming including 1% digoxigenin-dUTP.

Addition of internal MW standards to plant genomic DNA

The appropriate quantities of internal standards should be added to each genomic DNA for fingerprinting analysis. The easiest procedure consists of adding these when re-suspending the DNAs after restriction digestion.

13. Prepare a working bulk of the fragments according to the following:

		Amount to add per single gel lane	
Fragment	**Stock**	**ng/lane**	**μl stock/lane**
24.5 kb	10 ng/μl	0.25 ng	0.025μl
1.5 kb	10 ng/μl	0.50 ng	0.050μl

Do not forget to add the right amount of 5X SGB to complete the loading mixture of DNA, TE, and internal MW standards.

Internal Molecular Weight Markers for Fingerprinting with SSRs

A "top" molecular weight standard is routinely used in every lane of SSR fingerprinting gels, both agarose and polyacrylamide. It is PCR amplified from the Phi plasmid (ϕX174RF) and simple to prepare. It is not possible to use a"bottom" fragment, since fragments smaller thanabout 80 base pairs show up in both agarose and polyacrylamide gels as a smear, if they show up at all. Larger "bottom" standards would interfere with the SSR alleles themselves, which can often be as small as 80-100 base pairs.

1. Obtain the following primers from any source that manufactures oligonucleotides (we frequently use Operon for this purpose):

   Forward primer (5′-3′): CGCCAAATGACGACTTCTAC

   Reverse primer (5′-3′): GCGCATAACGATACCACTGA

   These primers correspond to position 1547 and 2050, respectively, of the Phi plasmid, and amplify a fragment 523 base pairs in length.

2. Run the following PCR reaction using uncut Phi (ϕX174RF) plasmid DNA. We recommend you do several reactions, as you will need a lot of product.

Stock	**Final or amount**	**25 μl RXN**	**100 μl RXN**
ddH2O	--	10.3μl	41.2μl
10X *Taq* buffer	1X	2.5μl	10.0μl
dNTP (2.5mM each)	50μM each	0.5μl	2.0μl
MgCl2 (50 mM)	1.2 mM	0.6μl	2.4μl
Taq polymerase (5U/μl)	0.5 U	0.1μl	0.4μl
Phi DNA (5 ng/μl)	25 ng	5.0μl	20.0μl
Forward primer (2.0 μM)	0.24μM	3.0μl	12.0μl
Reverse primer (2.0 μM)	0.24μM	3.0μl	12.0μl

3. Amplify using the following program:

1 cycle of:	**30 cycles of:**	**1 cycle of:**
93°C for 1 min	93°C for 30 sec	72°C for 5 min
62°C for 1 min		
72°C for 1 min		

4. Run a minigel to check for amplification and correct size on some of the reactions; if there has been amplification of a single 523 bp fragment, combine all the reactions into one tube for storage.

5. Use about 200 ng of molecular weight standard in each lane of a polyacrylamide fingerprinting gel; you can add it directly to the reaction mixture with the loading buffer.

   (Source: Laboratory protocols; CIMMYT Applied Molecular Genetics Laboratory)

7

Polyacrylamide Gel Electrophoresis (PAGE)

Polyacrylamide

Polyacrylamide is a cross-linked polymer of acrylamide. The length of the polymer chains is dictated by the concentration of acrylamide used, which is typically between 3.5 and 20%. Polyacrylamide gels are significantly more annoying to prepare than agarose gels. Because oxygen inhibits the polymerization process, they must be poured between glass plates (or cylinders).

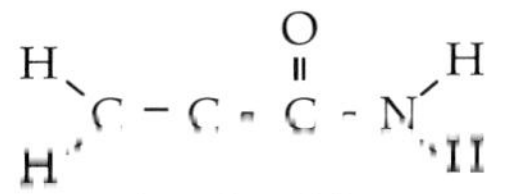

Acrylamide

Acrylamide is a potent neurotoxin and should be handled with care! Wear disposable gloves when handling solutions of acrylamide, and a mask when weighing out powder. Polyacrylamide is considered to be non-toxic, but polyacrylamide gels should also be handled with gloves due to the possible presence of free acrylamide. Polyacrylamide gels have a rather small range of separation, but very high resolving power. In contrast to agarose, polyacrylamide gels are used extensively for separating and characterizing mixtures of proteins

Polyacrylamide is the same material that is used for skin electrodes and in soft contact lenses. Polyacrylamide gel may be prepared so as to provide a wide variety of electrophoretic conditions. The pore size for the gel may be varied to produce different molecular seiving effects for separating proteins of different sizes. In this way, the percentage of polyacrylmide can be controlled in a given gel. By controlling the percentage (from 3% to 30%), precise pore sizes can be obtained, usually from 5 to 2,000 kdal. In the case of DNA, polyacrylamide is used for separating fragments of less than about 500 bp. However, under appropriate conditions, fragments of DNA differing is length by a single base pair are easily resolved. This is the ideal range for gene sequencing, protein, polypeptide, and enzyme analysis. Polyacrylamide gels can be cast in a single percentage or with varying gradients. Gradient gels provide continuous decrease in pore size from the top to the bottom of the gel, resulting in thin bands. Because of this banding effect, detailed genetic and molecular analysis can be performed on gradient polyacrylamide gels. Polyacrylamide gels offer greater flexibility and more sharply defined banding than agarose gels.

Polyacrylamide Gel Electrophoresis Apparatus (BIO-RAD)

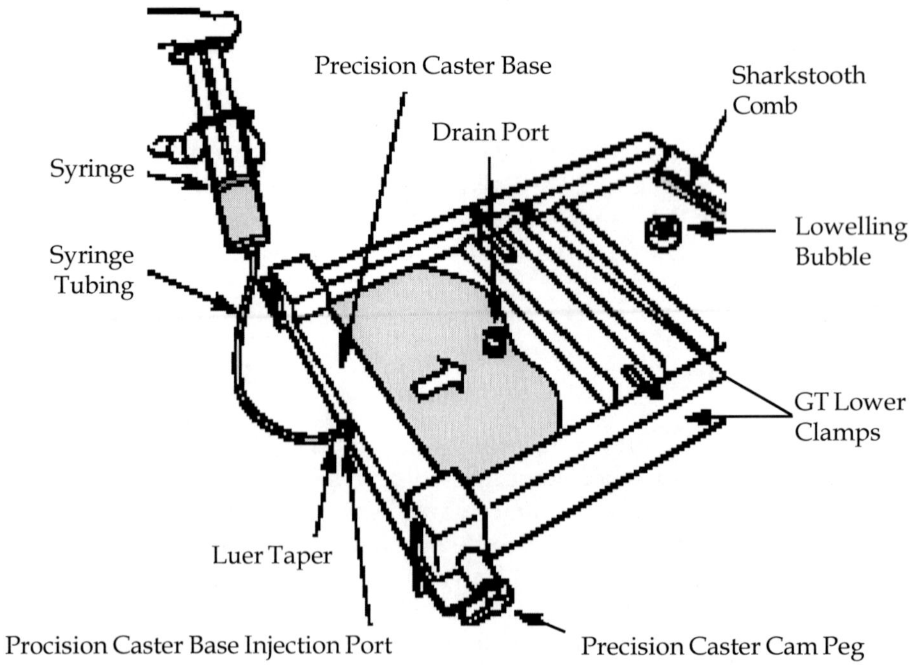

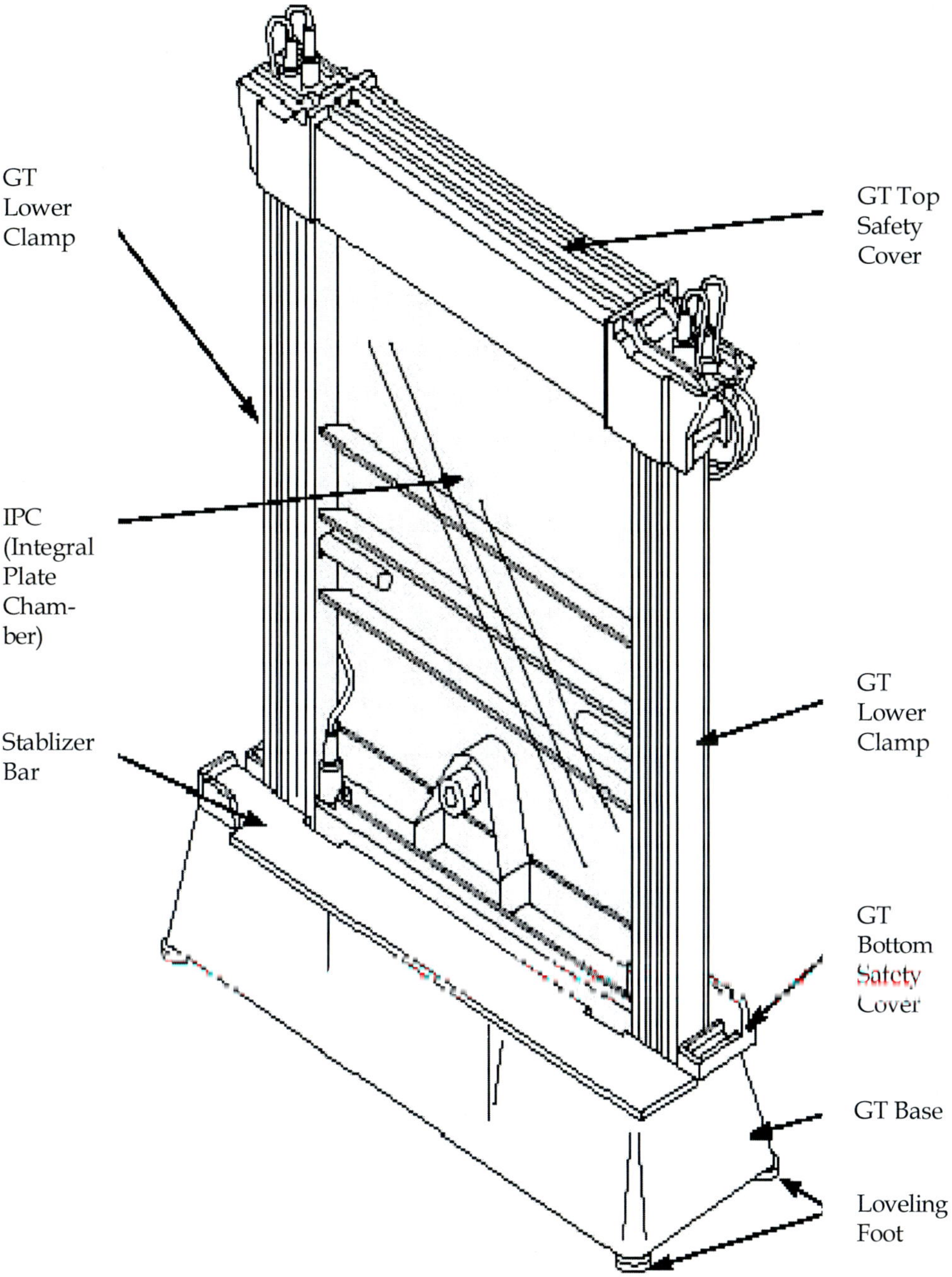
GT Lower Clamp
GT Top Safety Cover
IPC (Integral Plate Chamber)
GT Lower Clamp
Stablizer Bar
GT Bottom Safety Cover
GT Base
Loveling Foot

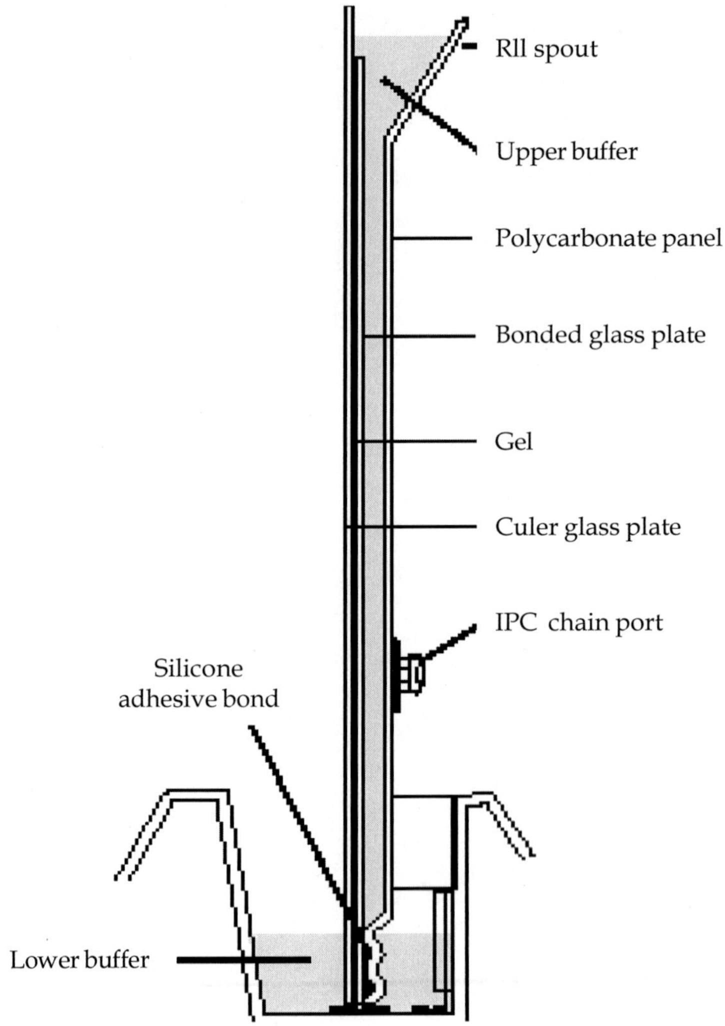

Polyacrylamide Gel Electrophoresis (PAGE) System

The polyacrylamide gel electrophoresis (PAGE) technique was introduced by Raymond and Weintraub (1959). Polyacrylamide gels are more cumbersome to prepare and run than agarose gels. They are almost always poured between two glass plates that are held apart by spacers and sealed by electrical tape. In this arrangement, most of the acrylamide solution is shielded from exposure to the air, so that inhibition of polymerization by oxygen is confined to a narrow layer at the top of the gel. Polyacrylamide gels can range in length from 10 cm to 100 cm, depending on the separation required;

they are invariably run in a vertical position. However, they have three major advantages over agarose gels: (1) Their resolving power is so great that they can separate molecules of DNA whose lengths differ by as little as 0.2% (i.e., 1 bp in 500 bp). (2) They can accommodate much larger quantities of DNA than agarose gel: Up to 10mg of DNA can be applied to a single slot (1 cm x 1 mm) of a typical polyacrylamide gel without significant loss of resolution. (3) DNA recovered from polyacrylamide gels is extremely pure and can be used for the most demanding purpose.

Different samples are loaded in wells or depressions at the top of the polyacrylamide gel. The proteins move into the gel when an electric field is applied. The gel minimizes convection currents caused by small temperature gradients, and it minimizes protein movements other than those induced by the electric field. Proteins can be visualized after electrophoresis by treating the gel with a stain such as Coomassie blue, which binds to the proteins but not to the gel itself. Each band on the gel represents a different protein (or a protein subunit); smaller proteins are found near the bottom of the gel.

Two types of Polyacrylamide Gels are in Common Use

(i) *Nondenaturing polyacrylamide gels for the separation and purification of fragments of double-stranded DNA*. These gels are poured and run in 1 x TBE at low voltage (1-8 V/cm) to prevent denaturation of small fragments of DNA by heat generated by the passage of electric current. Most species of double-stranded DNA migrate through nondenaturing polyacrylamide gels at a rate that is approximately inversely proportional to the $\log_{10}$ of their size. However, their electrophoretic mobility is also affected by their base composition and sequence, so that DNAs of exactly the same size can differ in mobility by up to 10%. This effect is believed to be caused by kinks that form at specific sequences in double-stranded DNA. Because it is impossible to know whether or not the migration of an unknown DNA is anomalous, electrophoresis through nondenatruing polyacrylamide gels cannot be used to determine the size of double-stranded DNAs.

(ii) *Denaturing polyacrylamide gels for the separation and purification of single-stranded fragments of DNA*: These gels are polymerized in the presence of an agent (urea or, less frequently, formamide) that suppresses base pairing in nucleic acids.

Effective Range of Separation of DNAs in Polyacrylamide Gels

Acrylamide (% [w/v])[a]	Effective Range of separation (bp)	Xylene cyanol FF[b]	Bromophenol blue[b]
3.5	1000-2000	460	100
5.0	80-500	260	65
8.0	60-400	160	45
12.0	40-200	70	20
15.0	25-150	60	15
20.0	6-100	45	12

a N, N'- methylenebisacrylamide is included at 1/30th the concentration of acryl amide.
b The number given are the approximate sizes (in nucleotide pairs) of fragments of double-stranded DNA with which the dye co-migrates.

2-D Gel Electrophoresis

When screening cell extracts for individual proteins, separation based on size alone is generally not sufficient to identify a protein of interest. Since it is improbable that any two proteins will share the same size and net charge, it is also useful to separate complex mixtures of proteins based on charge. As you know, the net charge of a protein depends on the number of its basic and acidic residues. You should also know that the charge character of a protein is given by its isoelectric point (pI) - the pH at which the protein has no net charge. Thus acidic proteins have a pI below 7, and basic proteins have a pI above 7. Separation according to pI is accomplished if the protein is loaded onto a gel containing a stable pH gradient. Upon induction of an electric field, proteins will migrate until it reaches a pH in the gel that is equal to its pI (i.e. the protein will have no net charge). This process is called **isoelectric focusing (IEF**; See Fig.) and is most commonly performed in conjunction with SDS page in a method known as **2D electrophoresis**.

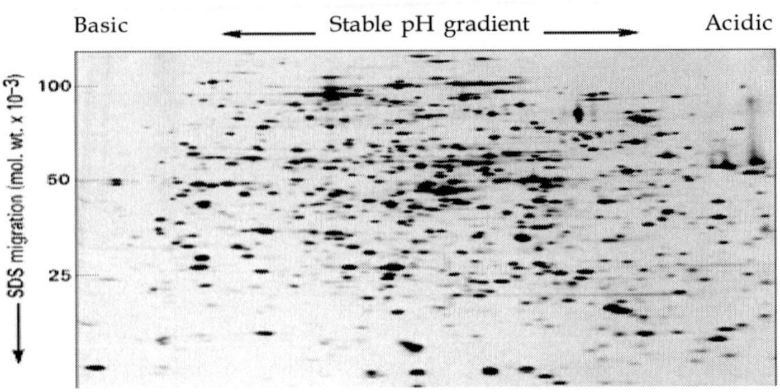

Two Dimensional separation of protein

This is a method for the separation and identification of proteins in a sample by displacement in 2 dimensions oriented at right angles to one another. This allows the sample to separate over a larger area, increasing the resolution of each component. 2D gel electrophoresis is generally used as a component of proteomics and is the step used for the isolation of proteins for further characterisation by mass spectroscopy. . The second use of this technique is differential expression; this is when you compare two or more samples to find differences in their protein expression.

The two steps involved in this are

Isoelectric focusing (IEF) is used in the 1st Dimension (Righetti, P.G., 1983). This separates proteins by their charge (pI).

SDS-PAGE in the 2nd Dimension. This separates proteins by their size (molecular weight, MW). The procedure is known as ISO-DALT: iso for isoelectric focusing and dalt for dalton weight.

Pulse Filed Gel Electrophoresis

It is the sieving properties of agarose gel that determines the degree of separation between DNA molecules of different sizes. In general, linear DNA molecules of smaller size percolate through the pores/channels of the agarose matrix with less drag than the larger molecules. Thus, over a certain size range there is a linear, inverse correlation between the size of the DNA molecule and ????????????. However, when the size of the DNA (>30 kb) exceeds the pore size, the sieving properties of the gel matrix are lost. In 1983 and 1984, Schwartz and Cantor described the separation of large DNA molecules (up to 2 Mb) by periodic alteration of non-uniform electric fields. This technique was called **pulsed field gradient** electrophoresis.

Principle of Pulse Field Gel Electrophoresis : Since the electric field gradient was thought to be the critical parameter in achieving separation. There was considerable variability in the mobility and trajectories of identical DNA molecules in non-uniform electric fields. Multiple technological improvements of the basic design ultimately demonstrated that the angle between the alternating electrical fields, rather than the non-uniform field strength, was the critical determinant in achieving separation of large DNA molecules.

Steps in PAGE

- Assemble the gel plates following the manufacture's instructions. The plates should be clean and free of detergents. Silicon grease can be applied to spacer to make a good, waterproof seal.

- Prepare the acrylamide polymerization solution for the separating gel using the instructions given in the table supplied along with the reagents.
- Pour the acrylamide solution between the glass plates. The meniscus of the acrylamide solution should be far enough below the top of the notched plate to allow for the length of the teeth on the comb plus 1 cm. Carefully overlay the acrylamide solution with isobutanol or water. For gels made with acrylamide concentrations lower than 8% use water; for gels of 10% or greater use isobutanol (or water-saturated isobutanol). Place the gel in a vertical position at room temperature.
- After the gel has set (about-20-30 min), pour off the overlay and wash the top of the separating gel several times with distilled water. Drain well.
- If required, prepare the acrylamide solution for the stacking gel from the values given in the table. Pour the solution for the stack directly onto the polymerized separting gel. Place the appropriate comb into the gel solution, being careful not to trap any bubbles. To clean the comb, wash with H_20 and ethanol. Place the gel in a vertical position at room temperature. The stacking gel will set in approximately 10min.
- After the stacking gel has set, carefully remove the comb. Wash the wells immediately with distilled water to remove unploymerized acrylamide. Straighten the teeth of the stacking gel, if necessary, and place in an appropriate gel box with running buffer in the bottom reservoir. Add appropriate electrophoresis buffer to top reservoir. And bubbles caught between the plates at the bottom of the gel can be removed by squirting running buffer .through a syringe fitted with a bent needle.
- Load samples into the bottom of the wells. The samples can be conveniently loaded using either: (1) a microliter syringe (washed by pipetting in the bottom reservoir buffer between each sample), (2) a micropipetter fitted with a long narrow tip.
- Start electrophoresis at 100-125 V. After the dye front has moved into the separating gel, increase the voltage to 200 V. If the plates get too warm, lower the voltage.
- When the dye front reaches the bottom of the gel, turn off the power pack. Remove the gel plates and gently put the plates apart. Use a spatula or similar tool to separate the plates. Cut a corner from the bottom of the gel that is closest to the number 1 well.

SDS-PAGE of Proteins

Separation of Proteins Under Denaturing Conditions

Sodium dodecyl sulphate (SDS) is an anionic detergent which denatures proteins by "wrapping around" the polypeptide backbone - and SDS binds to proteins fairly specifically in a mass ratio of 1.4:1. In so doing, SDS confers a negative charge to the polypeptide in proportion to its length - ie: the denatured polypeptides become "rods" of negative charge cloud with equal charge or charge densities per unit length. It is usually necessary to reduce disulphide bridges in proteins before they adopt the random-coil configuration necessary for separation by size: this is done with 2- mercaptoethanol or dithiothreitol. In denaturing SDS-PAGE separations therefore, migration is determined not by intrinsic electrical charge of the polypeptide, but by molecular weight.

Determination of Molecular Weight

This is done by SDS-PAGE of proteins - or PAGE or agarose gel electrophoresis of nucleic acids - of known molecular weight along with the protein or nucleic acid to be characterised. A linear relationship exists between the logarithm of the molecular weight of an SDS-denatured polypeptide, or native nucleic acid, and its *Rf*. The Rf is calculated as the ratio of the distance migrated by the molecule to that migrated by a marker dye-front. A simple way of determining relative molecular weight by electrophoresis (*M*r) is to plot a standard curve of distance migrated vs. log10MW for known samples, and read off the log*Mr* of the sample after measuring distance migrated on the same gel.

Principle

Crosslinked polyacrylamide gels are formed by the polymerization of acrylamide monomers in the presence of smaller amount of N, N′-methylene bis-acrylamide (generally referred as bis-acrylamide) which serves as a cross linking agent. The polymerization of acrylamide catalyzed by N′, N′, N′, N′ tetramethylenediamine (TEMED) and initiated by ammonium persulfate.

Protocol for Running Protein

Assembling Gel Apparatus

1. Assemble two glass plates (one notched) with two side spacers, clamps, grease, etc. as shown by demonstrators or instructions. Stand assembly upright using clamps as supports, on glass plate. Pour some pre-heated 1% agarose onto glass plate, place assembly in pool of agarose: this seals the bottom of the assembly.

2. Gel concentration of 12.5% in 0.25 M Tris-HCl pH 8.8

Reagent:	**Volume (ml: to make 30 ML)**	**Volume (ml: to make 10 ML)**
40% Acrylamide stock	9.4	3.1
water (distilled)	12.3	3.8
1 M Tris-HCl pH 8.8	7.5	2.5
10% SDS	0.3	0.1
Peroxydisulphate 1%	0.5	0.5
TEMED (added last)	20 ul	20 ul

19:1 - 38:1 w:w ratio of acrylamide to N,N′-methylene bis-acrylamide

Stacking Gels

Gel concentration of 4.5% in 0.125 M Tris-HCl pH 6.8

Reagent:	**Volume ml to make 15 ML)**	**Volume (ml to make 10 ML)**
40% Acrylamide stock	1.7	1.1
water	10.8	7.1
1 M Tris-HCl pH 6.8	1.9	1.25
10% SDS	0.15	0.1
Peroxydisulphate 1%	0.5	0.5
TEMED (stir quickly)	20 ul	20 ul

Mix as before, then pour onto top of set resolving gel, insert comb, allow to set, remove comb, and fill with electrophoresis buffer. Assemble top tank onto glass plate assembly. Fill with electrophoresis buffer.

3. *Electrophoresis buffer* : The final TANK buffer composition is 196mM glycine / 0.1% SDS / 50mM Tris-HCl pH 8.3, made by diluting a 10x stock solution. This goes in both top and bottom tanks.

4. *Sample Preparation*: Grind a little leaf material (e.g. 2 grams) in a mortar. Centrifuge in an Eppendorf tube for 3 min. Take supernatant and mix 100ul 1:1 (v:v) with SDS-PAGE disruption mix: this is 125mM Tris-HCl pH 6.8 / 10% 2-mercaptoethanol / 10% SDS / 10% glycerol, containing a little bromophenol blue. For liquid / purified samples, take e.g. 100 ul After the stacking gel has set, carefully remove the comb wash the wells immediately with distilled water to remove non-polymerized acrylamide. Straighten PAGE apparatus with running buffer in the bottom reservoir.

Any bubbles caught between the plates at the bottom of the gel can be removed by squirting running buffer through the syringe fitted with bent needle. Add running buffer to top reservoir.

5. Load sample into the bottom of the wells. The sample can be conveniently loaded using either a micro syringe (washed by pipetting in the bottom reservoir buffer between each sample) or a micropipette fitted with a long narrow tip.

6. Start electrophoresis at 100 volts.

7. When the dye front comes to 0.5 cm above the bottom of the gel, turn off the power pack. Remove the gel plates and gently dry the plates apart. Use a spatula or similar tool to separate the plates. Cut a corner from a bottom of the gel that is closet to the number 1 well.

8. Transfer the gel to a glass or plastic tray containing a minimum of 5 gel volumes of brilliant blue stain at 1 × concentration. Shake at room temperature for 30-60 minutes.

9. Remove the stain and add 5 volumes of 1 × Destainer when it gets dark. Continue the process till the background is clear view the gel against a right backgroundand add 50 - 100 µl of disruption mix.

Preparation of Staining of Gels

1. Coomassie Brilliant Blue/Page-Blue 83

Make up stain: 0.2% CBB in 45:45:10 % methanol:water:acetic acid. Cover gel with staining solution, seal in plastic box and leave overnight on shaker (RT) or for 2 to 3 hours at 37°C also with agitation. Destain with 25% 65% 10% methanol water acetic acid mix, with agitation.

2. Copper Chloride (0.3M CuCl2)

Rinse gel in distilled water, immerse in copper chloride solution with agitation for about 20 minutes (RT), rinse with distilled water and immerse in sufficient fresh distilled water to cover the gel (this acts as the destaining step). Seal in a plastic box.

8

Gel Electrophoresis of Protein

Proteins

Proteins are of central importance in the structure and function of all living organisms. Some proteins serve as structural components while others function in communication, defense, and cell regulation. Some proteins serve as enzymes that act as biological catalysts which control the pace and nature of essentially all biochemical events. Although DNA serves as the genetic blueprint of a cell, none of the life processes would be possible without the proteins.

Electrophoresis of Proteins

Proteins can be separated and purified. Methods for separating proteins take advantage of properties such as charge, size, and solubility, which vary from one protein to the next. Because many proteins bind to other biomolecules, proteins can also be separated on the basis of their binding properties. The source of a protein is generally tissue or microbial cells. The cell must be broken open and the protein must be released into a solution called a crude extract. If necessary, differential centrifugation can be used to prepare subcellular fractions or to isolate organelles. Once the extract or organelle preparation is ready, a variety of methods are available for separation of proteins. Ion-exchange chromatography can be used to separate proteins with different charges (similar to the way amino acids are separated). Other chromatographic methods take advantage of differences in size, binding affinity, and solubility. Nonchromatographic methods include the selective precipitation of proteins with salt, acid, or high temperatures. In addition to chromatography, another important set of methods is available for the separation of proteins, based on the migration of charged proteins in an electric field, a process called (gel) electrophoresis. Gel electrophoresis is especially useful as an analytical method. Its advantage is that proteins can be visualized as well as separated, permitting a researcher to estimate quickly the number of proteins in a mixture or the degree of purity of a particular protein preparation. Also, gel electrophoresis allows determination of crucial properties of a protein such as its isoelectric point and approximate molecular weight.

Amino acids differ not only in R-group characteristics but also in molecular weight. Different amino acids are linked together in a linear chain by peptide bonds in various combinations and sequences to form specific proteins. A protein may be comprised of amino acids from all of the categories. The net charge of a protein will depend on its amino acid composition. If it has more positively charged amino acids such that the sum of the positive charges exceeds the sum of the negative charges, the protein will have an overall positive charge and migrate to the cathode (negatively charged electrode) in an electrical field. Proteins even with a variation of one amino acids will have a different overall charge, and thus are electrophoretically distinguishable.

If a protein is electrophoresed before the disulfide bonds are broken, it will yield only one band of 125,000 Daltons molecular weight on the gel. That is, the protein will move as one entity and the peptide chains cannot be distinguished. However, after treatment with a reducing agent, the protein will yield four distinct bands with a combined molecular weight of 125,000 Daltons—the same as the original protein.

The gel results will show that some of the high molecular weight bands from samples not treated with the disulfide reducing agent is missing in the samples treated with the disulfide reducing agent. However, two or more bands are imaged on the gel lane containing the treated samples, thereby replacing each of the missing high molecular weight bands from the samples that were not treated. The additional bands that appear in the treated samples represent the individual polypeptides that make up complex proteins. The break up of complex proteins into their respective polypeptides allows us to study the structure of proteins that result from the interaction of several genes. A gene is a discrete unit of hereditary information that usually specifies a protein. A single gene provides the genetic code for only one polypeptide.

Thus, a protein consisting of four polypeptides requires the interaction of four genes to synthesize that specific protein. A molecular weight protein marker is used to prepare a standard separation curve with which various unknown proteins or polypeptide fractions can be identified.

Raising of source material for Electrophoresis : Etiolated seedlings are raised in the germinator or fresh young seedlings grown in the green house may also be used to extract the source material for protein analysis. Seeds can also be used but need proper sterilization in 0.2% $HgCl_2$ followed by thorough washings with distilled water.

Extraction of Proteins

Add 0.4 ml of extraction buffer to 30 mg of sample and mix on vortex shaker for 2 min. Keep it in water bath at 80°C for 30 min. Cool the samples at room temperature. The samples are then centrifuged at 10,000 rpm for 15 min and supernatant is used for electrophoresis. For studying polypeptide pattern denaturing discontinuous polyacrylamide gel electrophoresis, method of Laemmli (1970) is followed.

Procedure

Electrophoresis is carried out using standard electrophoretic apparatus, for example, the apparatus of Hoefer Scientific Instruments, San Francisco, California may be used. The glass plates are thoroughly washed with water followed by ethanol and air dried. The plates are fitted in gel casting assembly and then clamped. Mix all the components of resolving gel solution as described below. The solution mixture is poured immediately between the plates. A layer of water is then gently poured over using a syringe to obtain a smooth layer of running gel. The gel is set to polymerize for 30-40 minutes. After polymerization remove water using filter paper strips and stacking gel mixture is prepared as per composition of gel as described below and running gel is immediately poured above it. The comb is inserted immediately with care so that no air bubble is trapped. It is allowed to polymerize for 15-30 min.

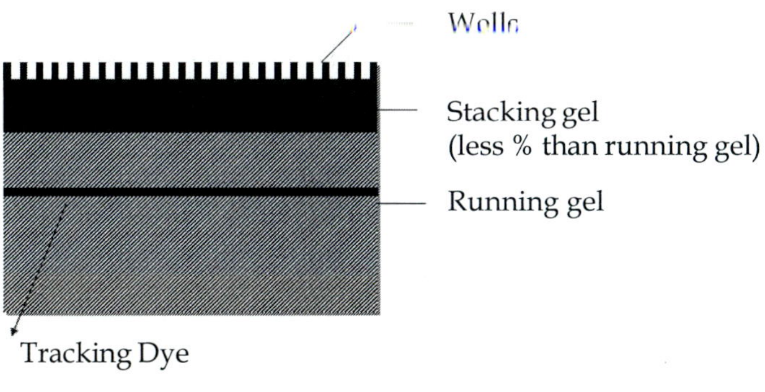

Comb and tygon tube are removed after polymerization of the stacking gel and the gel assemble is fitted in the electrophoretic apparatus. Reservoir buffer is paved in the upper and lower tanks of the electrophoretic apparatus. The wells are washed with reservoir buffer before sample loading. Thirty - forty µl of the sample of each genotype is loaded directly into the wells with the help of fine pipette. Electrophoresis is carried out at a constant current of 10 mA for first 30 min and subsequently at 20 mA till the tracking dye migrates to the lower end of the gel. However, current can be adjusted as per requirement in terms of time for run and enzyme(s) being investigated. At this stage, the electrophoresis is terminated. Gel is removed from glass plates and stained overnight with the staining reagent. Excess of stain is removed next day by diffusion using destaining solution. After proper destaining, gels are photographed using any good quality digital camera or using gel documentation system. Gels are stored in 7 per cent acetic acid solution.

Scoring of Gels

Gels are visually scored by putting the gel on a light box. Bands with dark to very light intensities are scored and are used to construct the zymograms. Rf (=Rm=Relative mobility) value of each band is calculated using the following formula :

$$Rm = \frac{\text{Distance traveled by the band}}{\text{Distance traveled by the tracking dye}}$$

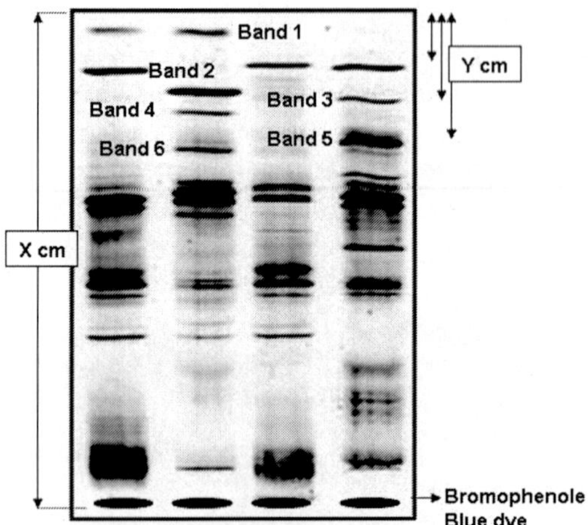

Sample of Protein Gel
Source: http://www.magazine.noaa.gov/stories/images/ief.jpg

For Example: Rf or Rm (Band 5) = Y/x

Bands are numbered on the basis of increasing Rf value or according to the distance traveled in the gel. Alternatively, the molecular weight of different polypeptide can be determined from the calibration curve which is obtained by plotting Rf value vs log molecular weight of standard proteins.

Construction of Zymogram

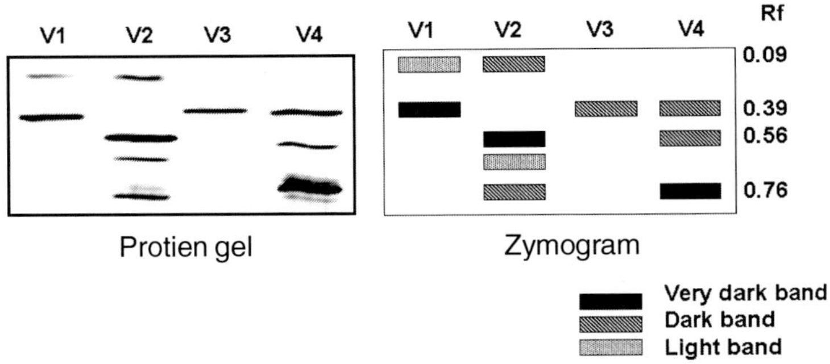

Similarity Index

Similarity index values are calculated based on proportion of common bands between two lanes by using the following formula (Nei and Li, 1979) :

$$F = \frac{2\,Mxy}{Mx + My}$$

Where, F is the similarity index, Mx is the number of bands in variety x, My is the number of bands in variety Y and Mxy is the number of bands common to both X and Y. F × 100 gives the per cent similarity between two variety, thus F=1.0 would mean that the patterns in the two varieties/genotypes are identical.

Reagents

All reagents are prepared in distilled water and stored in brown bottles in refrigeration.

Stock Solutions

A. Acrylamide-bisacrylamide solution

Acrylamide	30 g
Bis acrylamide	0.8 g
volume	100 ml

B. Stacking Gel buffer

Tris	6.0 g
IMHCl	48.0 ml
pH	6.8
Volume	100 ml

C. Resolving gel buffer

Tris	36.3 g
IMHCl	48.0 ml
pH	8.8
Volume	100 ml

The solutions A, B and C were filtered through Whatman No. 1 filter paper and stored at 4°C.

D. Ammonium per sulfate (APS) solution

APS	0.15g
Volume	10 ml

Solution D was prepared fresh before use.

E. N, N, N, N′ - tetramethyl ethylene diamine (TEMED) - TEMED used as supplied.

F. SDS solution

SDS	10 g
Volume	100 ml

G. Reservoir buffer

Tris	3.0 g
Glycine	14.4 g
SDS	1 g
pH	8.3
Volume	1 L

H. Staining solution

Coomassie Brilliant Blue R 250	1.25 g
Methanol	200 ml
Glacial acetic acid	35 ml
Volume	500 ml

The dye was dissolve in the methanol and water components first, and then glacial acetic acid was added. Filtered the final solution with Whatman No. 1 filter paper to remove insoluble material. It is stable for several months at room temperature.

I. Destaining solution

Glacial acetic acid	75 ml
Methanol	50 ml
Volume	1 L

J. 2×sample buffer

1 M Tris HCl, pH 6.8	2.5 ml
SDS	0.4 g
2-mercaptoethanol	1.0 ml
Glycerol	2.0 ml
1% bromophenol blue	0.4 ml
Volume	10 ml

Preparation of gel

(a) Running/separating gel (1 mm)

Acrylamide-bisacrylamide	7.5 ml
Resolving gel buffer	2.8 ml
SDS solution	0.225 ml
Ammonium persulphate (APS)	1.125 ml
Water	10.84 ml
TEMED	0.012 ml

TEMED was added just before pouring the gel mixture. APS was prepared fresh just before use.

(b) Stacking gel (1 mm)

Acrylamide-bisacrylamide	0.625 ml
Stacking gel buffer	1.250 ml
SDS solution	0.050
Ammonium persulphate	0.250
Water	2.800
TEMED	0.004

9

Isoenzyme

Hunter and Mohler discovered the isozymes. In 1959 Markert and Moler introduced the concept of isozymes, which they defined as the different molecular forms in which proteins may exist with the same enzymatic specificity (Buth 1984). This means that different variants on the same enzymes have identical or similar functions and are present in the same individual. As such, their importance for understanding gene action in development and differentiation was exploited during the1960s in both animals and plants. Nevertheless, isozymes played a minor role in research on plant biochemistry until 1966 when genetic polymorphism for isozymes within the same population was discovered (Stebbins 1989, Wendel 1989). That revealed the possibility for population genetics to make precise quantitative estimates of genetic variability based upon one parameter of the molecular structure of the primary products of the genes themselves.

Plant population genetics were not long in following their zoological colleagues and the investigation of both animals and plants increased explosively. This application, however, necessitated a partitioning of the isozyme concept, since the relevant isozyme subset, defined by Prakash (Buth 1984) as the variant proteins produced by allelic forms of the same locus, to avoid the now common confusion with isozymes which are the various polymer produced from monomers specified by different loci. The development of genetic distance or similarity coefficients (Lukasová 1985) allowed the summarization of isozyme data for intersample comparisons and so quantified isozyme data were soon applied to comparative studies of taxa.

Polymorphism and Basic Principles of Isozymes

Polymorphism may be defined as simultaneous occurrence within or between populations of multiple phenotypic forms of a trait attributable to the allels of a single gene or the homologs of a single chromosome (Acquaah 1992). In natural populations recurrent mutations of genes produce variability. There are polymorphic loci, variable in the sense described above, and monomorphic or nonvariant. The main points of the concept of isozymes perceived by researchers are summarized below:

1. Multiple molecular forms of enzymes (isozymes) are common in organisms.
2. Isozymes share a common catalytic activity. Each isozyme has a specific role in the metabolic pathway and functions in harmony with other enzymes within the organizational framework of cells.
3. Isozymes often exhibit tissue or cell specificity.
4. Molecular heterogenity of enzymes confers flexibility, versatility and precision upon an organism in terms of metabolic functions.
5. Molecular multiplicity is desirable for biological efficiency. Isozymes arise in nature by two general mechanisms, i.e. genetic and epigenetic.

The source of gene multiplicity is duplication through mutation, polyploidization and chromosomal aberrations (Hoelzel 1991). Those events constitute the present epigenetic origin of isozymes. Epigenetically formed enzymes are not considered isozymes by some researchers. Epigenetic mechanisms may be divided into

a) post-translational addition,
b) post-translational deletion and
c) post-translational conformation.

On the other hand there are four genetic mechanisms (Acquaah 1992):

a) Multilocus system I – different genes code for independent proteins with the same enzymatic activity. The various genes are nuclear in origin but their protein products are located in different parts of the cell.

b) Multilocus system II – is similar to system I except that the enzymes involved arepolymeric and the subunits are encoded by more than one locus.

c) Multilocus-polymeric system – enzymes display a series of polymers that consist of identical subunits.

d) Isozyme system – the term isozyme describe isozymes encoded by allelic genes.

Alleles at various loci may be modified to produce isozymes that are distributed in a population according to Mendelian laws of inheritance (Weeden 1983).

Application Isozymes as a Tool in Research

Genetic markers generally have contributed to the study of plant biology by providing methods for detecting genetic differences among individuals. There are some important ecological topics which often use isozymes as powerful markers:

a) Genetic relatedness within and among populations, often with relations to geographic structure.

b) Patterns across a broad range of taxa are mostly consistent with our understanding of the effects of the breeding system (species with selfing tend to poses lower levels of genetic variation within populations), life history (longer-lived perennials tend to be more variable) and the distribution of genetic diversity within and among populations (Clegg 1990, Hamrick 1990, Kudoh 1997, Mahy 1997, Tarayre 1997). There is a narrow linkage between geography and the spatial patterns of genetic variation (Epperson 1990, Newton 1999) as well as the genetics of plant migration and colonization (Barrett 1989,1990; Sun 1997).

c) Mating system estimation. An alternative approach of the study to plant mating uses classifications of mating events to characterize levels of inbreeding and patterns of gene dispersal in a population and other consequences (Holm 1997, Soltis 1987, Brown 1990, Soltis 1989, Sun 1996, Mitchell 1998, Noyes 1996, Wang 1996).

d) Genetic diversity in clonal plant species. Asexual reproduction is relatively common in plant species and can occur through a number of modes, e.g. vegetative spread, production of vegetative propagules, apomixis (Cruzan 1998). One difficulty with the study of genotypic diversity in clonal species is the inadequacy of isozyme markers to reliably identify all genotypes present (Ellstrand 1987, Etoh 1981, Lehman 1997, Ollejonsson 1996, Pooler1993, Sipes 1997, Zeidler 1999).

e) Forestry and agriculture (Bliss 1990, Hamrick 1989, 1990, 1997; Èurn 1995; Doebley 1989; Gärtner 1996; Kara 1997; Löchelt 1995, Marschal 1990, Muona 1990, Weber 1990). It is often necessary to test paternity (Cruzan 1998) or chromosomal locations and mapping (Peffley 1988, Satovic 1996, Shigyo 1995a, 1995b, 1994a,).

f) Selection and linkage in plant populations and their evolutional consequences (Cruzan 1998, Ennos 1990, Hartl 1988, Hastings 1990, Sun 1996) and some tasks of polyploidy (Chase 1988, Kirschner 1995, Maki 1996, Sun 1996).

g) Conservation biology is a rapidly rising field because of effective molecular tools (Gärtner 1996, Gemmill 1998, Godt 1996, Haig 1998, Newton 1999).

h) Other studies: Seed bank populations (Cabin 1998, McCue 1998), environmental changes and habitat heterogenity (Lehman 1997), combination with cytological studies and other methods (Anderson 1995, Noyes 1995), plant pathology (Leuchtman 1996), germaplasm collections (Lamboy 1996), mycorrhizal genetic variation (Karkouri 1996, Martin 1998) and phytopathology (Ylimattila 1997, Forbres 1997).

Analytical Procedures

- Preparation of tissue homogenates
- Separation of the polymorphisms by polyacrylamide or starch gel-electrophoresis
- Visualization of the polymorphisms by specific enzymatic staining.

Main Requirements

- Laboratory setup for gel-electrophoresis
- Facilities/equipment to maintain gels at 2-5 °C during electrophoresis

A comparison of the main features between isozymes and DNA markers are shown in Table.

Allozymes	**DNA methods**
A limited number of protein-coding genes (-)	Detection is not limited on coding regions of DNA, so there is a potentially unlimited number of markers (+)
The number of allozyme that can be screened is limited (-)	Nearly unlimited number of DNA markers, mainly in non coding area of DNA (+)
Obtaining data is relatively inexpensive (+)	Special laboratory construction is often necessary that is not always very cheap (-)
Large numbers of samples can be processed with far less training and time per sample (+)	DNA analysis requires sophisticated equipment and generally more time per sample (-)
Standard statistical procedure used for fineand broad- scale genetic variation (+)	There are a lot of specific statistical procedures for analysing (-)

Advantages

- No DNA extraction involved: Comparable data from previous studies and a wealth of standard statistical methods make isozymes appealing for studies of genetic variation. Most isozymes represent codominant loci distributed according to Mendelian laws of inheritance and many loci express at all stages of the life cycle (Hamrick 1989). So it is feasible to use variation in isozymes rather than nucleic acids for a particular question. With RAPDs and sequencing isozymes are considered excellent for population structure analysis for sufficiently polymorphic taxa.
- No primers or probes required
- Quick and easy to assay
- Low costs involved
- Zymograms can be interpreted in terms of loci and alleles
- Codominance of alleles
- High reproducibility

Disadvantages

- Low abundance
- Low level of polymorphism
- Zymograms sometimes difficult to interpret due to complex banding profiles arising from polyploidy or duplicate genes and the formation of intergenic heterodimers
- Proteins with identical electrophoretic mobility (co-migration) may not be homologous
- Selective neutrality dubious.

Separation of Plant Proteins Using Agarose Gels

The agarose gels can be used for a number of plant proteins. The first three applications suggested utilize the protocols presented here. The study of plants from the Brassicacae family, a comparison of native grasses and herbs, and the comparison of C3 and C4 plants all employ the agarose gel system. Students can employ a variety of stains to detect the presence of particular proteins (isozymes) in these plant tissues.

Follow the protocols presented below for the preparation of the gel, protein extractions, and loading of the gel. Then select the appropriate staining gel protocol for the enzyme you wish to study. You will be able to detect heterozygosity or **variation in isozymes** in plants from the same population. You will be able to observe the presence of particular isozymes in comparative studies.

Methods

Preparation of Source Material

Raising of Seedlings

1. Surface sterilize the seeds with 0.1% $HgCl_2$ for 10 minutes followed by 3-4 washing with distilled water to remove the traces of $HgCl_2$.
2. Sow them in the plastic boxes between two filter papers. Keep at 37°C in the germinator (i.e. in dark). Five days old etiolated seedlings (7-9 cm) are harvested (except roots) for getting the crude extracts to be used for isozyme study.

Preparation of Imbibed Seeds

Imbibitions of seeds is done by putting the surface sterilized seeds on water-soaked filter papers in sterilized glass petriplates at room temperature for the required duration to enhance the enzymatic activity.

Dry Seeds

Dry seeds can also be used to extract samples for enzyme detection after sterilization.

One buffer will be used to grind the plants, run the samples on agarose gels, and to assist in the staining process at the end. This buffer is called: *Protein Extraction Buffer/Electrophoresis Running Buffer:*

Tris/Glycine pH ~ 6.0*
3.0g Tris
14.4 g Glycine
up to one liter with distilled water

*The buffer Tris is normally at a pH of 8.0-8.3. For best results, use 1.0 M HCl to lower the pH of the extraction buffer. Better results have been obtained at pH 5 or 6 as compared to pH 8.3. If you are planning to view protein bands at a pH other than 8.3, be sure to change the pH of the extraction/running buffer NOW before proceeding with the lab.

Hint : Make the agarose gel first, before choosing plant samples to grind. As the gel is cooling, you can grind your plant samples. The gel will be ready when your plant extract samples are ready.

Casting the Agarose Gel

1. Make a 1.5% agarose gel using the Tris/Glycine buffer:

   For 100 mls: 1.5 g agarose

   100 ml Tris/Glycine buffer

2. Microwave or heat the agarose and Tris/Glycine buffer in an Erlenmeyer flask until liquid boils and agarose is completely dissolved. Be careful, the glass is hot! Use a paper towel or mitt to handle the hot flask.

3. Allow the melted agarose to cool to the touch before pouring into the casting tray.

4. Place the gel comb in the **middle** of the gel tray, as enzymes will migrate towards both the negative and positive poles of the gel. Properly tape the gel tray in preparation for the agarose.

5. Once the melted agarose has cooled somewhat, pour it into the casting tray and allow the gel to harden. (NOTE: gel thickness should be about 0.3-0.5 cm: thinner gels seem to work better and are easier to stain and interpret). Take care not to bump or hit the gel box during this time.

Protein Extraction Protocol

1. Obtain a small amount of plant matter such as grass, clover, herbaceous leaves, etc. (Tree and other woody plant leaves do not work well with this procedure as they are hard to grind).
2. Put plant matter into a mortar and pestle with a very small volume (1-2 ml) of Extraction Buffer. Grind the plant matter until the buffer is dark green. If the material is too dry, add another ml of buffer. You want the extract to be liquid, but concentrated, so use the grinding buffer sparingly.
3. Transfer 1 ml of dark green extract into a microcentrifuge tube and spin for three minutes to pellet the plant debris. (If no microcentrifuge is available, use a clinical centrifuge. More sample can be spun, and increase spin time to seven minutes on top speed).
4. Transfer the liquid extract to a fresh microcentrifuge tube. Discard the plant debris pellet.

   Note : This is a stop point. The extract can be stored in the refrigerator overnight if desired.
5. Increase the density of the extract before loading into the agarose gel. You will want to fill the wells of the gel. Typically, a gel well will hold 15-25 μL. Choose ONE of the following methods to increase extract density:

   a. Dilute 3 parts extract with 1 part 50% glycerol/Bromophenol Blue.

   b. Dilute 3 parts extract with 1 part 40% sucrose.

   c. Dilute 3 parts extract with 1 part loading dye (i.e. DNA loading dye)
6. The sample is now ready to load on an agarose gel.

Crude extracts for electrophoresis are prepared either from etiolated seedlings (SL) or imbibed seeds (IS) or dry seeds (DS). A sample of 50-100 mg of tissue, including coleoptile, plumule, and mesocotyle, are homogenized in grinding buffer solution (@ 1 ml buffer per 1 g of tissue) with a prechilled mortar and pestle (Appendix). Thereafter, samples are transferred in prechilled 1.5 ml polyethylene microcentrifugation tubes (kept on ice). These are centrifuged at 12,000 rpm for 20-25 minutes and supernatant is collected in 0.5 ml microcentrifugation tubes. Samples are either used immediately or tored at -70°C for later use.

Loading Agarose Gel

Possible Alternative Gel Loading Options

a. When working in groups of four with an eight-well gel, make two identical gels on the same gel. Load samples 1, 2, 3, and 4 in wells 1-4. Then load the same samples 1, 2, 3, and 4 in wells 5-8. When the gel is completed, slice the gel carefully down the middle and stain in two different enzyme stains. **Note :** this will work when viewing different proteins of the same pH range.

b. Make three identical gels on the same gel. Load samples 1 and 2 in wells 1-2, skip well 3 and load samples 1 and 2 in wells 4-5, skip well 6 and load samples 1 and 2 in wells 7-8. When the gel is completed, slice it into three pieces by slicing down through wells 3 and 6. The three pieces now contain the same samples and can be stained in three different enzyme stains.

1. Load 15-25 µL (fill wells) of increased density protein extract (from #5 above) into each well of gel.
2. Connect gel electrophoresis leads and run gel at 100 volts for approximately 30 minutes. Check to see that the dye has migrated about two-thirds down the gel.
3. When using stains that are pH sensitive (pH 5 or 6), allow gel to soak in 100 mM NaOAc for 10 minutes. This step is omitted when banding is not dependent upon pH.

Staining Gel

Use one of the following recipes for each gel to be stained. Only one enzyme can be stained for each gel, therefore it may be desirable to use one of the alternative loading techniques described in the previous section. Each enzyme stain will form a non-soluble precipitate with the bands IN the gel. This means the gel can be stored in the refrigerator (minus the stain) for up to several weeks and the banding pattern will be distinct.

All gels should be bathed in stain for a minimum of 10 minutes. For best results, gels may be left in stain for longer periods of time (i.e., over night). (Special procedures follow for some stains.)

Staining Method

A. Enzyme: Alcohol Dehydrogenase (for corn raised in drought and flood conditions)

1. Combine the following ingredients and pour over gel:
   50 ml 50 mM Tris-HCl pH <7
   1 ml NAD soln. (10 mg/1 ml)
   0.2 ml Ethanol
   1 ml NBT (Nitro blue tetrazolium) or MTT (tetrazolium thiazolyl blue) soln. (10 mg/ml)
   2 mg (0.4 ml) PMS (phenazine methosulfate) soln. (5 mg/ml)
2. Incubate at 37°C until bands are optimally developed. Rinse and store in water (if NBT is used) or fixative.

B. Enzyme: Catalase (CAT) EC 1.11.1.6. Anodal (Cardy *et al.* 1983)

1. Combine the following ingredients and pour over gel:

Hydrogen peroxide, 0.01% (fresh)	50 ml
H_2O	50 ml
Ferric chloride	500 mg
Pottasium ferricyanide	500 mg

2. Pour H_2O_2 on gel slice and leave for 5 min. Meanwhile, mix remaining ingredients. Pour off peroxide and add stain solution. Agitate the gel gently until bands are developed. Bands appear as achromatic zones on dark green background. Rinse and fix.

   Note : Bands disappeared very soon and whole of the gel turned dark bluish green.

C. Enzyme: Esterases Same banding pattern regardless of pH used.

Method I

0.1M sodium phosphate buffer (pH 7.2)	100 ml
α - napthyl acetate	20 mg (dissolve in 2 ml acetone)
Fast blue RR salt	50 mg

Add fast blue salt to the phosphate buffer and stirr till fully dissolved. Just before staining add α - napthyl acetate to the buffer, mix and pour it on the top of the gel. Incubate for 30-40 min in dark at 37°C. Wash and fix the gels when the bands are clearly visible.

Method II

0.1M potassium phosphate buffer (pH 6.0)	100 ml (adjust pH with NaOH)
Fast blue RR salt	80 mg
α - napthyl acetate	200 mg (in 10 ml acetone)

Procedure for staining is same except for the time of incubation. Bands fully appear within 15 min of incubation at 37°C.

D. Glutamate dehydrogenase (GDH) EC 1.4.1.2.

Method I

0.1M Tris-HCl (pH 8.5)	100 ml
L-Glutamic acid	300 mg
NAD	40 mg
NBT/MTT	30 mg
PMS	10 mg
$CaCl_2$	100 mg

Mix ingredients and pour over the gel. Incubate until blue bands appear. Rinse and store in water (when NBT used).

Method II

All ingrediants same except that $CaCl_2$ is not included.

E. Glutamate oxaloacetate transaminase (GOT)

AAT substrate solution*	50 ml
H_2O	50 ml
Fast blue BB salt	100 mg

* AAT substrate solution (pH 7.4 with NaOH):

H_2O	800 ml
α - ketoglutaric acid	292 mg
L - aspartic acid	1.07 g
PVP-40	4.00 g
EDTA, Na_2 salt	400 mg
Sodium phosphate, dibasic	11.36 g

Add fast BB to substrate solution and incubate at room temperature in the dark until blue bands appear. Rinse and fix.

F. Lactate dehydrogenase (LDH)

0.05M Tris HCl (pH 8.0)	100 ml
NAD	20 mg
Lactic acid, lithum salt	200 mg
MTT/NBT	20 mg
PMS	4 mg

Combine all the ingredients and pour over the gel. Incubate until blue bands appear. Rinse and fix.

G. Malate dehydrogenase (MDH) EC 1.1.1.37. Anodal (Cardy *et al.* 1983)

0.1M Tris HCl (pH 9.1)	100 ml
Neutral malic acid	200 mg
NAD	40 mg
NBT	20 mg
PMS	2 mg

Dl-malic acid was added as a neutralized (with NaOH) aqueous solution. Ingredients were combined and poured over the gel. Incubated until blue bands appeared. Rinsed and stored in water.

H. Malic enzyme (ME) EC 1.1.1.40. Anodal (Soltis *et al.* 1983)

0.1M Tris HCl (pH 8.6)	100 ml
L-malic acid	280 mg
NADP	20 mg
NBT	30 mg
PMS	5 mg
IM $MgCl_2$	0.1 ml

Mix the above chemicals and pour over the gel. Incubate in dark at room temperature until blue bands appear (may take 15 h).

I. Phosphoglucoisomerase (PGI) or Glucose-6-phosphate isomerase EC 5.4.2.2. (formerly EC 2.7.5.1.) Cathodal & Anodal (O. Malley *et al.* 1980)

0.1M Tris - HCl (pH 7.5)	100 ml
Fructose-6-phosphate	15 mg
NADP	7 mg
MTT	15 mg
Glucose-6-phosphate dehydrogenase	20 units
IM $MgCl_2$	1.0 ml

Combine all the ingredients and pour over the gel. Incubate until blue bands develop at 32°C. Rinse and fix the gel.

J. 6-phosphogluconate dehydrogenase (6-PGD) EC 1.1.1.44. Anodal (Modified O.Malley *et al.* 1980)

0.05M Tris-HCl (pH 7.5)	100 ml
6-phosphogluconate (in 1 ml H_2O)	40 mg
NADP	10 mg
$MgCl_2$	100 mg
MTT	10 mg
PMS	2 mg

Combine all the ingredients and pour over the gel. Incubate at 37°C until blue bands appear. Rinse and fix.

K. Shikimate dehydrogenase (SKDH) EC 1.1.1.25. Anodal (Soltis *et al.* 1983)

0.1M Tris-HCl (pH 8.5)	100 ml
Shikimic acid	100 mg
NADP	20 mg
MTT	20 mg
PMS	4 mg

Combine all the ingredients and incubate the gel at 37°C for 1 h and keep at room temperature overnight in the dark. Rinse and fix.

L. Superoxide dismutase (SOD) EC 1.15.1.1. Anodal (Wendel and Weeden 1989)

0.05M Tris-HCl (pH 8.2)	100 ml
Riboflavin	2 mg
EDTA	1 mg
NBT	10 mg

Combine ingredients and pour over the gel. Incubate for 30 min in dark at 37°C. Remove from incubator and illuminate on a light box. Zones of SOD activity are revealed as achromatic regions on a dark blue background. Rinse with water and fix.

M. Enzyme: Acid Phosphatase

For a four-well gel

1. Place gel in ~ 30 ml Tris/Glycine Buffer (use same pH as used for grinding and gel)
2. Dissolve 10 mg alpha-Naphthyl phosphate in 1 ml Tris/Glycine buffer and add to the gel.
3. Dissolve 50 mg Fast Garnet GBC (salt) in 2-3 ml of 1M NaOAc, pH 4.
4. Add small amounts of the Fast Garnet solution to gel; add more solution as needed.
5. View gel with white light box. Compare banding patterns of plants.

N. Enzyme: Peroxidases

1. True Blue Peroxidase Substrate, Add the pre-mixed substrate directly to the gel in the volume needed to cover and stain the gel.

   OR

2. Make your own solution of True Blue Peroxidase Substrate with the following recipe:
   20-50 mg 4-chloro-1-naphthol dissolved in methanol (do this step first)
   200-600 μL of 3.0% Hydrogen Peroxide
   add to a final volume of 100 ml distilled water
3. View the gel with a white light box to see the banding pattern.

O. Enzyme: Esterases Same banding pattern regardless of pH used.

For a ~4 well gel,

1. Place gel in ~30 ml Tris/Glycine Buffer (use same pH as used for grinding and gel)
2. Dissolve 10 mg Naphthylacetate in 1 ml DMSO (dimethylsufoxide) OR in 1 ml DMF (Dimethylformamide). **Note :** DMSO is preferred over DMF for less toxicity.

3. Add the dissolved Naphthylacetate solution to the gel.
4. Dissolve 50 mg Fast Garnet GBC (salt) in 2-3 ml 1M NaOAc pH 4.
5. Add small amounts of the Fast Garnet solution to gel; add more solution as needed.
6. View the gel with a white light box to see the banding patterns.

P. Enzyme: Ribulose-bisphosphate carboxylase (for C3 and C4 plants)

1. Dissolve the following:

   25 mg Amido Black

   50 ml Fixative consisting of: 5 parts water, 5 parts methanol, and 1 part acetic acid
2. Dissolve the Amido Black in the fixative solution. Pour over gel and stain for 15-30 minutes. Pour off stain and rinse three times with fixative for 15 minutes each to remove background staining. You may vary staining by adjusting time.

Q. **Enzyme**: Malate dehydrogenase (for C3 and C4 plants)

1. Dissolve the following:

   50 ml 50 mM Tris-HCL, pH 8.5

   1 ml NAD soln. (10 mg/1 ml)

   1 ml Malic acid (150 mg/1 ml) Use L or DL, which is less expensive.

   1 ml NBT (Nitro blue tetrazolium) or MTT (tetrazolium thiazolyl blue) soln. (10 mg/ml)

   0.4 ml PMS (phenzaine methosulfate) soln. (5 mg/ml)
2. After ingredients above are dissolved pour over gel. Incubate until blue bands appear. Rinse and store in water (if NBT is used) or fixative.
3. Note: Both C3 and C4 plants have these specific enzymes you will test, but as different isozyme forms.

Notes : In order to visualize protein bands, Fast Garnet GBC is used as a coupler. The coupler binds to the enzyme/protein substrate and produces a color reaction, hence the bands become distinct. Any commercial dye containing dizonium salt can be used as a coupler. It is not recommended that Fast Green be used, but any other (colour) Fast dye can be used.

To change the pH of the extract and visualize different bands at various pH's:

Mix up the Protein Extraction/Running Buffer as directed, but change to the desired pH by adding HCl. Use this "corrected pH" buffer for protein extraction, making of agarose gel, and gel running buffer. Remember to let the finished gel soak in 100 mM NaOAc for 10 minutes before staining. This step will ensure the gel is at a lowered pH and thus different protein bands will appear.

10

Extraction of DNA Fragments from Agarose Gel

There are many different methods of extracting DNA bands from an agarose gel. Which one you choose will probably depend on the consumables you have available in your lab. Another important consideration is the yield/ purity of the DNA after extraction. Agarose contains various impurities which may inhibit downstream reactions if not efficiently removed from the DNA. We used to give the generic term 'hoonose' to the unspecified carbohydrates that co-purified along with the DNA and inhibited enzyme reactions. Modern spin-column kits are very good at removing impurities, old-fashioned 'kitchen sink' methods are less good.

Cutting Out the DNA Band

Most of the methods require you to cut out the band. You must visualise the band, in an ethidium bromide stained gel, in a dark-room on a UV light-box (a trans-illuminator). UV is dangerous, wear gloves, long-sleeves and face protection. Never look at UV with unprotected eyes. If possible, set the trans-illuminator to long-wavelegth UV (or low-power) and minimise the amount of time the DNA is exposed. This is because the UV mutagenises the DNA at a measurable rate. Use a scalpel blade to cut around the band of interest. Switch off the transilluminator, switch on the white light and carefully remove the band from the gel and place it on the glass. It is good to trim off as much empty agarose as possible so go back to UV illumination briefly to do this. Place the excised band in a 1.5mL microfuge tube.

Extracting DNA

There are different methods for the extraction of DNA from the gel. Here we are mentioning few methods which can be used according to the facility available.

1. DNA Fragment Purification from Agarose or Acrylamide

For fragments from 200 bp to 10 kb the agarose purification is ideal. For smaller fragments (20 bp to 400 bp) the acrylamide purification is preferred.

Solutions

Crush and Soak Solution

500 mM NH_4OAc 3.3 g NH_4OAc

0.1% SDS 0.1 g SDS

0.1 mM EDTA 20 l 500 mM EDTA

up to 100 ml with Q

store at room temperature

3 M NaOAc pH 5.2

24.6 g anhydrous sodium acetate

pH to 5.2 with acetic acid and bring up to 100 ml with Q

store at room temperature

Other Reagents

DMCS treated glass wool (Alltech Assoc. Inc. #4037, 50 g)

0.22 m disposable micro tip filters (syringe type)

blue tips with melted tips to serve as pestle for crushing acrylamide

Procedure

Agarose Gels

1. Prepare spin columns by cutting off the cap of a 0.5 ml eppendorf tube and forming a hole in the bottom with a hot 18 ga needle. Fill this "mini-column" with a small ball of DMCS treated glass wool and pack down with a pipet tip.

2. Cut out the desired band from an agarose gel and place in a spin column inside a 1.5 ml eppendorf tube with the top cut off.
3. Spin at 6,000 rpm in a microfuge for 10 minutes.
4. Phenol/chloroform extract the flow through and EtOH precipitate with glycogen or tRNA and 10% v/v of 3M NaOAc pH 5.2.
5. Wash and dry, resuspend in 20 microliters TE, run 10 microliters on a gel and use 1-2 microliters for a ligation.

Acrylamide

1. Run a 4-6% acrylamide gel in 1X TBE, stain in EthBr (1-10 ?g/ml) and cut out the desired band.
2. Crush the acrylamide with a p1000 tip with a melted end to resemble a pestle for the eppendorf "mortar."
3. Add 1 ml crush and soak solution and incubate overnight at 37° C.
4. Spin in the microfuge for 10 minutes at 14,000 rpm. Remove as much liquid as possible and add another 500 microliters of crush and soak solution.
5. Repeat the spin and pool the recovered supernatant.
6. Add 0.1 volume of 3M NaOAc, 2.5 volumes of EtOH and carrier (see above).
7. Spin as usual, wash and dry. Resuspend in 20 microliters TE.

2. Spin-columns (Nucleic Acid Purification Columns)

These are excellent for extracting DNA if you can afford them. They cost 1–2 US$ each. Manufacterers include Qiagen, Sigma, Novagen. Everything you need comes with the kit, including protocol, but to summarise,

Procedure

1. Dissolve the gel-slice in 3 volumes of chaotropic agent at 50C for 10 minutes.
2. Apply the solution to a spin-column and spin for 1 minute (the DNA remains in the column).
3. Wash the column by passing 70% ethanol through (the DNA remains in the column, salt and impurities are washed out).

4. Elute the DNA in a small volume (30µL) of water or buffer, spin to collect.

3. Dialysis Tubing (Semi-permeable Membrane, Visking Tubing)

Procedure

1. Freeze the gel slice at -20C for 30 minutes. This is to make it easier to handle the gel slice
2. Cut a 5cm length of dialysis tubing and rinse it inside and out with distilled water. Then rinse it with the same buffer used for the gel (eg. 0.5 x TBE) and leave it submerged in a small beaker of this buffer. Seal one end with a dialysis clip Dialysis tubing (Sigma D9777) is purchased in rolls (dried), prepared by boiling and stored submerged in buffer at 4C. Dialysis clips are Sigma Z37,096-7.
3. Insert the frozen gel-slice into the tubing and add 200–400µL of buffer (eg. 0.5 x TBE). Seal the other end of the tubing with a second dialysis clip. The buffer around the gel-slice must be the same as the buffer inside the gel
4. Immerse the sealed tubing in an electrophoresis tank so that the DNA band is parallel to the electrodes and apply 5V/cm electric field. The DNA will migrate out if the gel towards the positive electrode. It will be retained by the dialysis tubing. You can see this happening under long-wavelength UV if you like. It takes about 10–15 minutes.
5. Remove the buffer from the tubing and place into a 1.5mL microfuge tube. Phenol/chloroform extract and ethanol precipitate the DNA. Re-dissolve the DNA pellet in an appropriate volume of water or TE buffer (e.g. 10µL). The pellet is often so small that it is invisible

4. Paper Strip Method

Procedure

1. Using a scalpel blade, cut a slit immediately in front of the band to be extracted. Do not remove the band from the gel
2. Cut a piece of filter paper (e.g. 3MM paper) to size to fit inside the slit. For example, 3mm x 10mm
3. Place the paper strip in the slit, return the gel to the electrophoresis tank (submerged in buffer) and switch on the current for 2–5 minutes. The DNA runs onward into the paper and is delayed in the smaller mesh size of the paper. Eventually the DNA will pass through so you have to

keep checking it under long-wavelength UV so as not to leave it too long.

4. Remove the strip of paper (carrying at least some of the DNA) and place into a 0.5ml microfuge tube, *DNA side down*. Make a tiny hole in the bottom of the tube using a needle (CAREFUL!)

5. Place the 0.5mL tube inside a 1.5mL tube and spin for 30 seconds. You may have to remove the lid of the 1.5mL tube. The buffer and DNA are retained in the larger tube. You can add 100µL of TE to the paper and re-spin to get a little more DNA out.

6. Phenol/chloroform extract and ethanol precipitate the DNA. Re-dissolve the DNA pellet in an appropriate volume of water or TE buffer (e.g. 10µL). The pellet is often so small that it is invisible

5. Isolation of DNA from Agarose Gels with DEAE Paper

Materials

The DEAE paper can be used as supplied by Schleicher and Schuell, however the binding capacity can be increased by washing the paper in 10mM EDTA, pH 7.5 for 10 minutes, 0.5M NaOH for 5 minutes, followed by several rinses in distilled water. The paper can be stored at 4°C for several months.

NET buffer
0.15M NaCl
0.1mM EDTA
20mM Tris-HCl, pH 8.0

High salt NET buffer
1.0M NaCl
0.1mM EDTA
20mM Tris-HCl, pH 8.0

Procedure

1. After the DNA bands have been separated electrophoretically (at around 80v), the gel is viewed under long wave UV light and a slit cut in the gel, just ahead of the desired band. A piece of DEAE paper is inserted in the slit. Another piece of DEAE paper can be inserted behind the desired band, to prevent unwanted bands being trapped by the paper.

2. Electrophoresis is continued for approximately 10 to 15 minutes with the voltage increased to 100-120V.

3. The DEAE paper is removed and binding of the DNA checked by viewing the paper with long wave UV light. The paper is then placed in an eppendorf tube containing 500ul NET buffer to wash away remaining agarose. The bound DNA can now be stored at 4°C for several days. The DNA should always remain covered with buffer to prevent irreversible binding which may occur if the paper is allowed to dry out.
4. To elute the bound DNA, place the DEAE paper in an eppendorf tube containing 250ul of high salt NET buffer so that the paper is completely covered. Make sure that the side of the paper containing the DNA faces the buffer and not the side of the tube. If necessary, centrifuge briefly to submerge the paper.
5. Incubate the tube at 55-68°C for 45 minutes with occassional swirling (or until all the DNA has been eluted as checked under long wave uv light).
6. Remove the buffer to a fresh tube and add 50ul high salt NET buffer to wash the membrane. Remove the additional buffer and add to the fresh tube.
7. Centrifuge the buffer for 30 seconds to pellet any remaining DEAE paper fibres and remove the buffer to a fresh tube (leaving approx. 20 ul in the bottom of the tube).
8. Extract the residual ethidium bromide with 3 volumes of water saturated butanol.
9. Remove and retain the bottom layer and add 1/20 volume of 3.0M sodium acetate, pH 5.25 and 2.5 volumes of cold absolute ethanol. Incubate at -70°C for at least one hour. Ethanol precipitate and vacuum dry.

6. General Method

Materials and Reagents

1. Low gelling agarose
2. 1 x TBE buffer
3. Buffered phenol
4. TE

Procedure

1. Digest DNA sample to completion. Pour, load and electrophorese on 1 % low gelling/melting temperature agarose gel in 1 x TBE buffer.

2. Stain the gel using ethidium bromide and cut out the desired band with a scalpel.
3. Melt the gel slice at 65°C and add enough TE buffer to decrease agarose percentage to 0.4% or less.
4. Add an equal volume of buffered phenol; mix vigorously for 5 to 10 min; and centrifuge at 10,000 rpm for 10 min at room temperature.

   Note : Use buffered phenol and not the phenol: chloroform : isoamyl alcohol mixture. A thick white interface will be visible after centrifugation.
5. Collect the aqueous phase and set aside. Re-extract phenol phase and interface with an equal volume of TE buffer. Centrifuge as above.
6. Collect second aqueous phase. If large interface still appears re-extract once more.
7. Combine all the aqueous phases and ethanol precipitate.

Example of DNA Quantitation

Volume of DNA: 100 µl

Dilution : 10 µl DNA sample + 390 µl distilled water (1:40 dilution) A260 of diluted sample (1 cm path length) = 0.60

DNA concentration (µg/ml) = 50 x A_{260} x dilution factor = 50 x 0.60 x 40 = 1200 µg/ml

Total amount of DNA = concentration x volume of sample in ml = 1200 µg/ml x 0.1 ml = 120 µg of DNA.

Determination of Nucleic Acid Purity

- Determine DNA or RNA purity by measuring the A_{260}/A_{280} ratio.
- Measure A_{260}/A_{280} ratios in low-salt buffer. Buffered solutions provide more accurate and reliable values than water since A_{260}/A_{280} ratios are influenced considerably by pH. Lower pH results in lower A_{260}/A_{280} values and reduced sensitivity to protein contamination.
- Pure DNA has an A_{260}/A_{280} ratio of 1.8-2.0 in 10 mM Tris-CI, pH 8.5.
- DNA with protein as impurity would have an A_{260}/A_{280} ratio < 1.8
- DNA with RNA as impurity would have an A_{260}/A_{280} ratio >2.0

- Conservative estimates indicate that value between 1.8-1.9 are appropriate for pure DNA and pure RNA is 1.9-2.0.

Absorbance by Contaminants

- Strong absorbance at 280 nm, resulting in a low A_{260}/A_{280} ratio (1.8), indicates the presence of contaminants such as proteins.
- Strong absorbance at 270 and 275 nm may indicate the presence of contaminating phenol.
- Absorbance of 325 nm suggests contamination by particulates in the solution or dirty cuvettes.

11

Why Molecular Markers ?

Agriculture Scenario

It has been seen since last few decades there is huge decrease in the production of the agricultural production. Three factors which can be held responsible for the increase of agricultural production per hectare:

1. Overall increase in efficiency due to the merger of farms and a more efficient redistribution of farm land.
2. More efficient cultural practises and better quality and availability of inputs (fertilizer, pesticides).
3. Genetic improvement of crops.

At this moment over 6 billion people need to be fed and it is expected that in the year 2050 more than 10 billion people will inhabit the earth (FAO, 1996). Moreover, demands per capita will rise when standards of living in developing counties improve. As a consequence there will be a huge increase in the demand for food, and production will need to triple in the coming 40 years. A continued effort to secure agricultural production in the future will therefore be vital. Reduction of losses caused by a-biotic and biotic stress will be a key issue. Scenario studies on world food security generally assume that a large increase in production will be achieved by genetic improvement of crop species and biotechnology. History has indeed shown a continuous increase in crop yields, resulting from plant breeding efforts. However, in the light of the speed at which the human population develops, and taking into account the expected reduction of available arable land due to climate change and human intervention, it may be necessary to accelerate the rate at which genetic improvement is achieved. Modern biotechnology provides new tools that can facilitate development of improved plant breeding methods and augment our knowledge of plant genetics. The knowledge that is obtained with these new tools can be used to contribute to an enhanced food security throughout the world.

Plant Breeding

Plant Breeding is a dynamic area of applied science. It relies on genetic variation and uses selection to gradually improve plants for traits and characteristics that are of interest for the grower and the consumer. Practical breeding of many economic important crops is performed by commercial companies that strive in a fierce competition for the favour of agricultural producers and consumers. This competition has ensured a continued improvement of cultivated varieties (cultivars) over the past decades. These improvements were partly realised through an efficient use of existing variability, present within the available material. Another important way of improvement is the introduction of new genetic material (*e.g.* genes for disease resistance) from other sources, such as gene bank accessions and related plant species. Although current breeding practises have been very successful in producing a continuous range of improved varieties, recent developments in the field of biotechnology and molecular biology can be employed to enhance plant breeding efforts and to speed up the creation of cultivars. Also, new ways and methods that allow an easier introduction of genetic material from related and unrelated plant species, without the drawbacks that are normally associated with the introduction of "wild genes" through conventional methods, become feasible.

Quantitative Traits

Some of the most difficult tasks of plant breeders relate to the improvement of traits that show a continuous range of values. Among such quantitative traits are important traits like yield, plant length and days to flowering (speed of plant development). Selection for quantitative traits is difficult, because the relation between observed trait values in the field (the phenotype) and the underlying genetic constitution (the genotype) is not straightforward. Quantitative traits are typically controlled by many genes that each contribute only a small part to the observed variation. The environmental variance resulting from differences in growing conditions further obscures the relation between phenotype and genotype. In practice, this problem is typically dealt with by evaluating large, replicated trials, which allow identification of genotypic differences through statistical analysis. Plant breeders would like to get a better grip on quantitative traits by direct selection for the genetic factors that are responsible for the observed variability in quantitative traits. This can be achieved through indirect selection: selection for other, well recognisable factors, that are linked to the target genes. Molecular markers, derived from recent bio-technological developments, can be used for this purpose.

Molecular Markers

The discovery of restriction enzymes (Smith & Wilcox, 1970) and the polymerase chain reaction (PCR; Mullis & Faloona, 1987) have created the opportunity to visualise the composition of organisms at the DNA level, and obtain a so-called genetic fingerprint (*e.g.* Kearsey & Pooni, 1996). The visualisation is routinely performed by the separation, on a gel, of DNA-fragments that result from a selective digestion with enzymes or fragments that result from a selective amplification using PCR. DNA-fragments that result in different gel patterns between samples or individuals are called polymorphic markers.

The visible differences on the gel result from differences at the DNA level. Not all types of markers are the same, the information content depends on the method that was used to obtain the marker data and the population in which the markers were 'scored'. For instance, it is not always possible to distinguish genome fragments that are present in homozygous condition from heterozygous fragments. In a heterogeneous population like an F2, co-dominant markers like RFLPs (Botstein *et al.* 1980) and co-dominantly scored AFLPs (Vos *et al.* 1995) yield more information than dominant markers like RAPDs (Welsh & McClelland, 1990) and dominantly scored AFLPs. Advanced tools for the retrieval of marker data and the subsequent analysis have been developed and allow a quick and reliable analysis in most plant species. These developments have opened up a new era for genetics and selection. Important information on the genetic background of individual plants and populations can be derived from linkage that is observed between markers.

Genetic Linkage Maps

Segregation analysis can be applied to a segregating population that is derived from a common set of ancestors. Markers that co-segregate (are always present or absent together) must be linked, i.e. they must be located in each others vicinity on the genome. In some cases however, due to recombination events, the linkage between the markers may be lost. The frequency with which the linkage between co-segregating markers is broken is an indication of the genetic distance between the markers. An extensive analysis of the linkage between a large number of molecular markers yields information on their arrangement on the genome. Such analysis can finally result in the construction of a genetic map, on which all markers are arranged in separate linkage groups or chromosomes. On such a map the distances between markers reflect the degree of observed linkage. Genetic linkage maps should not be confused with physical genomic maps, which can be obtained by determining the DNA sequence of chromosomes, as is currently being done in several genome mapping projects. Linkage maps and physical maps are related, but this relation is usually not linear. Nowadays, software for the calculation of genetic maps has brought marker analyses, aimed at the construction of genetic maps, within the reach of many scientists.

QTL Analysis

Genetic factors that are responsible for a part of the observed phenotypic variation for a quantitative trait can be called quantitative trait loci (QTLs). Although similar to a gene, a QTL merely indicates a region on the genome, and could be comprised of one or more functional genes. In a process called QTL-mapping association between observed trait values and presence/absence of alleles of markers that have been mapped onto a linkage map is analysed. When it is significantly clear that the correlation that is observed did not result from some random process, it is proclaimed that a QTL is detected. Also the size of the allelic effect of the detected QTL can be estimated. A breeder can analyse QTL occurrences and use this knowledge to his advantage, for instance by using indirect selection. When selection is (partly) based on genetic information retrieved through the application of molecular markers this is called marker-assisted selection.

Marker-assisted Selection

Marker-assisted selection (MAS), sometimes also called marker-aided selection, is a relative new tool for plant breeders. In its simplest form it can be applied to replace evaluation of a trait that is difficult or expensive to evaluate. When a marker is found that co-segregates with a major gene for an important trait, it may be easier and cheaper to screen for the presence of the marker allele linked to the gene, than to evaluate the trait. From time to time the linkage between the marker and the gene should then be verified. When more complex, polygenic controlled traits are concerned, the breeder is faced with the problem how to combine as many as possible beneficiary alleles for the QTLs that were detected. In this case the breeding material can be screened for markers that are linked to QTLs. Based on such an analysis, specific crosses can be devised for the creation of an optimal genotype, combining beneficiary QTL alleles from different sources. Marker-assisted selection, when applied within the current breeding material to enhance a breeding program, does not solve the problem of limited genetic variability that is often seen in breeding stocks. A different application of marker-assisted selection could contribute to a genetic enrichment of breeding material. Marker-assisted selection may be used to facilitate a controlled inflow of new genetic material. 'Wild' or unadapted material often carries desired components that may be missing in cultivated material. Such components can be transferred to elite cultivated material by repeated backcrossing. However, breeders are often reluctant to apply this method because of unpredictable side-effects. These are caused by other genes, which are unintentionally transferred along with the genes that control the target trait. It may take considerable effort and screening to get rid of the unwanted genes and return the material to an acceptable agronomic value. Markers can be used to pinpoint the genetic factors that are responsible for the desired characteristics in the unadapted material. In a backcross program, the presence of the desired QTL alleles can be verified continuously by observing linked markers. At the same time, and with little extra effort, markers provide information on the origin of the remaining genome, allowing selection within the backcross material for genotypes that have lost the majority of unwanted donor DNA. Usually the application of this marker-assisted backcross procedure will also result in a reduction of the number of backcross generations that are required, thereby speeding up the breeding program.

Here a table has been presented which explains the limitation of quantitative genetic selection and opportunities for the use of molecular data.

Characteristic Features of Molecular Markers-Comparisons

The potential benefits that a breeding program can obtain from the use of molecular markers are:

- A significant reduction in the time to develop a new variety. The use of molecular markers can lead to more direct identification of the desirable traits than is possible through phenotypic selection, so that the time to develop a new variety (generally 10 years or more) can be reduced.

Limitations of quantitative genetic selection and opportunities for the use of molecular data

Limit to quantitative selection	Example trait(s)	Help provided by molecular data	Possible breeding solution	Economic merit of molecular data
Phenotype is a poor predictor of breeding value (low heritability).	Reproduction in animals, yield in plants.	Better estimate of breeding value at identified QTL	Select on molecular score and phenotype	Depends on requirements for QTL detection. Difficult to prove.
Phenotype is difficult or expensive to record.	Disease-related traits.	Markers are easier or cheaper to score than phenotype.	Select on molecular score.	Proportional to cost of phenotyping versus genotyping. Easy to prove.
Phenotype expressed subsequent to reproductive age. Long generation interval.	Reproduction traits in animals, grain yield in plants. Tree breeding.	Molecular score is available at earlier stage, resulting in faster selection.	Select on molecular score in combination with phenotype of ancestors.	Allows more rapid genetic gain and earlier release of improved genetic material.
Individual has to be sacrificed to score its phenotype.	Meat quality in animals, malting quality in barley.	Molecular score is available on all selection candidates.	Select on molecular score in combination with phenotype of relatives.	Substantial increase in genetic gain expected.
Traits observed only in one gender.	Milk yield in dairy cattle.	Molecular score is available at an early age on both genders.	Select on molecular score and phenotype. Pre-select on molecular score for further phenotypic testing (for example, progeny test).	Moderate, depending on opportunity for and costs of pre-selection.
Genetic potential is masked by epistatic interactions between QTL, or by linked QTL that are in repulsion phase.	Many traits.	Dissect and break down unfavourable interactions at the genetic level.	Select on molecular score and phenotype.	Difficult but can be spectacular if successful.
Genotype–environment interactions.	Many traits.	Predict interactions at the genetic level	Select on molecular score and phenotype	Unknown. Difficult to prove.

- An increase in the overall rate of genetic gain in a breeding cycle. By allowing the more ready introduction of desirable traits in combination, the varieties produced using molecular markers can be more productive than those developed by conventional breeding methods.
- The possibility of improving traits that were not possible to measure using traditional phenotypic screening. Some problems have been extremely difficult to assess using phenotypic screening, such as: (i) frost tolerance; (ii) pre-harvest sprouting; (iii) late maturity amylase in wheat; (iv) black point; (v) drought tolerance; (vi) tolerance to exotic pests and diseases such as Russian wheat aphid and Karnal bunt; and (vii) out of-region breeding for constraints not applicable in the region where breeding is taking place.
- Faster access to new genes to provide greater genetic diversity. By enabling screening for the presence of particular segments of DNA, molecular markers can allow the breeder to identify diversity at an early stage, which can be maintained through the breeding cycle.

An ideal marker should have the following characteristics:

1. It should be present in abundance
2. It should be uniformly distributed throughout the genome.
3. It should be co-dominant.
4. It should give high level of polymorphism.
5. It may be random / anonymous byt gene specific markers are desired to fish out a gene.
6. It should be cost-effective.
7. It should be high throughput.
8. It should be amenability to automation.
9. It should be highly reproducible.
10. It should involve less development and operational cost.
11. It should be less labour-intensive.

Frequency of Occurrence

The number of markers which can be generated is determined mainly by the frequency at which the sites of interest occur within the genome. RFLPs and AFLPs are abundant markers due to the large number of restriction endonucleases available and the frequent occurrence of their recognition sites within genomes. However AFLPs have been found to cluster near the centromere and the telomeres. They are also found in other areas of the genome but not in abundance. Within the eukaryotic genome also microsatellites have been found to occur frequently. RAPD markers are even more abundant because numerous random sequences can be used for primer construction. In contrast, the number of allozyme markers is restricted due to the limited number (about 30) of enzyme systems available for analysis. To investigate specific genomic regions by PCR-sequencing, SSCP, CAPS, SCAR, RGAs, cDNA, EST or TaqMan, sequence data of the sites of interest (structural genes mainly) are required for primer construction. Although, in principle, many sites of interest may occur within genomes, the proportion of the genome covered by these markers in reported studies to date is limited. Genomic abundance is essential to studies where a large fraction of the genome needs to be covered, e.g. for the development of high-density linkage maps in gene mapping studies, however studies like MAS do not require high density maps rather require makers in close relation to the target loci.

Degree of Polymorphism

The resolving power of genetic markers is determined by the level of polymorphism detected, which is primarily affected by the mutation rate at the genomic sites involved. Variation at allozyme loci is caused by point mutations, which occur at low frequency ($<10^{-6}$ per meiosis). Moreover, only mutations altering the net electric charge of proteins can be detected, thereby even reducing the resolving power of allozymes. In contrast, mutations at minisatellite and microsatellite loci, mainly due to changes in the number of repeat units of the core sequence, have been estimated to occur at the relatively high frequency of 10^{-3}-10^{-2} and 10^{-5}-10^{-2} per meiosis, respectively. The other markers presented generally show intermediate levels of polymorphism, resulting from base substitutions, insertions or deletions which may alter primer annealing sites and recognition sites of restriction endonucleases or change the size of restriction fragments and amplified products. In choosing the appropriate technique, the level of polymorphism generally detected by the marker needs to be considered in relation to the presumed degree of genetic relatedness within the material to be studied. In general, higher resolving power is required when samples are more closely related.

Anonymous / Locus Specific

Genetic markers using multilocus probes or primers benefit from the fact that multiple polymorphisms, representing various genomic regions, are generated simultaneously. However, major drawback is that the band profiles generally can not be interpreted in terms of loci and alleles, but can only be scored as the presence or absence of bands of particular size. Main consequence is that similar sized fragments may represent alleles from different loci (non-homology). The probability of non-homology increases when comparisons are made across higher taxonomic levels. Therefore, locus specificity of the marker should be considered particularly when studies are carried out above the species level. In general, polymorphisms of known identity are generated by locus specific markers but in most cases genomic data are needed for their development.

Codominance of Alleles

Alleles at a locus of a codominant marker are expressed when co-occurring in an individual. Therefore, heterozygotes (hybrids) can be distinguished from homozygotes (parents) which allows the determination of genotypes and allele frequencies at loci. In contrast, band profiles of dominant markers are scored as the presence or absence of fragments of particular size from which heterozygosity can not be derived and allele frequencies can only be estimated. Therefore, codominant markers are preferred for many applications, for example in genetical studies based on the statistical analysis of allele frequencies (e.g. estimation of genetic differentiation and genetic distances).

Reproducibility

To obtain reproducible results, the extraction of purified, high quality DNA is a prerequisite for the majority of the marker techniques. For example, degraded and/or unpurified DNA may affect the amplification or restriction digestion of fragments resulting in nonspecific (non-reproducible) polymorphisms. Even when purified, high molecular weight DNA is used, RAPDs often fail to show reproducible results because they have small primers. Even small variations affect their annealing. Therefore, highly standardized experimental procedures are required when RAPD markers are being used.

Industriousness Involved

RFLPs and minisatellites are quite laborious markers because their analysis includes the time-consuming steps of Southern-blotting, radioactive labeling of probes and hybridization. Therefore, PCR-based techniques are currently preferred. Unless the process is automated, throughput of PCR-sequencing samples is still low because four separate sequence reactions are needed per sample and data processing is time-consuming. The labor-intensity of the other PCR-based techniques presented varies from low to medium, depending on the methodological procedures required in addition to PCR.

Technical Skill Involved

RFLPs, minisatellites and PCR-sequencing require technical skills and facilities to carry out radioactive labeling. In addition, Southern blot hybridizations are part of the RFLP and minisatellite analysis. These techniques are therefore among the more technically demanding markers. Allozymes, TaqMan and PCR-based markers analyzed on agarose gels (e.g. RAPD and SCAR) are the least technically demanding.

Operational Costs

Wages, laboratory facilities, technical equipment and chemicals all contribute to the operational costs of the technologies. Relatively expensive chemicals include Taq-polymerase needed for PCR, restriction enzymes (particularly, frequently cutting endonucleases used in AFLP-analysis) and radioactive label. Polyacrylamide gels are more expensive to run than agarose gels and require visualization of polymorphisms by autoradiography or silver staining which are more costly compared to ethidium-bromide staining. Laborious and technically demanding markers, such as RFLPs, minisatellites and PCR-sequencing are therefore quite expensive. Compared to these techniques the operational costs of the other markers presented vary from low to medium, depending on the methodological procedures followed.

Cost Involved

Marker development may be very time-consuming and costly when necessary probes or sequence data for primer construction are unavailable. In addition, sufficient technical skills and facilities need to be present. Development of suitable probes for Southern blot hybridizations (e.g. for RFLP-analysis) requires the construction of a genomic library, the isolation of DNA fragments and the examination of various probe/restriction enzyme combinations for the ability to detect polymorphisms. Development of site specific PCR-primers (*e.g.* for microsatellite analysis) also requires the construction of a genomic library, which then needs to be screened to identify the fragments of interest. Subsequently, the identified fragments need to be sequenced in order to obtain the necessary data for primer construction. Therefore, the investment required for marker development should be evaluated in relation to the intended range of application of the technique. Alternatively, probes, primers and sequence data may be obtained from genome databases of other species, although the usefulness of this approach decreases with increasing evolutionary distance.

Cost Components Involved in MAS are:

- Once a marker has been developed and validated for a given environment and gene pool, it is ready for implementation into the breeding program, if it is advantageous. The costs of the implementation of a marker in a breeding program depend on a number of factors:
- the cost and conditions of the intellectual property relating to that marker;
- the costs of tissue collection and storage;
- the method of extraction of DNA/protein for analysis;
- the extent to which extracted DNA is used for a number of different markers;
- the analytical method used;
- the rate of throughput of the analytical system;
- the degree to which the analysis is combined with that for other markers
- ("multiplexing");
- the ease of scoring the alleles after the analysis;
- the costs of the labour required for the different operations (Dreher *et al.* 2003);
- the capital cost of the laboratory equipment required;
- the utilisation of the equipment required (that is, the overall throughput of the laboratory).

Quantity of DNA Required

Because only small quantities of template DNA (5-100 ng per reaction) are required, techniques which are based on the Polymerase Chain Reaction (PCR) currently are preferred. RFLPs and minisatellites require the largest amount of DNA (5-10 μg per reaction) but Southern blots may be reprobed several times. Intermediate quantities of DNA are needed for AFLP-analysis (0.3-1 μg per reaction) because endonuclease restriction of the DNA template precedes the PCR-reaction. Application of PCR-based markers may be relevant when only small amounts of DNA can be extracted, e.g. when working with tiny organisms.

Amenability to Automation

Currently, techniques which can be automated are preferred because they enable increased sample throughput. Although considerable financial investment still is required, automation may be cost-effective when techniques are applied on a routine basis. Nearly all techniques which are based on the Polymerase Chain Reaction (PCR) are amenable to automation.

Comparison of molecular markers for important characteristics

Marker System	Abundance	Level of polymorphism	Locus specificity	Codominance of alleles	Reproducibility	Labor-intensity	Technical demands	Operational costs	Development costs	Quantity of DNA required	Amenability to automation
Allozymes	low	low	yes	yes	high	low	low	low	low	-	no
RFLP	high	medium	yes	yes	high	high	high	high	medium-high	high	no
Minisatellites	med	high	no/yes	no/yes	high	high	high	high	medium-high	high	no
PCR-sequencing	low	low	yes	yes	high	high	high	high	high	low	yes
RAPD	high	medium	no	no	low	low	low	low	low	low	yes
Microsatellites	high	high	yes	yes	high	low	low-medium	low-medium	Initial high	low	yes
ISSR	medi-high	medium	no	no	medium-high	low	low-medium	low-medium	low	low	yes
SSCP	low	low	yes	yes	medium	low-medium	medium	low-medium	high	low	no
CAPS	low	low-medium	yes	yes	high	low-medium	low-medium	low-medium	high	low	yes
SCAR	low	medium	yes	yes/no	high	low	low	low	high	low	yes
AFLP	high	medium	no	no/yes	high	medium	medium	medium	low	medium	yes
TaqMan	low		yes	no/yes	high	low	low		high	low	yes

Modified from : http://www.cgn.wageningen-ur.nl/pgr/research/molgen/overview.htm

Considering all the factors together, SSRs seem to be the most ideal marker system as they are codominant, show high level of polymorphism, less labour intensive, highly reproducible, locus specific, abundant, uniformly distributed through out the genome, require less amount of DNA and are amenable to automation.

12

Restriction Enzyme Digestion of DNA and Its Agarose Gel Electrophoresis

Restriction Enzymes

Restriction enzymes are enzymes isolated from bacteria that recognize specific sequences in DNA and then cut the DNA to produce fragments, called restriction fragments. Restriction enzymes play a very important role in the construction of recombinant DNA molecules, as is done in gene cloning experiments. Another application of restriction enzymes is to map the locations of restriction sites in DNA.

Restriction endonucleases, often called restriction enzymes, cut DNA within the molecule by hydrolyzing the phosphodiester bonds between the nucleotides. Restriction enzymes were discovered in bacteria and there are now more than 1200 known restriction enzyme types. These enzymes are named using a simple system. EcoRI, for example, was isolated from E. coli and was the first enzyme isolated from a particular strain, hence the designation of I. HaeIII was isolated from Haemophilus aegyptius. HindIII was isolated from Haemophilus influenzae, and was the third enzyme discovered in a particular strain.

Usually, restriction enzymes only cut the DNA at or near a very specific nucleotide sequence known as a recognition site. This "restrictive" nature of these enzymes allows molecular biologists to pin-point exactly where the DNA is to be cut. In many cases, the recognition site of a restriction enzyme consists of a palindromic sequence where the order of the bases is read the same on each side of the helix. The two sides of the helix are antiparallel and the recognition sequence reads from 5′ to 3′ on either side of the helix. For example the recognition site for EcoRI is as follows;

GAATTC
CTTAAG

Some restriction enzymes cut both sides of the helix at the same point. HaeIII is such an enzyme and the fragments produced are "blunt" ended. Many restriction enzymes make asymetric cuts in the helix. EcoRI cuts DNA at the recognition site in the following fashion;

G?AATTC
CTTAA?G.

This type of asymetric cut leaves a single stranded sequence hanging on either 5′ or 3′ end. Such a trailing end is often called a "sticky" end since it can bond by hydrogen bonding to a complementary single stranded sequence. The sticky ends make it rather easy to insert DNA into a plasmid. If the plasmid and DNA to be inserted are cut with the same restriction enzyme, the sticky ends will match and will anneal using DNA ligase enzyme. The result is a recombinant plasmid with a foreign fragment incorporated. Cleavage of DNA by a restriction enzyme is called digestion and is usually carried out at 37°C in a small volume (e.g., 20 μl) of buffer and salt. Although restriction enzymes are active over a fairly broad range of conditions, optimal conditions have been determined for each restriction enzyme using a buffer which typically contains 10-100mM Tris-HCl pH 8.0, a salt (usually 0-150 mM NaCl), and 10 mM of Magnesium Chloride. Restriction enzyme digestion involves adding enough restriction enzyme to the substrate DNA in the appropriate buffer and letting the digest incubate at the appropriate temperature for at least one hour. Restriction enzymes are measured functionally in units (μ), which are defined as the amount of restriction enzyme that will completely digest a defined substrate DNA in a defined time at a specific temperature. When stored at the recommended conditions (-20°C in storage buffer containing 50% glycerol), restriction enzymes can remain active for many months. However, once they are diluted into digestion buffer and incubated at 37°C, they lose activity within hours.

Restriction Enzyme Digestion of DNA

Restriction endonucleases (RE) are of three types, Type I, Type II and Type III. Type II REs recognize specific sequences and cut double stranded DNA within their recognition sequence to produce fragments with protruding or blunt ends, however, Type I and Type III are non-specific in identifying and cutting the DNA sequences, i.e they identify another region of DNA and make a cut at a different site. Fragments thus produced get separated on agarose gel electrophoresis and are visualized after staining with ethidium bromide, a dye which intercalates between bases of DNA and gives orange coloured fluorescence under UV light.

Flow Diagram showing the process of restriction enzyme digestion of Plant DNA and its agarose gel electrophoresis

SET UP DIGEST
↓
SPIN SAMPLES
↓
DIGEST AT 37°C
↓
PREPARE FOR GEL LOADING
→
MAKE AGAROSE
↓
COOL TO 65°C, POUR GEL
↓
LOAD SAMPLES INTO GEL
↓
RUN GEL AT 80 Volts 60 MIN.
↓
STAIN 5 MIN
↓
VISUALIZE AND PHOTOGRAPH GEL
Determine size of fragments

Materials and Reagents

1. Microfuge
2. Water bath set at 37°C
3. Eppendorf tubes (1.5 ml capacity) and their rack
4. Adjustable micropipette (0-20 µl capacity)
5. DNA samples (1 µg/µl): Dissolve DNA samples in a suitable buffer (10 mM Tris-HCI, pH 8.0) at concentration of 1 µg DNA/µl .
6. Restriction enzyme: EcoRI or Hind III

7. Restriction buffer (10x): As supplied by the supplier along with the enzyme.
8. Sterile water or sterile ddH_2O
9. Loading dye

NOTE

1. Restriction enzymes are very sensitive to temperature changes. Keep them on ice and avoid warming and contamination by handling the middle of the tube, not at the bottom or top.
2. While 50% glycerol is present to stabilize the enzyme and to prevent freezing, the final glycerol concentration in the digest mixture must be less than 5%. Therefore, the total enzyme volume must be less than 10% of the total digest volume.
3. One unit of enzyme is defined as the amount of enzyme required to digest 1 ug of lambda DNA to completion in one hour at optimum conditions. (Conditions vary for each enzyme, and it is important to note the specific requirements of each before setting up reactions).
4. Ethidium bromide is a mutagen, so wear gloves whenever handling solutions or the stained gel.
5. UV light causes severe eye and skin damage. You must wear UV shields or goggles when viewing the transilluminator.

Procedure

A. Set up Digestion

Set up the digest reaction as detailed below. Use a fresh tip when adding a different solution to the tube. Mix the following constituents in the order shown below in a sterile 1.5 ml of Eppendorf tube

DNA	5	μg
10x buffer	5.0	μl
Restriction enzyme	2.5	μl (10 units/μl)
Sterile water to	50	μl

B. Spin Samples

Mix contents of the tube carefully and spin at full speed (14,000 rpm) in a microfuge for 1 min.

C. Digest at 37°C

Incubate the mixture at 37°C overnight for the restriction digestion of DNA samples to occur.

D. Prepare for Gel Loading

1. Stop the reaction by adding 5 μl of the loading dye.
2. Mix and follow Step B.
3. Load the entire sample on 0.8% agarose gel slab to separate fragments and analyse them

Interpretations: Undigested DNA would show single band at the top of the gel near the well. It indicates complete failure of RE digestion. Several clear bands indicate proper digestion of prokaryotic DNA (plasmid and λDNA), however, a continuous smear in a lane would show complete digestion of eukaryotic DNA. Presence of both thick band(s) and smear at upper end only indicates partial digestion of DNA.

Precautions

1. Restriction enzymes should be taken out of -20°C freezer only at the time of use and should be immediately kept on ice. Store them back at -20°C immediately after use.
2. Ethidium bromide is a powerful mutagen and carcinogenic. Gloves should be worn while working with solutions containing this dye.
3. Follow all the general precautions for handling nucleic acids.
4. The concentration of restriction enzyme and buffer in the final volume of reaction mixture should be 1x. So, the reaction volumes should be adjusted accordingly

Agarose Gel Electrophoresis

Agarose gel electrophoresis is used to separate DNA digested by restriction enzyme. Agrarosegel seperates fragments of DNA molecules according to their size. The negatively charged DNA molecules migrate in an electrical field from negatively charged cathode to positively charged anode, with smaller molecules migrating faster than the bigger ones through the pores of the matrix. After electrophoresis, DNA fragments are visualized by staining with ethidium bromide which gives orange coloured fluorescence in UV.

Material and Reagents

1. 10 x TBE

Tris base	108 g
Boric acid	55 g
0.5 M EDTA (pH 8.0)	40 ml
Water to	1 liter

2. 1x TBE
3. Take 10 ml of 10x TBE and dilute to 100 ml with DDW
4. Agarose (0.8 % in 1 x TBE)
5. Restriction digest of plant DNA (from the above experiment)
6. 10 x gel loading dye

Glycerol	
10 x TBE	5ml
Bromophenol blue (Saturated)	1 ml
Xylene cyanol 10%	1 ml

   Mix well autoclave and store in aliquots.
7. 10 mg/ml ethidium bromide in sterile water
8. Horizontal slab gel electrophoresis apparatus
9. UV transilluminator

Procedure

1. Weigh 0.8 g agarose and put in a 250 ml conical flask. Add 100 ml of 1 x TBE buffer and gently boil the solution in a microwave oven with occasional mixing until all agarose particles are completely dissolved. Allow it to cool to 50°C. Add 10 μl of ethidium bromide (10 mg/ml) and mix well. Prepare the gel mould and keep the comb in position. Pour the cooled gel solution into the gel mould and allow the gel to set for 20 min.
2. Fill the horizontal electrophoresis chamber with 1xTBE. Remove the comb from the gel and place gel with the tray in the electrophoresis chamber.
3. Load the digested DNA sample carefully into the wells. In one well load a standard marker (Lambda DNA restricted with Hind III).

4. Run the gel at 20 mA overnight.
5. Visualize the DNA bands on a UV transilluminator and determine the fragment sizes by placing a ruler next to the gel. Take a picture using Polaroid camera with a red filter or digital camera fitted with gel documentation system with UV filter.

13

Thermocyclers (PCR Machines)

Polymerase Chain Reaction

The Polymerase Chain Reaction (PCR) technique, invented in 1985 by Kary B. Mullis, allowed scientists to make millions of copies of a scarce sample of DNA. The technique has revolutionized many aspects of current research, including the diagnosis of genetic defects and the detection of the AIDS virus in human cells. The technique is also used by criminologists to link specific persons to samples of blood or hair via DNA comparison. PCR also affected evolutionary studies because large quantities of DNA can be manufactured from fossils containing but trace amounts.

Kary Mullis invented the PCR technique in 1985 while working as a chemist at the Cetus Corporation, a biotechnology firm in Emeryville, California. The procedure requires placing a small amount of the DNA containing the desired gene into a test tube. A large batch of loose nucleotides, which link into exact copies of the original gene, is also added to the tube. A pair of synthesized short DNA segments, that match segments on each side of the desired gene, is added. These "primers" find the right portion of the DNA, and serve as starting points for DNA copying. When the enzyme *Thermus aquaticus* (Taq) is added, the loose nucleotides lock into a DNA sequence dictated by the sequence of that target gene located between the two primers.

The test tube is heated, and the DNA's double helix separates into two strands. The DNA sequence of each strand of the helix is thus exposed and as the temperature is lowered the primers automatically bind to their complementary portions of the DNA sample. At the same time, the enzyme links the loose nucleotides to the primer and to each of the separated DNA strands in the appropriate sequence. The complete reaction, which takes approximately five minutes, results in two double helices containing the desired portion of the original. The heating and cooling is repeated, doubling the number of DNA copies. After thirty to forty cycles are completed a single copy of a piece of DNA can be multiplied to hundreds of millions.

When completed manually, Mullis' PCR technique was slow and labor-intensive. Therefore, Cetus scientists began looking for ways in which to automate the process. Before the discovery of the thermostable *Taq* enzyme, scientists needed to add fresh enzyme to each cycle. The first thermocycling machine, "Mr. Cycle" was developed by Cetus engineers to address that need to add fresh enzyme to each test tube after the heating and cooling process. And the purification of the *Taq* polymerase resulted in the need for a machine to cycle more rapidly among different temperatures. In 1985, Cetus formed a joint venture with the Perkin-Elmer Corporation in Norwalk, Connecticut, and introduced the DNA Thermal Cycler.

Thermocycler Models

Block thermocyclers house the sample in a conductive block to assure uniformity and supply or reject heat to the sample via the block. All but one of the commercially manufactured thermocyclers are of this type. Samples are usually placed inside tubes or tube racks for transport.

1. Capillary thermocyclers hold the sample within a glass capillary and supply heat conductively or convectively to the exterior of the capillary. There are two models of this type available commercially.

2. Micromachined thermocyclers are simply any thermocycler constructed of micromachined components; these are generally etched structures in glass or silicon with heat supplied by integral resistive heating and rejected passively to ambient through the structure. These thermocyclers have not yet been commercialized.

Block Thermocyclers

The earliest commercial thermocyclers offered an aluminum block with holes machined to fit common labware forms. At the time (1988), 1.5 ml microfuge tubes were common and researchers rarely did more than a few dozen samples at a time. These were accommodated by the first thermocyclers commercialized by the Perkin-Elmer Corporation which were programmable, held 24 microfuge tubes, and were heated and cooled using a vapor compression heat pump to heat and cool the block. Soon after this introduction, many companies began offering similar devices with programmability and a variety of tube formats. Devices using Peltier elements for heat transfer to the block were introduced.

Commonly available thermocyclers implement a 'hot-plate' design. That is a metal plate, which is designed to accept a 96-well micro-strip. This plate is actively heated and cooled to cycle through the appropriate temperatures (MJ Research) While such designs have proven reliable they do suffer from a few fundamental limitations. First, sample volume must be above some minimum (usually approx 10μl) for the reactions to proceed reliably. Second, in the case of the MJ, thermal cycling is achieved by heating and cooling a massive metal block. The heat capacity of the block slows the temperature ramping time.

In recent days several foreign companies have started manufacturing thermocycler viz.,

1. M. J. Research, USA,
2. Eppendorf, Germany
3. Bio-Rad, USA. Etc.

Gradient PCR (to Determine the Optimum Annealing Temperature)

Problems with the amplification of a specific DNA fragment using the PCR are an everyday occurrence in the lab. For example, unspecific secondary bands may form after the PCR reaction, which hinder, or even prevent, further analyses (cycle sequencing, mutation detection, etc.) or an unequivocal assessment of the PCR result. In such cases, PCR conditions must be optimized. This is normally achieved by titrating the magnesium-, template-, primer-, dNTP- and *Taq* polymerase concentration, "Hot Start PCR", "Touch Down PCR", adding detergents, reducing the PCR cycles or by gradually increasing the annealing temperature.

The selection of the annealing temperature is possibly the most critical component for optimizing the specificity of a PCR reaction. In most cases, this temperature must be empirically tested. The PCR is normally started at 5°C below the calculated temperature of the primer melting point (Tm). However, the possible formation of unspecific secondary bands shows that the optimum temperature is often much higher than the calculated temperature (>12°C).

Further PCR reactions with gradually increasing temperatures are required until the most stringent conditions have been found. When a standard PCR cycler is used, this method is the most time-intensive optimization strategy.

The Mastercycler® gradient now enables rapid testing of the optimum temperature conditions on one block and in one experiment. During the PCR, a temperature gradient, which can be programmed between 1°C and 20°C, is built up across the thermoblock. This allows the most stringent parameters for every primer set to be calculated with the aid of only one single PCR reaction.

PCR Principles and Procedure

PCR is used to amplify specific regions of a DNA strand (the DNA target). This can be a single gene, a part of a gene, or a non-coding sequence. Most PCR methods typically amplify DNA fragments of up to 10 kilo base pairs (kb), although some techniques allow for amplification of fragments up to 40 kb in size.

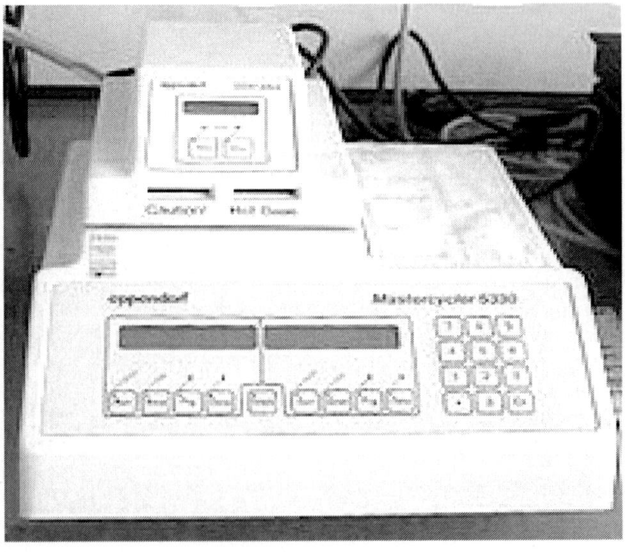

A basic PCR set up requires several components and reagents. These components include:

- DNA template that contains the DNA region (target) to be amplified.
- Two primers that are complementary to the 3' (three prime) ends of each of the sense and anti-sense strand of the DNA target.
- Taq polymerase or another DNA polymerase with a temperature optimum at around 70°C.
- Deoxynucleoside triphosphates (dNTPs; also very commonly and erroneously called deoxynucleotide triphosphates), the building blocks from which the DNA polymerases synthesizes a new DNA strand.
- Buffer solution, providing a suitable chemical environment for optimum activity and stability of the DNA polymerase.
- Divalent cations, magnesium or manganese ions; generally Mg^{2+} is used, but Mn^{2+} can be utilized for PCR-mediated DNA mutagenesis, as higher Mn^{2+} concentration increases the error rate during DNA synthesis.
- Monovalent cation potassium ions.

The PCR is commonly carried out in a reaction volume of 10-200 ìl in small reaction tubes (0.2-0.5 ml volumes) in a thermal cycler. The thermal cycler heats and cools the reaction tubes to achieve the temperatures required at each step of the reaction *(see figure below)*. Many modern thermal cyclers make use of the Peltier effect which permits both heating and cooling of the block holding the PCR tubes simply by reversing the electric current. Thin-walled reaction tubes permit favorable thermal conductivity to allow for rapid thermal equilibration. Most thermal cyclers have heated lids to prevent condensation at the top of the reaction tube. Older thermocyclers lacking a heated lid require a layer of oil on top of the reaction mixture or a ball of wax inside the tube.

Procedure

1. Denaturing at 94-96°C.
2. Annealing at ~65°C
3. Elongation at 72°C.

Four cycles are shown here. The blue lines represent the DNA template to which primers (red arrows) anneal that are extended by the DNA polymerase (light green circles), to give shorter DNA products (green lines), which themselves are used as templates as PCR progresses.

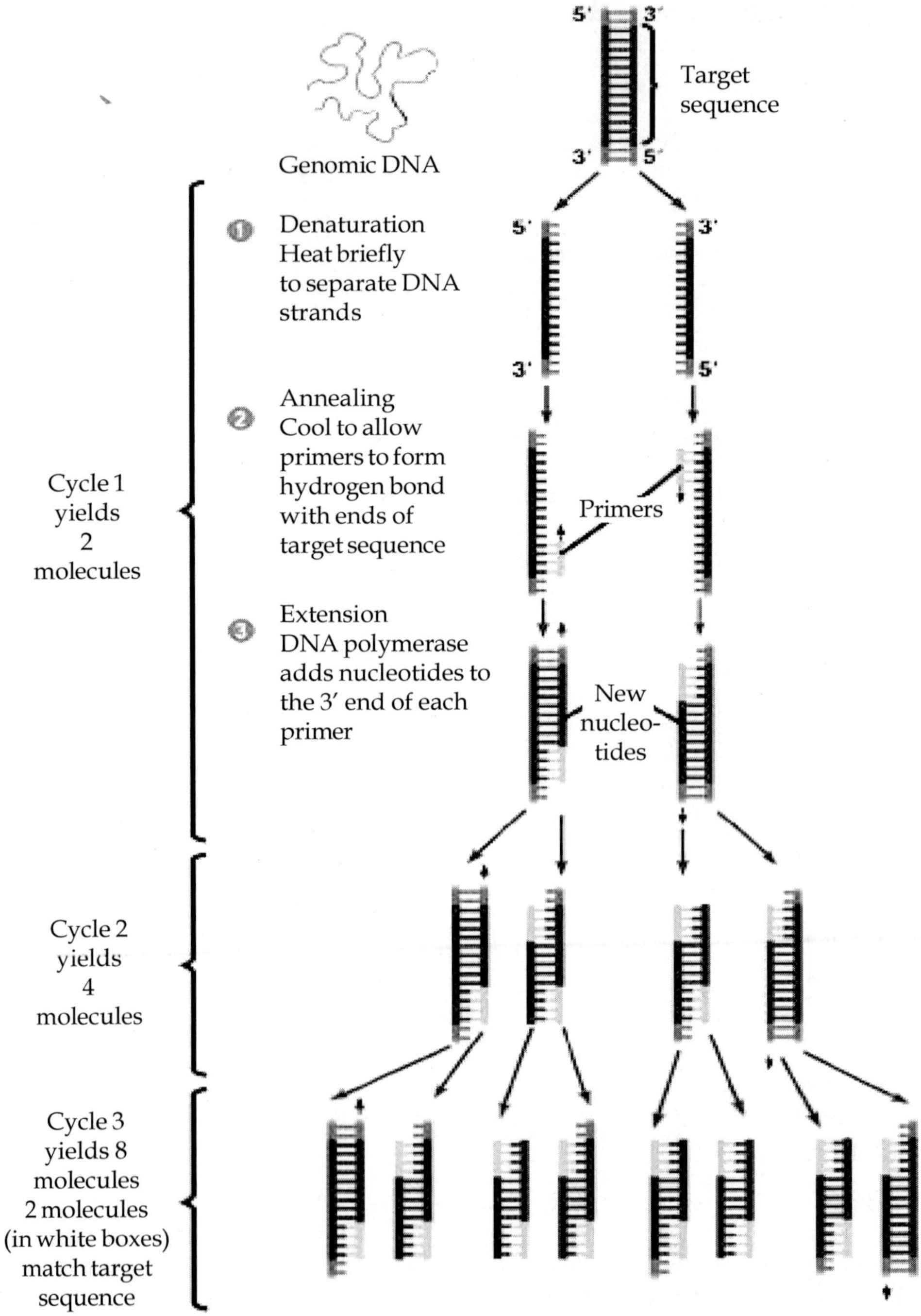
5'
3'
Target sequence
3'
5'
Genomic DNA
Cycle 1 yields 2 molecules
1 Denaturation Heat briefly to separate DNA strands
5'
3'
3'
5'
2 Annealing Cool to allow primers to form hydrogen bond with ends of target sequence
Primers
3 Extension DNA polymerase adds nucleotides to the 3′ end of each primer
New nucleo-tides
Cycle 2 yields 4 molecules
Cycle 3 yields 8 molecules 2 molecules (in white boxes) match target sequence

The PCR usually consists of a series of 20 to 40 repeated temperature changes called cycles; each cycle typically consists of 2-3 discrete temperature steps. Most commonly PCR is carried out with cycles that have three temperature steps The cycling is often preceded by a single temperature step (called *hold*) at a high temperature (>90°C), and followed by one hold at the end for final product extension or brief storage. The temperatures used and the length of time they are applied in each cycle depend on a variety of parameters. These include the enzyme used for DNA synthesis, the concentration of divalent ions and dNTPs in the reaction, and the melting temperature (Tm) of the primers.

- *Initialization step*: This step consists of heating the reaction to a temperature of 94-96°C (or 98°C if extremely thermostable polymerases are used), which is held for 1-9 minutes. It is only required for DNA polymerases that require heat activation by hot-start PCR.
- *Denaturation step*: This step is the first regular cycling event and consists of heating the reaction to 94-98°C for 20-30 seconds. It causes melting of DNA template and primers by disrupting the hydrogen bonds between complementary bases of the DNA strands, yielding single strands of DNA.
- *Annealing step*: The reaction temperature is lowered to 50-65°C for 20-40 seconds allowing annealing of the primers to the single-stranded DNA template. Typically the annealing temperature is about 3-5 degrees Celsius below the Tm of the primers used. Stable DNA-DNA hydrogen bonds are only formed when the primer sequence very closely matches the template sequence. The polymerase binds to the primer-template hybrid and begins DNA synthesis.
- *Extension/elongation step*: The temperature at this step depends on the DNA polymerase used; Taq polymerase has its optimum activity temperature at 75-80°C, and commonly a temperature of 72°C is used with this enzyme. At this step the DNA polymerase synthesizes a new DNA strand complementary to the DNA template strand by adding dNTPs that are complementary to the template in 5' to 3' direction, condensing the 5'-phosphate group of the dNTPs with the 3'-hydroxyl group at the end of the nascent (extending) DNA strand. The extension time depends both on the DNA polymerase used and on the length of the DNA fragment to be amplified. As a rule-of-thumb, at its optimum temperature, the DNA polymerase will polymerize a thousand bases per minute. Under optimum conditions, i.e., if there are no limitations due to limiting substrates or reagents, at each extension step, the amount

of DNA target is doubled, leading to exponential (geometric) amplification of the specific DNA fragment.

- *Final elongation* : This single step is occasionally performed at a temperature of 70-74°C for 5-15 minutes after the last PCR cycle to ensure that any remaining single-stranded DNA is fully extended.
- *Final hold*: This step at 4-15°C for an indefinite time may be employed for short-term storage of the reaction.

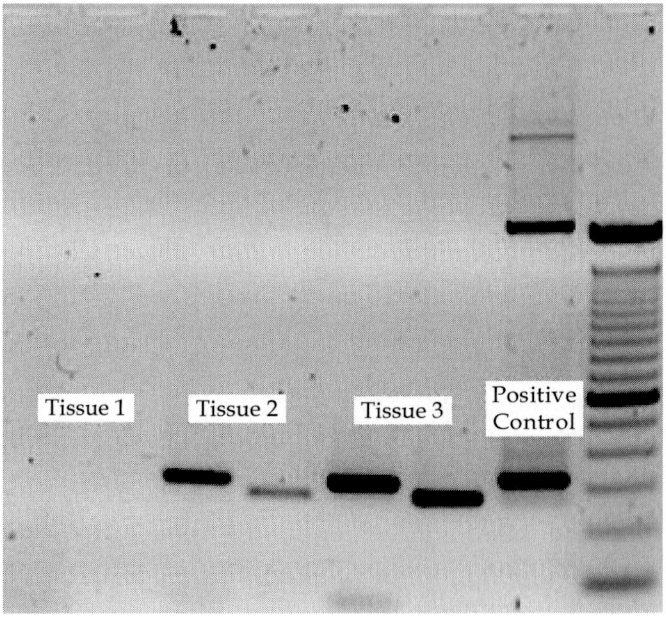

Ethidium bromide-stained PCR products after gel electrophoresis. Two sets of primers were used to amplify a target sequence from three different tissue samples. No amplification is present in sample #1; DNA bands in sample #2 and #3 indicate successful amplification of the target sequence. The gel also shows a positive control, and a DNA ladder containing DNA fragments of defined length for sizing the bands in the experimental PCRs.

To check whether the PCR generated the anticipated DNA fragment (also sometimes referred to as the amplimer or amplicon), agarose gel electrophoresis is employed for size separation of the PCR products. The size(s) of PCR products is determined by comparison with a DNA ladder (a molecular weight marker), which contains DNA fragments of known size, run on the gel alongside the PCR products.

PCR Reaction Conditions and Components

The PCR process is widely employed in a tremendous variety of experimental applications to produce high yields of specific DNA target sequences. Since no single set of conditions can be applied to all PCR amplifications, individual reaction component concentrations (and time and temperature parameters) must be adjusted within suggested ranges for efficient amplification of specific targets. While there are a number of possible concentration parameters, logical titrations of interrelated reaction components can be readily defined. In addition, the time and temperature optima can often be determined within a few experiments.

Sample Volume and Reaction Tubes

Most PCR protocols are performed at the 25 µL - 50 µL scale in 0.2 mL or 0.5 mL microcentrifuge tubes; however, 5 µL reactions have also been successful. Larger (> 100 µL) volumes are usually not recommended, but if they are used, longer incubation times may be needed to assure adequate thermal equilibration of the reaction mixture. Reaction Tubes are designed for optimal fit in the each thermal cycler.

Vapor Barrier and Thermal Transfer Fluid

In the recently developed PCR, the need for mineral oil as a vapor barrier, and mineral oil or glycerol as a thermal transfer fluid in the wells, is eliminated since there is a tight fit between the thin walls of the Reaction Tubes in the block, and the heated cover of each instrument.

Template DNA or RNA

The PCR sample may be single- or double-stranded DNA or RNA. If the starting sample is RNA, total RNA, poly (A+) RNA, viral RNA or rRNA can be used. A reverse transcriptase (for example MuLV or r*Tth* DNA Polymerase) is used to prepare first-strand cDNA prior to conventional PCR amplification. For RNA templates with high G+C content or complex secondary structure, the high-temperature reverse transcriptase activity of thermostable r*Tth* DNA Polymerase is effective. When using DNA as the starting template, nanogram amounts of cloned template, up to microgram amounts of genomic DNA, or up to 20,000 target copies are chosen to start optimization trials. However, even very low levels of sample (*i.e.,* mRNA from tens of cells, DNA from single cells or individual viral genomes) may be sufficient for PCR amplification.

Primers

PCR primers are oligonucleotides, typically 15-30 bases long, hybridizing to opposite strands and flanking the region of interest in the target DNA. When choosing two PCR primers, it is important that they not contain bases complementary to themselves or with each other. Complementarity at the 3' ends should especially be avoided to minimize the formation of an artifactual product, often called "primer-dimer" or "primer-oligomer." A 40% - 60% G+C content is recommended for each primer, avoiding internal secondary structure and long stretches of any one base. Also, primers should not sit on regions of secondary structure (within the target) having a higher melting point than the primer. Non-template, complementary 5' extensions may be added to primers to allow a variety of useful post-amplification manipulations of the PCR product without significant effect on the amplification itself. These 5' extensions can be restriction sites, promotor sequences, etc.

Optimal annealing temperatures and primer concentrations must be determined empirically. *Taq* DNA Polymerase has activity in the 25 °C-72 °C range. Therefore, primer extension will occur during the annealing step and the hybrid will be stabilized. Primers are always present at an excess and equal concentration in conventional (symmetric) PCR amplification and, typically, are within the range of 0.1 μM to 1 μM. It is generally advisable to use purified oligomers of the highest chemical integrity.

Deoxynucleoside Triphosphates

In the standard PCR protocol, each deoxynucleoside triphosphate (dNTP) concentration is 200 ìM. It is important to keep the four dNTP concentrations above the estimated K_m of each dNTP (10 μM-15 μM) and balanced for best base incorporation. However, deviations (from these standard recommendations) may be beneficial in certain applications. For example, when random mutagenesis of a specific target is desired, unbalanced dNTP concentrations promote a higher degree of misincorporation by *Taq* DNA Polymerase. Lowering the dNTP and magnesium ion by an equal molar concentration can improve fidelity. Modified dNTPs (dig-11-dUTP, 5-bromo-dUTP, inosine, biotin-11-dUTP, biotin-16-dUTP and 7-deaza dGTP) and dUTP are useful substrates for DNA Polymerase.

PCR Buffers

Standard buffers for PCR amplification with DNA Polymerase are the 10X PCR Buffer and the PCR Buffer II. The 10X PCR Buffer is composed of 500 mM potassium chloride, 100 mM Tris-HCl (pH 8.3 at room temperature), 15 mM magnesium chloride and 0.01% (w/v) gelatin.

Magnesium Ion

In the presence of 0.8 mM total dNTP concentration, a magnesium chloride titration series in 0.5 mM increments over the 1 mM-4 mM range will locate the magnesium ion concentration, producing the highest yield of a specific PCR product.

DNA Polymerase

DNA Polymerase is an ultra-pure, modified form of DNA Polymerase designed to make the PCR process easy. DNA Polymerase is provided as an inactive enzyme, requiring heat activation to regenerate polymerase activity. The necessity for thermal activation provides highly specific PCR conditions resulting in lower background and increased yield of specific product (see Section below). The improvements in specificity and product yield, along with the increased sensitivity for low copy detection and easy room temperature set-up makes enzyme the enzyme of choice for the majority of PCR applications.

PCR Thermal Profiles

DNA denaturation is the critical step in the PCR process and is often the focus of attention if PCR experiments fail. The practical range of denaturation temperatures for most samples is 94°C-96°C.

Annealing temperature is based on the T_m (melting temperature) of the oligonucleotides chosen for PCR amplification. If unwanted bands are observed, the annealing temperature is raised in 2°C-5°C increments in subsequent optimization runs. While the primer annealing temperature range is often 37°C-55°C, it may be raised as high as the extension temperature in some cases. In fact, high-temperature annealing should result in enhanced specificity.

Primer extension, in most applications, occurs effectively at a temperature of 72°C and seldom needs optimization. ypically, 25-45 cycles are required for extensive amplification (i.e., 106-fold) of a specific target.

Materials

1. Thermocycler
2. Eppendorf tubes (0.5 ml capacity)
3. Micropipettes
4. Template DNA

5. 10 x PCR buffer: 100 mM Tris-HCI (pH 8.3) containing 500 mM KCI and
6. 15 mM $MgCl_2$
7. dNTPs mix: Mixture of dATP, dCTP, dTTP and dGTP each at a concentration of 1.25mM.
8. Primer pair: Dissolve each primer strand (forward and reverse) in 5 ml of distilled H_2O. Primers are available commercially eg OPERON Kits are available in alphabetical series.
9. Taq DNA polymerase
10. Mineral oil (not required in all thermocyclers)
11. Sterile distilled water
12. Hand gloves

Reagents required

1.	10X reaction buffer (with $MgCl_2$)	5ul
2.	25mM $MgCl_2$(optional)	3ul
3.	dNTP mix (10mM of each dNTP)	1ul
4.	Taq DNA Polymerase(5U/ul)	0.25ul
5.	downstream primer (10 um/ul)	1.5ul
6.	upstream primer (10 um/ul)	1.5ul
7.	DMSO	2.5ul
8.	template DNA (10-100ng/ul)	1-2ul
9.	nuclease-free water (adjust to a final 50 ul)	

Cycle condition

Denaturation	94°C	5 min	
Denaturation	94°C	1 min	30 CYCLES
Annealing	35°C	2 min	
Polymerization	72°C	3 min	

Note : These conditions may change slightly from material to material

Procedure

1. Thaw 10x buffer, dNTPs, and primers. Keep on ice.

   Note : Always Include A Reaction Tube "Blank" Which Will Have All Ingredients Except DNA. This Is An Essential Control For Each PCR Reaction.

2. Create a bulk reagent tube for each set of primers to be used. This will contain the 10x buffer, nucleotides, polymerase, primers, and water that will be required for this set of PCR reactions. For the purposes of determining amounts to use in the bulk reagent tube, consider that you will have 11 reactions for each 10 reactions you will be setting up. Label bulk reaction tubes with primers to be used and number or reactions you will be setting up. Mix well and place on ice.
3. Use 0.2 ml thin walled centrifuge tubes for reaction. Label top of tubes with sample
4. Pipet mix reagent (except template) into each tube.
5. Pipet DNA samples into appropriate tubes, using a new tip for each sample.
6. Mix tube well. Spin for 5 seconds in microfuge to remove liquid from sides of tube.
7. Place tubes in PCR machine, program machine, and begin PCR reaction.

Analysis

1. Analyze the PCR reaction products by agarose gel electrophoresis of a 5 ul aliquot from the total reaction. The products should be readily visible by UV transillumination of the ethidium bromide-stained gel.
2. Store reaction products at -20°C until needed. The reaction products can be further purified using a number of procedures, including the Qiagen PCR gel extract or column purification system

Effect of annealing temperature on PCR amplification

Since DNA-DNA hybridization is a temperature-dependent phenomenon, the annealing temperature is very important. The success of PCR depends on it.

- If annealing temperature is too high, pairing between the primers and template DNA will not take place and the PCR will fail i.e. there is no amplification even if the target sequence is present in the template DNA.
- If the temperature is too low, the primers would stably pair with even such sequences that are not perfectly complementary with them. As a result, some sequences other than the target sequence may also become amplified, thereby reducing the specificity of the PCR. Specificity of PCR

describes the probability of a non-target sequence being amplified; the lower this probability, the higher is the PCR specificity.

What is an Ideal Annealing Temperature?

An *ideal annealing temperature* must be low enough to enable hybridization between primer and template, but high enough to prevent amplification of' non-target sites. The ideal annealing temperature is usually 1-2°C lower than the melting temperature of the template-primer duplex. *Melting temperature (Tm)* of any DNA duplex is the temperature at which the two strands of the duplex dissociate, *i.e.,* denature. The Tm of primer-target sequence duplex can be determined either experimentally or, more generally, calculated using the following formula.

$$Tm = [4(G+C)] + [2\ (A+T)]$$

Where, (G+C) is the number of G and C nucleotides and (A+ T) is the number of A and T nucleotides in the primer sequence.

The annealing temperature for PCR experiment is, therefore, determined as follows: the Tm for each primer is calculated as above and a temperature 1-2°C below this value is used as annealing temperature. Therefore, it can not be over-emphasized that the two primers for a PCR must be designed in such a way that they both have an identical Tm. If this were not the case, the ideal annealing temperature for one primer would be either too low or too high for the other primer.

Factors Affecting Efficiency of PCR

The efficiency of PCR may be described as the number of copies of the target sequence generated after a given number of PCR cycles. PCR efficiency is influenced by several factors, some of which are listed below as explained by Singh (2005):

1. The efficiency of amplification declines with an increase in the length of the target sequence.
2. Efficiency of PCR is affected by primer length; it declines if the primers used for PCR are too long.
3. The annealing temperature has a marked influence on PCR efficiency; a temperature higher than the ideal annealing temperature reduces PCR efficiency.
4. Primer sequence also affects PCR efficiency. For example, if the primers used for PCR have complementary regions; primer dimers will be formed and PCR efficiency will markedly decline.

5. The sequence of the target DNA -itself can affect PCR efficiency. The GC-rich sequences in target DNA may form secondary structures in the single strands produced by denaturation; this could reduce PCR efficiency. A higher annealing temperature would minimize secondary structure and, thereby, improve efficiency. If required, 7-deaza guanosine triphosphate (7-deaza GTP) can be used in place of GTP; it reduces hairpin loop formation without affecting the Watson-Crick base pairing required for annealing and primer extension.
6. Addition of certain proteins, such as, bovine serum albumin enhances PCR efficiency by protecting the DNA polymerase and by binding to PCR inhibitors.
7. Certain compounds act as inhibitors of PCR. Presence of such a compound in the reaction mixture would, therefore, have an adverse effect on PCR efficiency.

Application of PCR

1. Isolation of Genomic DNA

PCR allows isolation of DNA fragments from genomic DNA by selective amplification of a specific region of DNA. This use of PCR augments many methods, such as generating hybridization probes for Southern or northern hybridization and DNA cloning, which require larger amounts of DNA, representing a specific DNA region. PCR supplies these techniques with high amounts of pure DNA, enabling analysis of DNA samples even from very small amounts of starting material.

Other applications of PCR include DNA sequencing to determine unknown PCR-amplified sequences in which one of the amplification primers may be used in Sanger sequencing, isolation of a DNA sequence to expedite recombinant DNA technologies involving the insertion of a DNA sequence into a plasmid or the genetic material of another organism. Bacterial colonies (*E.coli*) can be rapidly screened by PCR for correct DNA vector constructs. PCR may also be used for genetic fingerprinting; a forensic technique used to identify a person or organism by comparing experimental DNAs through different PCR-based methods.

Some PCR 'fingerprints' methods have high discriminative power and can be used to identify genetic relationships between individuals, such as parent-child or between siblings, and are used in paternity testing. This technique may also be used to determine evolutionary relationships among organisms.

2. Amplification and Quantitation of DNA

Because PCR amplifies the regions of DNA that it targets, PCR can be used to analyze extremely small amounts of sample. This is often critical for forensic analysis, when only a trace amount of DNA is available as evidence. PCR may also be used in the analysis of ancient DNA that is tens of thousands of years old. These PCR-based techniques have been successfully used on animals, such as a forty-thousand-year-old mammoth, and also on human DNA, in applications ranging from the analysis of Egyptian mummies to the identification of a Russian Tsar.

Quantitative PCR methods allow the estimation of the amount of a given sequence present in a sample – a technique often applied to quantitatively determine levels of gene expression. Real-time PCR is an established tool for DNA quantification that measures the accumulation of DNA product after each round of PCR amplification.

3. PCR in Diagnosis of Diseases

PCR allows early diagnosis of malignant diseases such as leukemia and lymphomas, which is currently the highest developed in cancer research and is already being used routinely. PCR assays can be performed directly on genomic DNA samples to detect translocation-specific malignant cells at a sensitivity which is at least 10,000 fold higher than other methods.

PCR also permits identification of non-cultivatable or slow-growing microorganisms such as mycobacteria, anaerobic bacteria, or viruses from tissue culture assays and animal models. The basis for PCR diagnostic applications in microbiology is the detection of infectious agents and the discrimination of non-pathogenic from pathogenic strains by virtue of specific genes.

Viral DNA can likewise be detected by PCR. The primers used need to be specific to the targeted sequences in the DNA of a virus, and the PCR can be used for diagnostic analyses or DNA sequencing of the viral genome. The high sensitivity of PCR permits virus detection soon after infection and even before the onset of disease. Such early detection may give physicians a significant lead in treatment. The amount of virus ("viral load") in a patient can also be quantified by PCR-based DNA quantitation techniques.

4. PCR Enables Rapid Amplification of Template DNA for Screening of Uncharacterized Mutations

Because of its rapidity and simplicity, PCR is ideally suited to providing numerous DNA templates for mutation screening. Partial DNA sequences, at the genomic or the cDNA level, from a gene associated with disease, or some other interesting phenotype, immediately enable gene-specific PCR reactions to be designed. Amplification of the appropriate gene segment then enables rapid testing for the presence of associated mutations in large numbers of individuals. By contrast, cell-based DNA cloning of the gene from numerous different individuals is far too slow and labor-intensive to be considered as a serious alternative.

Typically, the identification of exon-intron boundaries and sequencing of the ends of introns of a gene of interest offers the possibility of genomic mutation screening. Individual exon-specific amplification reactions are developed by designing primers which recognize intronic sequences located close to the exon-intron boundary. The resulting PCR products are then analyzed by rapid mutation-screening methods, in which the optimal size for mutation screening is usually about 200 bp. Conveniently, the average size of a human exon is about 180 bp but, in the case of very large exons, it is usual to design a series of primers to generate overlapping exonic products. PCR can also quickly provide amplified cDNA sequences for mutation screening. Such cDNA mutation screening may be the only way in which mutations can be screened if the exon-intron organization of a gene has not been established. To do this, mRNA is isolated from a convenient source of tissue, such as blood cells, converted into cDNA using reverse transcriptase and the cDNA is used as a template for a PCR reaction. This version of the standard genomic PCR reaction is consequently often referred to as RT-PCR (**reverse transcriptase-PCR**). Clearly, the method is ideally suited to genes expressed at high levels in easily accessible cells, such as blood cells. However, as a result of low level ectopic transcription of genes in all tissues, it has also been applied to transcript analysis of genes which are not significantly expressed in blood cells, such as the dystrophin (*DMD*) gene.

5 PCR Permits Rapid Genotyping for Polymorphic Markers

Restriction site polymorphisms (RSPs) result in alleles possessing or lacking a specific restriction site. Such polymorphisms can be typed using Southern blot hybridization. A DNA probe representing the locus is hybridized against genomic DNA samples that have been digested with the appropriate restriction enzyme and size-fractionated by agarose gel electrophoresis. The resulting RFLPs have two alleles corresponding to the presence or absence of the restriction site. As a convenient alternative to RFLPs, PCR can type RSPs by simply designing primers using sequences which flank the polymorphic restriction site, amplifying from genomic DNA, then cutting the PCR product with the appropriate restriction enzyme and separating the fragments by agarose gel electrophoresis.

Short tandem repeat polymorphisms (STRPs), also called microsatellite markers, consist of a short sequence, typically from one to four nucleotides long, that is tandemly repeated several times, and often characterized by many alleles. For example, $(CA)_n/(TG)_n$ repeats are often polymorphic when n exceeds 12, and have been widely used as polymorphic markers in the human genome. Increasingly, however, trinucleotide and tetranucleotide marker polymorphisms are being typed. In each case the STRPs can be typed conveniently by PCR. Primers are designed from sequences known to flank a specific STRP locus, permitting PCR amplification of alleles whose sizes differ by integral repeat units. The PCR products can then be size-fractionated by polyacrylamide gel electrophoresis. The PCR normally includes a radioactive or fluorescent nucleotide precursor which becomes incorporated into the small PCR products and facilitates their detection. To ensure adequate size fractionation of alleles, the PCR products are denatured prior to electrophoresis.

6. A Wide Variety of PCR-based Methods can be used to Assay for known Mutations

PCR is a very rapid and valuable tool for detecting pathogenic mutations and other mutations of interest. The examples below illustrate some popular methods.

a. Allelic Discrimination by size or Susceptibility to Restriction Enzyme

Small insertions or deletions (such as the three nucleotide deletion in the common cystic fibrosis (CFTR) allele, F508del) can be simply detected by designing primers from regions closely flanking the mutation site and distinguishing the normal and mutant alleles by size on polyacrylamide or agarose gels. If the mutation changes a restriction site, mutant and normal alleles can be distinguished by amplifying across the mutant site and digesting the PCR product with relevant restriction endonuclease.

b. Allelic Discrimination by Susceptibility to an Artificially Introduced Restriction Site

Even if the mutation does not result in a restriction site difference, it may be possible to exploit the difference between normal and mutant alleles by amplificationcreated restriction site PCR. This is a form of mismatched primer mutagenesis in which a primer is deliberately designed from sequence immediately adjacent to, but not encompassing, the restriction site. The primer is deliberately designed to have a mismatched nucleotide which together with the sequence of the mutant site creates a restriction site not present in normal alleles.

c. Allele-specific PCR (ARMS Test)

Oligonucleotide primers can be designed so as to discriminate between target DNA sequences that differ by a single nucleotide in the region of interest. This is a form of allele-specific PCR, the PCR equivalent of the allele-specific hybridization which is possible with ASO probes. In the case of allele-specific hybridization, alternative ASO probes are designed to have differences in a central segment of the sequence (to maximize thermodynamic instability of mismatched duplexes). However, in the case of allele-specific PCR, ASO primers are designed to differ at the nucleotide that occurs *at the extreme 3' terminus*. This is so because the DNA synthesis step in a PCR reaction is crucially dependent on correct base-pairing at the 3' end. This method can be used to type specific alleles at a polymorphic locus, but has found particular use as a method for detecting a specific pathogenic mutation, the so-called amplification refractory mutation system.

d. Mutation Detection Using the 5' → 3' Exonuclease Activity of Taq DNA Polymerase (TaqMan™ Assay)

Taq polymerase does not possess a proofreading 3' → 5' exonuclease activity but does possess a 5' → 3' exonuclease activity. This property can be exploited to facilitate detection of specific alleles. Such an assay involves hybridization of three primers, the third primer being intended to bind just downstream of one of the conventional primers which should be allele-specific. The additional primer carries a blocking group at the 3' terminal nucleotide so that it cannot prime new DNA synthesis and at its 5' end carries a labeled group. In modern versions of the assay, the label is a fluorogenic group and the third primer also carries a quencher group. If the upstream primer which is bound to the same strand is able to prime successfully, *Taq* DNA polymerase will extend a new DNA strand until it encounters the third primer in which case its 5' → 3' exonuclease will degrade the primer causing release of separate nucleotides containing the dye and the quencher, and an observable increase in fluorescence.

7. Degenerate Oligonucleotide Primers and Primers Specific for Ligated Linker Sequences Permit co-amplification of Sequence Families, or even Indiscriminate Amplification

a. *DOP-PCR (degenerate oligonucleotide-primed PCR)* is a form of PCR which is deliberately designed to permit possible amplification of several products. The two primers may be *partially degenerate* oligonucleotides, composed of panels of oligonucleotide sequences that have the same base at certain nucleotide positions, but are different at others. As a result, there may be comparatively many primer binding sites in the source DNA. This provides a means of searching for a new or uncharacterized DNA sequence that belongs to a family of related sequences either within or between species. **Note** : the use of such primers also provides a way of cloning a gene when only a limited portion of amino acid sequence is known for the product.

DOP-PCR can also be used to permit comparatively indiscriminate amplification of target DNA. Primer sequences with random sequences can bind to numerous locations in the template DNA and permit a form of whole-genome amplification. This can be advantageous where the amount of starting DNA may be limiting (as in the case of extracts from ancient DNA samples, microdissected chromosome bands, single cell typing, etc.), and PCR amplification of essentially all sequences increases the amount of DNA for study.

b. *Linker-primed PCR (ligation adaptor PCR)* : Another way of enabling amplification of essentially all DNA sequences in a complex DNA mixture involves first ligating a known sequence to all fragments. To do this, the target DNA population is digested with a suitable restriction endonuclease, and double-stranded oligonucleotide linkers (also called **adaptors**) with a suitable overhanging end are ligated to the ends of target DNA fragments. Amplification is then performed using oligonucleotide primers which are specific for the linker sequences. In this way, all fragments of the DNA source which are flanked by linker oligonucleotides can be amplified.

8. Anchored PCR uses a Target-specific Primer and a Universal Primer for Amplifying Sequences Adjacent to a known Sequence

It is often desirable to be able to amplify previously uncharacterized DNA sequences that neighbor a known DNA sequence, either at the genomic or cDNA level. To do this a form of anchored PCR is used. One of the primers is specific for the target sequence and the second primer is specific for a common sequence that can be introduced in different ways, such as by using a linker-primer method as described in the previous section, or by using primers that are modified at the 5' end so as to introduce a novel sequence.

Ten Things That Can Kill Your PCR

(Source: http://www.bioron.net/PDS%20files/Ten% 20Things %20That %20Can %20Kill %20Your%20PCR.pdf)

A blank PCR gel has got to be one of the most aggravating things about molecular biology. We've all had PCRs work one day, then fail inexplicably the next. And we've all banged our heads against a wall trying to figure out what went wrong. If your gels are turning up blank, check out this list of things that can kill your PCR — and get your experiments working!

Nowhere is the old adage "an ounce of prevention is worth a pound of cure" more applicable than to PCR. Depending on your level of PCR expertise, some of these hints may seem obvious. But how many times have you kicked yourself because you overlooked the obvious? Whether you're an expert or a novice, any of these may be just what you're searching for to get your PCR working.

1. *Too much dNTP, or degraded dNTP.* Too much dNTP can actually inhibit your PCR reaction. Between 40 - 200 uM is the optimal range. Also, dNTPs are sensitive to repeated freeze-thaw cycles. Make small aliquots when you get a fresh batch, and turn over your stock frequently since dNTPs frozen at -20° C will eventually go bad.
2. *Not mixing $MgCl_2$.* Magnesium chloride solutions form a concentration gradient when frozen and need to be vortexed prior to use.
3. *Wrong $MgCl_2$ concentration.* Every PCR reaction has an optimal $MgCl_2$ concentration range, usually between 1 - 4 mM. Mg2+ ions form complexes with dNTPs and can also act as a cofactor for polymerases, so you'll need to try several conditions to optimize your concentration.
4. *Inhibitors in the reaction mix.* Make sure you know how you got your source DNA. Chloroform, phenol, EDTA, ionic detergents (SDS and Sarkosyl), xylene cyanol, bromophenol blue and ethanol — among many other things — can inhibit PCR. An extra clean-up step on your template may do the trick. Also, certain polymerases can be more susceptible to certain substances, so be sure to check your polymerase for possible inhibitors.
5. *Poor quality mineral oil.* Lower-grade preparations may contain nucleases that can kill PCR. Also, avoid autoclaving mineral oil if possible. Exposure to high heat may cause reaction-inhibiting hydrocarbons to form. Similarly, do not irradiate mineral oil with UV for long periods.

6. *Too much enzyme.* Excess enzyme in the PCR can lead to smearing of PCR products. Most people seem to use 0.5 ul of their stock Taq per reaction, but that may contain way more than necessary for a particular reaction.

7. *Wrong primer concentration.* If you have too little primer you won't see any product. Too much primer and you may get primer dimerization and not enough amplification. Stay within 0.1 - 1.0 uM of primer.

8. *Wrong PCR program.* Make sure the program you selected on your PCR machine is actually the one you want! It only takes a slip of a finger, or some klutz, to alter your personal program on a common PCR machine. Check your program while it's cycling to make sure it's what you wanted.

9. *Excess or insufficient template.* Too much template can inhibit PCR by binding all the primers. Too little template, and amplification may not be detectable. For 25 - 30 cycles, 104 copies of the target sequence are sufficient.

10. *Poor primer design.* While primer design can seem like a black art, avoid obvious errors like self-complementarity, complementarity between paired primers, or excessively long oligos (>30 bp). Often, making a new primer next to a suspect one can solve the problem and can be faster and cheaper than trying numerous variations in reaction conditions.

14

Modification of Basic PCR Technique

Allele-Specific PCR

This diagnostic or cloning technique is used to identify or utilize single-nucleotide polymorphisms (SNPs) (single base differences in DNA). It requires prior knowledge of a DNA sequence, including differences between alleles, and uses primers whose 3' ends encompass the SNP. PCR amplification under stringent conditions is much less efficient in the presence of a mismatch between template and primer, so successful amplification with an SNP-specific primer signals presence of the specific SNP in a sequence.

Amplification Refractive Mutation System (ARMS) - PCR

In ARMS, the primer pair is designed so that one of the 3' ends coincides with a variant nucleotide in the target sequence. When the primer mismatches the template the frequency of extension is very low and consequently the effective number of sequence copies available for amplification is greatly reduced. ARMS PCR exploits a thermostable polymerase that lacks 3' exonuclease activity (usually *Taq* polymerase). Such enzymes extend primers bound to their target sequences very inefficiently when the 3' base is mismatched. Because the 3' exonuclease activity required for mismatch repair is not present, the extension of such primers in PCR is a rare event and amplification is retarded. However, once mismatch extension has occurred amplification proceeds normally from the newly synthesised target strand. Thus an important principle of ARMS is that non-matching (*i.e.* non-matching at only the 3' end of one primer) template may be amplified. The factors that contribute to this apparent 'failure' of ARMS are that when very high numbers of template molecules are added to the reaction, amplification is likely to proceed because the small proportion of mismatched primer extensions reach the sensitivity threshold of the reaction. Second, certain mismatches are extended more efficiently than others so that proportion extended in each PCR cycle varies.

Anchored PCR

In the basic PCR technique and the inverse PCR, one has to use two primers representing the sequences lying at the ends of sequences LO be amplified. But sometimes, we may have knowledge about sequence at only one of the two ends of the DNA sequence to be amplified.

In such cases, anchored PCR may be used, which will utilize only one primer instead of two primers. In. this technique, due to the use of one primer, only one strand will be copied first, after which a poly G tail will be attached at the end of the newly synthesized strand.

This newly synthesized strand with poly G tail at its 3' end will then become template for the daughter strand synthesis utilizing an anchor primer with which a Poly C sequence is linked to complement with poly G of the template. In the next cycle, both the original primer and anchored primer will be used for gene amplification.

Assembly PCR or Polymerase Cycling Assembly

Assembly PCR can be used to assemble two gene-sized pieces of DNA into one piece for easier cloning of fusion genes/parts. Briefly, it essentially involves PCR'ing the two pieces separately with primers that have a 20bp overlap and then doing an extra PCR step using the two products as the template. This is essentially just for ease of cloning. Instead of trying to PCR or cut out of a vector two separate pieces and then assemble them by endonuclease digestion and ligation (aka 3-way ligation), it can be easier simply to PCR the first piece w/ a reverse primer that overlaps with the forward primer of the second piece, and then use the product of the first PCR reactions as a template for the assembly reaction. If the reverse primer for the 5' piece and the forward primer for the 3' piece overlap by ~20bp, the product of the first PCR reactions should anneal in the overlapping region and create a full length (gene fusion) product. Using the forward primer for the first piece and the reverse primer for the second piece in the assembly reaction then amplifies the desired full-length product. The merits for this technique are that it's arguably faster than standard 3-way ligation assembly, and it's more reliable.

Protocol *(by Felix Moser)*

Remember, that this technique is good if: You want to assemble in series two long pieces of DNA from PCR product.

1. Design the reverse primer for the DNA that will be 5' w/ significant overlap w/ the forward primer for the 3' piece. Essentially, as long as one of the primers has ~20bp overlap w/ the 'reverse complement' of the other primer, the products should anneal in the assembly reaction.
2. Do PCR as normal for the two (5' and the 3') pieces using the longer primers that correspond to each piece.
3. Check on a gel to make sure you got product from the first PCR reaction. Some people like to cut out the product band and use the purified products as template for the next reaction. I just use the PCR product straight from the first reactions w/o any purification; the logic is that the undesired primers/templates will be in such low concentrations that the intended reaction will be highly favoured. Besides, if the residual "middle" primers did create product, they would just be making more starting template, which shouldn't hurt your rxn.
4. Set up the assembly reaction like a regular PCR, except: 1) as template use equal amounts of product from the first reactions, and 2)use the Forward primer for the 5' piece and the Reverse primer for the 3' piece to amplify the annealed template. (I use 45μl of Invitrogen Taq HIFI supermix, 2μl of 5μM primer each, and 0.5μl of each PCR product as template). Cycle like you did for the first reactions, except w/ longer extension time corresponding to the length of your product.
5. Run the product on a gel. If the reaction worked, you should see a band the size of the sum of your two templates.
6. Purify the product (I use the Quiagen PCR pur. kit), cut w/ desired endonucleases, and clone away. The quality of the product from this reaction is usually very good and very plentyful and I can get up to >100 transformants.

7. An arguable disadvantage of this technique, besides slightly higher up-front cost for primers, is that it requires sequencing following assembly to make sure the PCR rxn hasn't produced mutations. Use HiFi polymerase and you shouldn't really have a problem, though... but don't be lazy: you should still get your clones sequenced.

Asymmetric PCR

Asymmetric PCR is used to preferentially amplify one strand of the original DNA more than the other. It finds use in some types of sequencing and hybridization probing where having only one of the two complementary stands is ideal. PCR is carried out as usual, but with a great excess of the primers for the chosen strand. Due to the slow (arithmetic) amplification later in the reaction after the limiting primer has been used up, extra cycles of PCR are requiredAsymmetric PCR is used to preferentially amplify one strand of the target DNA more than the other.

Applications of Asymmetric PCR

The Asymmetric PCR is useful in some sequencing and hybridization probing applications where having only one of the two complementary stands is sufficient or required.

Asymmetric PCR Method

The asymmetric PCR method is conducted as the standard pcr protocol, however a great excess of the primers for the chosen strand is used.

Due to the slow (arithmetic) amplification later in the reaction after the limiting primer has been used up, extra cycles of PCR are required.

Primers for Asymmetric PCR

A PCR in which the predominant product is a single-stranded DNA, as a result of unequal primer concentrations. As asymmetric PCR proceeds, the lower concentration primer is quantitatively incorporated into double-stranded DNA. The higher concentration primer continues to primer synthesis, but only of its strand.

Colony PCR

Colony PCR the screening of bacterial (*E.Coli*) or yeast clones for correct ligation or plasmid products. Selected colonies of bacteria or yeast are picked with a sterile toothpick or pipette tip from a growth (agarose) plate. This is then inserted into the PCR master mix or pre-inserted into autoclaved water.

PCR is then conducted to determine if the colony contains the DNA fragment or plasmid of interest.

Bacterial *E.Coli* Colony PCR Protocol

1. Pick a bacterial colony with an autoclaved toothpick or pipette tip and swirl it into 25 µl of TE autoclaved dH2O in an autoclaved microfuge tube. Do not remove the entire colony, and mark the area you picked preferrably number the bottom of the plate (and the microfuge tube).
2. Heat the mix in a boiling water bath or hot (90-100C) heating block for 2 minutes
3. Spin sample for 2 minutes high speed in centrifuge.
4. Transfer 20 µl of the supernatant into a new microfuge tube.
5. Take 1-2 µl of the supernatant as template in a 25 µl PCR standard PCR reaction.

Yeast Colony PCR Protocol

Take yeast cells which were grown on a plate fresh from the 37°C incubator (freshly grown yeast are better). Take an autoclaved toothpick or pipette tip and touch a bit of the yeast colony. Do not scrape away a whole colony. Suspend it into a standard PCR mastermix.

PCR Cycling Conditions

First Denaturing Step: 95°C , 5 min

Denaturing Step: 95°C, 30 sec

Annealing Step: 50-55°C, 30 sec

Extension Step: 72°C, 1min/kb

Repeat Steps 2-4: 34 Times

Final Extension: 72°C, 3 min

Load the entire PCR reaction onto 1% agarose gel.

Helicase-dependent Amplification

Helicase-dependent amplification (HDA) is a method for *in vitro* DNA amplification like the polymerase chain reaction (PCR).

The disadvantage of PCR is that it consumes a lot of time with uncoiling the double-stranded DNA with heat into single strands (a process called denaturation) and copying the single strands to create new double-stranded

DNA (DNA synthesis). Instead of these thermocycles, HDA mimics nature's method of replicating DNA by using helicase (an enzyme) to denature the DNA at a constant temperature of 37°C.

The advantages of HDA over PCR are increased speed, reduced costs and the prospect of creating a hand-held DNA diagnostic device. The disadvantage is that HDA still requires a relatively large amount of DNA, so it presumably will not be used in cases where high sensitivity is needed.

Hotstart PCR

Hot Start PCR allows the inhibition of polymerase activity during PCR reaction preparation. By limiting polymerase activity prior to PCR cycling, Hot Start PCR reduces non-specific amplification and increases PCR product target yield.

Hot Start PCR is commonly performed by using included chemical modifications, wax-barrier methods, and inhibition by a taq-directed antibody.

PCR or polymerase chain reaction uses DNA polymerases which were isolated from thermostable organisms such as the Taq enzyme, a eubacterial type I DNA polymerase, and Pfu, an archaeal type B DNA polymerase.

PCR amplifies DNA by multiple cycles of melting DNA at high temperature, annealing primers at a lower temperature, and then allowing polymerase extension at an intermediate temperature. Unfortunately, these thermophilic DNA polymerases show a very small but measurable polymerase activity at room temperature during assembly of the experiment.

This in-efficient DNA polymerase activity will catalyze the extension of any annealed 3' ends. After PCR amplification, the resulting product contains a mixture of specific and non-specific bands. Moreover, the 5'-3' exonuclease activity of the DNA polymerasel degrades any free 5' ends of partially annealed nucleic acids, destroying the primer and template substrates of the polymerase reaction. This results in a lower yield of the desired pcr product.

In Situ PCR

During the initiation and progression of disease, minute quantities of a product in small populations of cells or tissues may be vital for the pathogenesis of the disease.

In many slowly-evolving diseases which require months or even years to manifest themselves clinically, it has been shown that the majority of the affected cell population is in a transcriptionally inactive state, and at a level of one gene per host cell.

Nucleic acid hybridization methods and the polymerase chain reaction (PCR) have both been employed to examine the expression and detection of such affected genes during pathogenesis. While both these techniques are quite useful, the disadvantage of these techniques is that they are essentially conducting cell expression and population studies. Nucleic acids are isolated from a population of cells which contains either a sufficient number of molecules to detect directly by standard hybridization techniques, or, when a subpopulation contains as little as a single copy of nucleic acid, that molecule amplified by the PCR, and detected after amplification.

In situ hybridization (ISH) applies the methodology of the nucleic acid hybridization technique to the cellular level. Combining cytochemistry and immunocytochemistry, *In Situ* PCR allows the identification of cellular markers to be identified, and further permits the localization of to cell specific sequences within cell populations, such as tissues and blood samples.

In Situ PCR is limited to the detection of non-genomic material such as RNA, genes or genomes, as the detection limit in most conditions is several copies of the target nucleic acid per cell. Therefore, due to copy number limitations, hybridization of RNA is more sensitive than DNA detection.

Factors affecting In Situ PCR sensitivity include:

1. The strandedness of the target molecule
2. The lack of a complementary sequence proximal to the target sequences

Reverse transcriptase-catalyzed *in situ* transcription has even been used to detect RNAs which occur at relatively high copy number.

Intersequence-specific PCR

Intersequence-specific PCR or ISSR PCR is a method based on the polymerase chain reaction. A PCR method for DNA fingerprinting that amplifies regions between some simple sequence repeats to produce a unique fingerprint of amplified fragment lengths

Applications of ISSR PCR

The ISSR PCR method is used for DNA fingerprinting by amplifying regions between some simple sequence repeats to produce a unique fingerprint of amplified fragment lengths.

Inverse PCR

Inverse PCR also called IPCR, and was first described by Ochman et al. in 1988. A limitation of standard PCR is that 5' and 3' flanking regions of your DNA fragment of interest must be known. Inverse PCR allows you to conduct PCR when you only have the information of one internal sequence.

Inverse polymerase chain reaction is a variant of PCR, and is used when only one internal sequence of the target DNA is known. It is therefore very useful in identifying flanking DNA sequences of genomic inserts. Similar to other PCR methods, inverse PCR amplifies target DNA using DNA polymerase.

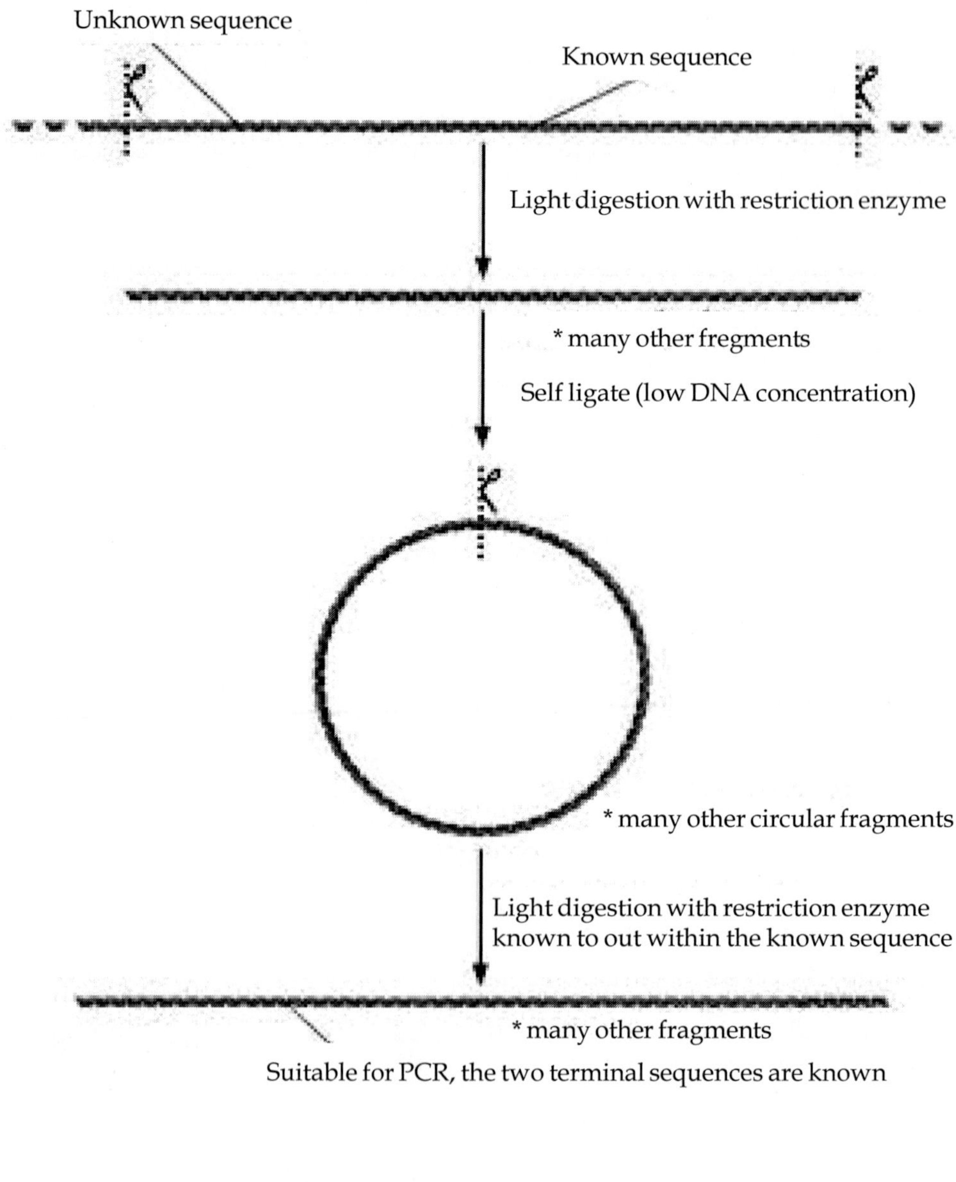

Inverse PCR uses standard PCR (polymerase chain reaction), however it has the primers oriented in the reverse direction of the usual orientation. The template for the reverse primers is a restriction fragment that has been ligated upon itself to form a circle.

The Inverse PCR Method

The inverse PCR method includes a series of digestions and self-ligations with the DNA being cut by a restriction endonuclease. This cut results in a known sequence at either end of unknown sequences.

Inverse PCR Steps

1. Target DNA is lightly cut into smaller fragments of several kilobases by restriction endonuclease digestion.
2. Self-ligation is induced under low concentrations causing the phosphate backbone to reform. This gives a circular DNA ligation product.
3. Target DNA is then restriction digested with a known endonuclease. This generates a cut within the known internal sequence generating a linear product with known terminal sequences. This can now be used for PCR (polymerase chain reaction).
4. Standard PCR is conducted with primers complementary to the now known internal sequences.

Protocol

1. Digest 2.5 µg of genomic DNA with restriction enzymes. Stop reaction at 65°C for 20′.
2. Fractionate on a 1% agarose gel to check DNA has been cut.
3. Set up self-ligation reaction:
   40µl 10 × DNA Ligase buffer
   1.5µl NEB concentrated DNA ligase
   358.5µl 1µg digested DNA plus dH_2O
   Incubate overnight at room temperature. Do not put in more that 1 µg of DNA as fragments may ligate to each other rather than to themselves.
4. Add 400 µl 25:24:1 phenol:chloroform: IAA. Vortex and spin for 5′ in microfuge. Remove aqueous layer to fresh tube.
5. Add 2.5 volumes ethanol and 0.1 volume 3M NaOAc pH4.5.
   Precipitate at –20°C overnight or at –80°C for 1 hr.

6. Spin in microfuge for 10′ at full speed. Wash pellet with 70% ethanol, spin again and drain off supernate.
7. Resuspend pellet in 30 μl dH2O and check concentration on a gel or by OD260
8. Set up PCR with 50ng DNA using extend taq for long products:
   37.5 μl Ligated DNA plus dH2O
   5 μl 10 X PCR buffer
   4 μl 25 mM MgCl2
   1 μl 10 mM dNTP's
   1 μl Primer 1 (10 μM)
   1 μl Primer 2 (10 μM)
   0.5 μl Taq DNA Polymerase
   50 μl TOTAL
9. Cycle:
   94°C (45s) X 1
   94°C (15s), 50°C (30s), 68°C (4′) X 10
   94°C (15s), 50°C (30s), 68°C (4′ plus 5s per cycle) X 20
   72°C (7′) X 1
10. 25 μl of the final PCR product should be run out on a 1% agarose gel, viewed and blotted for band verification by hybridization. The remaining 25 μl of the PCR reaction can then be used, if positives are forthcoming, to clone and sequence the band of interest.

Applications of Inverse PCR

Inverse PCR has numerous applications in molecular biology including the amplification and identification of sequences flanking transposable elements, and the identification of genomic inserts.

Ligation-mediated PCR

This method uses small DNA linkers ligated to the DNA of interest and multiple primers annealing to the DNA linkers; it has been used for DNA sequencing, genome walking, and DNA footprinting.

Methylation-specific PCR

This alternative method of methylation analysis also uses bisulfite-treated DNA but avoids the need to sequence the area of interest. Instead, primer pairs are designed themselves to be "methylated-specific" by including sequences complementing only unconverted 5-methylcytosines, or conversely "unmethylated-specific", complementing thymines converted from unmethylated cytosines. Methylation is determined by the ability of the specific primer to achieve amplification. This method is particularly useful to interrogate CpG islands with possibly high methylation density, as increased numbers of CpG pairs in the primer increase the specificity of the assay. Placing the CpG pair at the 3'-end of the primer also improves the sensitivity. The initial report using MSP described sufficient sensitivity to detect methylation of 0.1% of alleles. In general, MSP and its related protocols are considered to be the most sensitive when interrogating the methylation status at a specific locus.

The MethyLight method is based on MSP, but provides a quantitative analysis using real-time PCR. Methylated-specific primers are used, and a methylated-specific fluorescence reporter probe is also used that anneals to the amplified region. Alternatively, the primers or probe can be designed without methylation specificity if discrimination is needed between the CpG pairs within the involved sequences. Quantitation is made in reference to a methylated reference DNA. A modification to this protocol to increase the specificity of the PCR for successfully bisulfite-converted DNA (ConLight-MSP) uses an additional probe to bisulfite-unconverted DNA to quantify this non-specific amplification.

Further methodology using MSP-amplified DNA analyzes the products using melting-curve analysis (Mc-MSP). This method amplifies bisulfite-converted DNA with both methylated-specific and unmethylated-specific primers, and determines the quantitative ratio of the two products by comparing the differential peaks generated in a melting-curve analysis. A high resolution melting analysis method which using both real time quantification and melting analysis has been introduced particularly for sensitive detection of low level methylation.

Miniprimer PCR

Miniprimer PCR uses a novel thermostable polymerase (S-Tbr) that can extend from short primers ("smalligos") as short as 9 or 10 nucleotides, instead of the approximately 20 nucleotides required by Taq. This method permits PCR targeting smaller primer binding regions, and is particularly useful to amplify unknown, but conserved, DNA sequences, such as the 16S (or eukaryotic 18S) rRNA gene. 16S rRNA miniprimer PCR was used to characterize a microbial mat community growing in an extreme environment, a hypersaline pond in Puerto Rico. In that study, deeply divergent sequences were discovered with high frequency and included representatives that de?ned two new division-level taxa, suggesting that miniprimer PCR may reveal new dimensions of microbial diversity. By enlarging the "sequence space" that may be queried by PCR primers, this technique may enable novel PCR strategies that are not possible within the limits of primer design imposed by Taq and other commonly used enzymes.

Multiplex Ligation-dependent Probe Amplification

Multiplex ligation-dependent probe amplification (MLPA) is a variation of the polymerase chain reaction that permits multiple targets to be amplified with only a single primer pair. Each probe consists of a two oligonucleotides which recognise adjacent target sites on the DNA. One probe oligonucleotide contains the sequence recognised by the forward primer, the other the sequence recognised by the reverse primer. Only when both probe oligonucleotides are hybridised to their respective targets, can they be ligated into a complete probe. The advantage of splitting the probe into two parts is that only the ligated oligonucleotides, but not the unbound probe oligonucleotides, are amplified. If the probes were not split in this way, the primer sequences at either end would cause the probes to be amplified regardless of their hybridisation to the template DNA, and the amplification product would not be dependent on the number of target sites present in the sample DNA. Each complete probe has a unique length, so that its resulting amplicons can be separated and identified by (capillary) electrophoresis. This avoids the resolution limitations of multiplex PCR. Since the forward primer used for probe amplification is fluorescently labelled, each amplicon generates a fluorescent peak which can be detected by a capillary sequencer. Comparing the peak pattern obtained on a given sample with that obtained on various reference samples, the relative quantity of each amplicon can be determined. This ratio is a measure for the ratio in which the target sequence is present in the sample DNA.

Applications of MLPA

MLPA has a variety of applications including detection of mutations and single nucleotide polymorphisms, analysis of DNA methylation, relative mRNA quantification, chromosomal characterisation of cell lines and tissue samples, detection of duplications and deletions in human cancer predisposition genes such as BRCA1, BRCA2, hMLH1 and hMSH2 and aneuploidy determination. MLPA has potential application in prenatal diagnosis both invasive and noninvasive.

Multiplex-PCR

Multiplex PCR amplifies multiple DNA targets using more than one pair of primers in one reaction tube. Multiplex PCR has been successfully applied on genetics screening, genotyping, diagnositic analysis of infectious disease etc. It is also cost-saving for large scale PCR analysis. Quantitative multiplex PCR requires extensive multiplex PCR reaction optimization to determine the primer concentration, DNA template concentration, PCR cycling condition and PCR buffer components.

Nested PCR

Nested PCR is a variation of the polymerase chain reaction (PCR), in that two pairs (instead of one pair) of PCR primers are used to amplify a fragment.

The first pair of PCR primers amplify a fragment similar to a standard PCR. However, a second pair of primers called nested primers (as they lie / are nested within the first fragment) bind inside the first PCR product fragment to allow amplification of a second PCR product which is shorter than the first one.

The advantage of nested PCR is that if the wrong PCR fragment was amplified, the probability is quite low that the region would be amplified a second time by the second set of primers. Thus, Nested PCR is a very specific PCR amplification.

The Nested PCR Reaction

Nested PCR requires two sets of primers which are used to amplify a specific DNA fragment using two separate runs of PCR. The second pair of primers function to amplify a smaller specific DNA fragment located within the first PCR product.

Overlap-Extension PCR

The **Overlap extension polymerase chain reaction** (or **OE-PCR**) is a variant of PCR which can a) insert mutations at specific points in a sequence, further than ~55 nucleotides from either end and/or b) produce polynucleotides from smaller fragments[1].

To insert a mutation into a sequence, a primer is generated depending on requirements. A single substitution may be included or a totally new sequence added at the 5' end. If a deletion is required, then the continuation of the sequence 5' of the deletion effectively becomes an addition. The 3' end must be complementary to the template strand so that DNA polymerase can copy the template from the primer (DNA polymerase can only copy DNA in one direction; 5' to 3').

Once the primer has annealed to the template, the DNA polymerase proceeds to the end of the template. The duplex is denatured and the second primer anneals to the newly formed DNA strand, containing the first primer. Replication proceeds to produce a strand of the required sequence, containing the mutation.

The duplex is denatured again and the first primer can now bind to the latest DNA strand. The replication reaction continues to produce a fully dimerised DNA fragment. After further PCR cycles, to amplify the DNA, the sample can be separated by agarose gel electrophoresis, followed by electroelution for collection.

Efficiently generating oligonucleotides beyond ~110 nucleotides in length is very difficult, so to insert a mutation further into a sequence than a 110 nt primer will allow, it is necessary to employ overlap extension PCR. In OE-PCR the sequence being modified is used to make two modified strands with the mutation at opposite ends, using the technique described above. After mixing and denaturation, the strands are allowed to anneal to produce three different combinations as detailed in the diagram. Only the duplex without overlap at the 5' end will allow extension by DNA polymerase in 3' to 5' direction.

Following separation, the eluted fragments of appropriate size are subject to normal PCR, using the outermost primers used in the initial, mutagenic PCR reactions.

PCR for Site-directed Mutagenesis

This technique is used to introduce mutations at desired sites in the DNA. This is achieved by changing the sequences of the primers. *In vitro* site-directed mutagenesis is an invaluable technique for studying protein structure-function relationships, gene expression and vector modification. Several methods have appeared in the literature, but many of these methods require single-stranded DNA as the template. The reason for this, historically, has been the need for separating the complementary strands to prevent reannealing. Use of PCR in site-directed mutagenesis accomplishes strand separation by using a denaturing step to separate the complementing strands and allowing efficient polymerization of the PCR primers. PCR site-directed methods thus allow site-specific mutations to be incorporated in virtually any double-stranded plasmid; eliminating the need for M13-based vectors or single-stranded rescue.

Several points should be mentioned concerning site-directed mutagenesis using PCR. First, it is often desirable to reduce the number of cycles during PCR when performing PCR-based site-directed mutagenesis to prevent clonal expansion of any (undesired) second-site mutations. Limited cycling which would result in reduced product yield, is offset by increasing the starting template concentration. Second, a selection must be used to reduce the number of parental molecules coming through the reaction. Third, in order to use a single PCR primer set, it is desirable to optimize the long PCR method. And fourth, because of the extendase activity of some thermostable polymerases it is often necessary to incorporate an end-polishing step into the procedure prior to end-to-end ligation of the PCR-generated product containing the incorporated mutations in one or both PCR primers.

A protocol is provided as a facile method for site-directed mutagenesis and accomplishes the above desired features by the incorporation of the following steps: (i) increasing template concentration approximately 1000-fold over conventional PCR conditions; (ii) reducing the number of cycles from 25-30 to 5-10; (iii) adding the restriction endonuclease *Dpn*I (recognition target sequence: 5-Gm6ATC-3, where the A residue is methylated) to select against parental DNA (note: DNA isolated from almost all common strains of *E. coli* is *Dam*-methylated at the sequence 5-GATC-3); (iv) using Taq Extender in the PCR mix for increased reliability for PCR to 10 kb; (v) using *Pfu* DNA polymerase to polish the ends of the PCR product, and (vi) efficient intramolecular ligation in the presence of T4 DNA ligase.

Protocol

1. Plasmid template DNA (approximately 0.5 pmole) is added to a PCR cocktail containing, in 25 ul of 1x mutagenesis buffer: (20 mM Tris HCl, pH 7.5; 8 mM MgCl2; 40 ug/ml BSA); 12-20 pmole of each primer (one of which must contain a 5-prime phosphate), 250 uM each dNTP, 2.5 U *Taq* DNA polymerase, 2.5 U of *Taq* Extender (Stratagene).

2. The PCR cycling parameters are 1 cycle of: 4 min at 94 C, 2 min at 50 C and 2 min at 72 C; followed by 5-10 cycles of 1 min at 94 C, 2 min at 54 C and 1 min at 72 C (step 1).

3. The parental template DNA and the linear, mutagenesis-primer incorporating newly synthesized DNA are treated with *Dpn*I (10 U) and *Pfu* DNA polymerase (2.5U). This results in the *Dpn*I digestion of the *in vivo* methylated parental template and hybrid DNA and the removal, by *Pfu* DNA polymerase, of the *Taq* DNA polymerase-extended base(s) on the linear PCR product.

4. The reaction is incubated at 37°C for 30 min and then transferred to 72 C for an additional 30 min (step 2).
5. Mutagenesis buffer (1x, 115 ul, containing 0.5 mM ATP) is added to the *Dpn*I-digested, *Pfu* DNA polymerase-polished PCR products.
6. The solution is mixed and 10 ul is removed to a new microfuge tube and T4 DNA ligase (2-4 U) added.
7. The ligation is incubated for greater than 60 min at 37 C (step 3).
8. The treated solution is transformed into competent *E. coli* (step 4).

Quantitative PCR (Q-PCR)

Quantitative PCR (Q-PCR): is used to measure the quantity of a PCR product (preferably real-time). It is the method of choice to quantitatively measure starting amounts of DNA, cDNA or RNA. Q-PCR is commonly used to determine whether a DNA sequence is present in a sample and the number of its copies in the sample. The method with currently the highest level of accuracy is Quantitative real-time PCR. It is often confusingly known as RT-PCR (Real Time PCR) or RQ-PCR. QRT-PCR or RTQ-PCR are more appropriate contractions. RT-PCR commonly refers to reverse transcription PCR (see below), which is often used in conjunction with Q-PCR. QRT-PCR methods use fluorescent dyes, such as Sybr Green, or fluorophore-containing DNA probes, such as *Taq* Man, to measure the amount of amplified product in real time.

RT (Reverse Transcribed) -PCR

Reverse transcription polymerase chain reaction (RT-PCR) is based on the polymerase chain reaction (PCR). More importantly it is based on the process of reverse transcription, which reverse transcribes RNA into DNA and was initially isolated from retroviruses.

The techniques of RT-PCR allows the formation of cDNA (complementary or copy DNA) from RNA, which stores the sequence of RNA (such as messenger RNA, mRNA) in the more stable form of nucleic acid, DNA. This reverse transcription from RNA into its reverse complement DNA (cDNA) is the first step of a usually two-step process of RT-PCR. Furthermore, by copying the RNA into DNA, one can then amplify the cDNA sequence by using primers specific for the DNA sequence. This amplification is the final second major step of the two-step process of RT-PCR.

The Process of RT-PCR

The First step of RT-PCR is referred to as the "first strand reaction". In the first-strand reaction, complementary DNA also termed cDNA, is made from the messenger RNA template of interest using oligo dT (oligonucleotide poly-dTs act similar to primers and bind to the 3' polyA sequence located at the 3' UTR - untranslated region, which are present in most mRNAs), dNTPs, and an RNA-dependent DNA polymerase, reverse transcriptase, through the process of reverse transcription.

These factors are combined in a reverse transcriptase buffer for 1 hour at 37°C. After reverse transcriptase reaction is complete, and the cDNA has been synthesized, RNaseH is added (an RNA digestion enzyme) which digests the RNA away from the RNA-cDNA hybrid. After incubation with RNaseH, standard PCR or polymerase chain reaction is conducted using DNA oligo primers specific for the sequence of interest. This second step is referred to as the "second strand reaction".

Thus by adding the thermostable DNA polymerase, upstream and downstream DNA primers, the single stranded DNA becomes double stranded and is amplified, allowing the detection of even rare or low copy mRNA sequences by amplifying its complementary DNA.

RT(Real Time)-PCR

Although the traditional methods of quantitating mRNA are fairly good such as northern blotting and in situ hybridization, they do not approach the ease and speed of Real Time PCR.

RT-PCR or reverse transcriptase PCR is semi-quantitative due to need to load samples on a gel and the insensitivity of ethidium bromide. Thus, real time PCR was developed out of the need to quantitate differences in mRNA expression in a easy and quick manner, and due to the need to use of small amounts of mRNA such as those obtained by small tissue samples, and LCM (laser capture microdissection) isolated cells.

Real-time reverse-transcriptase (RT) PCR is different from other quantitative PCR as it quantitates the initial amount of the template instead of detecting the amount of final amplified product.

Real Time PCR is characterized by the point in time during cycling when amplification of the PCR product of interest is first detected rather than the amount of the PCR product of interest which is accumulated at the end-point after PCR which contained a large number of cycles. Real Time PCR does this by monitoring the amount of fluorescence emitted during the PCR reaction, and this acts as an indicator of the amount of PCR amplification that occurs during each PCR cycle. Thus, in newer Real Time PCR machines, one can visually see the progress of the reaction in "real time".

Real Time PCR also has a much wider dynamic range of up to 107-fold (compared to 1000-fold in conventional RT-PCR). The dynamic range of an assay determines how much the target concentration can vary and yet still be quantified. This wide dynamic range also results in a more accurate quantitation.

Solid Phase PCR

Solid Phase PCR encompasses multiple meanings, including Polony Amplification (where PCR colonies are derived in a gel matrix, for example), 'Bridge PCR' (the only primers present are covalently linked to solid support surface), conventional Solid Phase PCR (where Asymmetric PCR is applied in the presence of solid support bearing primer with sequence matching one of the aqueous primers) and Enhanced Solid Phase PCR (where conventional Solid Phase PCR can be improved by employing high Tm solid support primer with application of a thermal 'step' to favour solid support priming).

TAIL-PCR: Thermal Asymmetric Interlaced

TAIL-PCR is a powerful tool for the recovery of DNA fragments adjacent to known sequences. Basically, TAIL-PCR utilises three nested primers in consecutive reactions together with an arbitrary degenerate (AD) primer having a lower Tm (melting temperature) so that the relative amplification frequencies of specific and non-specific products can be thermally controlled.

Primary Reaction

In the primary reaction, one low stringency PCR cycle is conducted to create one or more annealing sites for the AD primer in the targeted sequence. Specific products are then amplified over non-specific ones by interspersion of two high-stringency PCR cycles with one reduced-stringency PCR cycle.

1. Set up 4 reactions as follows (one with each AD primer):

2μl	10 X PCR buffer
1.2μl	25 mM $MgCl_2$
0.2μl	10 mM dNTP's
0.2μl	100 ngμl-1 specific primer 1 (furthest away from AD) (0.15μM final)

2μl	20μM AD primer (2μM final)
0.2μl	Taq DNA polymerase
0.4μl	DMSO
1μl	DNA (1-20ngμl^{-1})
12.8μl	H2O

2. Cycle as follows: (if doesn't work, try dropping annealing step to 60°C)

92°C (3′), 95°C (1′)	X 1
94°C (30s), 65°C(1′), 72°C (2′)	X 5
94°C (30s), 25°C (2′), ramping to 72°C over 2′, 72°C (2′)	X 1
94°C (30s), 65°C (1′), 72°C (2′)	
94°C (30s), 65°C (1′), 72°C (2′)	
94°C (30s), 44°C (1′), 72°C (2′)	X 15
72°C (5′)	X1
cool to 4°C	

There is no need to run out this primary reaction. It should contain a medium yield of specific products, a high yield of non-targeted products, and a low yield of non-specific products. The nested primers used in the secondary and tertiary reactions result in very low yields of non-specific products, very high yields of specific products and no amplification of non-targeted products.

Secondary reaction

For the secondary reaction, a 1/40 dilution of the primary PCR product is used as template, and the specific primer is the middle one of the three specific primers.

1. Set up reaction as follows:

2.5μl	10 X PCR buffer
1.5μl	25 mM MgCl2
0.25μl	10 mM dNTP's
0.3μl	100 ngμl-1 specific primer 2 (middle nested) (0.2 μM final)
2.5μl	20 μM AD primer (2 μM final)
0.2μl	Taq DNA polymerase
0.5μl	DMSO
1μl	DNA (1/40 dilution of primary PCR products)
16.25μl	H2O

2. Cycle as follows: (if doesn't work, try dropping annealing step to 60°C)
   94°C (30s), 65°C (1′), 72°C (2′)
   94°C (30s), 65°C (1′), 72°C (2′)
   94°C (30s), 45°C (1′), 72°C (2′)   X 12
   72°C (5′)   X 1
   cool to 4°C

Tertiary Reaction

For the tertiary reaction, the SAME template (i.e primary PCR product) is used but this time in a 1/10 dilution. I usually simply add 4 X of the 1/40 dilution used for the secondary reaction. This removes the possibility of getting false positives. The specific primer used is the primer nearest the unknown sequence.

1. Set up reaction as follows:

5µl	10 X PCR buffer
3µl	25 mM MgCl2
0.5µl	10 mM dNTP's
0.6µl	100ngµl-[1] specific primer 3 (closest to AD) (0.2 µM final)
5µl	20 µM any one AD primer (2 µM final)
0.4µl	Taq DNA polymerase
1µl	DMSO
4µl	DNA (1/40 dilution of primary PCR products)
31µl	H2O

2. Cycle as follows:
   94°C (30s), 45°C (1′), 72°C (2′)   X 20
   72°C (5′)   X 1
   cool to 4°C

Agarose Gel Analysis

The secondary and tertiary products are run in adjacent lanes on a 1.2% agarose gel. The specificity of the products is confirmed by the expected size change between the secondary and tertiary products.

Touchdown PCR

Touchdown polymerase chain reaction or touchdown style polymerase chain reaction is a method of polymerase chain reaction by which primers will avoid amplifying nonspecific sequence. The temperature at which primers anneal during a cycle of polymerase chain reaction determines the specificity of annealing. The melting point of the primer sets the upper limit on annealing temperature. At temperatures just below this point, only very specific base pairing between the primer and the template will occur. At lower temperatures, the primers bind less specifically. Nonspecific primer binding obscures polymerase chain reaction results, as the nonspecific sequences to which primers anneal in early steps of amplification will "swamp out" any specific sequences because of the exponential nature of polymerase amplification.

The earliest steps of a touchdown polymerase chain reaction cycle have high annealing temperatures. The annealing temperature is decreased in increments for every subsequent set of cycles (the number of individual cycles and increments of temperature decrease is chosen by the experimenter). The primer will anneal at the highest temperature which is least-permissive of nonspecific binding that it is able to tolerate. Thus, the first sequence amplified is the one between the regions of greatest primer specificity; it is most likely that this is the sequence of interest. These fragments will be further amplified during subsequent rounds at lower temperatures, and will out compete the nonspecific sequences to which the primers may bind at those lower temperatures. If the primer initially (during the higher-temperature phases) binds to the sequence of interest, subsequent rounds of polymerase chain reaction can be performed upon the product to further amplify those fragments.

Universal Fast Walking

This method allows genome walking and genetic fingerprinting using a more specific 'two-sided' PCR than conventional 'one-sided' approaches (using only one gene-specific primer and one general primer - which can lead to artefactual 'noise') by virtue of a mechanism involving lariat structure formation. Streamlined derivatives of UFW are LaNe RAGE (lariat-dependent nested PCR for rapid amplification of genomic DNA ends), 5'RACE LaNe and 3'RACE LaNe.

15

Random Amplified Polymorphic DNA (RAPD)

RAPD stands for Random Amplification of Polymorphic DNA. It is a type of PCR reaction, but the segments of DNA that are amplified are random. The scientist performing RAPD creates several arbitrary, short primers (8-12 nucleotides), then proceeds with the PCR using a large template of genomic DNA, hoping that fragments will amplify. By resolving the resulting patterns, a semi-unique profile can be gleaned from a RAPD reaction.

No knowledge of the DNA sequence for the targeted gene is required, as the primers will bind somewhere in the sequence, but it is not certain exactly where. This makes the method popular for comparing the DNA of biological systems that have not had the attention of the scientific community, or in a system in which relatively few DNA sequences are compared (it is not suitable for forming a DNA databank). Because it relies on a large, intact DNA template sequence, it has some limitations in the use of degraded DNA samples. Its resolving power is much lower than targeted, species specific DNA comparison methods, such as short tandem repeats. In recent years, RAPD is used to characterize, and trace, the phylogeny of diverse plant and animal species.

Introduction

Random Amplified Polymorphic DNA (RAPD) markers are decamer (10 nucleotide length) DNA fragments from PCR amplification of random segments of genomic DNA with single primer of arbitrary nucleotide sequence and which are able to differentiate between genetically distinct individuals, although not necessarily in a reproducible way.

How It Works

Unlike traditional PCR analysis, RAPD (pronounced "rapid") does not require any specific knowledge of the DNA sequence of the target organism: the identical 10-mer primers will or will not amplify a segment of DNA, depending on positions that are complementary to the primers' sequence. For example, no fragment is produced if primers annealed too far apart or 3' ends of the primers are not facing each other. Therefore, if a mutation has occurred in the template DNA at the site that was previously complementary to the primer, a PCR product will not be produced, resulting in a different pattern of amplified DNA segments on the gel.

Advantages

- Low quantities of template DNA required (5-50 ng per reaction)
- No sequence data for primer construction required
- Easy and quick to assay
- Low costs involved
- Very high genomic abundance
- Random distribution throughout the genome
- Generation of multiple bands per reaction
- Amenable to automation

Disadvantages

- Purified, high molecular weight DNA required
- Precautions are needed to avoid contamination of DNA because short, random primers are used that are able to amplify DNA fragments in a variety of organisms
- Highly standardized experimental procedures are needed because of sensitivity to reaction conditions

- Band profiles can not be interpreted in terms of loci and alleles
- Dominance of alleles
- Low reproducibility
- Similar sized fragments may not be homologous

Limitations of RAPD

- Nearly all RAPD markers are dominant, *i.e.* it is not possible to distinguish whether a DNA segment is amplified from a locus that is heterozygous (1 copy) or homozygous (2 copies). Co-dominant RAPD markers, observed as different-sized DNA segments amplified from the same locus, are detected only rarely.
- PCR is an enzymatic reaction, therefore the quality and concentration of template DNA, concentrations of PCR components, and the PCR cycling conditions may greatly influence the outcome. Thus, the RAPD technique is notoriously laboratory dependent and needs carefully developed laboratory protocols to be reproducible.
- Mismatches between the primer and the template may result in the total absence of PCR product as well as in a merely decreased amount of the product. Thus, the RAPD results can be difficult to interpret.

Developing Locus-specific, Co-Dominant Markers from RAPDs

- The polymorphic RAPD marker band is isolated from the gel.
- It is amplified in the PCR reaction.
- The PCR product is cloned and sequenced.
- New longer and specific primers are designed for the DNA sequence, which is called the Sequenced Characterized Amplified Region Marker (SCAR).

Materials

1. *Equipment:* Thermocycler, Power supply Unit
2. *Water:* Sterile de-ionized or distilled water should be used for preparing all the reagents and pre-mixes
3. *Reaction buffers:* Assay buffer for Taq DNA polymerase (supplied by the manufacturer of Taq DNA polymerase)
4. *Deoxynucleoside triphosphates (dNTP'S):* 2.5 mM each of dCTP, dATP, dTTP, dGTP. Readymade solutions of dNTPS are available from many manufactures. Store at –20°C

5. *Magnesium chloride:* 25mM stock and store at –20°C
6. Taq DNA polymerase Genomic DNA 5-25 ng/ml stocks. DNA of sufficient quality can be obtained in coconut by using SDS protocol.

Methods

Assemble RAPD reactions as follows:

2.5μl DNA stock (25ng/μl)

2.5μl Assay buffer containing 2.5mM $MgCl_2$ (2.5mM)

1μl $MgCl_2$ stock (1.5mM) 1μl primer stock (25pmol)

4μl dNTPS (400μM)

1μl Taq polymerase (1U)

Sterile water to make 25 μl

- Wear gloves throughout RAPD reaction preparation procedure. Assay buffer, dNTPs, $MgCl_2$ and primer solution are thawed from frozen stock. Keep the assembled reaction in themocycler for amplification.
- Amplify DNA in themocycler: Cycling conditions may be modified depending on the thermocycer used.

Temperature Profile

General cycling steps followed are:

Step 1: Initial denaturation at 94°C for 5.00 min

Step 2: Denaturation at 94°C for 1.00 min

Step 3: Primer annealing at 55°C for 1.00 min

Step 4: Primer extension at 72°C for 2.00 min

Step 5: Go to 2, 39 times

Step 6: Final extension at 72°C for 10min

Step 7: 4°C for ever

After the reaction, DNA is analyzed through gel electrophoresis.

Agarose Gel Electrophoresis

Reagents for Agarose Gel Electrophoresis

1. Agarose, TBE/TAE buffer, ethidium bromide, gel loading dye,
2. To prepare 100ml of a 0.7% agarose solution, measure 0.7g agarose into a glass beaker or flask and add 100ml 1 X TBE or TAE.
3. Microwave or stir on a hot plate until agarose is dissolved and solution is clear.

4. Allow solution to cool to about 55°C before pouring. (ethidium bromide can be added at this point to concentration of 0.5μg/ml
5. Place the comb in gel tray.
6. Pour 50°C gel solution into tray to a depth of about 5mm. Allow the gel to solidify for about 20 min at room temperature.
7. To run, gently remove the comb, place the tray in electrophoresis chamber, and cover (just until wells are submerged) with electrophoresis buffer (the same buffer used to prepare the agarose).
8. To the RAPD sample from refrigerator, add 1μl of 6%gel loading dye for every 5μl of DNA solution. Mix well. Load 20μl of DNA per well. Load also the DNA size standards along side RAPD reactions.
9. Connect the electrodes to the power pack. And electrophoresis at 50-150Volts until the bromophenol blue dye has reached three fourth of the gel length.
10. Stain the gel with ethidium bromide (if not already included in the gel).
11. Examine the gel under UV light (transilluminator).
12. Depending on the objective of the experiment make a note of polymorphism, segregating bands, and appearance of overall pattern with in fingerprint.
13. Bands may be sized by comparison to molecular weight standards. The standards should be used to generate a standard curve for interpolation.
14. After you have run the gel, obtain a photograph, and label and measure the migration of the DNA bands. Make a standard curve plot of the known size markers, and determine the size of the marker bands.

Gel Interpretation

- Bands are sized and matched directly on gels, or photographic films, or photocopies on transparency overlays.
- Note the presence and absence of bands.
- Analyze the data using computer software NTSYS / RAPD Distance.

Applications of RAPD

RAPD protocol can be used in genetic mapping and finger printing application in plant as well as in micro organisms.

Example

Here is an example of RAPD, which was carried out by us in wheat.

Genomic DNA isolation

Genomic DNA was isolated using CTAB method of Saghai-Maroof *et. al.* (1984) from young leaf samples of ten wheat genotypes. Leaf samples were taken from 2-3 week old seedlings and ground into fine powder using liquid nitrogen in a sterile pestle and mortar. Approximately 5 g of the ground leaf tissue powder was transferred into a 50 ml polypropylene tube. To this, 15 ml of preheated (65°C) CTAB buffer was added and the samples were thoroughly mixed by inverting the tubes several times. Incubation was done at 65°C for 90 minutes with regular gentle mixing of samples.

After incubation, samples were cooled to room temperature and 15 ml of Chloroform: Isoamyl alcohol (24:1) mixture was added. Contents were mixed well by inverting the tubes gently for 10-15 min. Tubes were then centrifuged at 10,000 rpm for 10 min. and the upper aqueous layer was transferred into new sterilized 50 ml tubes using a micropipette.

Equal volume of ice-cold isopropanol was subsequently added to precipitate the DNA and the tubes were kept undisturbed for 15 min. The DNA was then spooled out and put into 1.5 ml centrifuge tubes (Eppendorf-tubes). Centrifugation was done at 12,000 rpm for 5 min to pellet down the DNA. The supernatant was discarded and the DNA pellet was washed with 70% ethanol for 1-2 min. Ethanol was then poured off and the DNA pellet was dried overnight at room temperature.

DNA was subsequently dissolved in appropriate volume of T.E. buffer and samples were stored at -20°C till further use.

Reagents

Stock Solution Required for DNA Isolation

1M Tris	12.11 g Tris + dH_2O	
	pH 8.0 adjusted with HCl	
	Final made volume	100 ml
0.5 M EDTA	18.6 g EDTA + dH_2O	
	pH 8.0 adjusted with NaOH	
	Final volume made	100 ml
5 M NaCl	58.44 g + dH_2O	
	Final volume made	100 ml

100 ml of CTAB buffer	1M Tris	20 ml
	0.5 M EDTA	4.32 ml
	5M NaCl	28 ml
	CTAB	2 g
	Sterile d H_2O	45.8 g
	β-mercaptoethanol	1 ml

CTAB Extraction Buffer

Tris-HCl (pH 8.0)	1 M
EDTA (pH 8.0)	0.5 M
NaCl	5 M
β-mercaptoethanol	2%
CTAB	2%

T.E. Buffer

Tris (pH 8.0)	10 mM
EDTA (pH 8.0)	1 mM

Quality and Quantity of DNA

When DNA was completely dissolved in T.E., its quality and concentration were checked by running DNA samples on 0.8% agarose gels along with a standard marker. RNA, if any was also detected on the gel.

RNAase Treatment

To remove RNA contamination, DNA samples were treated with RNAase. 5μ of 10 mg/ml RNAase was added; samples were mixed gently and then incubated at 37°C for 30-40 min.

DNA was again extracted by adding equal volume of chloroform: isoamyl alcohol (21:1) mixture. Samples were mixed well and then centrifuged at 10,000 rpm for 10 min. Supernatant was transferred to new Eppendorf-tubes. 1/10 Vol. of 3M sodium acetate (pH 6.8) and 2 Vol. of chilled absolute alcohol was added, followed by centrifugation to pellet down the DNA. Pellet was then washed with 70% alcohol, air-dried and finally dissolved in T.E. buffer.

Polymerase Chain Reaction (PCR)

PCR amplifications were performed using Eppendorf Master Cycler Gradient. The PCR reaction was conducted in a reaction volume of 20μl containing 1 x PCR buffer, 200μM dNTPs, 0.25μM of primer, 2 mM $MgCl_2$, 1 unit *Taq* DNA polymerase and 50-60 ng template DNA. Thirty RAPD primers were used and PCR amplification was performed as follows:

1. Initial denaturation	94°C for 4 min.	40 cycles
2. Denaturation	94°C for 1 min.	
3. Annealing	40°C for 1 min.	
4. Extension	72°C for 2 min.	
5. Final extension	72°C for 2 min.	

Cycles were set by repeating steps II to IV- 40 times. Amplified products were stored at -20°C till further use.

Agarose Gel Electrophoresis

PCR amplified DNA fragments for RAPD markers were resolved by submerged horizontal agarose gel electrophoresis (1.5% w/v). Gel casting tray was washed, air-dried and its ends were sealed with tape. Agarose was melted in 1X TBE buffer and ethidium bromide (5 µg/ml) was added. Gel solution was then poured into gel casting tray with an appropriate comb inserted. Gel was allowed to set for 30 min. After agarose polymerization, sealing tapes were removed from both the ends and the tray was placed in electrophoresis chamber. 1X TBE buffer was added to submerge the gel and comb was removed gently. Samples were prepared by adding 6X loading dye and pulse centrifuged for proper mixing. DNA samples were then loaded in the wells and electrophoresis was carried out at constant voltage (3V/cm of gel) till the dye migrated to the other end of the gel. PCR amplified products were visualized under U.V. light and photograph was taken.

Solution

10X TBE Buffer

Tris base	108 g
Boric acid	55 g
EDTA	4.6 g
Final vol	1000 ml

6X Loading Dye

Sucrose	4 g
BPB	0.025 g
Xylene cyanol	0.025 g
Final volume	10 ml

Allele Scoring / Gel Scoring

Bands for RAPD analysis were scored based on the presence (taken as 1) or absence (taken as 0) of bands for each wheat genotype. For each RAPD primer used. Only clear and unambiguous bands were scored. The size (in nucleotide base pairs) of the amplified bands was observed based on its migration and conformation relative to molecular size marker.

Data Analysis

Polymorphism was calculated based on the presence or absence of bands. The 0/1 matrix was used to calculate the similarity index genetic distance using 'simqual' sub-program of software NTSYS-PC (numerical taxonomy and multivariate analysis system program) (Rohlf, 1990). Dendrogram was constructed by using distance matrix by the unweighted pair-group method with arithmetic average (UPGMA) sub-program of NTSYS-PC.

16

Restriction Fragment Length Polymorphism (RFLP)

In 1985, Sir Alec Jeffreys developed restriction fragment length polymorphism (RFLP), which quickly became the standard technique for DNA testing throughout the 1980s. RFLP provided the world with the first form of genetic testing based on DNA, the body's genetic material.

Definition : RFLP analysis is a difference in homologus DNA sequences that can be detected by the presence of fragments of different lengths after digestion of the samples in question with specific restriction endonucleases (RE).

A *restriction fragment length polymorphism*, or RFLP, (commonly pronounced "rif lip"), is a variation in the DNA sequence of a genome that can be detected by breaking the DNA into pieces with restriction enzymes and analyzing the size of the resulting fragments by gel electrophoresis. It is the sequence that makes DNA from different sources different, and RFLP analysis is a technique that can identify some differences in sequence (when they occur in a restriction site). Though DNA sequencing techniques can characterize DNA very thoroughly, RFLP analysis was developed first and was cheap enough to see wide application. Analysis of RFLP variation was an important tool in genome mapping, localization of genetic disease genes, determination of risk for a disease, genetic fingerprinting, and paternity testing.

Analysis Technique

The basic technique for detecting RFLPs involves fragmenting a sample of DNA by a restriction enzyme, which can recognize and cut DNA wherever a specific short sequence occurs, in a process known as a restriction digest. The resulting DNA fragments are then separated by length through a process known as agarose gel electrophoresis, and transferred to a membrane via the Southern blot procedure. Hybridization of the membrane to a labeled DNA probe then determines the length of the fragments which are complementary to the probe. A RFLP occurs when the length of a detected fragment varies between individuals. Each fragment length is considered an allele, and can be used in genetic analysis.

Types

Usually there are two types of RFLP analysis.

1. MLP (Multi locus polymorphism) and
2. SLP(Single locus polymorphism).

Usually, SLP method is most preffered than MLP method.SLP method is most sensitive,easy to interpret and can analyse mixed DNA samples. Moreover data can be generated even when the DNA is too degraded. *e.g.* DNA can be analysed from bone remains.

Applications

Analysis of RFLP variation in genomes was vital tool in genome mapping and genetic disease analysis. If researchers were trying to initially determine the chromosomal location of a particular disease gene, they would analyze the DNA of members of a family afflicted by the disease, and look for RFLP alleles that show a similar pattern of inheritance as that of the disease (see Genetic linkage). Once a disease gene was localized, RFLP analysis of other families could reveal who was at risk for the disease, or who was likely to be carriers of the mutant gene.

RFLP analysis was also the basis for early methods of Genetic fingerprinting, useful in the identification of samples retrieved from crime scenes, in the determination of paternity, and in the characterization of genetic diversity or breeding patterns in animal populations.

Alternatives

The technique for RFLP analysis is, however, slow and cumbersome. It requires a large amount of sample DNA, and the combined process of probe labeling, DNA fragmentation, electrophoresis, blotting, hybridization, washing, and autoradiography could take up to a month to complete. A limited version of the RFLP method that used oligonucleotide probes was reported in 1985[1]. Fortunately, the results of the Human Genome Project have largely replaced the need for RFLP mapping, and the identification of many single-nucleotide polymorphisms (SNPs) in that project (as well as the direct identification of many disease genes and mutations) has replaced the need for RFLP disease linkage analysis (see SNP genotyping). The analysis of VNTR alleles continues, but is now usually performed by polymerase chain reaction (PCR) methods. For example, the standard protocols for DNA fingerprinting involve PCR analysis of panels of more than a dozen VNTRs.

RFLP is still a technique used in marker assisted selection. Terminal restriction fragment length polymorphism (TRFLP or sometimes T-RFLP) is a molecular biology technique initially developed for characterizing bacterial communities in mixed-species samples. The technique has also been applied to other groups including soil fungi.

TRFLP works by PCR amplification of DNA using primer pairs that have been labeled with fluorescent tags. The PCR products are then digested using RFLP enzymes and the resulting patterns visualized using a DNA sequencer. The results are analyzed either by simply counting and comparing bands or peaks in the TRFLP profile, or by matching bands from one or more TRFLP runs to a database of known species. The technique is similar in some aspects to DGGE or TGGE.

The sequence changes directly involved with an RFLP can also be analyzed more quickly by PCR. Amplification can be directed across the altered restriction site, and the products digested with the restriction enzyme. This method has been called Cleaved Amplified Polymorphic Sequence (CAPS). Alternatively, the amplified segment can by analyzed by Allele specific oligonucleotide (ASO) probes, a process that can often be done by a simple Dot blot.

Analytical Procedures

- Extraction of DNA and purification
- Digestion of DNA by endonuclease restriction enzyme (DNA fragmentation)

- Separation of the fragments by agarose gel-electrophoresis
- Transference of the fragments to a nylon filter by Southern blotting
- Hybridization of fragments to a locus specific radiolabeled DNA probe
- Visualization of the polymorphisms by autoradiography

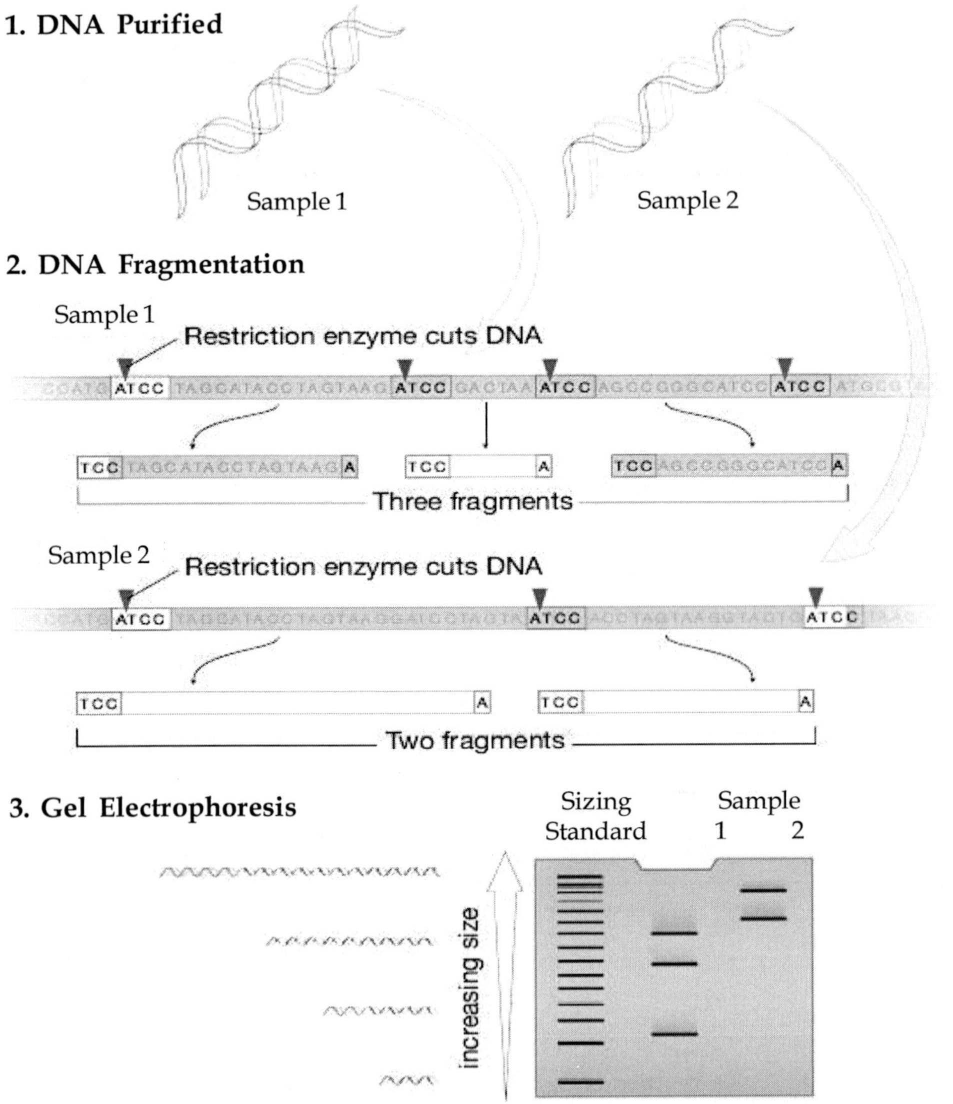

RFLP Flow Chart

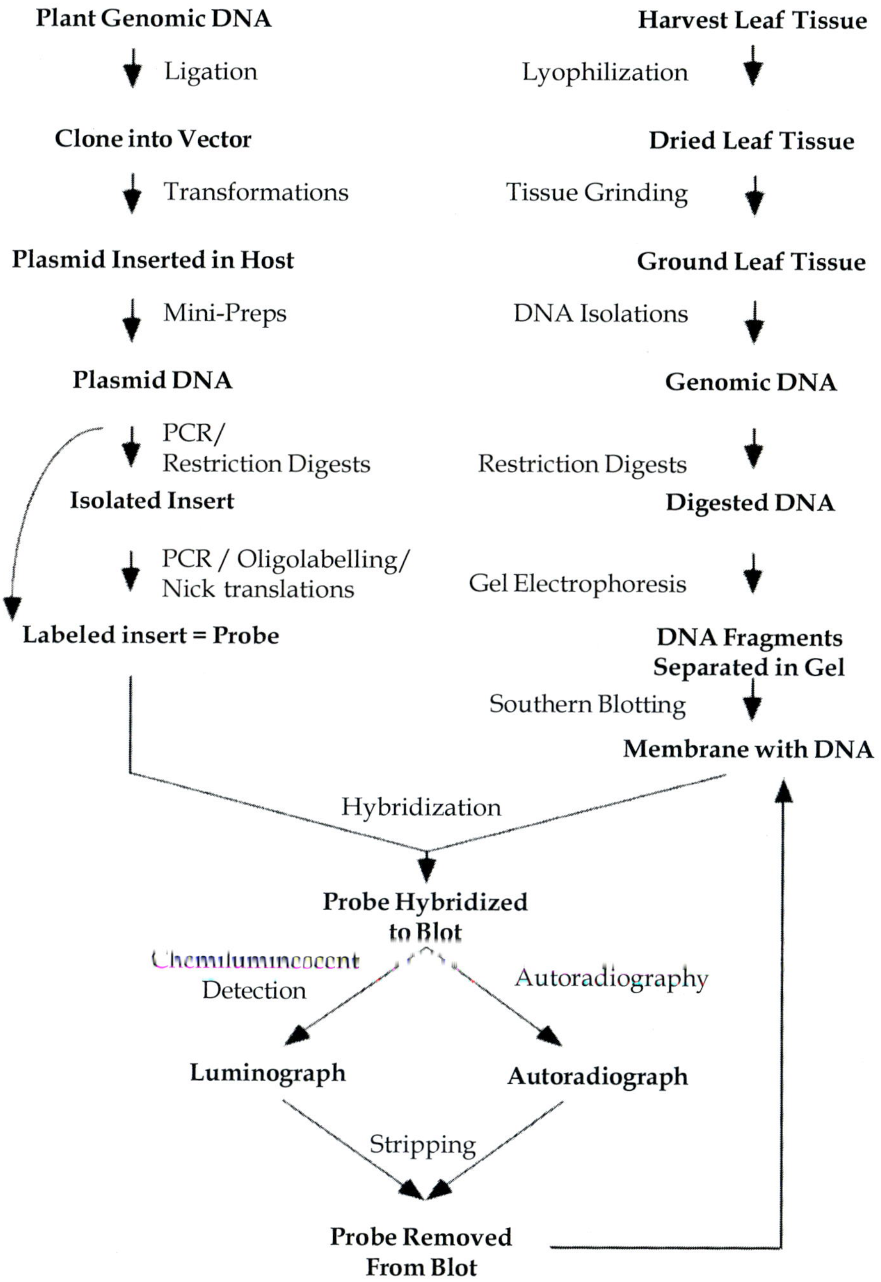

Main Requirements

- Laboratory setup for gel-electrophoresis
- Technical equipment to carry out Southern blot hybridizations
- Laboratory facilities for radioactive labeling

PROTOCOL

Digestion, Run and Gel Blotting

- Digest 3-5ug genomic DNA with 10Units of enzyme per ug of DNA. (Try to do reactions in a final volume of 15ul)
- Incubate samples at the suitable temperature for minimum 5 hours.
- At least 30′ before use, prepare a 0.6-0.8% agarose gel in 1x TBE, without EtBr. Use wide but thin combs. Don't load the 2 most external lanes.
- Add blue juice to each sample (3ul) and load as much as you can(18ul).
- You can use 5ul of HindIII digested Lambda DNA (250ng) as size marker.
- Run the gel overnight at 70Volts (about 16 hours) in 1x TBE running buffer. No cooling is needed.
- Stain the gel in EtBr for 20′, wash with tap water and take a picture putting a ruler beside the gel in order to estimate the distance run by the size marker.
- Put the gel in 0.25M HCl for 20′.
- Meanwhile set up the vacuum blot apparatus. Wash the porous support and wet the sealing rubber. Connect the vacuum pump to the vacuum trap and then to the apparatus.
- Wash the gel with millipore water.
- Put the nylon membrane onto the porous support, then the plastic mask (the mask must be slightly smaller then the gel), being very careful slide the gel onto the mask.
- Close the apparatus and start the pump (you should be able to see if vacuum is formed, if not look for the problem). Close the screws tighty and pour 1 litre of 0.4N NaOH. Apply a vacuum pressure of -50mm/ H_2O.
- Leave for 1 hour, checking from time to time that there is no leakage. If there is a small leakage add more NaOH to maintain the gel always submersed.

- Suck out the remaining NaOH and mark with a pencil the slots onto the nylon membrane.
- Remove the gel, and wash the membrane with 2X SSC to clean from agarose. Air dry the filter. Note with a pencil the date of the blot on the filter in an area covered by the mask. Cut to reduce the size of the membrane.

Labelling of the Probe

- Boil for 5′, 25 to 50μg of the DNA fragment to be labelled.
- Place on ice for 1′
- Set up the following labelling reaction. (**Pharmacia oligolabelling kit**)
- 25 to 50ng denatured DNA (max 36ml).
- 10ul provided reagent mix.
- 3ul [alpha-32P]dCTP
- 1ul Klenow enzyme
- Distilled water to 50ul
- Incubate at 37°C for 60′ at least.
- If the probe is not needed immediately it could be stored in the freezer for few days.
- Before use boil the probe for 5′. (Normally it is not necessary to get rid of the nucleotides.)

Pre-hybridisation and Hybridisation

- If the membrane to be used is employed for the first time then an overnight pre-hybridisation is needed, otherwise 4-5 hours should be enough.
- Wet the membrane with either distilled water or 2x SSC. Drain off the excess of water or SSC and roll the membrane into the hybridisation tube. Add 20ml of hyb solution and 600ug of sonicated and denatured (boil for 5′) salmon sperm DNA (usually stock is 10ug/ul)
- Incubate in the rotating hybridisation oven at 65°C.
- After pre-hybridization, start hybridisation by adding the boiled labelled probe (Not Directly on the membrane) to the pre-hybridisation.
- Incubate overnight.

Hyb Solution

1% SDS

1M NaCl

10% Dextrane Sulphate

Make 1 litre with 50mM Tris-Cl pH 7.5

Resuspend in waterbath at 65°C (You can store the solution in the freezer)

Washing of the Membranes

- Add just a little (50-100ml) of 500ml of 2 X SSC to the hyb solution.
- Mix for a couple of minutes then pour the solution into the radioactive waste tank.
- Put the hybridised membrane in a large box and pour the rest of the 500ml of 2 X SSC.
- Shake at room temperature for 10 min (Be careful radioactivity is still high).
- Pour off the solution and replace with 500ml of pre-heated (65°C) 1X SSC + 0.1% SDS. Incubate for 20′ at 65°C.
- Substitute the solution with pre-heated (65°C) 0.2X SSC + 0.1% SDS. Incubate while shaking for 20′ at 65°C.
- Now you can measure the radioactivity level of your filter using the hand counter. The filter should not count more than 5 cpm. If it counts too much, wash at 65°C with 0.1 X SSC +0.1% SDS for 20′.
- Wrap the membrane with Saran Wrap and expose.Leave at -80°C for 1 to 3 days.

Removal of the Probe

- Put the membrane into boiling water +0.1% SDS, let slowly cool down at room temperature.
- Put the membrane into a plastic bag or wrap it with Saran Wrap and store in the fridge. (Don't Let the Membrane Dry Out).

Advantages

- High genomic abundance
- Random distribution throughout the genome.

Disadvantages

- Large quantities of purified, high molecular weight DNA required (5-10μg)
- Laborious and technically demanding
- High costs involved
- Additional development costs in case suitable probes are unavailable
- Not amenable to automation.

17

Amplified Fragment Length Polymorphism (AFLP)

Introduction

The AFLP amplified fragment polymorphism also known as selective restriction fragment amplification (SRFA) technique is used to visualize hundreds of amplified DNA restriction fragments simultaneously. The AFLP band patterns, or fingerprints, can be used for many purposes, such as monitoring the identity of an isolate or the degree of similarity among isolates. Polymorphisms in band patterns map to specific loci, allowing the individuals to be genotyped or differentiated based on the alleles they carry. produces highly complex DNA profiles by arbitrary amplification of restriction fragments ligated to double-stranded adaptors with hemi-specific primers harboring adaptor-complementary 5' termini.The technique has been widely used in the construction of genetic maps containing high densities of DNA marker loci.

The AFLP protocol amplifies restriction fragments obtained by endonuclease digestion of target DNA using "universal" AFLP primers complementary to the restriction site and adapter sequence. However, not all restriction fragments are amplified because AFLP primers also contain selective nucleotides at the 3' termini that extend into the amplified restriction fragments.

These arbitrary terminal sequences result in the amplification of only a small subset of possible restriction fragments. The number of amplified fragments can therefore be "tailored" by extending the number of arbitrary nucleotides added to the primer termini. Alternatively, the use of endonuclease combinations that vary in their restriction frequency can also be used to tune the number of amplicons. Generally, the abundant restriction fragments produced from complex genomes require of AFLP primers with longer selective regions. Conversely, analysis of small genomes require of only few arbitrary nucleotides added at the primer 3' termini. The resulting AFLP fingerprints are usually a rich source of DNA polymorphisms that can be used in mapping and general fingerprinting endeavors.

The AFLP can be divided in the following steps:

- DNA is digested with two different restriction enzymes
- Oligonucleotide adapters are ligated to the ends of the DNA fragments
- Specific subsets of DNA digestion products are amplified, using combinations of selective primers
- Polymorphism detection is possible withradioisotopes, fluorescent dyes or silver staining

The DNA being examined is digested with two different restriction enzymes, one of which is a frequent cutter (the four-base restriction enzyme) and the other a rare cutter (the six-base restriction enzyme). Various enzyme/ primer combinations can be used. MseI and EcoRI are best used in AT-rich genomes as they give fewer fragments in GC-rich genomes. Specific synthetic adapters for each restriction site are then ligated to the digested DNA. Both the restriction and ligation steps can be performed in a single reaction.

Amplification of very small "genomes" requires of primers with no selective nucleotides. AFLP fingerprinting of plants requires primers with 2 selective bases. Complex genomes require the use of more than 2 selective bases in one or both primers. In the case of complex genomes it is suggested to carry the amplification in two consecutive steps to increase specificity and the amount of initial template. The AFLP fragments are usually detected by labeling one of the two AFLP primers. For example, radioactively labeled primers can be obtained by phosphorylating the 5' ends with g-33P-ATP and polynucleotide kinase. Do not label the two primers if the generation of doublets resulting from the different mobility of complementary strands in sequencing gels wants to be avoided. Finally, the labelled reaction products are separated by electrophoresis using denaturing polyacrylamide gels and exposed to X-ray films to visualize the AFLP fingerprints.

Materials

1. *Equipment:* Refrigerator and freezer, Centrifuge, Thermocycler, Power supply units, Hotplate or microwave, pH meter, Standard balance, Vertical gel electrophoresis units, UV transilluminator, Automatic sequencer Disposables: PCR tubes, X-ray film
2. *Genomic DNA:* For genomes smaller or larger than 500 Mb use 50 ng or 100 ng DNA for template preparations, respectively. Determine DNA concentrations by measuring OD260, and confirm the measurement and the integrity of DNA by electrophoresing the sample together with a series of phage l DNA dilutions ranging from 50 ng to 500 ng in agarose gels
3. AFLP primers (50 ng/μL). Primers are named "+0" when having no selective bases, "+1" when having a single selective base, "+2" for having two selective bases, and so on.

   *Eco*RI-primer+0: 5'-GACTGCGTACCAATTC-3'
   *Eco*RI-primer +1: 5'-GACTGCGTACCAATTCA-3'
   *Eco*RI-primers +2: 5'-GACTGCGTACCAATTCAN-3'
   *Eco*RI-primers +3: 5'-GACTGCGTACCAATTCANN-3'
   *Pst*I-primer +0: 5'-GACTGCGTACATGCAG-3'
   *Pst*I-primer +1: 5'-GACTGCGTACATGCAGA-3'
   *Pst*I-primers +3: 5'-GACTGCGTACATCGAGANN-3'
   *Mse*I-primer +1: 5'-GATGAGTCCTGAGTAAC-3'
   *Mse*I-primers +2: 5'-GATGAGTCCTGAGTAACN-3'
   *Mse*I-primers +3: 5'-GATGAGTCCTGAGTAACNN-3'
   *Taq*I-primer +0: 5'-GATGAGTCCTGAGCGAA-3'
   *Taq*I-primers +3: 5'-GATGAGTCCTGAGCGAANN-3'
4. Corresponding adaptors (5 or 50 pmol/μL):

   EcoRI-Adapter: 5' CTCGTAGACTGCGTACC
   CAT CTGACGCATGG-5'
   Pst- Adapter, *Mse*I- Adapter, Taq- Adapter,
5. Double-distilled water.
6. Buffers: 1 M Tris.HAc pH 7.5; 1 M Tris.HCl pH 8.0 and pH 8.3.
7. Magnesium: 0.1 mM MgCl2.
8. TE (10x): 100 mM Tris.HCl, 10 mM EDTA pH 8.0.

9. 100 mM DTT.
10. 5 mM of dNTPs
11. Gamma-32P-ATP (~3000 Ci/mmol) or gamma-33P-ATP (~2000 Ci/mmol).
12. 10 mM ATP.
13. T4-buffer (10x): 250 mM Tris.HCl pH 7.5, 100 mM MgCl2, 50 mM DTT, 5 Mm spermidine
14. Restriction-ligation buffer (5x): 50 mM Tris-HAc, 50 mM MgCl2, 250 mM KAc, 25 mM DTT, 250 ng/μL, pH 7.5.
15. Restriction endonucleases: *Eco*RI, *Pst*I, *Mse*I, *Taq*I Enzymes: T4 DNA ligase, T4 polynucleotide kinase, *Taq* DNA polymerase
16. PCR buffer (10x): 100 mM Tris.HCl pH 8.3, 15 mM MgCl2, 500 mM KCl.
17. Molecular weight standards
18. General reagents for polyacrylamide gel electrophoresis.

Methods

Depending on the size of the genome to be analyzed a different set of primers will have to be used. High complexity genomes require of a pre-amplification step. The following table shows primer combinations required for the amplification of different genomes. Note that numbers depict the number of selective nucleotides at the 3' terminus of the individual primers.

Template Preparation

1. *Adapters*: Prepare double-stranded (ds) adapters by mixing individual synthetic oligonucleotides. No denaturing-renaturation step is required. EcoRI, MseI, PstI, and TaqI adapters have double-stranded regions of 14, 12, 14 and 12 base pairs, respectively. Mix 1500 pmoles of each oligonucleotides to produce 5 pmol/μL solutions. Generally, use 8.5μg, 10.5μg, 8 μg or 8μg of the top strand oligonucleotide with 9.0μg, 7μg, 7μg or 7μg of the bottom strand oligonucleotidein 300μL of water for the EcoRI, PstI, MseI and TaqI adapter, respectively.

2. Digestion of DNA: Digest genomic DNA (50-100 ng) with restriction endonucleases in a 40μL reaction containing DNA, 8μL 5x RL buffer, 5 units EcoRI and 2 units MseI. Mix well and incubate for 2 h at 37°C. For genomes larger than 5000 Mb use 5 units *Pst*I and 2 units *Mse*I. For mammalian and vertebrate genomes use 5 units *Taq*I and 5 units *Eco*RI

in subsequent incubation steps at 65°C and 37°C. DNA preparations need to be of sufficient quality to allow complete digestion, since this step is crucial for the production of good quality AFLP fingerprints. Often, contaminating agents can interfere with digestion.

3. *Adaptor ligation:* Ligate adaptors to the digested genomic DNA by adding 1 μL *Eco*RI adapter (5 pmol), 1μL *Mse*I adapter (50 pmol), 1μL PstI adapter (5 pmol) or 1μL TaqI adapter (50 pmol), 1 μL 10 mM ATP, 2μL 5¥ RL buffer, 1 unit T4 DNA ligase and 5μL water to the digestion mix. Incubate another 2 h at 37°C. Overall, DNA is incubated for a total of 4 h with endonucleases, the last 2 h in the presence of T4 DNA ligase and oligonucleotide adapters. Avoid longer incubation periods because of possible "star" activity of *Eco*RI that gives reduced cleavage specificity and aberrant AFLP fingerprints. When using *Taq*I, use high concentration of adapters; if not *Taq*I will not efficiently re-digest aberrant fragment-ligation products at 37°C.

4. *Dilution*: Dilute the ligation reaction mixture 10 times with TE (usually 10μL in 100μL) and use the diluted reaction mixture directly as template DNA for the AFLP reactions. Store diluted DNA at -20°C.

AFLP Pre-Amplification

1. *Pre-amplification:* Check above for the right combination of primers for your individual application. If genomic complexity is sufficiently low, AFLP preamplification is not required(see Table above); therefore, skip this preamplification section. Assemble the preamplification reaction (50 μL total volume) with following components: 5μL ligated DNA, 1.5 μL EcoRI-primer +0 (75 ng), 1.5μL MseI-primer +C (75 ng), 1.5μL of the PstI one-base extension primer -A (75 ng), 1.5μL MseI-primer +C (75 ng) and/or 1.5μL TaqI-primer +A (75 ng), 2μL 5 mM dNTPs, 0.2μL Taq polymerase (1 unit), 5μL 10x PCR-buffer and 34.8μL water. Preamplify the mix for 20 cycles using the following regime: 30 s at 94°C; 60 s at 56°C; 60 s at 72°C. For mammalian genomes use 30 cycles. For genomes larger than 5000 Mb modify the amplification regime as follows: a first cycle of 30 s at 94°C, 60 s at 65°C, 60 s at 72°C, followed by 12 cycles with a stepwise decrease of the annealing temperature in each subsequent cycle by 0.7°C, and 23 cycles of 30 s at 94°C, 30 s at 56°C and 60 s at 72°C. In this step it is advisable to assemble one reaction mix containing primer and dNTPs and another containing the *Taq* polymerase and its buffer.

2. *Dilution:* Preamplification, 10μL of the reaction is diluted with 190μL of TE0.1 to 100μL which is sufficient for 40 AFLP-reactions +2/+3. The diluted reaction mix and the rest of the preamplification reaction is stored at -20°C. If necessary new dilutions of the preamplification reactions may be made to give additional template for the AFLP reactions.

AFLP Amplification

1. *Preparation of labelling mix:* Label primers for selective AFLP amplification by phosphorylating the 5' end of the primers with gamma-32P-ATP or gamma-33P-ATP and polynucleotide kinase. Check above for the right primer combination to use in this step. Only one of the two primers of the AFLP reaction should be labelled (*e.g.,* the EcoRI-primer). When possible use the more expensive 33P-labelled primers because they give better product resolution in polyacrylamide gels, and are less prone to degradation due to autoradiolysis. Prepare the following primer labelling mixes (40μL) for 100 AFLP reactions. either 20μL gamma-32P-ATP (~3,000 Ci/mmol) or 10μl gamma- 33P-ATP (~2,000 Ci/mmol), 5μL 10xT4-buffer, 2μL T4-kinase (10 units/μL) and water to 40μL.

2. *Primer labelling:* Add 10μl of primer (either EcoRI- or PstI-primers at 50 ng/μL) to 40μL labelling mix and incubate 60 min at 37°C, followed by incubation at 70°C for 10 min for the inactivation of the kinase. This gives a labelled primer with a concentration of 10 ng/μL.

3. *Prepare AFLP reaction mixes:* Prepare reaction mixes for a minimum of 10 reactions. Working with AFLP reaction mixes is important for the reliability and reproducibility of AFLP reactions and because it facilitates reaction assembly. *Primer* and *dNTPs* mix (50μL): 5 μL labelled primer (10 ng/μL), 6μL unlabelled primer (50 ng/μL), 8 μL 5 mM dNTPs and 31μL water. Taq polymerase mix (100μL): 20μL 10x PCR-buffer, 0.8μL Taq polymerase (4 units) and 79.2μL water.

4. *AFLP amplification:* Assemble the reaction by adding 5μL of the primers and dNTPs mix and 10μL of the Taq polymerase mix to 5μL of pre-amplified ligated DNA. The template DNA should be pipetted first followed by the two mixes. The reagents should be mixed by tapping the base of the tubes on the bench. Pipetting mixes is essential for the rapid start of the AFLP reactions that are assembled at room temperature (to avoid loss of AFLP fingerprint quality). Tubes are amplified in a thermocycler with the following cycle regime a first cycle of 30 s at 94°C, 30 s at 65°C and 60 s at 72°C, followed by 12 cycles with a stepwise decrease of the annealing temperature in each subsequent cycle by 0.7°C,

and 23 cycles of 30 s at 94°C, 30 s at 56°C and 60 s at 72°C. The reaction is started at a high annealing temperature to obtain optimal primer selectivity. In the following steps the annealing temperature is lowered gradually to a temperature for optimal primer annealing.

Polyacrylamide Gel Electrophoresis of AFLP Products

1. *General:* Amplification products are analyzed on 4.5% denaturing polyacrylamide sequencing gels (see the Separation section for more details). Treat back glass plate of the gels with 2 mL of repel silane, and the front plate with 10 mL of bind silane solution (30μL acetic acid and 30μL bind silane in 10 mL ethanol, freshly made immediately before use). The gels should be cast at least 2 h before use and should be prerun for 0.5 h just before loading the samples Perform pre-running and running electrophoretic steps at 110 W. Use TBE (1x) as running buffer.
2. *Sample loading:* Mix AFLP reaction products with an equal volume (20μL) of loading dye. Heat the samples for 3 min at 90°C, and then quickly cool on ice. Rinse the the gel wells with running buffer and push carefully two 24-well sharktooth combs about 0.5 mm into the gel surface to create the gel slots. Rinse the gel slots formed in this way with TBE and load 2μL of each sample per well.
3. *Post-electrophoretic procedures:* Disassemble the gel cassette and remove the front glass plate with the silane-attached gel to the front. Fix the gel by soaking in 10% acetic acid for 30 min and dry it subsequently at room temperature in a fume hood for 10-20 h. Autoradiographic exposure of the 32P-gels to standard X-ray film overnight, without intensifying screens. Exposure of the 33P-gels takes 2 to 3 d in order to generate similar band intensities.

Aplication

- Genetic diversity assessment
- Genetic distance analysis
- Genetic fingerprinting
- Analysis of germplasm collections
- Genome mapping
- Monitoring diagnostic markers

Advantages

- No sequence data for primer construction required
- High genomic abundance

- Random distribution throughout the genome, although clustering around centromers has been reported
- Generation of many informative bands per reaction
- High reproducibility
- Wide range of applications
- Amenable to automation

Disadvantages

- Purified, high molecular weight DNA required
- Band profiles can not be interpreted in terms of loci and alleles
- Dominance of alleles
- Similar sized fragments may not be homologous

18

Simple Sequence Repeats (Microsatellite)

Also Referred to as Microsatellites

SSR markers are based short sequences of nucleotides (2-6 units in length) that are repeated in tandem:

Di-nucleotide repeat: CACACACA

Tri-nucleotide repeat: ATGATGATGATG

It is believed that when DNA is being replicated, errors occur in the process and extra sets of these repeated sequences are added to the strand. Over time, these repeated sequences vary in length between one cultivar and another.

Example of Allelic Variation in SSRs

Allele A: CACACACA (4 repeats of the CA sequence)
Allele B: CACACACACACA (6 repeats of the CA sequence)

These variations in length are easy to trace in the lab and allow us to track genotypic variation in breeding programs.

A short tandem repeat (STR) in DNA occurs when a pattern of two or more nucleotides are repeated and the repeated sequences are directly adjacent to each other. The pattern can range in length from 2 to 10 base pairs (bp) (for example (CATG)n in a genomic region) and is typically in the non-coding intron region, making it junk DNA. By identifying repeats of a specific sequences, it is possible to create a genetic profile of an individual. There are currently over 10,000 published STR sequences in the human genome. STR analysis has become the prevalent analysis method for determining genetic profiles in forensic cases.

The DNA sequences flanking SSRs are known to be conserved and that have been exploited to design suitable primers for amplification of the SSR loci using PCR. Differences in the length of the PCR amplified product from different genotypes reveal the SSR polymorphism. The length differences are attributed to the variation in the number of repeat units at a particular SSR locus possibly caused by slippage during DNA replication.

Microsatellites and their flanking sequences can be detected by constructing a small-insert genomic library, screening of the library with a synthetic labeled oligonucleotide repeat and sequencing of the positive clones. Alternatively, primers designed for closely related species may be used. Polymerase slippage during DNA replication (or slipped strand mispairing) is considered to be the main cause of variation in the number of repeat units, resulting in length polymorphisms that can be detected by gel-electrophoresis. Due to their high level of polymorphism, microsatellites are informative markers that can be used for many population genetic purposes, ranging from the individual level (e.g. clone and strain identification) to closely related species. In addition, microsatellites are considered ideal markers in gene mapping studies.

SSR Loci in DNA Fingerprinting

SSR markers can be employed in the development of unique allelic profiles for establishing individual identity. Distinctive profiles can be readily generated by defining the allelic constitution of individuals at relatively few loci, each of which is multi-allelic. Such a system is open-ended in that additional loci can be added if those already in use are inadequate to produce a unique profile for all individuals. With the advent of the Plant Variety Protection Act such a definitive cultivar identification system would be extremely useful.

Do SSRs Occur in Plants?

Can SSR's be used in plant genetic studies? Of particular interest in this regard is the occurence of SSR sequences in higher plants. Tautz et al. (1986) examined the European Molecular Biology Laboratory DNA sequence library for the presence of di-trinucleotide repeats. A comparison of very limited plant and algal sequence data with those of vertebrates indicated a similar frequency of the two types of SSR's in the two groups of organisms.

Sarkar et al. (1991) reported a search of GenBank sequences for purine/ pyrimidine repeats greater than 13 units in length. Calculations from their data indicate 2.3 such SSR loci per 100 kb in primates and 1.8 in yeast. While these reports are not clearly indicative of SSR's in higher plants, they do suggest the possibility of their occurrence. The only published report clearly documenting SSR's in higher plants is that of Condit and Hubbell (1991). They screened DNA libraries of five tropical tree species as well as *Zea mays* for the presence of clones containing (AC)n and (AG)n repeat sequences. They estimated a total of (AC)n + (AG)n SSR sequences ranging from 5 x 103 to 3 to 105 among six species examined.

It seemed appropriate to undertake a further investigation of SSR sequences in plant genomes. Therefore, a search of GenBank was completed with the assistance of Dr. Susan McCarthy, Coordinator for the National Agricultural Library Plant Genome Data and Information Center. Despite the limited number of plant sequences available in GenBank, a number of di- and tri-nucleotide SSR's were identified (Table below). The data supports the presence of SSR sequences in numerous plant species, including crop plants.

Do Plant SSR's Exhibit Length Polymorphism?

At this time no published information is available to answer this question. However, Weber (1990) suggested that human (CA)n sequences with n of 10 or less are unlikely to exhibit length polymorphism, whereas sequences with n greater than 15 are consistently polymorphic.

A more detailed look at the data obtained from the search of GenBank, reported in Table below, indicates at least one instance of a SSR with greater than 15 tandem repeats in each of the following higher plant species: *Arabidopsis thanliana, Daucus carota, Glycine max, Hordeum vulgare, Nicotiana tabacum, Pisum sativum, Oryza sativa, Solanum tuberosum*, and *Zea mays*. This finding suggest that plant SSR's have a high probability of exhibiting length polymorphism.

Applications

Marker assisted selection in plant breeding : The ability to distinguish between closely related individuals is important for crop species, which tend to have a narrow genetic base. To date, diversity detected with microsatellites reveal more polymorphism compared with other assay procedures

The number and average distance (in kilobase pairs) between all possible dimeric, trimeric, and tetrameric Simple Sequence Repeat DNA sequences in plant species determined from a search of GenBank.

Plant Species	Kilobases searched kbp	Dimeric repeats No.	Dimeric repeats Distance between repeats kbp	Trimeric repeats No.	Trimeric repeats Distance between repeats kbp	Tetrameric repeats No.	Tetrameric repeats Distance between repeats kbp
Saccharomyces cervisiae	2288	40	57	29	79	2	1144
Nicotiana tabacum	118	4	29	0	-	0	-
Glycine max	212	6	35	1	212	4	53
Lycopersicon esculentum	135	2	68	0	-	1	135
Triticum aestivum	151	1	151	43	4	0	-
Medicago sativa	30	0	-	1	30	0	-
Pisum sativum	129	3	43	3	43	2	64
Zea mays	683	3	123	4	91	6	61
Arabidopsis thaliana	247	4	62	1	247	0	-
Oryza sativa	137	2	68	2	68	6	23

Dimeric repeats such as (AT) with $n>9$.
Trimeric repeats such as (ATT) with $n>7$.
Tetrameric repeats such as (AGAT) with $n>4$.

Determination of hybridity : The codominant nature of the microsatellite primers allows the allelic contribution of each parent to be detected in sexual or somatic hybrids. Recently microsatellites have been used to characterize and determine the hybridity of intraspecific somatic hybrids of potato.

Genome mapping : Microsatellites can provide reference (anchor) points for specific regions of the genome. Hence, creation of microsatellite index is possible although a high-resolution linkage map based exclusively on SSR is not possible.

Gene tagging : One of the reasons for developing microsatellites is their potential use as diagnostic markers for important traits in plant breeding programs (Marker – assisted programmes). Several resistant genes for soybean mosaic virus and peanut mottle virus have been identified using microsatellite primers.

Cultivar identification, estimation of genetic relatedness and germplasm conservation : The unique mechanism responsible for generating high levels of microsatellite-allele variability-by "replication slippage" may reduce the usefulness of microsatellite for assessing genetic similarity between genotypes. However, this phenomenon is exploited to describe phenetic relationship (i.e. reflecting phenotypic character) between accessions that correspond to their geographical origins.

Analytical Procedures

- Extraction of DNA
- Amplification of DNA fragments by PCR
- Separation of the polymorphisms by polyacrylamide or agarose gel-electrophoresis
- Visualization of the polymorphisms by autoradiography, silver staining or fluorescence (polyacrylamide gels), or ethidium-bromide staining and ultraviolet light (agarose gels).

Main Requirements

- Thermocycler
- Laboratory setup for gel-electrophoresis
- (Laboratory facilities for radioactive labeling)
- (Ultraviolet light source)

Protocol

Microsatellite PCR reaction mix (50ml):

DNA (50ng/ml)	4 ml
Microsatellite primer - Forward (100ng/ml)	2 ml
Microsatellite Primer - Reverse (100ng/ml)	2 ml
1.0 mM dNTPs	4 ml
Taq DNA polymerase	0.6 ml
10xPCR buffer with $MgCl_2$	5.0 ml
Sterile double distilled water	32.4 ml

Microsatellite PCR reaction conditions:

Initial denaturation	94°C for 3 min	45 cycles
Denaturation	94°C for 1min	
Annealing	50/55/60°C for 1 min	
Extension	72°C for 2 min	
Final Extension	72°C for 10 min	
Storage	4°C till further use.	

Advantages

- Co-dominant (more informative when dealing with heterozygotes)
- Highly variable (important for species with narrow gene pools)
- Widely used
- Excellent for use in marker assisted selection, fingerprinting and marker assisted backcrossing
- Low quantities of template DNA required (10-100 ng per reaction)
- High genomic abundance
- Random distribution throughout the genome
- High level of polymorphism
- Band profiles can be interpreted in terms of loci and alleles
- Codominance of alleles
- Allele sizes can be determined with an accuracy of 1 bp, allowing accurate comparison across different gels
- High reproducibility

- Different microsatellites may be multiplexed in PCR or on gel
- Wide range of applications
- Amenable to automation

Disadvantages

- High development costs in case primers are not yet available
- Heterozygotes may be misclassified as homozygotes when null-alleles occur due to mutation in the primer annealing sites
- Stutter bands on gels may complicate accurate scoring of polymorphisms
- Underlying mutation model (infinite alleles model or stepwise mutation model) largely unknown
- Homoplasy due to different forward and backward mutations may underestimate genetic divergence.

Steps Involved in SSR

(http://www.dnalandmarks.com/english/marker_systems_ssr.html)

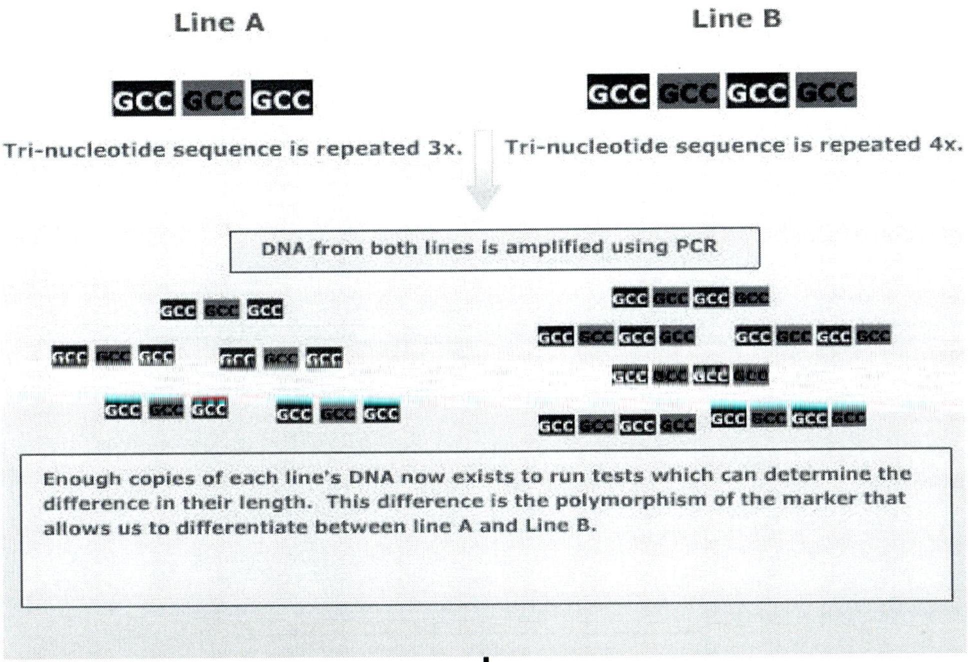

↓

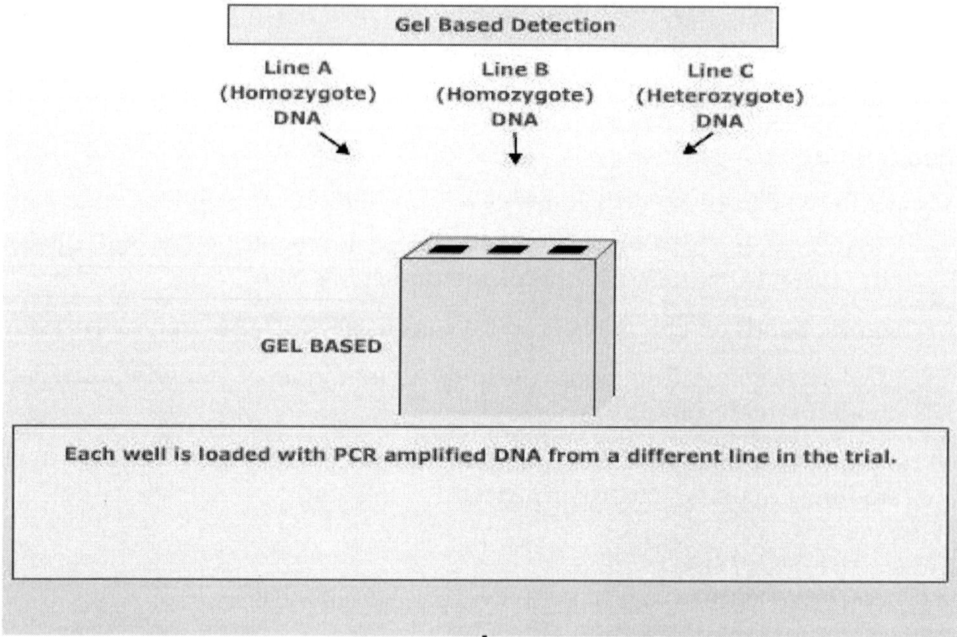
Gel Based Detection
Line A (Homozygote) DNA
Line B (Homozygote) DNA
Line C (Heterozygote) DNA
GEL BASED
Each well is loaded with PCR amplified DNA from a different line in the trial.

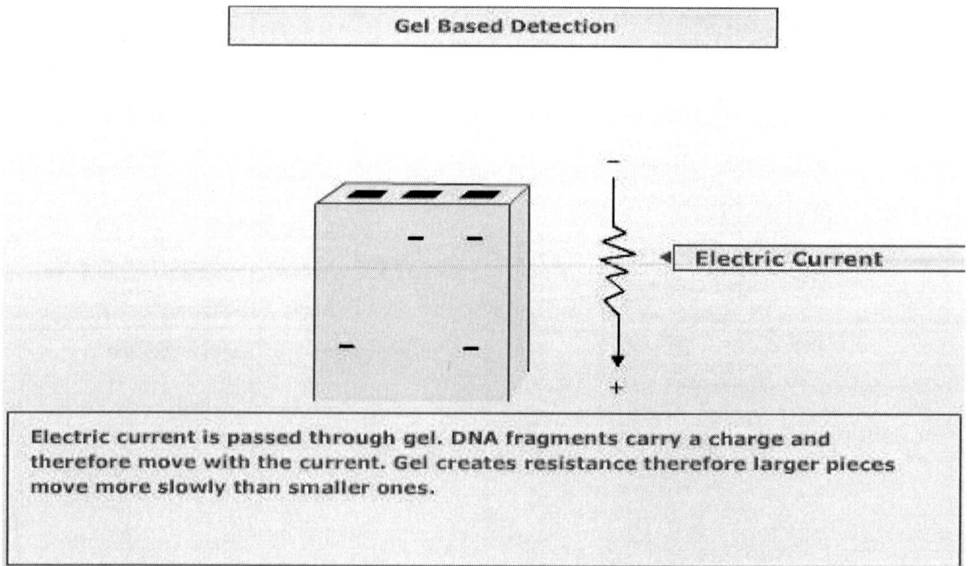
Gel Based Detection
Electric Current
Electric current is passed through gel. DNA fragments carry a charge and therefore move with the current. Gel creates resistance therefore larger pieces move more slowly than smaller ones.

GEL BASED

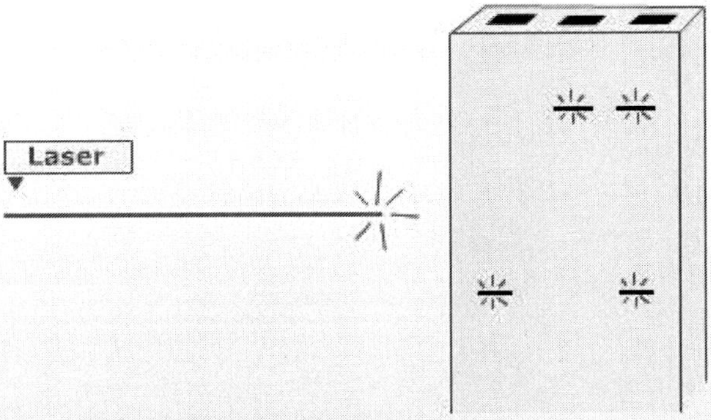

DNA fragments were labelled with a flourescent tag during the PCR. Now the gel is read with a laser which casuses the fragments to flouresce.

↓

RESULTS - GEL PHOTO

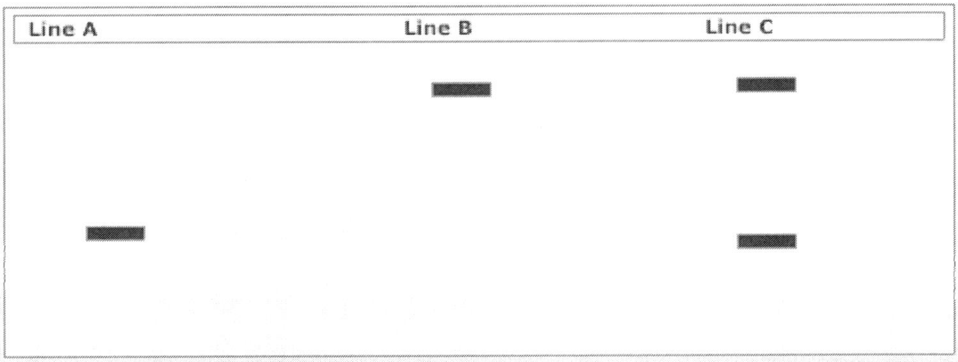

Line A is homozygos for the allele that contains 3 copies of the repeat sequence. Line B is homozygous for the allele that contains 4 copies of the repeat. Therefore, Line A's Shorter DNA fragments have travelled further along the gel than line B's. Line C is Heterozygous for the allele and therefore shows fragments at both Positions

↓

Microcapillary Detection

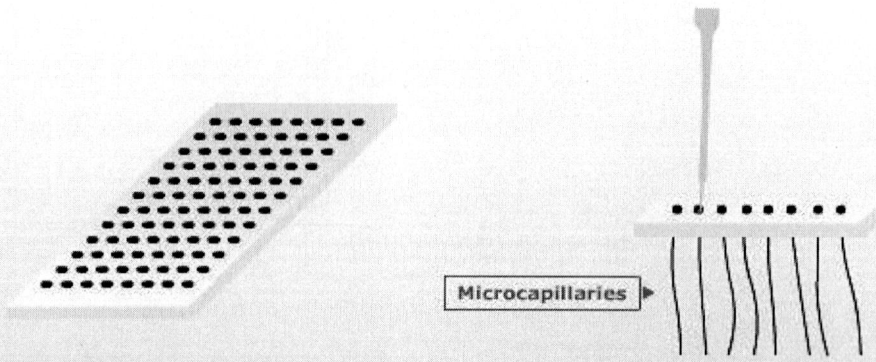

Each capillary is loaded automatically by an robotic arm, with PCR amplified DNA from a different line in the trial.

↓

Microcapillary Detection

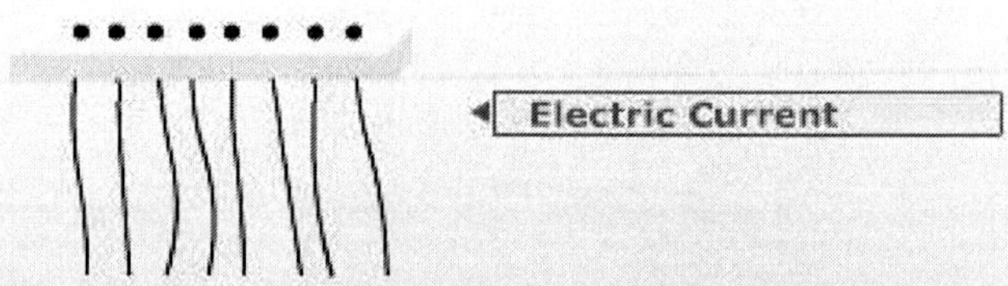

Electric current is passed through the microcapillaries. DNA carries a charge and therefore moves with the current. The solution in the capillaries provides resistance so smaller fragments travel faster than larger fragments.

↓

Microcapillary

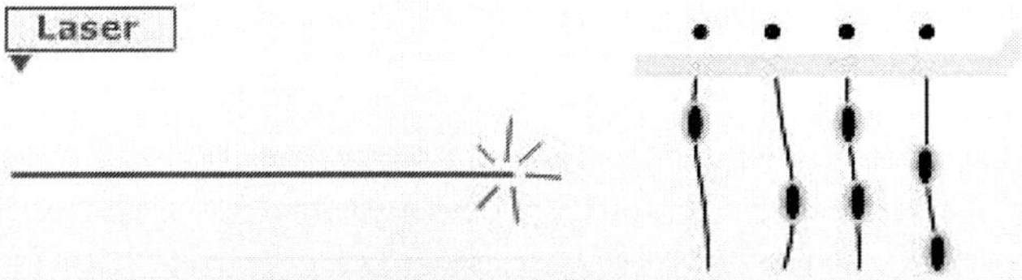

DNA fragments were labelled with a flourescent tag during the PCR. Now the microcapillaries are read with a laser which causes the fragments to flouresce.

↓

RESULTS - Microcapillary Detection

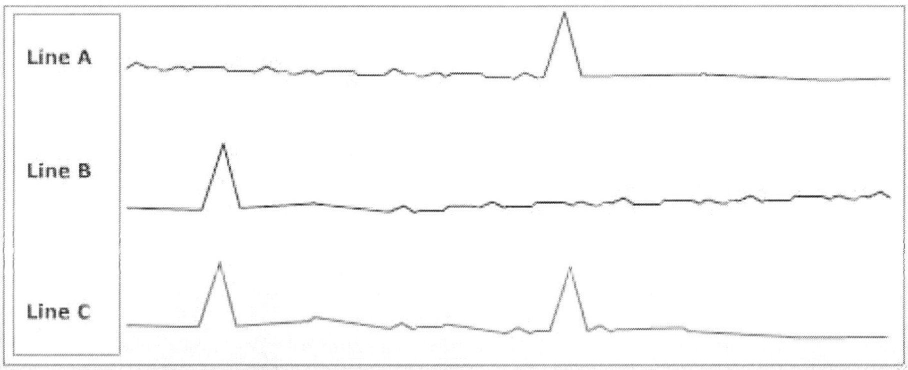

Line A is homozygos for the allele that contains 3 copies of the repeat sequence. Line B is homozygous for the allele that contains 4 copies of the repeat. Therefore, Line A's Shorter DNA fragments have travelled further along the microcapilary than line B's. Line C is Heterozygous for the allele and therefore shows peaks at both positions.

Microsatellite

One common example of a microsatellite is a $(CA)_n$ repeat, where n is variable between alleles. These markers often present high levels of inter- and intra-specific polymorphism, particularly when tandem repeats number ten or greater. The repeated sequence is often simple, consisting of two, three or four nucleotides (di-, tri-, and tetranucleotide repeats respectively), and can be repeated 10 to 100 times. CA nucleotide repeats are very frequent in human and other genomes, and are present every few thousand base pairs. As there are often many alleles present at a microsatellite locus, genotypes within pedigrees are often fully informative, in that the progenitor of a particular allele can often be identified. In this way, microsatellites are ideal for determining paternity, population genetic studies and recombination mapping. It is also the only molecular marker to provide clues about which alleles are more closely related.

Microsatellites owe their variability to an increased rate of mutation compared to other neutral regions of DNA. These high rates of mutation can be explained most frequently by slipped strand mispairing (slippage) during DNA replication on a single DNA strand. Mutation may also occur during recombination during meiosis. Some errors in slippage are rectified by proofreading mechanisms within the nucleus, but some mutations can escape repair. The size of the repeat unit, the number of repeats and the presence of variant repeats are all factors, as well as the frequency of transcription in the area of the DNA repeat. Interruption of microsatellites, perhaps due to mutation, can result in reduced polymorphism. However, this same mechanism can occasionally lead to incorrect amplification of microsatellites; if slippage occurs early on during PCR, microsatellites of incorrect lengths can be amplified.

Amplification of Microsatellites

Microsatellites can be amplified for identification by the polymerase chain reaction (PCR) process, using the unique sequences of flanking regions as primers. DNA is repeatedly denatured at a high temperature to separate the double strand, then cooled to allow annealing of primers and the extension of nucleotide sequences through the microsatellite. This process results in production of enough DNA to be visible on agarose or polyacrylamide gels; only small amounts of DNA are needed for amplification as thermocycling in this manner creates an exponential increase in the replicated segment[5]. With the abundance of PCR technology, primers that flank microsatellite loci are simple and quick to use, but the development of correctly functioning primers is often a tedious and costly process.

Development of Microsatellite Primers

If searching for microsatellite markers in specific regions of a genome; for example within a particular exon of a gene, primers can be designed manually. This involves searching the genomic DNA sequence for microsatellite repeats, which can be done by eye or by using automated tools such as repeat masker. Once the potentially useful microsatellites are determined (removing non-useful ones such as those with random inserts within the repeat region), the flanking sequences can be used to design oligonucleotide primers which will amplify the specific microsatellite repeat in a PCR reaction.

Random microsatellite primers can be developed by cloning random segments of DNA from the focal species. These are inserted into a plasmid or bacteriophage vector, which is in turn implanted into Escherichia coli bacteria. Colonies are then developed, and screened with fluorescently-labelled oligonucleotide sequences that will hybridise to a microsatellite repeat, if present on the DNA segment. If positive clones can be obtained from this procedure, the DNA is sequenced and PCR primers are chosen from sequences flanking such regions to determine a specific locus. This process involves significant trial and error on the part of researchers, as microsatellite repeat sequences must be predicted and primers that are randomly isolated may not display significant polymorphism. Microsatellite loci are widely distributed throughout the genome and can be isolated from semi-degraded DNA of older specimens, as all that is needed is a suitable substrate for amplification through PCR.

ISSR-PCR

ISSR (for inter-simple sequence repeat) is a general term for a genome region between microsatellite loci. The complementary sequences to two neighboring microsatellites are used as PCR primers; the variable region between them gets amplified. The limited length of amplification cycles during PCR prevents excessive replication of overly long contiguous DNA sequences, so the result will be a mix of a variety of amplified DNA strands which are generally short but vary much in length.

Sequences amplified by ISSR-PCR can be used for DNA fingerprinting. Since an ISSR may be a conserved or nonconserved region, this technique is not useful for distinguishing individuals, but rather for phylogeography analyses or maybe delimiting species; sequence diversity is lower than in SSR-PCR, but still higher than in actual gene sequences. In addition, microsatellite sequencing and ISSR sequencing are mutually assisting, as one produces primers for the other.

Limitations of Microsatellites

Microsatellites have proved to be versatile molecular markers, particularly for population analysis, but they are not without limitations. Microsatellites developed for particular species can often be applied to closely related species, but the percentage of loci that successfully amplify may decrease with increasing genetic distance. Point mutation in the primer annealing sites in such species may lead to the occurrence of 'null alleles', where microsatellites fail to amplify in PCR assays. Null alleles can be attributed to several phenomena. Sequence divergence in flanking regions can lead to poor primer annealing, especially at the 3′ section, where extension commences; preferential amplification of particular size alleles due to the competitive nature of PCR can lead to heterozygous individuals being scored for homozygosity (partial null). PCR failure may result when particular loci fail to amplify, whereas others amplify more efficiently and may appear homozygous on a gel assay, when they are in reality heterozygous in the genome. Null alleles complicate the interpretation of microsatellite allele frequencies and thus make estimates of relatedness faulty. Furthermore, stochastic effects of sampling that occurs during mating may change allele frequencies in a way that is very similar to the effect of null alleles; an excessive frequency of homozygotes causing deviations from Hardy-Weinberg equilibrium expectations. Since null alleles are a technical problem and sampling effects that occur during mating are a real biological property of a population, it is often very important to distinguish between them if excess homozygotes are observed.

When using microsatellites to compare species,homologous loci may be easily amplified in related species, but the number of loci that amplify successfully during PCR may decrease with increased genetic distance between the species in question. Mutation in microsatellite alleles is biased in the sense that larger alleles contain more bases, and are therefore likely to be mistranslated in DNA replication. Smaller alleles also tend to increase in size, whereas larger alleles tend to decrease in size, as they may be subject to an upper size limit; this constraint has been determined but possible values have not yet been specified. If there is a large size difference between individual alleles, then there may be increased instability during recombination at meiosis. In tumour cells, where controls on replication may be damaged, microsatellites may be gained or lost at an especially high frequency during each round of mitosis. Hence a tumour cell line might show a different genetic fingerprint from that of the host tissue.

19

Variable Number of Tandem Repeats (Minisatellites)

Minisatellite is a section of DNA that consists of a short series of bases 10–60bp. These occur at more than 1000 locations in the human genome. Some minisatellites contain a central (or "core") sequence of letters "GGGCAGGANG" (where N can be any base) or more generally a strand bias with purines (Adenosine (A) and Guanine (G)) on one strand and pyrimidines (Cytosine (C) and Thymine (T)) on the other. It has been proposed that this sequence per se encourages chromosomes to swap DNA. In alternative models, it is the presence of a neighbouring cis-acting meiotic double-strand break hotspot which is the primary cause of minisatellite repeat copy number variations. Somatic changes are suggested to result from replication difficulties (which might include replication slippage, among other phenomena). When such events occur, mistakes are made, this causes minisatellites at over 1000 locations in a persons genome to have slightly different numbers of repeats, thereby making them unique. The most highly mutable minisatellite locus described so far is CEB1 (D2S90) described by Vergnaud.

Structure

"Minisatellites" consist of repetitive, generally GC-rich, variant repeats that range in length from 10 to over 100 bp. These variant repeats are tandemly intermingled, which makes minisatellites ideal for studying DNA turnover mechanisms.

Application

Since the fortuitous discovery of the first human minisatellite in 1980 by A.R. Wyman and R. White and especially the discovery that the extreme polymorphism of minisatellites made them superb for DNA fingerprinting by Alec Jeffreys, this class of repeats has been an intense focus of studies that have addressed the turnover mechanisms that provoke their instability. Due to their high level of polymorphism, minisatellites have been extensively used for DNA fingerprinting as well as for genetic markers in linkage analysis and population studies.

Minisatellites have also been implicated as regulators of gene expression (e.g., at levels of transcription, alternative splicing, or imprint control) or as part of bona fide open reading frames.

Pathology

Minisatellites have been associated with chromosome fragile sites and are proximal to a number of recurrent translocation breakpoints.

Some human minisatellites (~1%) have been demonstrated to be hypermutable, with an average mutation rate in the germline higher that 0.5% up to over 20%, making them the most unstable region in the human genome known to date. It is interesting to note that while other genomes (mouse, rat and pig) contain minisatellite-like sequences, none was found to be hypermutable. Since all hypermutable minisatellites contain internal variants, they provide extremely informative systems for analyzing the complex turnover processes that occur at this class of tandem repeat. Minisatellite variant repeat mapping by PCR (MVR-PCR) has been extensively used to chart the interspersion patterns of variant repeats along the array, which provides details on the structure of the alleles before and after mutation.

Studies have revealed distinct mutation processes operating in somatic and germline cells. Somatic instability detected in blood DNA shows simple and rare intra-allelic events two to three orders of magnitude lower than in sperm. In contrast, complex inter-allelic conversion-like events occur in the germline.

Additional analyses of DNA sequences flanking human minisatellites have also revealed an intense and highly localized meiotic crossover hotspot that is centered upstream of the unstable side of minisatellite arrays. Repeat turnover therefore appears to be controlled by recombinational activity in DNA that flanks the repeat array and results in a polarity of mutation. These findings have suggested that minisatellites most probably evolved as bystanderd of localized meiotic recombination hotspots in the human genome.

Evolutionary Aspect

Finally, studies have shown that the evolutionary fate of minisatellites tends towards an equilibrium distribution in the size of alleles, until mutations in the flanking DNA affect the recombinational activity of a minisatellite by suppressing DNA instability. Such event would ultimately lead to the extinction of a hypermutable minisatellite by meiotic drive.

Types

Hypervariable Minisatellites

These have core units 9–24 bp long and are found mainly at the centromeric regions.

Telomeric Minisatellites

These have core units 6 bp long, and have thousands of repeated sequences at the telomeres.

Location

In humans, 90% of minisatellites are found at the sub-telomeric region of chromosomes. The telomere sequence itself is a tandem repeat: TTAGGG TTAGGG TTAGGG.

A **Variable Number Tandem Repeat** (or **VNTR**) is a location in a genome where a short nucleotide sequence is organized as a tandem repeat. These can be found on many chromosomes, and often show variations in length between individuals. Each variant acts as an inherited allele, allowing them to be used for personal or parental identification. Their analysis is useful in genetics and biology research, forensics, and DNA fingerprinting.

Schematic of a Variable Number of Tandem Repeats in 4 alleles.

VNTR Structure and Allelic Variation

In the schematic above, the rectangular blocks represent each of the repeated DNA sequences at a particular VNTR location. The repeats are tandem - they are clustered together and oriented in the same direction. Individual repeats can be removed from (or added to) the VNTR via recombination or replication errors, leading to alleles with different numbers of repeats. Flanking the repeats are segments of non-repetitive sequence (shown here as thin lines), allowing the VNTR blocks to be extracted with restriction enzymes and analyzed by RFLP, or amplified by the polymerase chain reaction (PCR) technique and their size determined by gel electrophoresis.

Use of VNTRs in Genetic Analysis

VNTRs were an important source of RFLP genetic markers used in linkage analysis (mapping) of genomes. Now that many genomes have been sequenced, VNTRs have become essential to forensic crime investigations, via DNA fingerprinting and the CODIS database. When removed from surrounding DNA by the PCR or RFLP methods, and their size determined by gel electrophoresis or Southern blotting, they produce a pattern of bands unique to each individual. When tested with a group of independent VNTR markers, the likelihood of two unrelated individuals having the same allelic pattern is extremely improbable. VNTR analysis is also being used to study genetic diversity and breeding patterns in populations of wild or domesticated animals.

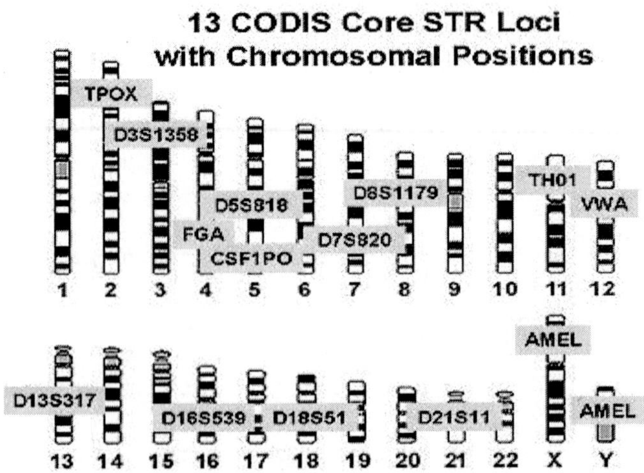

Chromosomal locations of the 13 VNTR loci in the CODIS panel.

VNTR Inheritance

In analyzing VNTR data, two basic genetic principles can be used:

- *Identity Matching*- both VNTR alleles from a specific location must match. If two samples are from the same individual, they must show the same allele pattern.
- *Inheritance Matching*- the VNTR alleles must follow the rules of inheritance. In matching an individual with his parents or children, a person must have an allele that matches one from each parent. If the relationship is more distant, such as a grandparent or sibling, then matches must be consistent with the degree of relatedness.

Relationship to Other Types of repetitive DNA

Repetitive DNA, representing over 40% of the human genome, is arranged in a bewildering array of patterns. Repeats were first identified by the extraction of Satellite DNA, which does not reveal how they are organized. The use of restriction enzymes showed that some repeat blocks were interspersed throughout the genome. DNA sequencing later showed that other repeats are clustered at specific locations, with tandem repeats being more common than inverted repeats (which may interfere with DNA replication). VNTRs are the class of clustered tandem repeats that exhibit allelic variation in their lengths.

Classes of VNTRs

There are two principal families of VNTRs: microsatellites and minisatellites. The former are repeats of sequences less than about 5 base pairs in length (an arbitrary cutoff), while the latter involve longer blocks. Confusing this distinction is the recent use of the terms Short Tandem Repeat (STR) and Simple Sequence Repeat (SSR), which are more descriptive, but whose definitions are similar to that of microsatellites. VNTRs with very short repeat blocks may be unstable - dinucleotide repeats may vary from one tissue to another within an individual, while trinucleotide repeats have been found to vary from one generation to another (see Huntington's disease). The 13 assays used in the CODIS database are usually referred to as STRs, and most analyze VNTRs that involve repeats of 4 base pairs.

Main Requirements

- Laboratory setup for gel-electrophoresis
- Technical equipment to carry out Southern blot hybridizations
- Laboratory facilities for radioactive labeling

Analytical Procedures

Extraction of DNA
↓
Digestion of DNA by endonuclease restriction
↓
Separation of the fragments by agarose gel-electrophoresis
↓
Transference of the fragments to a nylon filter by Southern blotting
↓
Hybridization of fragments to a radiolabeled minisatellite probe
↓
Visualization of the polymorphisms by autoradiography

Advantages

- High level of polymorphism
- Generation of many informative bands per reaction (in case of multilocus probes)
- High reproducibility
- Filters can be washed and reprobed several times

Disadvantages

- Large quantities of purified, high molecular weight DNA required (5-10 μg)
- Laborintensive and technically demanding
- High costs involved
- Distribution across the genome may be non-random (unlike SSRs)
- Band profiles can not be interpreted in terms of loci and alleles (in case of multilocus probes)
- Dominance of alleles (in case of multilocus probes, SSRs are codominant)
- Similar sized fragments may not be homologous (in case of multilocus probes)
- Difficulties in comparing polymorphisms across different gels, so only small number of samples are compareable that can be run in one gel (depending upon No. of wells a gel has).

  Not amenable to automation (PCR is not used).

20

Inter Simple Sequence Repeat (ISSR)

Since the mid 1980s, genome identification and selection has progressed rapidly with the help of PCR technology. A large number of marker protocols that are rapid and require only small quantities of DNA have been developed. Three widely-used PCR-based markers are RAPDs (Williams *et al.,* 1990), SSRs or microsatellites (Tautz, 1989), and AFLPs (Vos *et al.,* 1995). Each marker technique has its own advantages and disadvantages. RAPD markers are very quick and easy to develop (because of the arbitrary sequence of the primers) but lack reproducibility (Karp *et al.,* 1997; Hansen *et al.,* 1998; Jones *et al.,* 1999; Virk *et al.,* 2000). AFLP has medium reproducibility but is labour intensive and has high operational and development costs (Karp *et al.,* 1997). Microsatellites are specific and highly polymorphous (Karp *et al.,* 1997, Jones *et al.,* 1999), but they require knowledge of the genomic sequence to design specific primers and, thus, are limited primarily to economically important species.

The choice of a molecular marker technique depends on its reproducibility and simplicity. The best markers for genome mapping, marker assisted selection, phylogenic studies, and crop conservation have low cost and labour requirements and high reliability. Since 1994, a new molecular marker technique called inter simple sequence repeat (ISSR) has been available (Zietkiewicz *et al.,* 1994). ISSRs are semiarbitrary markers amplified by PCR in the presence of one primer complementary to a target microsatellite. Amplification in the presence of nonanchored primers also has been called microsatellite-primed PCR, or MP-PCR, (Meyer *et al.,* 1993). Such amplification does not require genome sequence information and leads to multilocus and highly polymorphous patterns

(Zietkiewicz *et al.,* 1994; Tsumara *et al.,* 1996; Nagaoka *et al.,* 1997). Each band corresponds to a DNA sequence delimited by two inverted microsatellites. Like RAPDs, ISSRs markers are quick and easy to handle, but they seem to have the reproducibility of SSR markers because of the longer length of their primers. We now report the effect of various PCR parameters on the specificity and reproducibility of nonanchored ISSR amplifications.

ISSR (for inter-simple sequence repeat) is a general term for a genome region between microsatellite loci. The complementary sequences to two neighboring microsatellites are used as PCR primers; the variable region between them gets amplified. The limited length of amplification cycles during PCR prevents excessive replication of overly long contiguous DNA sequences, so the result will be a mix of a variety of amplified DNA strands which are generally short but vary much in length.

Sequences amplified by ISSR-PCR can be used for DNA fingerprinting. Since an ISSR may be a conserved or nonconserved region, this technique is not useful for distinguishing individuals, but rather for phylogeography analyses or maybe delimiting species; sequence diversity is lower than in SSR-PCR, but still higher than in actual gene sequences. In addition, microsatellite sequencing and ISSR sequencing are mutually assisting, as one produces primers for the other.

The ISSR technique is a PCR- based multi-locus marker system that employs oligonucleotide primers homologous to microsatellite or SSR sequences such as $(GATA)_n$ to amplify mainly the inter-SSR regions. *Amplification products are obtained only if SSRs in opposite orientation are found within a PCR amplifiable distance, with flanking sequences matching the oligonucleotide primers.* ISSR markers are multilocus, unmapped dominant markers. Therefore, ISSRs are DNA fragments (100-3000 bp) located between adjacent, oppositely oriented microsatellites (SSRs), that are amplified by the Polymerase Chain Reaction (PCR) using microsatellite core sequences and a few selective nucleotides as a single primer (16-18 bp).

About 10-60 fragments from multiple loci are generated (contrary to SSRs which produce 1-2 fragments only) simultaneously, separated by gel-electrophoresis, and scored as the presence or absence of fragments of particular size. Because of the multilocus fingerprinting profiles obtained, ISSR-analysis can be applied in studies involving genetic identity, parentage and clone and strain identification. In addition, ISSRs are considered useful in gene mapping studies.

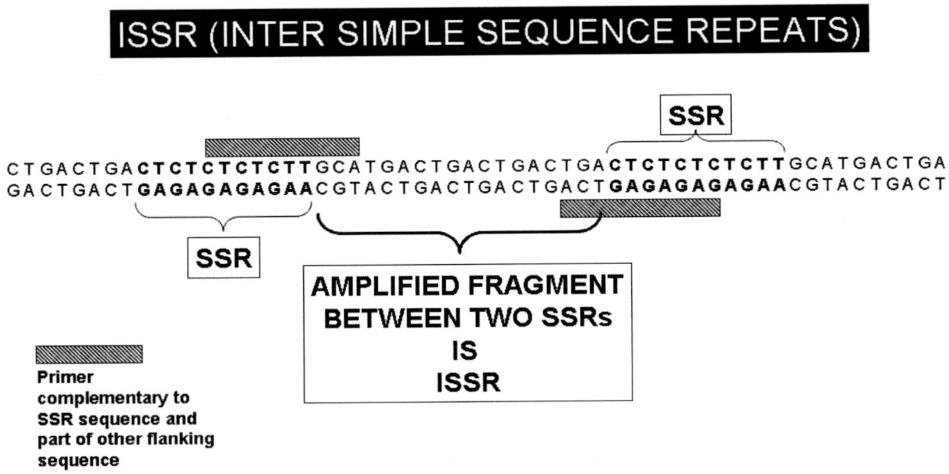

Synonyms: Various techniques have been developed to use microsatellite sequences as PCR primers with minor variations. This has resulted in many synonyms of ISSR techniques viz.

- SSR-anchored PCR;
- Anchored SSR-PCR (ASSR-PCR);
- Microsatellite primed PCR (MP-PCR);
- Anchored microsatellite primed PCR (AMP-PCR);
- Single primer amplification reaction (SPAR);
- Randomly amplified microsatellites (RAM) and
- Randomly amplified multiple polymorphism (RAMP).

Comparison Between ISSR and STMS Markers

STMS (Sequence Tagged Microsaleltite Site) is also a PCR-based marker, where unique DNA sequences flanking microsatellites are used to construct oligonucleotide primers. Resulting amplification products spanning the microsatellite region show rich allele divergence (No. of repeats are depicted in the gel). However, ISSR polymorphisms are not influenced by repeat polymorphisms within the SSR. This is because primers of STMS are purely based on sequence of the flanking regions of SSRs, whereas, primers have some complementary portion of repeat units and their flanking regions in ISSRs.

Comparison Between ISSR and RAPD Markers

ISSR technique is nearly identical to RAPD technique except that ISSR primer sequences are designed from microsatellite regions. Higher stringency of amplification in the form of longer primers (16-25bp) and elevated annealing temperatures (45-60°C) makes ISSR markers more reproducible than RAPD markers.

Factors Affecting Polymorphism of ISSR Markers

Polymorphism levels and extent of utility in ISSR experiments depend upon (1) nature of primers used (unanchored, 3′ anchored or 5′ anchored), (2) motif of the repeats targeted (di-, tri-, tetra-, penta-nucleotides or higher), (3) sequence of the primer (AT rich, GC rich or balanced), (4) length of the primer, (5) extent of optimization in the PCR (both reaction composition and reaction condition-mainly annealing temperature), (6) method of detection (agarose gel electrophoresis ethidium bromide or PAGE-silver staining/ radioactive detection).

Unanchored primers show a tendency to slip within the repeat units during PCR, which may result in background smear. To prevent internal priming, slippage and to avoid stuttering, ISSR primers can be anchored to unique genomic sequences flanking the repeat by extending them at either of ends using degenerate nucleotides up to 5 bases. 3′-anchoring is shown to give clearer banding pattern and 5′-anchoring is reported to produce more bands and higher polymorphism.

Procedure

DNA Extraction

Plant genomic DNA can be extracted by a CTAB (cetyltrimethylammonium bromide) protocol. Ground leaf tissues (100 mg) in 1000 mL of CTAB extraction buffer (100 mM Tris [pH 8.0], 1.4 M NaCl, 20 mM EDTA [pH 8.0], 0.2% (p/v) β-mercaptoethanol, 2% [p/v] CTAB) and heated at 60°C for 30 min. DNA can be extracted with one volume of a chloroform:isoamyl alcohol mix (24:1) and precipitated in presence of isopropanol (40% [v/v] final concentration). The DNA pellet washed with 5 mM ammonium acetate and 70% ethanol, dried, and dissolved in 100 mL of TE (10 mM Tris-HCl [pH 8.0], 1 mM EDTA [pH 8.0]). After addition of 1 mL of RNase (10 mg/mL), DNA concentrations can be determined with a fluorometer using bisbenzimide as the fluorescent dye.

PCR Conditions

Use ISSR primer for amplifications.

Electrophoresis and Analysis of Amplification Products

The PCR products then can be analyzed by electrophoresis using a 1.5% agarose gel. DNA can be stained by soaking the gel in a 0.5 mg/mL ethidium bromide solution.

Visualization

Visualization of the polymorphisms by ethidium-bromide staining and ultraviolet light (agarose gels), silver staining or autoradiography (polyacrylamide gels).

Main Requirements

- Thermocycler
- Laboratory setup for gel-electrophoresis
- Ultraviolet light source
- Laboratory facilities for radioactive labeling

Reaction Mixture for 1SSR

1. Sterile dH_2O = x ml (to make total 50ml)
2. PCR Buffer (10x) = 2.5 ml
3. dNTP (lmM) mix = 2.5 ml
4. Spermidine = 0.5 ml
5. Formamide = 0.5 ml
6. Taq polymerase (5u/m l) = 0.16 ml
7. Primer = 0.4 ml
8. DNA template = 3.0ml (25-30ng)

PCR Amplification

Step	Temp.	Time	No. of Cycles
Denaturation	94°C	5.00 min	
Denaturation	94°C	30 sec	
Annealing	50°C	45 sec	45 cycle
Extension	72°C	90 sec	
Final Extension	72°C5.00 min		

Related Techniques

- DAMD (Directed Amplification of Minisatellite-region DNA): uses a single primer consisting of the core motif of minisatellites only
- SPAR (Single Primer Amplification Reaction): uses a single primer consisting of the core motif of microsatellites only.

Advantages

- Low quantities of template DNA required (5-50 ng per reaction).
- No sequence data for primer construction required
- Random distribution throughout the genome
- Generation of many informative bands per reaction
- Amenable to automation

Disadvantages

- Band profiles can not be interpreted in terms of loci and alleles
- Dominance of alleles
- Similar sized fragments may not be homologous

21

Sequence Characterized Amplified Region (SCAR)

SCARs are DNA fragments amplified by the Polymerase Chain Reaction (PCR) using specific 15-30 bp primers, designed from nucleotide sequences established in cloned RAPD (Random Amplified Polymorphic DNA) fragments linked to a trait of interest. By using longer PCR primers, SCARs do not face the problem of low reproducibility generally encountered with RAPDs. Obtaining a codominant marker may be an additional advantage of converting RAPDs into SCARs. However, SCARs may exhibit dominance when one or both primers partially overlap the site of sequence variation. SCARs are locus specific and have been widely applied in gene mapping studies and marker assisted selection (Paran & Michelmore, 1993).

Characteristic Features of SCAR

- Less abundant in genome
- Show medium polymorphism
- Are locus specific
- Are codominant
- Their reproducibility is high

- Less labour intensive.
- Less technical demand
- Low operational cost.
- Development cost is high.
- Low quantity of DNA is required.
- Are amenable to automation.

Analytical Procedures

- Extraction of DNA
- Amplification of DNA fragments by PCR
- Separation of the polymorphisms by agarose gel-electrophoresis
- Visualization of the polymorphisms by ethidium-bromide staining and ultraviolet light.

Main Requirements

- Thermocycler
- Laboratory setup for gel-electrophoresis
- Ultraviolet light source

Advantages

- Low quantities of template DNA required (10-100 ng per reaction)
- Easy and quick to assay
- Band profiles can be interpreted in terms of loci and alleles
- High reproducibility
- Amenable to automation

Disadvantages

- Sequence data required for primer construction

22

Cleaved Amplified Polymorphic Sequence (CAPS)

Synonym : PCR-RFLP (PCR- Restriction Fragment Length Polymorphism)

The **Cleaved Amplified Polymorphic Sequence** or **CAPS** method is a technique in molecular biology for the analysis of genetic markers. It is an extension to the Restriction Fragment Length Polymorphism (RFLP) method, using polymerase chain reaction (PCR) to more quickly analyse the results.

Like RFLP, CAPS works on the principle that genetic differences between individuals can create or abolish restriction endonuclease restriction sites, and that these differences can be detected in the resulting DNA fragment length after digestion.

In the CAPS method, PCR amplification is directed across the altered restriction site, and the products digested with the restriction enzyme. When fractionated by agarose or acrylamide gel electrophoresis, the digested PCR products will give readily distinguishable patterns of bands. Alternatively, the amplified segment can by analyzed by Allele specific oligonucleotide (ASO) probes, a process that can often be done by a simple Dot blot.

In comparison with RFLP-analysis, polymorphisms are more difficult to identify because of the limited size of the amplified fragments (300-1800 bp). CAPS-analysis, however, does not require time-consuming Southern blot hybridization and radioactive detection. CAPS have been applied predominantly in gene mapping studies.

Main Requirements

- Thermocycler
- Laboratory setup for gel-electrophoresis
- Ultraviolet light source
- Gel documentation system

Advantages

- Low quantities of template DNA required (50-100 ng per reaction)
- Band profiles can be interpreted in terms of loci and alleles
- Codominance of alleles, thus heterozygotes can be differentiated from homozygotes.
- High reproducibility

Protocol

DNA Extraction

↓

Amplification of DNA fragments by PCR

↓

Restriction Endonuclease Digestion of the amplified fragments

↓

Separation of the fragments by Agarose or Polyacrylamide gel-electrophoresis

↓

Visualization of the polymorphisms by ethidium-bromide staining and ultraviolet light (agarose gels) or silver staining (polyacrylamide gels)

Disadvantages

- Sequence data required for primer construction

23

Single-Strand Conformation Polymorphism (SSCP)

Single strand conformation polymorphism (SSCP), or single strand *chain* polymorphism, is defined as conformational difference of single stranded nucleotide sequences of identical length as induced by differences in the sequences under certain experimental conditions. This property allows to distinguish the sequences by means of gel electrophoresis, which separates the different conformations.

Background

A single nucleotide change in a particular sequence, as seen in a double stranded DNA, cannot be distinguished by electrophoresis, because the physical properties of the double strands are almost identical for both alleles. Contrarily, after denaturation, single stranded DNA undergoes a 3-dimensional folding and may assume a unique conformational state based on its DNA sequence. The difference in shape between two single-stranded DNA strands with different sequences can cause them to migrate differently on an electrophoresis gel, even though the number of nucleotides is the same. Predictability, however, is a shortfall of SSCP - since conformational states are subject to many experimental conditions and sequence differences that cause minimal changes in strand conformation may not be detected.

The mobility of double-stranded DNA in gel electrophoresis is dependent on strand size and length but is relatively independent of the particular nucleotide sequence. The mobility of single strands, however, is noticeably affected by very small changes in sequence, possibly one changed nucleotide out of several hundred. Small changes are noticeable because of the relatively unstable nature of single-stranded DNA; in the absence of a complementary strand, the single strand may experience intrastrand base pairing, resulting in loops and folds that give the single strand a unique 3D structure, regardless of its length. A single nucleotide change could dramatically affect the strand's mobility through a gel by altering the intrastrand base pairing and its resulting 3D conformation (Melcher, 2000).

Single-strand conformation polymorphism analysis takes advantage of this quality of single-stranded DNA. First announced in 1989 as a new means of detecting DNA polymorphisms, or sequence variations, SSCP analysis offers an inexpensive, convenient, and sensitive method for determining genetic variation (Sunnucks *et al.,* 2000).

Like restriction fragment length polymorphisms (RFLPs), SSCPs are allelic variants of inherited, genetic traits that can be used as genetic markers. Unlike RFLP analysis, however, SSCP analysis can detect DNA polymorphisms and mutations at multiple places in DNA fragments (Orita *et al.,* 1989). As a mutation scanning technique, though, SSCP is more often used to analyze the polymorphisms at single loci, especially when used for medical diagnoses (Sunnucks *et al.,* 2000).

Applications in Molecular Biology

SSCP used to be a way to discover new DNA polymorphisms apart from DNA sequencing, but is now being supplanted by sequencing techniques on account of efficiency and accuracy. These days SSCP is most applicable as a diagnostic tool in molecular biology. It can be used in genotyping to detect homozygous individuals of different allelic states, as well as heterozygous individuals which should each demonstrate distinct patterns in an electrophoresis experiment.

SSCP is also widely used in virology to detect variations in different strains of a virus. The idea being that a particular virus particle present in both strains will have undergone changes due to mutation, and that these changes will cause the two particles to assume different conformations and thus be differentiable on an SSCP gel.

Related Techniques

- Denaturing Gradient Gel Electrophoresis (DGGE): separation of DNA fragments due to mobility differences under increasing denaturing conditions (formamide/urea concentrations)
- Thermal Gradient Gel Electrophoresis (TGGE): similar to DGGE, but subjected to heat-denaturing conditions.

Analytical Procedures

- Extraction of DNA
- Amplification of DNA fragments by PCR
- Denaturation of the amplified fragments to a single-strand form
- Separation of the polymorphisms by polyacrylamide gel-electrophoresis
- Visualization of the polymorphisms by silver staining or autoradiography.

Main Requirements

- Thermocycler
- Laboratory setup for gel-electrophoresis
- (Laboratory facilities for radioactive labeling).

Advantages

The major advantage of SSCP is that many individual PCR products may be screened for variation simultaneously. Forty nine to 97 PCR products may be analyzed on each full-size sequencing gel (depending upon comb size used). Most researchers use SSCP to reduce the amount of sequencing necessary to detect new alleles at loci of interest (*e.g.* Sweetman *et al.* 1992) or to better estimate allele frequencies of populations (*e.g.* Aguadé *et al.* 1994).

Two additional uses of SSCP have been advocated. First, SSCP should be used to screen PCR products of genes that researchers intend to sequence for phylogenetic analysis. This SSCP screening will allow researchers to determine 1) if the gene contains sufficient polymorphism that sequencing will be worthwhile, 2) which portion of the gene is most polymorphic, 3) what level of intraspecific variation exists, and 4) if there is polymorphism among multicopy genes within individuals (*e.g.* rDNA). Second, SSCP is a much more efficient method of obtaining information about levels of polymorphisms within anonymous nuclear loci than the restriction enzyme protocol originally described (Karl *et al.* 1992). The advantages of using SSCP in this context are: 1) the PCR amplicons are smaller and thus easier to amplify, 2) because the amplicons are only 200-300 bp, the entire sequence needed for PCR primer design can be obtained from a single sequencing reaction, 3) similar amounts of information are obtained by simply running the amplicons on 2 SSCP gels (*e.g.* one with glycerol and one without), rather than doing 50 restriction digests and assumming that the restriction sites are independent (a problematic assumption when dealing with a 1 kb amplicon), and 4) when polymorphisms are detected, the amplicons are of optimal size to sequence using an ABI automated sequencer (so getting sequence level information is easy).

Other advantages are:

- Low quantities of template DNA required (10-100 ng per reaction)
- Band profiles can be interpreted in terms of loci and alleles
- Codominance of alleles

Disadvantages

There are 2 major disadvantages of SSCP. First, the amount of mobility differences have little if any correlation to the amount of sequence differences. For example, under some SSCP conditions, cytochrome b PCR products from one subspecies of Red Panda are closer in mobility to Raccoons than to the other subspecies of Red Panda. Thus, the only information that can be gained from SSCP is if PCR amplicons are "identical" or different. Second, the optimal amplicon size for detection of most point mutations is rather small, around 200 bp. The strategies to deal with this limitation (*e.g.* dideoxy fingerprinting or cutting amplicons with restriction enzymes) are often more trouble than they are worth.

Other disadvantages are:

- Sequence data required for primer construction
- Highly standardized electrophoretical conditions are needed to obtain reproducible results
- Absence of mutation cannot be proven because some mutations may remain undetected.

SSCP Limitations and Considerations

1. Single-stranded DNA mobilities are dependent on temperature. For best results, gel electrophoresis must be run in a constant temperature.
2. Sensitivity of SSCP is affected by pH. Double-stranded DNA fragments are usually denatured by exposure to basic conditions: a high pH. Kukita

et al. found that adding glycerol to the polyacrylamide gel lowers the pH of the electrophoresis buffer—more specifically, the Tris-borate buffer—and the result is increased SSCP sensitivity and clearer data (1997).

3. Fragment length also affects SSCP analysis. For optimal results, DNA fragment size should fall within the range of 150 to 300 bp, although SSCP analysis of RNA allows for a larger fragment size (Wagner, 2002). Tthe presence of glycerol in the gel may also allow a larger DNA fragment size at acceptable sensitivity (Kukita *et al.*, 1997).
4. Under optimal conditions, approximately 80 to 90% of the potential base exchanges are detectable by SSCP (Wagner, 2002).
5. If the specific nucleotide responsible for the mobility difference is known, a similar technique called Single Nucleotide Polymorphism (SNP) may be applied.

Other Methods

Other methods of quickly screening amplicons have been described (e.g. Kirkpatrick *et al.* 1993). For a general review, Lessa and Applebaum is recommended (1993; see also Langemeier *et al* 1994 and references therein for dideoxy fingerprinting). Lessa and Applebaum (1993) recommend *Denaturing Gradient Gel Electrophoresis* (DGGE) as the most widely useful method. The essential trade-offs between DGGE and SSCP are: 1) DGGE can detect single-base substitions more efficiently in longer amplicons (about twice the optimal length for SSCP), 2) DGGE requires that one of the PCR primers include a "GC-clamp" (about 40 bp of G's and C's; which is about as expensive as getting 2 additional primers to either screen a second fragment of the same locus or get another locus -> thus primer costs per bp screened are similar), 3) DGGE requires one to determine the optimal gradient conditions before screening individuals for variation, 4) DGGE requires additional specialized equipment. Therefore, if one is concentrating on finding intraspecific variation within one genus (insects) or family (vertebrates) of organisms, a small number of primers and conditions will be highly useful and thus DGGE is the method of choice. However, when one wants to screen many loci or the organisms of interest are quite dissimilar at the sequence level (e.g. raccoons vs. red pandas) then SSCP will usually be better.

Finally, many methods of screening for SSCP variants have been described. For a general review, Sheffield *et al.* (1993) and Vidal-Puig and Moller (1994) can be consulted. Among non-radioactive methods, both ethidium bromide (Yap and McGee 1993) and silver-staining can be used, rather than radiation to detect SSCP alleles. However, Ethidium bromide and Silver staining are not as reliable as using radiation.

SSCP Procedure

- The procedure used during the development of SSCP was as follows:
  - Digestion of genomic DNA with restriction endonucleases
  - Denaturation in an alkaline (basic) solution
  - Electrophoresis on a neutral polyacrylamide gel
  - Transfer to a nylon membrane
  - Hybridization with either DNA fragments or more clearly with RNA copies synthesized on each strand as probes (Orita *et al.*, 1989).
- Since then, more convenient procedures have been developed, taking into account other molecular techniques, although sometimes it is simpler to amplify the double strand and then denature it into single strands instead of trying to find suitable primers for the below PCR method if the targeted sequence is unknown.
- Most experiments involving SSCP are designed to evaluate polymorphisms at single loci and compare the results from different individuals.

Step by Step SSCP

(*Source*: Travis Glenn, Laboratory of Molecular Systematics(LMS), Smithsonian Institution, Washington, DC 20560)

Warning : This protocol is currently being updated. It is dangerous to use 35S in PCR reactions. It forms a volatile gas that WILL escape from the PCR tubes (*see Clinton & Scougall 1995*).

Presented below is the protocol for Single-Strand Conformation Polymorphism (SSCP) analysis used at LMS. This protocol is basically the same as any SSCP protocol available in the literature. The only major difference is that doing a two reaction PCR (unless you are screening more than 48 individuals, then you should optimize PCR conditions and amplify hot the first time around) is recommended. The first reaction is a "normal" cold PCR to amplify the intended products. Check that these reactions were successful by running out on a minigel. The second reaction is a hot PCR to label the products produced in the first PCR reaction. The hot product will be diluted in sequencing buffer, denatured, and loaded onto a native polyacrylamide gel.

Detailed Protocol

I. Initial PCR Reaction

1. Set-up a "normal" cold PCR reaction. The following recipe is generally used:

   Master Mix

   2.5 µL 10x Thermo Buffer

   2.5 µL Primer 1 (5 µM)

   2.5 µL Primer 2 (5 µM)

   1.5 µL $MgCl_2$ (25 mM)

   1.5 µL dNTPs (2.5 mM each)

   11.8 µL dH_2O

   0.2 µL Taq DNA Polymerase (1 unit)

   Multiply each component by the number of reactions you plan to do + 1 for a negative control + 10% (e.g. to do nine experimental reactions - multiply each component by 11, so the master mix would contain 27.5 µL 10x Thermo Buffer, ...). **ENSURE THAT EACH COMPONENT IS THOROUGHLY MIXED BEFORE ADDING IT TO THE MASTER MIX.** Aliquot 22.5 µL of the Master Mix into a tube for each reaction. Then, add 2.5 µL DNA (@20 ng/µL) to each tube, Ensuring that the DNA Is thoroughly Mixed. Add 1 drop of mineral oil to each reaction tube (unless you plan to use the hot lid - Cycler-Mate, BioLogic Engineering, Inc.). If you plan to use the hot lid when screening your samples, then you must use it while optimizing the PCR conditions.

2. Thermocycle with the parameters established during optimization. In general, the following parameters are used: 1 cycle of 94° for 2 min., then 25 to 30 cycles of 94° for 1 min, 50°* for 30 sec., 72° for 30** sec. **Note:** *50° is a standard annealing temperature, use temperature established during optimization; ** 30 seconds is usually sufficient to synthesize PCR products less than 400 bp, use 1 min. per kb if synthesizing longer products.

3. Examine your results on a 1.5% agarose minigel. Use Gel Marker I (Research Genetics, Huntsville, AL) mixed with Light Loading Buffer (NOT the buffer that comes with the Gel Marker I) as a size marker. Use light loading buffer to screen 10 µL of each sample.

4. If the reactions contain anything other than bands of approximately the correct size, try the following to re-optimize conditions:

   1. Change the annealing temperature
   2. Add 2.5μL of BSA (2.5 mg/mL - Non-acetylated)
   3. Increase the $MgCl_2$ concentration 2X or 3X
   4. Add 2.5μL of DMSO
   5. Add 2.5μL of Glycerol

Note: If you add anything to the PCR reaction (e.g. BSA) be sure to subtract an equal amount of water, so the total reaction volume remains constant.

II. Making the SSCP Gel

Note: Always be very careful when handling acrylamide. It is a potent neurotoxin. Use gloves, glasses and lab coat when handling.

A) Gel Mix (15 - 30 min.)

1. Make gel solution:

For one MDE gel:	For five gels:
15 mL 2 x MDE	75 mL 2 x MDE
3.6 mL 10 x TBE	18 mL 10 x TBE
41.1 mL dH_2O	205.5 mL dH_2O

For Gels with 10% Glycerol:

For one gel:	For five gels:
16.5 mL 2 x MDE	82.5 mL 2 x MDE
4 mL 10 x TBE	20 mL 10 x TBE
6.6 mL Glycerol	33 mL Glycerol
38.7 mL dH_2O	193.5 mL dH_2O

For Glycerol Gels, it is usually convenient to measure the glycerol in a 25 mL graduated cylindar, then fill the cylindar with dH_2O (18.4 mL dH_2O), cover with parafilm, and mix by inversion. This way, the glycerol is diluted & easier to manipulate. The cylindar is then "washed" with 20.3 mL dH_2O.

If making enough for multiple gels, then store the mix in a brown plasic bottle. The mix should be used within 1 month of being made.

Note: Each MDE gel will also contain 330μL APS + 30μL TEMED and 10% Glycerol gels will contain 400μL APS + 36μL TEMED added immediately prior to pouring the gel (see below).

2. When making enough mix for multiple gel, it is usually a good idea to filter the Gel Mix. This should be done with the reusable magnetic filter units and the 0.22 micron AcetatePlus filters (MSI cat. # A02SP04700).
3. Get frozen aliquot of 10% Ammonium Persulfate (APS) or make fresh 10% APS solution by adding 1 gram of APS to a 14mL snap cap tube and then filling to 10 mL with dH_2O. [Note: the excess 10% APS can be aliquoted and frozen for future use.]

B) Setting up and Pouring a SSCP Gel (30min.-1hour)

Setting up the gel mold (glass plate sandwich):

1. Take one long glass plate and one short glass plate from the drying racks to the Acrylamide Gel Pouring Area.
2. Place the plates on the table with the inside surfaces of the plates facing up (inside surface can be identified by etched notation on outside surface of plate). Treat the *short plate only* with rainex, then spray and wipe dry both plates with deionized water, and then 95% ethanol.
3. Make "glass plate sandwich" which rests on two empty pipette racks by first placing long plate on pipette racks inside surface facing up. Then wipe two plastic spacer strips with ethanol and place one along each of the long edges of the long plate with the rubber cushions at the top and facing up. Place the short plate directly on top with the inside surface down and make sure the plates and spacers are lined up precisely with the rubber cushions of the spacer strips fitting snugly against the top edge of the short plate.
4. Place three clamps along each long side of glass plate sandwich leaving about 6 inches without clamps at the bottom. Pressure points of clamps should be directly over the center of the spacers.
5. **Optional:** Insert a strip of 3MM filter paper 1 cm thick into the bottom of the gel sandwich.
6. Carefully place one long strip of sequence gel tape along the bottom edge of the sandwich and coming about 5 inches up the side edges. Make sure the tape is sealed tightly along the bottom and side by firmly pressing it along the edges of the glass (the depth) before folding it and sealing along the length & then width of the glass.

7. Put a fourth clamp on each side of the sandwich near the bottom (on the taped area). Adding polymerizing agents and pouring gel (to be done only when gel mold is ready)

8. Ensure the following items are ready to use: 60cc syringe; 18 gauge, 1.5 inch needle; 100 mL graduated cylinder; 150 mL beaker; P1000 pipetter set to 330μL; P200 pipetter set to 30μL; 10% APS; TEMED; gel mix; and gel mold

9. Measure out 55-60 mL of gel mix. Add 330 μL of 10% APS and then 30μL of TEMED to the gel mix.

   Note: a minimum of 45 ml is enough for a gel but it is a good idea to have at least some extra in case of some drippage and to monitor polymerization.

10. Using 60cc syringe without needle, suck up solution. Wipe the end of the syringe and put the 18 Guage 1.5 inch needle on.

11. Pick up gel mold and tilt it up (45 degree angle with table or greater) and to one side (slightly). Begin squirting gel mix into the opening between the plates at a point near the low side edge (near the spacer cushion).

12. Squirt in gel mix slowly and steadily. Try to avoid forming bubbles or spilling. Spilling can be avoided by squirting slowly. Bubbles can usually be avoided by increasing the angle of tilt to the side or more vertically as you pour.

13. When the fluid is about three inches from reaching the top of the mold, stop squirting and place the mold, long plate down, flat on the empty pipette racks again. The fluid should come up to fill the entire gap between the plates. Squirt remainder of fluid back into bottom of the 150 mL beaker so that polymerization may be monitored.

14. Slide the two plastic combs in between the plates into the gel mix with the flat sides in (teeth out). Flat surfaces of combs should form an even line, about a half centimeter deep into the gel (the top of the oblong holes should be even with the top of the short glass plate).

15. Place a large piece of saran wrap around top of gel mold and avoid disturbing combs.

16. Take one clamp from each side of the gel mold and place them clamps along top of gel mold. Spread the 3 remaining clamps along each side evenly.

17. Let gel sit on empty pipette racks for a minimum of an hour (up to overnight)

III. Preparing and Loading Samples

If you skipped step one, then increase the number of cycles to the number determined during primer optimization and ensure that the $MgCl_2$ concentration is the same as determined during optimization rather than what is given below (it may also be ok to double the dNTP concentration).

A) Making "Hot" PCR Amplicons (1 hour => 2 hours)

1. Label one tube for each PCR product. Then make the following:

The New Less Dangerous Way

1. A. End label one of the primers using the following recipe (enough for 50 individuals/reactions):

5.0µL 10x Kinase Buffer

20.0µL Primer (5 µM)

20.0µL dH_2O

2.5µL PNK (PolyNucleotide Kinase - Promega - 8 units/µL)

2.5µL gamma ^{33}PdATP (10 mCi/mL; 2000 Ci/mmol = 5 pmol/µL)

Incubate at 37° for 30+ min.

1. B. Label the PCR products with the hot primer by making new reactions according to the following recipe:

Master Mix

1.00µL 10x Thermo Buffer

0.75µL Hot Primer (2µM)

0.30µL Cold Primer (5µM)

0.6µL $MgCl_2$ (25 mM)

0.6µL dNTPs (2.5 mM each)

5.65µL dH_2O

0.1µL Taq DNA Polymerase (0.5 unit)

The Old Dangerous Old Way:

Master Mix

1.0µL 10x Thermo Buffer

1.0µL Primer 1 (5µM)

1.0µL Primer 2 (5µM)

0.6μL $MgCl_2$ (25 mM)
0.3μL dNTPs (2.5 mM each)
3.5μL dH_2O
0.5μL ^{35}S dATP (Note: use alpha labelled nucleotide)
0.1μL Taq DNA Polymerase (0.5 unit)

Note: Handle all radioactive materials with much care to avoid contamination. Monitor everything frequently to avoid spreading radiation.

2. Aliquot 8μL of master mix to each reaction tube. (Mix well by gently pumping pipette before making aliquots!)
3. Add 2μL of previous PCR products (or DNA if you did not do step I) to each reaction tube.
4. Add 1 drop of mineral oil to each tube and then close.
5. Using the same thermocycling parameters as used in step I, cycle 10-15 times (the intial denaturation isn't necessary).
6. Turn on the radioactive-use water bath to 90°.

B) Preparing Samples for Loading into SSCP Gel (30 min)

1. While the thermocycler is still running, add 18 μL of SSCP Loading Buffer (USB Stop Solution + 2 μL of 10 M NaOH per mL) to 0.7 mL tubes (one for each reaction done in part III A). [Note: the P250 electronic pipetters are great for this.] If you have many samples, it is easier to use a microtiter plate here, rather than tubes. [Note: adding 1 μL of 1 M CH3HgOH per sample will ensure denaturation - *see Weghorst & Buzard (1993)*].
2. Label the tubes with SSCP Loading Buffer, with the appropriate sample numbers. Centrifuge the tubes to ensure the drops of Loading Buffer are on the bottom.
3. After the amplicons have been labelled (the thermocycler is done cycling), add 2 μL of the Hot amplicon to the appropriate tube made in step 1 (above).

C) Preparing and Loading SSCP Gel (1hour)

It is best to start after 14:00 HRS because the gel will run 14-18 hrs.

1. Make sure radioactive-use water bath is at 90-95°C.
2. Check small sample of gel mix (remainder from step B15 above) to make sure that gel polymerized correctly. If it is still mostly liquid, there may

have been a problem with the acrylamide mix or one of the polymerizing reagents and you should not waste your reactions on the gel; make a new gel mix, NEW APS & pour again.

3. Remove the tape from the bottom, then remove the saran wrap, then all the clamps from gel mold, and finally, take out combs carefully to avoid damaging top edge of gel.
4. Place gel, long-plate-out, on gel rig and tighten clamps to secure gel.
5. Close buffer drain valve by turning knob on right side of rig clockwise until tight.
6. Get 1 liter of 1x TBE** from carboy, and pour enough into bottom tray of gel rig to cover bottom of gel (cover about 1 cm of the bottom of the glass plates). Pour the remainder in the top compartment of the gel rig.
   ** The concentration of TBE may be varied in this step.
7. Suck up some of the buffer, and squirt all along top edge of gel to remove air bubbles and gunk. You may use a Pasteur pipette, or a syringe (with needle) for this step.
8. Shut plastic lids on top and bottom compartments of gel rig and use electrical cords to connect gel rig with power source (make sure positive charge is at the bottom of the rig).
9. Turn on power source and set to desired Wattage (usually 4-6). Run gel for about 5 minutes before loading to ensure that it will carry current and there are no huge leaks.
10. Float reactions (either microtiter plate or eppendorf tubes) in 90°C water bath for 2 minutes to denature the amplicons. If using microtiter plates, make sure water contacts the bottoms of all of the wells with reactions in them.
11. While reactions are incubating, get ice bucket ready and bring the following materials to the gel rig: magic marker, P10 pipette man, P10 pipette tips, 60cc syringe with needle or Pasteur pipette with bulb, and a labelled beaker for temporary storage of used radioactive pipette tips.
12. After removing reactions from water bath, immediately place them on the ice.
13. Turn off electrical current on rig, open lid on top compartment and squirt buffer along the top of the gel to ensure that all the gunk is rinsed out.

14. Carefully slide toothed side of plastic combs into top of gel such that the tips of the teeth protrude only slightly into the gel (roughly 0.5-1mm). If comb slides in too easily, the wells will leak, so keep trying combs until you get two that fit snugly.
15. Using magic marker, mark lanes to be used. Select lanes with no bubbles and with a good contact with teeth. Samples should be run side-by-side as much as possible. It will be very difficult to judge differences if there are gaps between the samples.
16. Load 1-3 μL of each reaction into wells if using fine toothed combs, 3-5 μL if using larger toothed combs.
17. When all lanes are loaded, close lid, reconnect power source, and turn on power.
18. Mark your name, time of the start of the run, and times to stop the gel on the outside of the gel with a magic marker. Also mark a line at the level of the buffer near the top of the glass plate. This step will allow you to monitor the rig for buffer leaks. If you notice buffer leaking from the top compartment of the gel rig at this point or at any time during the run, disconnect the power and take steps to plug leak. Using two syringes with needles to poke a small piece of kimwipe into the corner where the rubber gasket touches the glass plate usually does the trick.
19. Make sure that the order of the samples, the primers, templates, reaction conditions and order of lanes on the gel are all written down on a data sheet. The left-right orientation of the gel can easily get confused in later gel handling steps. A good way to eliminate this confusion is to load the lanes in an asymmetrical configuration (*e.g.* 1,2,3,4,5, 6,7,8).
20. There is a fair amount of variation in the speed that individual gels run. For MDE gels, It is possible to run gels at 4 to 8 Watts for about 40 Watt hours (= 10 hours at 4 Watts or 5 hours at 8 Watts, or ...) when amplicons are 50 - 100 bp; 50-60 Watt hours with 100 - 200 bp amplicons; etc. Glycerol gels take considerably longer to run (usually a bit less than 50% longer). The cyanol blue (light green dye) usually runs at about 100 bp. Use that as a guide.

IV) Taking off, Drying and Exposing Gel

A) Taking Gel off rig (15 minutes)

1. Turn off power source and disconnect wires.
2. Unscrew buffer valve on right side of rig to allow buffer in top compartment to drain into bottom compartment.

3. When buffer is drained out of top, unscrew clamps, remove gel from rig and place, long-plate-down, on a flat surface with bench paper.

4. Allow gel to cool slightly while you remove buffer tray from bottom of gel rig and dump buffer from both compartments into radioactive liquid disposal carboy in sink (contains radiolabelled DNA fragments and unincorporated radioactive dATP. Rinse both compartments of the buffer tray with water and wipe down the broad flat vertical surface (heating plate) of the gel rig with a large kimwipe to remove any liquid buffer that has dripped down.

5. Using a spatula labeled for radioactive-use, gently pry top plate from bottom plate only slightly at first. This step can be very tricky, and if gel does not stick to correct plate, consult an experienced sequencer in the lab before separating plates very much. If gel is sticking to bottom plate, continue to separate plates and remove top plate and spacers entirely.

6. If using regular acrylamide gel, it may be desirable to fix the gel (though not imperative). If you are fixing the gel, go to "Fixing Gel" instructions. **If using "long Ranger" gel, you may not fix the gel (it will dissolve).** For non-fixed gels, label a 13"x15.5" sheet of filter paper with pencil to denote any desired information about the gel, and place it directly on top of gel making sure to cover the entire gel.Press down gently to make a good contact between gel and paper. Peel off paper with gel and go to "Drying gel" instructions.

B) Drying Gel (30-40 min)

1. Place saran wrap over surface of gel. Use two or more pieces if necessary to ensure that entire surface of gel is covered. Saran wrap should be laid out smoothly but a few wrinkles, folds, and regions of saran wrap overlap are fine.

2. Trim saran wrap so that no more than one inch is sticking out from the filter paper on any side.

3. Open gel dryer, lift up rubber gasket, and place the gel, saran wrap side up, on top of the filter paper.

4. Replace the gasket on top of the gel so that it is stretched out flat and close lid of gel dryer.

5. Follow directions on vacuum aspirator to start vacuum. Make sure valve to your gel dryer is open and is pulling vacuum. Reopen lid of gel dryer and check if gasket is holding vacuum.

6. Press "start" button to initiate gel heating program.
   - Gel dryer program settings:
   - Use the first (top) program (an even drying temp.).
   - Temp is usually set for 80°.

C) Removing Gel from Dryer, Checking and Exposing (15 minutes)

1. Remove gel from dryer, turn off aspirator according to directions on aspirator, and remove saran wrap from gel or from gel dryer gasket (whichever it is stuck to). **if saran wrap does not come off of gel easily, try to gently peel it off but if it really is stuck, don't rip it because it will rip the gel. You can expose the gel with the saran wrap still on if you allow for an exposure time roughly 25% longer.*

2. Check the dry gel with a geiger counter set to 1X, and record level of radioactivity on sequence data sheet. [Note: if the reading is greater than 4K, then reset the geiger counter to 10x to read the value.]

3. Bring gel and data sheet into dark room.

4. Select a metal cassette that is facing green side out (unused) and place on counter. (*see below if there are no metal cassettes available*).

5. Open cassette and place gel, gel-side-up, in cassette.

6. Turn on red light, turn off regular light, remove film from drawer, place one piece of film on top of gel inside cassette. Seal cassette, close up film packaging, replace film in drawer and turn on regular light. If your gel loading configuration was symmetrical (i.e. it would appear similar flipped over), you should take measures to ensure that you do not lose track of the correct left-right alignment of the film (making a small mark on one side of the film is usually sufficient).

7. Put a small piece of tape on cassette labeled with your name and the date of exposure.

8. Replace the cassette, red-side-out, in the rack and record which cassette your gel is in on data sheet.

   **If there are no metal cassettes available, use a cardboard cassette by placing gel and film in cassette with lights out as above, sealing cassette and film package as above, then place cassette in between two glass plates (in drawer) and clamp glass plates together along all edges.*

9. Leave the cassette in the rack to expose for 1-7 days (or more if you're desperate, it is difficult to over-expose when using ^{35}S) depending on

strength of radioactive signal on gel. A gel that reads 2K or more in step 2 (over the entire gel, not just the top) should yield readable sequence with an overnight exposure. If the gel reads 0.5-2K, a period of roughly 3 days is usually necessary for exposure.

D) Developing Gel (10 min. => 25 min.)

1. Make sure fluids in both fixer and developer tanks below automatic developing machine have at least 2 inches of fluid in them (if not, these fluids need to be replaced).
2. Turn water valve (blue handle near door) into vertical position (on) and turn on automatic developer by pressing two adjacent black switches on left (west) side of machine
3. Wait approximately 15 minutes for machine to warm up (indicated by red light flashing on and off above black switches).
4. Feed one piece of old film (sitting on top of machine) through feed tray on right (east) side of machine.
5. Place cassettes with films to be developed on counter and turn off lights (except for red light).
6. Open cassette and label the top left-hand corner with your intials and the gel number (e.g. TCG 94-15).
7. Take the film to the X-ray processor and feed the film into machine through feed tray. The rollers can be activated by pressing the black button next to the feed tray or by gently pushing the film into the processor.
8. When the machine buzzes, you may feed in a new film or turn on the lights again.
9. After three or four minutes, the machine will start spitting out the films in the order that you put them in. Remove films and label them with the order of samples in lanes, primers used, length of exposure, length of gel run,and other relevant information (date, experimental treatments, etc.).

Mutation Detection by Single-strand Conformational Polymorphism (SSCP)

(Source: http://www.ohsu.edu/nsi/faculty/reddyh/lab/protsscp.html)

PCR Protocol

10X PCR Buffer	1.0
$MgCl_2$ (2.0mM)	0.8μl

dNTPS mix*	0.8µl
Primer-F (µg/µl)	0.1µl
Primer-R (µg/µl)	0.1µl
Taq (5U)	0.1µl
^{32}P dCTP (10µci/µl)	0.1µl
ddH_2O	4.5µl
DMSO (100%)	0.5µl
Subtotal	**8.0µl**
Template DNA (50-100ng)	2.0µl

Note: Synthesize primers 21-30 nt in length for products of 200-350 bp.

Use standard PCR conditions, however the Tm has to be calculated from the specific primer pair.

*dNTPS mix (final concentration): dATP 2.5mM; dTTP 2.5mM; dGTP 2.5mM; dCTP 1.25mM

SSCP Gels

Prepare 0.5x MDE gel as follows:

MDE gel	16.0ml
ddH_2O	44.2ml
10X TBE	3.84ml
10% APS	256µl
TEMED	25.6µl

Pour sequencing gel format with appropriate sharkstooth comb. Gel will polymerize in about 1 hour.

Loading Buffer

- 95% formamide
- 10mM NaOH
- 0.025% Bromophenol Blue
- 0.025% Xylene Cyanol

Run gel in 0.6X TEB buffer.

Heat denature samples at 94°C for 5 minutes and place them on ice for 3-5 minutes. Load 2.0-4.0µl per sample. Include non-denatured controls.

Electrophoresis Conditions

1. Fragment Size: 150-200 bp
   - 6 Watts
   - 10-12 hours
   - Room temperature
2. Fragment Size: > 200 bp
   - 8 Watts
   - 10-12 hours
   - Room temperature

Exposure

Dry gel and expose either at -80°C for 2 hours or at room temperature for 16-18 hours.

Fixing Gel (25 minutes)

1. Place glass plate with gel on top in the fixing tub (gel side up).
2. Gently pour in 1 liter of fixing solution (5% acetic acid, 5% methanol) so as to cover the gel completely. Do not disturb gel by pouring in fixing solution too fast. Folds, bubbles, and floating portions of the gel may be eliminated by gently touching the gel while in the fixing solution.
3. Allow gel to sit 15 minutes in fixer.
4. Cut out a 13"x15.5" piece of filter paper, write any desired information about the gel in pencil on the filter paper, and place it in the fixing tub directly covering the entire gel.
5. Remove fixing solution with aspirator hooked up to faucet while holding filter paper in place over gel with other hand.
6. Lift plate out of fixing tub and place, gel-side-up, on a flat counter surface with bench paper.
7. Cover entire filter paper with small stacks of paper towels (0.5 inch stacks) to suck up moisture. Flip over paper towel stacks, press down again gently, and remove paper towels.
8. Gently peel up filter paper with the gel now stuck to it.

Single Strand Conformation Polymorphism (SSCP) Analysis by Nondenaturing PAGE

(Source: B.R. Yadav and D.S. Kale)

Principle

Single-strand conformation polymorphism (SSCP) technique is a simple and efficient means to detect any small alteration in PCR-amplified product. It is based on the assumption that subtle nucleic acid change affects the migration of single-stranded DNA fragment and, therefore, results in visible mobility shifts across a nondenaturing polyacrylamide gel (Orita *et al.*, 1989). Polyacrylamide gel is used for analysis of DNA with specialized buffer systems and without urea. In nondenaturing PAGE the components used to synthesize matrix are acrylamide monomers, N, N-methylene bisacrylamide (Bis), ammonium persulphate (APS) and N,N,N′,N′-tetramethylenediamine (TEMED). Ammonium persulphate (APS) when dissolved in water generates free radicals, which activate acrylamide monomers inducing them to react with other acrylamide molecules forming long chains. These chains cross-linked with Bis.TEMED act as catalyst for gel formation because of its ability to exist in free radical form. The acrylamide and bisacrylamide is used in 49:1 ratio using autoclaved HPLC water added to make 100 ml volume. This 49:1 acrylamide-bisacrylamide solution is dissolved completely using magnetic stirrer and kept it in refrigerator till used.

Procedure

Preparation of SSCP Gel

The percentage of gels used for SSCP analysis varies from 8-20 percent but most of the primers were optimized with good results in 12% PAGE solution. Composition of 12% nondenaturing PAGE Solution (For 100 ml)

Component	Volume
50% Acrylamide-bisacrylamide	24 ml
10 X TBE	5 ml
Glycerol	10 ml
Autoclaved HPLC water	61 ml

The gel mixture is kept dissolved completely and stored at -20°C till it is used.

PCR-SSCP

The PCR-SSCP procedure includes following steps viz; PCR amplification of the gene fragments, resolution in nondenaturing PAGE and visualization using silver staining. PCR conditions are optimised for PCR-SSCP by testing a number of variables such as concentration of DNA, Taq polymerase, dNTPs, $MgCl_2$ and temperature profile. The PCR amplification protocol for all the SSCP primers used is same except the annealing temperature, which varies between primers.

1. The Single Strand Conformation Polymorphism analysis of amplified gene fragments is carried out using Bio-Rad Protein II xi Cell vertical gel electrophoresis unit (Bio-Rad laboratories). The two glass plates are washed thoroughly using tap water with detergent and rinsed initially under running tap water till no remains of detergent are left. The plates are wiped two times with tissue paper soaked in distilled water first, 70 percent alcohol and then air-dried. The similar thorough cleaning treatment is given to spacers and comb to ensure proper alignment of 20 cm glass plates.

2. The gel sandwich is assembled on a clean surface laying down the long rectangular plate first, then two spacers of equal thickness along the long edges of plate and the short plate is placed on the rectangular plate. The two glass plates with spacers between them are fitted well with proper alignment by tightening the bulldog clamps. The sandwiched gel plates are fitted in the stand with screw clamps. The cleaned comb (20 wells) is inserted from the topside of the gel sandwich and immediately bulldog clamps are applied over the plates containing comb to create sharp wells.

3. The bottom side of the gel sandwich is sealed using 10 ml of 12% gel mix. The gel sandwich is kept in slanting position and the solution mixed with 50 µl APS and 20 µl TEMED is injected between the two glass plates using syringe fitted with 10 µl tip and allowed to polymerize for 10 minutes.

4. After polymerisation the assembled gel sandwich is placed in alignment slot of casting stand. The 12% native PAGE gel mix (25 ml) is prepared by adding APS (100 ul) and TEMED (40 µl) at a time and mixed well. This gel mix is filled from upper side of gel sandwich using syringe smoothly without any bubble and clamps are immediately applied over the comb to ensure sharp wells. The gel is kept undisturbed at least 45 minutes for polymerisation.

5. After polymerisation the comb is removed and wells are flushed with 0.5X buffer. The gel sandwich is placed in electrophoresis tank with notched plate facing towards the buffer reservoir. The reservoir of the

electrophoresis tank is filled with 0.5X TBE and the gel is given pre-run at 200 volts at constant temperature for minimum 45 minutes. Ice cooled water circulation with electric pump is applied to central cooling core of assembly to maintain constant temperature.

6. About 4 µl PCR product and 12 µl of a formamide dye is prepared in PCR tube and denatured at 95°C for 10 minutes in the Biometra PCR machine. After denaturation the samples are immediately kept in ice-chilled box and kept in -20°C deep freeze for 10 minutes.
7. After completion of pre-run the wells are flushed again using buffer. The samples are loaded on a nondenaturing 12% acrylamide: bis-acrylamide (49:l) gel with gel loading tip and immediately electrophoresis is performed in 0.5 X Tris borate (pH 8.3)-EDTA buffer at 10-12.5 volts/ cm for 3-24 hr at room temperature depending on the optimised conditions for each primer.
8. After completion of the electrophoresis for required time the glass plates are removed from the assembly. There after gel is subjected to silver staining to visualize SSCP band patterns.
9. In order to stain the gel it is immersed in a tray of appropriate size filled with 10% acetic acid (500ml) for at least 30 minutes for fixing DNA bands in gel so as to prevent diffusion of the DNA bands (Care has to be taken so that gel remains dipped well in solution). The gel is agitated slowly for 30 minutes or until the tracking dye is no longer visible.
10. The acetic acid is decanted and 500 ml of distilled water is poured in the tray and rinsed thoroughly by placing the tray on oscillatory automatic shaker for 20 minutes.
11. Meanwhile 500 ml of 0.1% silver nitrate solution is prepared in amber color bottle and 750 µl of 37% formaldehyde is added and mixed. Distilled water is gently decanted from tray. The gel is stained for 45 minutes in silver nitrate solution with constant shaking in a dark room or covering the tray with black cloth.
12. Then the gel is rinsed briefly for 25 seconds in distilled water.
13. Working quickly distilled water is decanted from the tray. The freshly prepared and chilled 3% sodium carbonate solution (3% Na_2CO_3 and 750µl of 37% formaldehyde + 1% sodium thiosulfate) is transferred to the tray. The gel is kept immersed until bands get developed sharply.

14. The gel is given 10 percent acetic acid (stop solution) treatment for 10 minutes. Then 500 ml distilled water is added to tray.
15. The gel is transferred gently on the transparency; excess water is soaked with tissue paper, and air dried for half an hour.

Observations

The silver stained gel is kept on transilluminator and SSCP variants are recorded. Then gel is labeled and scanned for computer image analysis and documentation.

24

Retrotransposon-Based Markers (S-SAP, IRAP, REMAP, RBIP, RGAS, SNP)

What are Retrotransposons?

The presence of mobile elements, capable of moving from one region of the genome to another, has been known and thoroughly studied since the 1950's. These elements are capable of exciding and inserting themselves from one region of the chromatin to another, through their peculiar terminal repeat regions. This process is mediated by enzymes encoded by the transposon itself.

Retrotransposons are a class of mobile elements, widespread in plant genomes, where they account for a significant part of noncoding DNA. Barley, for example, contains thousands of copies of the Bare-1 element, dispersed throughout the euchromatic regions of chromosomes and exhibiting sequence identity higher than 95% to each other.

Unlike other classes of mobile elements, retrotransposons insert into the genome via an RNA intermediate, synthesized by their own reverse transcriptase, without the loss of the parental copy. Unique sequence properties show that retrotransposons of the Ty1-copia group are close to retroviruses, and their mode of integration in the host genome is also very similar.

The number and exact point of insertion of retrotransposons may differ from one individual to another and this can be exploited to characterize individuals, varieties or species, depending on how recent the transpositional event is. Furthermore, insertion of a retrotransposon is an irreversible event, a characteristic that can be helpful when tracing an ancestry.

Retrotransposons are suitable to be used as molecular markers by means of their Long Terminal Repeat (LTR) sequences. These are stretches of DNA, variable in length and sequence, repeated at both ends of the retrotransposon. LTRs can be found by "mining" public databases with suitable search programs, but this method is only suitable for a restricted number of well-known species. Alternatively, it is possible to screen libraries for other, more conserved sequences closely related to LTRs, such as conserved domains of the transposon genes *rt* and *rnh*

These have proven superior to both biochemical (protein) and phenotypic markers, which suffer from a fairly low degree of polymorphism, lack of sufficient loci to generate dense maps, and environmentally variable expression. To generate a multiplicity of bands and maximize the potential to detect genetic polymorphisms, all molecular marker systems except RFLP (restriction fragment length polymorphism) are based on prevalent, widespread elements in the genome. Most of these systems rely on PCR amplification, either from random primers (RAPD), from primers to adapters matching particular restriction sites (AFLP, SSAP), from primers to the simple sequence repeats (SSRs) to amplify between microsatellites (ISSR), or from primers flanking specific microsatellite loci. The dispersion, ubiquity, and abundance of retrotransposons in plant genomes gives an excellent basis for development of marker systems.

Transposed sequences are "jumping genes" that are dispersed throughout the genome.Transposons move as DNA elements and retrotransposons move via an RNA intermediate which is reverse transcribed and reinserted into the genome. The analysis of SINEs - Short INterspersed Elements - LINEs - Long INterspersed Elements - or truncated LTRs - Long Terminal Repeats - as molecular cladistic markers represents a particularly interesting complement to DNA sequence and morphological data.

The target sites are relatively unspecific so that the chance of an independent integration of exactly the same element into one specific site in different taxa is not large and may even be negligible over evolutionary time scales. Retrotransposon integrations are currently assumed to be irreversible events; this might change since no eminent biological mechanisms have yet been described for the precise re-excision of class I transposons, but see van de Lagemaat *et al.* (2005). A clear differentiation between ancestral and derived character state at the respective locus thus becomes possible as the absence of the introduced sequence can be with high confidence considered ancestral.

In combination, the low incidence of homoplasy together with a clear character polarity make retrotransposon integration markers ideal tools for determining the common ancestry of taxa by a shared derived transpositional event (Hamdi *et al.* 1999; Shedlock and Okada 2000). The "presence" of a given retrotransposon in related taxa suggests their orthologues integration, a derived condition acquired via a common ancestry, while the "absence" of particular elements indicates the plesiomorphic condition prior to integration in more distant taxa. The use of presence/absence analyses to reconstruct the systematic biology of mammals depends on the availability of retrotransposons that were actively integrating before the divergence of a particular species.

REMAP (REtotransposon-Microsatellite Amplified Polymorphism) and IRAP (Inter-Retrotransposon Amplified Polymorphism) which require neither restriction enzyme digestion nor ligation to generate the marker bands. The IRAP bands are generated from two nearby LTRs using outward-facing primers annealing to LTR target sequences. In REMAP, amplification between LTRs proximal to simple sequence repeats such microsatellites produces the marker bands.

Retrotransposons have several advantages as molecular markers over the other systems described earlier. Their abundance and dispersion can yield many marker bands, the pattern possessing a high degree of polymorphism due to transpositional activity. The LTR termini are highly conserved even between families, yet longer primers can be tailored to specific families. *Unlike DNA transposons, the new copies are inserted but not removed.* Even intra-element recombination resulting in the conversion of a full-length element to a solo LTR either does not affect the marker band and can furthermore be exploited to produce a new marker band in RBIP. Retrotransposon families may vary in their insertional activity, as is seen clearly by the pattern of nested insertions in maize, allowing the matching of the family used for marker generation to the phylogenetic depth required. Furthermore, the length and conservation of primers to the LTRs facilitate cloning of interesting marker bands and the development of new retrotransposons for markers.

Retrotransposons have been found to comprise the most common class of transposable elements in eukaryotes, and to occur in high copy number in plant genomes. Several of these elements have been sequenced and were found to display a high degree of heterogeneity and insertional polymorphism, both within and between species. Because retrotransposon insertions are irreversible, they are considered particularly useful in phylogenetic studies. In addition, their widespread occurrence throughout the genome can be exploited in gene mapping studies. Moreover, they are frequently observed in regions adjacent to known plant genes. Retrotransposons consist *of long terminal repeats (LTR) with a highly conserved terminus, which is exploited for primer design in the development of retrotransposon-based markers.* Several techniques have emerged during the last few years and are briefly outlined below.

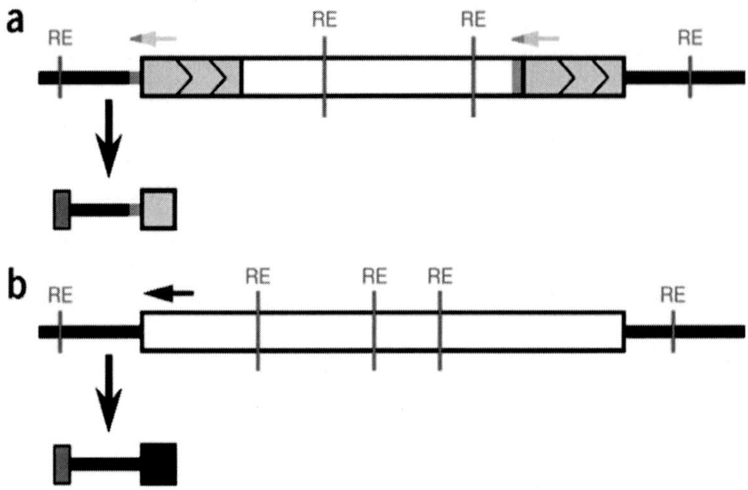

S-SAP (Sequence-Specific Amplified Polymorphism)

Dominant, multiplex marker system for the detection of variation in the DNA flanking the retrotransposon insertion site. Fragments are amplified by PCR, using one primer designed from the conserved terminus of the LTR and one based on the presence of a nearby restriction endonuclease site. Experimental procedures resemble those used for AFLP analysis. *Compared to AFLPs, S-SAP generally yields less fragments but higher levels of polymorphism* (Waugh *et al.*, 1997).

IRAP (Inter-Retrotransposon Amplified Polymorphism)

IRAP are an alternative valuable retrotransposon-based markers. In this method, PCR oligonucleotide primers facing outwards from the LTR or other regions of retrotransposons are made and amplify between two retroelements inserted into the genome. As discussed above, the insertion of elements into the genome mean that the number of sites amplified and sizes of inter-retroelement fragments differ between different lines, and can be used as markers to detect genotypes, measure diversity or reconstruct phylogeny (see Flavell *et al.* 1999; Kalendar *et al.* 1999; Kumar & Hirochika 2001).

Retrotransposons can be used for markers because integration of a daughter copy creates new joints between genomic DNA and the conserved LTRs. To detect polymorphisms for retrotransposon insertion, the marker systems that have been recently developed generally rely on PCR amplification between an LTR and some component of the flanking genomic DNA. Two methods have been developed,

Dominant, multiplex marker system that examines variation in retrotransposon insertion sites. IRAP fragments between two retrotransposons are generated by PCR, using outward-facing primers annealing to LTR target sequences. Fragments are separated by high-resolution agarose gel-electrophoresis (Kalendar *et al.*, 1999).

Procedure

1. Total genomic DNA samples were diluted with sterile double distilled water to 25 ng/ul.
2. IRAP was performed in a 25 ul reaction mixture containing 50 ng DNA, 1X Promega PCR buffer, 1.5 mM $MgCl_2$, 5 pmol of each primer, 200 μM dNTP mix, 1 U *Taq* polymerase (Promega, USA).
3. Amplification was performed using Biometra Tgradient (Goettingen) thermocycler or thermalcycler that can set the ramping time.
4. The PCR reaction parameters consisted of: 95°C, 5 min; 30 cycles of 95°C, 30s, annealing at the Ta specified, ramp 0.5°C s-1 to 72°C, and 72°C for 2 min + 3 s extension per cycle; a final extension at 72°C for 10 min
5. PCR products were analyzed by electrophoresis on 1.5 to 2 % (w/v) agarose gel and detected by ethidium bromide staining.
6. Polymorphic band patterns were scored and analysed using phylogenetic software such as PAUP*, powermarker, *etc.*

Recipes

2.5 ul 10 x PCR buffer (Promega)

1.5 ul 25mM MgCl2

0.5 ul 10mM dNTPs

2 ul outward LTR primer (10uM)
1 U Taq polymerase (Promega, USA)
sterile double distilled water, added up to 25 ul
2 ul 25 ng/ul genomic DNA

REMAP (Retrotransposon-Microsatellite Amplified Polymorphism)

Dominant, multiplex marker system that examines variation in retrotransposon insertion sites. REMAP fragments between retrotransposons and microsatellites are generated by PCR, using one primer based on a LTR target sequence and one based on a simple sequence repeat motif. Fragments are separated by high-resolution agarose gel-electrophoresis (Kalendar *et al.*, 1999). Retrotransposon-microsatellite amplified polymorphism amplified the region between simple sequence repeats (SSRs). The dispersion of retrotransposon and SSRs provided a good background for molecular marker application in Marker-Assisted Selection (MAS) and breeding.

Procedure

1. Total genomic DNA samples were diluted with sterile double distilled water to 25 ng/ul.
2. REMAP was performed in a 25 ul reaction mixture containing 50 ng DNA, 1X Promega PCR buffer, 1.5 mM MgCl2, 5 pmol of each primer, 200 μM dNTP mix, 1 U *Taq* polymerase (Promega, USA).
3. Amplification was performed using Biometra Tgradient (Goettingen) thermocycler or thermocycler that can set the ramping time.
4. The PCR reaction parameters consisted of: 95°C, 5 min; 30 cycles of 95°C, 30s, annealing temperature (depend on the SSRs and LTR primers) for 30 s, ramp 0.5°C s-1 to 72°C, and 72°C for 2 min + 3 s extension per cycle; a final extension at 72°C for 10 min.
5. PCR products were analyzed by electrophoresis on 1.5 t0 2 % (w/v) agarose gel and detected by ethidium bromide staining.
6. Polymorphic band patterns were scored and analysed using phylogenetic software.

Reagents

2.5 ul 10 x PCR buffer (Promega)

1.5 ul 25mM MgCl2

0.5 ul 10mM dNTPs

1 ul outward LTR primer (10 uM)

1 ul SSR primer

1 U Taq polymerase (Promega, USA)

sterile double distilled water, added up to 25 ul

2 ul 25 ng/ul genomic DNA

RBIP (Retrotransposon Based Insertional Polymorphism)

Codominant marker system that uses PCR primers designed from the retrotransposon and its flanking DNA to examine insertional polymorphisms for individual retrotransposons. Presence or absence of insertion is *investigated by two PCRs,* the first using one primer from the retrotransposon and one from the flanking DNA, the second using primers designed from both flanking regions. Polymorphisms are detected by simple agarose gel-electrophoresis or by dot hybridization assays. Drawback of the method is that sequence data of the flanking regions are required for primer design. Major advantage is that RBIP does not necessarily require a gel-based detection system but can easily be adopted to automated, gel-free procedures, such as *Taq* Man or DNA chip technology in order to increase sample throughput (Flavell *et al.,* 1998).

REMAP
(REtrotransposon-Microsatellite Amplified Polymorphism)

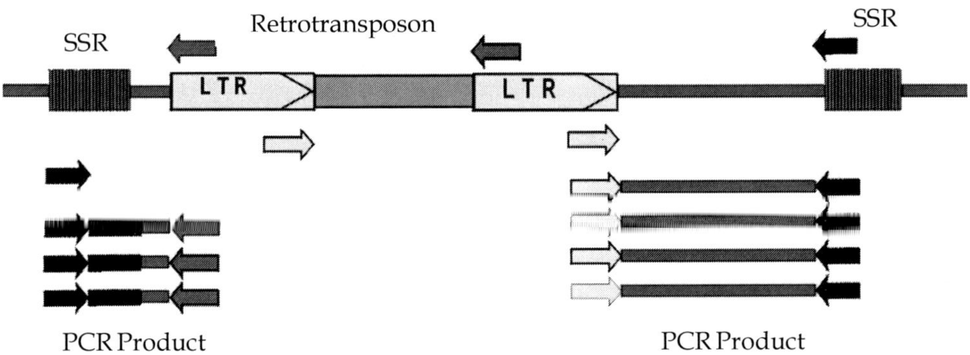

Functional Genomics

The sequence data that are becoming more and more available for increasing numbers of genes have stimulated the development of novel molecular marker technologies directed to the genes of interest. These technologies will allow the screening of germplasm for functional diversity in order to identify the variation relevant for conservation and exploitation. In addition, they may provide an efficient method to screen germplasm at an early developmental stage without the necessity of performing costly evaluation tests, such as pathogen resistance screening. Furthermore, they can be applied in marker-assisted selection for plant breeding purposes. Given the availability of gene sequences, such marker technologies can in principle be developed for any character. They are therefore, expected to become an important tool to identify and exploit the genetic diversity relevant to end-users. A number of evolving techniques are briefly outlined below.

Resistance-based Markers / Resistance Gene Analogues (RGAs)

Resistance Gene Analogues (RGA) are fairly recent class of markers that are very important for disease resistance. These are some sequences in a resistance gene that are conserved over species. Even in some cases different genera also have these conserved sequences. This property of these sequences which are actually not the resistance genes but are resistance gene analogues, help breeders to fish out/locate resistance genes over species and genera. These sequences are used as probes to identify similar sequences in other plants/varieties/species/genera. RGA in plants are related to the NBS-LRR class of disease resistance genes which are characterized by Nucleotide Binding Site (NBS) and Leucine Rich Repeat (LRR) regions. The NBS, in particular, contains several sequence motifs which are highly conserved among disease resistance genes and RGA from distantly related plants, thus facilitating their use for the isolation of new RGA and perhaps actual disease resistance genes.

RGAs have been successfully used in case of plants. One of the examples is being explained here (see figure on next page). Eight pairs of RGA primers were used, all of which are being maintained at the Genetics Laboratory of the Entomology and Plant Pathology Division, International Rice Research Institute. The primers were designed based on the conserved motif of various disease resistance genes. PCR amplification, polyacrylamide electrophoresis and silver staining were performed following the procedure described by Chen *et al.* (1998). All the eight primers detected polymorphisms between the parents and each generated 3-13 polymorphic bands. Altogether, 57 polymorphic RGA bands were scored. Most of them were inherited as dominant. All the eight primers detected polymorphisms between the parents and each generated 3-13 polymorphic bands. Altogether, 57 polymorphic RGA bands were scored. Most of them were inherited as dominant markers, but two pairs of bands appeared to be allelic and were suggested to be co-dominant markers by their segregation characteristics. Of the 55 RGA markers, 43 (76.4%) segregated in the expected 1:1 ratio, indicating that the majority of the RGA markers were inherited as a single locus. In comparison with parental survey using 190 RFLP probes and 280 RAPD primers, polymorphism was detected with only 35 probes (18%) and 69 RAPD primers (25%), and 63.2% of the RFLP markers and 84.6% of the RAPD markers segregated in the 1:1 ratio among the RILs.

How RGAs help in fishing out resistance genes

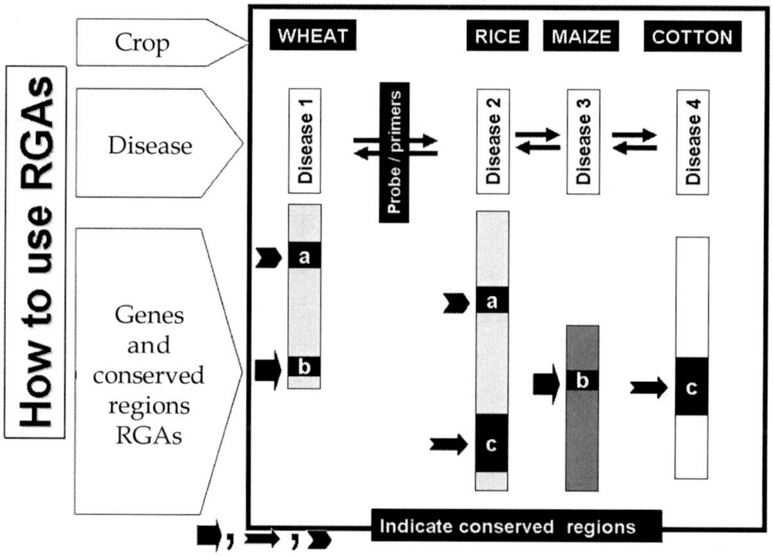

Linkage analysis was then conducted using the computer software MAPMAKER/ EXP version 3.0b (Lincoln et al. 1992). One resistance gene was mapped onto the interval of Pi-2 (t) and Pi-9 (t) on chromosome 6. Its resistance allele was derived from Gumei 2. Two RGA markers were found to be linked to this gene. Another gene was mapped onto the interval of Pi-4(t) and Pi-6(t) on chromosome 12. Its resistance allele was derived from Thong 156. Three RGA markers were found to be linked to this gene.

The finding that pathogen resistance genes share common sequence motifs, such as the nucleotide binding site (NBS) and leucine-rich repeat (LRR) regions, has resulted in the development of molecular markers derived from these conserved regions. Appealing feature of such marker systems is that they are not restricted to known genes, but in principle are able to detect variation in novel resistance genes and analogs as well. Marker development may be limited to the gene sequences (Chen *et al.*, 1998; Sicard *et al.*, 1999) or may be extended to the flanking DNA, e.g. by transformation into an S-SAP-like assay. This latter approach is provisionally termed NBS-directed profiling and is part of a project in which CGN is involved (GENE-MINE).

The Role of RGAs in Creating Disease Resistant Cultivars

The use of RGAs in creating disease resistant cotton genotype has genetic perspective. However, the big challenge ahead is to determine which of these cotton RGAs represent true resistance genes, and what is the nature of the organism they provide protection against in cotton. Research groups are initially targeting Verticillium wilt and Bacterial blight as examples of fungal and bacterial diseases of economic importance to the cotton industry. Cultivars of known disease response such as Pima S-7 (*G. barbadense,* Wilt resistant, Blight susceptible) and CS50 (*G. hirsutum,* Wilt susceptible, Blight resistant) are being employed. These cultivars allow researchers to compare the DNA fragment patterns of the RGAs in each plant and to relate presence or absence of a particular DNA fragment with disease response.

The presence of RGAs has been established in both *G. hirsutum* and *G. barbadense* and subtle differences have been detected that may account for the variation detected in disease resistance levels between the cotton species. Research groups engaged in RGAs work expect to find a RGA linked with Verticillium wilt resistance in Pima S-7, and a RGA linked with Bacterial blight resistance in CS 50. Linked RGAs could then be used as DNA markers to enable selection of plants carrying genes for resistance to these diseases.

Knowledge and understanding of the process of pathogen recognition and defence by plants holds great promise for the development of better disease control in cotton. If one can identify the resistance genes responsible for recognition of the organisms causing such diseases as Verticillium wilt, Fusarium wilt or Bacterial blight, one can ensure that all of these genes are included when breeding improved, more resistant cotton varieties.

SNP (Single Nucleotide Polymorphism)

Why SNPs?

- Highly abundant markers - allow very fine mapping of traits
- SNPs can occur within gene sequences and may even be the source of mutation behind the phenotypic variation
- Co-dominant - allows determination of zygosity, therefore more informative when screening for traits

- Various technologies exist for SNP genotyping - can assay SNPs from simplex reactions (ABI Taqman) to >50,000 SNPs simultaneously (Affymetrix).

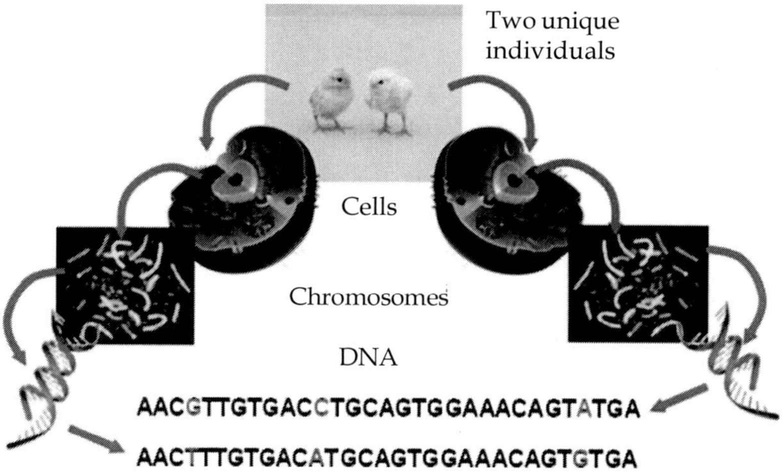

SNP Portability

- There exist numerous platforms for SNP genotyping
- SNPs are generally quite portable from one platform to another
- Adopting any set of SNPs to a given platform is never 100%
- Successful conversion rate varies from 80-95%
- SNP databases are also very robust and should remain relevant even when genotyping platforms are changed
- SNPs are likely to remain the predominant marker system for the foreseeable future

DNA LandMarks can carry out SNP analysis on a wide variety of genotyping platforms in order to best suit the needs of individual projects.

99.9% of one individuals DNA sequences are identical to that of another person. Over 80% of this 0.1% difference will be Single Nucleotide Polymorphisms, SNP's for short. A SNP is a single base substitution of one nucleotide with another. Both substitutions have to be observed in the general population at a frequency greater than 1%. An example of a SNP is individual "A" has a sequence GAACCT, while individual "B" has sequence GAGCCT, the polymorphism is a A/G.

Studying SNP profiles or haplotypes associated with a disease trait. Association study can detect and indicate which pattern is most likely associated with the disease-causing genes. Eventually, SNP profiles that are characteristic of a variety of diseases, will be established. Then, it will only be a matter of time before physicians can screen individuals for susceptibility to a disease just by analyzing their DNA samples for specific SNP patterns.

The race among pharmaceutical companies today, is to apply new system genomics approach to identify novel targets and validate these targets in the most efficient fashion. SNP research will provide fundamental understanding of many polygenic diseases, thus providing new therapeutic targets. Another significant goal is to identify those SNP's which are associated with significant biological effects in response to chemical drugs. A large percentage of people given a drug respond in the intended medically beneficial way, however some smaller percentage might either have no response or have a life threatening response and death. This adverse drug response (ADR) is believed to cause thousands of deaths annually. The SNP effort will serve as the bedrock of pharmacogenomics, the emerging field of personalized medicine: the right drug, in the right dose, to the right person, at the right time.

SNP study is also extremely important in organisms other than humans. Within agriculture, genetic modification of the agriculturally important crops (corn, wheat, rice, soybeans, etc.) could lead to improve crop yields at lower cost by reducing the amounts of fertilizer, insecticides, herbicides required. Within microorganisms and viruses, SNP's are known to cause increased drug resistance. Some of the recent *E. Coli* outbreaks are due to new evolving strains of the bacterium. HIV, the causative agent of AIDS, has historically been so difficult to treat with drugs due to very high mutation frequency primarily in the form of SNP's.

Recently, SNP's are found to be very useful for a complete different purpose. SNP's turned out to be valuable genetic markers, for revealing the evolutionary history of populations. Their occurrence throughout the genome also makes them ideal for analyses of specification and historical demography, especially in light of recent theory suggesting that many unlinked nuclear loci are needed to estimate population genetic parameters with statistical confidence. In spite of having lower variation compared with microsatellites, SNP's should make the comparison of genomic diversities and histories of different species (the core goal of comparative biogeography) more straightforward than has been possible with microsatellites. The most pervasive, but correctable, complication to SNP analysis is a bias towards analyzing only the most variable loci, an artifact that is usually introduced by the limited number of individuals used to screen initially for polymorphisms.

Distribution of SNP's

SNP's are not uniformly distributed over the introns and exons. There are one third as many SNP's within coding regions as non-coding regions. SNP's can be concentrated about a specific region, usually implying a region of medical or research interest. For instance, the sequence that encodes proteins that present antigens to the immune system in chromosome 6 displays very high nucleotide diversity compared to the other areas of that chromosome.

DNA Chip Technology

High throughput screening technique based on the hybridization between oligonucleotide probes and complementary DNA sequences in microarrays. Chips may consist of arrays of amplified DNA immobilized on miniature glass or nylon substrates that are then tested for hybridization to series of fluorescently-labeled oligonucleotide probes. Alternatively, arrays of oligonucleotide probes may be synthesized, followed by the exposure to fluorescently-labeled PCR samples.

Hybridization signals are determined by laser scan technology from which data on sequence variation are obtained. Microarrays may comprise up to 250,000 features per square centimeter, but technical advances are likely to further increase the extent of miniaturization in the future. Applications of the DNA chip technology include diagnostics (mutation and polymorphism detection), gene discovery, gene expression and gene mapping.

25

Silver Staining

Silver Staining for Protein

Silver staining is one of the procedures, in addition to Coomassie Blue, R and G types, and fluorescent dyes that are available for detecting proteins separated by gel electrophoresis. Switzer introduced silver staining in 1979, a technique that today provides a very sensitive tool for protein visualization with a detection level down to the 0.3-10 ng level.

The basic mechanisms underlying silver staining of proteins in gels are relatively well understood. Basically, protein detection depends on the binding of silver ions to the amino acid side chains, primary the sulfhydril and carboxiyl groups of proteins, followed by reduction to free metallic silver. The protein bands are visualized as spots where the reduction occurs and, as a result, the image of protein distribution within the gel is based on the difference in oxidation-reduction potential between the gel's area occupied by proteins and the free adjacent sites. A number of alteration in the silver staining procedure can shift the oxidation-reduction equilibrium in a way that gel-separated proteins will be visualized either as positively or negatively stained bands. Silver staining protocols can be divided into two general categories: 1) silver amine or alkaline methods, and 2) silver nitrate or acidic methods. In general, the detection levels' using the various procedures is determined by how quickly the protein bands develop in relationship to the background (e.g. signal-to-noise ratio). The silver amine or alkaline methods usually have lower background and as a result are most sensitive but require longer procedures. Acidic protocols, on the other hand, are faster but slightly less sensitive. Clearly, each protocol has different advantages regarding timing, sensitivity, cost, and compatibility with other analytical methods, especially mass-spectrometry (MS) a tool that is being used in combination with gel electrophoresis or chromatographic methods for rapid protein identification. Until recently most of the silver staining protocols used either glutaraldehyde-of formaldehyde-based sensitizes in the fixing and sensitization step, thus, introducing the chemical modifications into proteins. The utilization of those chemicals causes the cross-linking of two lysine residues within protein chain, thus, strongly affecting two most crucial steps for the subsequent MS analysis: hampering the trypsin digestion and highly reduces the protein extraction from the gel .

Protocol

Reagents

1. Acetic acid (1%) in H_2O
2. Developing solution
   2% (w/v) Sodium carbonate
   0.04% (v/v) formaldehyde in H_2O
3. Fixing solution (50% v/v methanol and 5% v/v glacial acid
4. Methnol (50%)
5. Silver nitrate (0.1% w/v) (chilled on ice) in H_2O
6. Sodium thiosulfate (0.02% w/v) in H_2O
7. Stopping solution (5% v/v glacial acid in H_2O

Equipment

Plastic container
Shaker

Biological Sample

Protein within a gel that have been separated by electrophoresis.

Procedure

1. Place the gel in a plastic container and add sufficient fixing solution to cove the gel. Fix the gel for 20 minutes with gentle shaking.
2. Remove the fixing solution and add enough 50% methanol to cover the gel. Shake the gel gently for 10 minutes.
3. Remove the 50% methanol and add deionized H_2O. Shake the gel gently for a further 10 minutes
4. Remove the deionized H_2O and soak the gel in 0.02% sodium thiosulfate for 1 minute.
5. Rinse the gel twice with deionized H_2O for 1 minute.
6. Submerge the gel in chilled 0.1% silver nitrate solution and incubate it for 20 minutes at 4C.
7. Rinse the gel twice with deionized H_2O for 1 minute.
8. Submerge the gel in developing solution and shake the container vigorously ntil the desired intensity of staining is reached. Discard the developing solution if it turns a yellow colour and replenish it with fresh developing solution.
9. When the desired intensity of staining is reached, stop the development by replacing the developing solution with stopping solution.
10. Store the silver stained gel in 1% acetic acid until further use.

Since this book is more concerned to molecular genetics, so here we will give more emphasis to silver staining of DNA.

Silver Staining of DNA

Silver staining was introduced more than a decade ago as a sensitive procedure to detect trace amounts of proteins in polyacrylamide gels. Since then, silver staining has been perfected and extended to the study of other biological molecules that have been separated by electrophoresis in a variety of supports. Silver staining follows generally one of two methods. One uses diamine or ammoniacal silver solutions for gel impregnation and dilute acid solutions of formaldehyde for image development. The other method impregnates with silver nitrate in a weakly acidic milieu and uses formaldehyde to reduce silver under alkaline conditions. Reportedly, diamine alkaline methods are less sensitive but better suited for thicker gels, while acidic methods are fast and work best with thin gels.

Of the many silver staining procedures available, some provide quite sensitive and reproducible stains. During image development, almost all staining procedures reduce silver to metallic silver, which is then deposited in the immediate vicinity of the staining substratum. For optimal image contrast, the level of silver reduction in the polyacrylamide matrix must be kept to a minimum, usually by silver complexation or by "enhancing" pre-treatments of the gel matrix that precede silver impregnation. This is accomplished by appropriate modulation of the speed of the reduction process, which depends mainly on pH, the absolute and relative concentrations of silver and reducing agent, and the rate constant of the reaction. For example, complexation of silver ions decreases both silver concentration and the rate constant, allowing for selective "priming" of separated bands with metallic silver. Note that this process is not thermodynamically driven, but rather depends on the kinetics of silver reduction.

Originally, silver staining was used effectively in the detection of small amounts of nucleic acids. However, more demanding applications, such as the analysis of complex DNA profiles generated in DNA amplification fingerprinting and DNA sequencing, or the use of polyesterbacked polyacrylamide gels, required a more versatile silver staining protocol. Complex nucleic acid mixtures usually are resolved in very thin polyacrylamide gels, producing bands that must be detected reliably with high sensitivity. Backed gels have the advantage of easy handling and preservation as permanent records but are especially difficult to stain with silver, since the backing film serves as a surface for silver deposition and restricts diffusion of reagents.

The following protocol can developed a fast and highly sensitive acidic silver stain which allows accurate detection of nucleic acids (down to 1pg DNA/mm^2 band cross-section) with minimum background staining. The method has fewer steps and chemicals, takes less time, and produces the least number of staining artifacts when compared to other published protocols. It has been applied successfully for the staining of DNA sequencing profiles and can also be used for the staining of oligonucleotides, RNA, proteins and polysaccharides.

Silver Staining of Polyacrylamide Minigels for Detection of DNA

Procedure

1. Remove the gel from between the electrophoresis plates as demonstrated by the instructor and place in a glass dish with Distilled Water. Rinse once in Distilled Water.

2. Fix the DNA in the gel by incubating in 10% Ethanol for 10 min.

3. Rinse once in Distilled Water.
4. Incubate in 1% Nitric Acid for exactly 3 min. Decant solution to nitric acid waste bottle. (Nitric acid is not to be disposed of down the drain.) Rinse 2 times in Distilled Water. Incubate 2 times for 5 min. in Distilled Water.
5. Incubate in 0.2% silver nitrate for 15 minutes. Decant silver nitrate solution back into bottle. (The silver nitrate solution can be reused several times, but mark bottle each time you use the solution.) Rinse 3 times with Distilled Water. Incubate for 2 min in Distilled Water.
6. Add Developing Solution and shake briefly, until brown or black precipitate forms in the solution. Decant and add fresh Developing Solution. Protect from fluorescent light. Incubate until desired intensity of bands and background is reached. When desired intensity is reached, i.e. best balance between band intensity and background, decant Developing Solution and add Stop Solution immediately. Incubate in Stop Solution for 5 min.
7. To preserve the gel, incubate overnight in Gel Preserving Solution.

Reagents

1. 10 % Ethanol

   10% (v/v) ethanol in H_2O

2. 1 % nitric acid

   1% (v/v) nitric acid in H_2O

3. 0.2 % silver nitrate

   0.2 % (w/v) silver nitrate in H_2O

4. Developing Solution

   3 % (w/v) sodium carbonate

   0.05 % (w/v) formaldehyde in H_2O

5. Stop Solution

   3% (w/v) acetic acid in H_2O

   note: glacial acetic acid is 100% (w/v)

6. Gel Preserving Solution

   5 % (w/v) glycerol

   35 % (v/v) methanol in H_2O

Source

This procedure is a modification of that published by Neilan *et al.*, Biotechniques 17 (4): 708-712.

Silver Staining of the Sequencing Gel for SSR Analysis

Principle

Highly sensitive detection of nucleic acid in the nanogram range has been achieved by the specific chemical reduction reaction of silver ions. The methods for silver staining nucleic acids employ either a histologically derived procedure that uses ammoniacal solutions of silver or a photo chemically derived reactions in which silver binds to nucleic acid bases and is then selectively reduced by chemical reagents or light. Bassam *et al* have modified the original method introduced by Merril *et al.*, for protein staining, which uses formaldehyde to selectively reduce silver ions to metallic silver under alkaline conditions. The suggested concentration of $AgNO_3$ (12mM) can be halved provided the concentration of formaldehyde is increased at least two-to three fold during image development. While band intensity increased with time, a minimum $AgNO_3$ impregnation of 30 minutes gives optimal sensitivity. The presence of formaldehyde in the silver stain improves both sensitivity and contrast.

Effect of Formaldehyde Concentration on Band Detection

Reduction of silver by formaldehyde is concentration dependent.

- An increase of formaldehyde in the range of 0.00185% (by volume) to 0.0222% increased sensitivity but also background staining, while reducing development time.
- Higher formaldehyde level produces dark brown-black bands as opposed to the orange -light brown colour obtained at lower concentrations.
- Optimal band intensity occurs at about 0.00555% (by volume) formaldehyde.

Reducing the Background by Thyosulphate

Thiosulphate dissolves insoluble silver salts by complex formation, removing silver ions from the gel surface which in turn decreases non-specific staining. A concentration of 4 mM thiosulphate is enough to reduce non-specific background staining without noticeably affecting DNA image development. Higher than this concentration have no noticeable benefit and can reduce the sensitivity of DNA detection. The presence of thiosulphate eliminates the formation of a dark precipitate in the gel and developer solution.

Effect of Formaldehyde Concentration and Temperature on Developing Time of the Gel

If higher formaldehyde concentration is used, the developing time shortens. This makes accurate image development difficult to control. Elevated temperatures enhance reduction of silver by formaldehyde. Conversely, decreasing the temperature of the developing solution below the recommended range of 18-25°C increases developing time. *A temperature range of 10°C approximately doubles the time to achieve the proper silver staining, allowing more accurate image development.*

Materials Required

A. Equipment

1. Orbital shaker/rocking platform (2-3)
2. 2-3 plastic trays with dimensions slightly larger than the glass plate

B. Solutions

1. Fix/stop solution
2. Staining solution
3. Developing solution (always use pre chilled)

Protocol

Step	Procedure	Reagents/Solution required	Description
1.	Fixing of gel	*Fix/Stop solution:* Add 200 ml of glacial acetic acid into 1800 ml of ultra pure or double distilled water.	• Fix the gel for 20 min. • Save this solution because it will be used again in Step 6.
2.	Washing of gel	Ultra pure water	• Rinse the gel 3 times (2 min. each wash) and agitate.
3.	Staining of gel	*Staining solution:* To 2L Ultra pure water, add 2g silver nitrate and 3 ml. 37% formaldehyde	• Agitate in staining solution for 30 min.
4.	Washing of gel	Ultra pure water	• *Very Critical step:* Dip the gel briefly into the tray containing ultra pure water.

contd...

			• Important: The time taken to do this should not exceed 10 sec. If it exceeds 10 sec, repeat Step 3.
5.	Developing of gel	*Developer:* In 2L ultra pure water, dissolve 60g sodium carbonate, 400 ml sodium thiosulfate and 3 ml 37% formaldehyde.	• Agitate the gel well until the first bands are visible.
6.	Fixing of gel	Fix/Stop solution	• Use the same solution used in Step 1 • Incubate with shaking for 2-3 min.
7.	Washing of gel	Ultrapure water	• Rinse the gel twice for 2 min each
8.	Drying of gel	–	• Dry the gel by either keeping it at room temperature or using convection heating
9.	Documentation	Camera / gel documentation system	• Make a permanent record of the gel by capturing the image in a digital camera or gel documentation system.

Troubleshooting and their Solutions in Silver Staining

S.N.	Troubleshooting	Reason(s)	Solution
1.	Faint or no sequencing bands	1. Improper rinsing following staining	Rinse for a shorter time as longer rinses (>5-10 seconds) remove(s) the silver deposited on DNA
		2. Poor water quality	Use ultra pure/ double distilled water
		3. Incorrect amount of sodium carbonate added to the developing solution	Add correct amount of the sodium carbonate i.e. 60 gm

contd...

		4. Too much sodium thiosulphate added to the developing solution	Add only 400ml aliquot of sodium thiosulphate to the developing solution
		5. Methanol present in the fix/stop solution	The fix/stop solution should only contain 10% acetic acid
2.	Low band intensity at the bottom of the gel	Poor quality water	Use ultra pure or double distilled water
3.	High background staining	1. Developing solution too warm	Cool the developing solution to 10°C in an ice bath.
		2. Development performed too long	Stop the development reaction after the bands near the bottom of the gel first appear. The bands should be light gray. They darken upon gel drying and exposure to APC film.
		3. Insufficient fixation	Fix the gel until the bromophenol blue and xyelene cyanol bands are no longer visible.
		4. Brown background	Detergent residue present on glass plate may result in a brown background. Rinse the cleaned plates thoroughly with de-ionized water to remove the detergent microfilms. Change detergent.
		5. Poor quality PAGE gel	Use freshly prepared solutions. Pre-mixed solutions may give higher background.
		6. Decomposition of solutions.	Use freshly prepared fix/stop solution, staining solution and developing solution.
		7. Poor quality sodium carbonate	Use only fresh ACS grade sodium carbonate.
4.	Dark swirling pattern of the gel surface	1. Inadequate agitation during staining steps	The gel must be agitated during all staining steps.

contd...

		2. Inadequate rinsing before the development step	Rinse the gel for 5-10 seconds prior to transferring it to development solution.
5.	Clear, unstained areas in the gel	Excess sigma cote on the long glass plate	Remove excess sigma cote with ethanol-saturated paper towels.
6.	Yellow gel	1. Improper fixing of the gel	Fix the gel for 20 minutes. To ensure that the urea is completely removed, fix the gel soon after electrophoresis.
		2. Poor quality sodium carbonate used	Use only fresh ACS grade sodium carbonate.
7.	Gray gel	The sodium thiosulphate was not added to the developing solution	Always add fresh 400ml aliquot of sodium thiosulphate to the developing solution.
8.	Yellowish-brown band with poor contrast	Dirty DNA template	Confirm by agarose gel electrophoresis. If impure, re-extract with phenol: chloroform or any other method.
9.	Gel adheres to both plates	Long glass plate contaminated with binding solution, or inadequate treatment of long plate with sigma cote solution	Wipe excessive binding solution from the short plate. Exercise care to avoid contaminating the long plate with binding solution. Ensure uniform coverage of long plate with sigma cote solution. If necessary, thoroughly clean the plates by soaking them in 10% NaOH for 30 minutes.
10.	Gel peels off the plate when dried	1. Build-up of binding solution (Bind silane) after multiple treatments	Wash plate with a window-cleaning agent and rinse with de-ionized water. Alternatively, soak the plates in 10% NaOH for 30 minutes.
		2. Acrylamide percentage of the gel was too high	Use a 4-6% polyacrylamide gel (19:1 acrylamide: bisacrylamide)

Limitations of Silver Staining

1. Sensitivity of this method is limited to 1 pg/mm^2 band.
2. The same gel cannot be reused.
3. The process/analysis cannot be automated and real time detection of the band is not possible.
4. The size of the gel limits the size of DNA to be sequenced/sized.
5. In PCR based microsatellite analysis the staining of both the DNA strands create(s) confusion as this method detects differential migration of both the DNA of same PCR product, especially in the analysis of mono- and di- nucleotide repeat loci.

Source : Manual on Molecular Markers: Theory and Practicals.

26

Basics of Marker Assisted Selection in Crop Plants

Deoxyribonucleic acid (DNA) is a molecule made up of pairs of building blocks called nucleotides. The four kinds of nucleotides that make up DNA are adenine (abbreviated as the single letter A), guanine (G), cytosine (C), and thymine (T). The DNA molecule has the shape of two intertwined spirals, referred to as a double helix. DNA is packaged into chromosomes that are located within the nucleus of all cells. These chromosomes are the same in every cell of an organism and together make up the organism's genetic information, its *genome*. Chromosomes contain stretches of DNA called genes that code for amino acids that make proteins. It is the proteins that are the foundation of life for all organisms. The interaction and structure of proteins determine the visible characteristics or *phenotype* of an organism, while the genetic makeup of an organism is called its *genotype*.

The sequence of nucleotides that make up a gene can differ among individuals. The different forms of a gene are called *alleles*. The alleles are the result of nucleotide differences in a gene that affect an amino acid sequence of a protein. This can result in a change, addition, or deletion of a protein that can affect the phenotype.

All organisms receive one copy of each gene from their mother and one from their father. The DNA sequence of a gene inherited from each parent may be identical, in which case the individual is said to be *homozygous* for that trait. Or the sequence of the gene from one of the parents may be different, in which case the individual is said to be *heterozygous*. Allele variations may differ in their DNA sequence by as little as a single nucleotide.

Differences among alleles caused by a single nucleotide, called SNPs, can be the basis of genotyping tests. Genotyping means using laboratory methods to determine the sequence of nucleotides in the DNA from an individual, usually a specific gene. Genetic tests based on SNPs utilize DNA derived from an individual to determine the nucleotide in the gene of interest.

Marker assisted selection is the process of using the results of DNA testing in the selection of individuals to become parents for the next generations. The information from the DNA testing, combined with the observed performance records for individuals, is intended to improve the accuracy of selection and increase the possibility of identifying organisms carrying desirable and undesirable traits at an earlier stage of development. Complex traits, including many of economic importance, are controlled by many genes and are influenced by the environment. When an animal has a favorable performance record for a certain trait, it means that based on edigree and phenotype, the animal has inherited a greater than average number of good alleles of each gene affecting that specific trait.

It is important to combine DNA results with performance and phenotype information to maximize the effectiveness of selection for traits of interest. Combining information from performance records and genetic tests into the selection process will be better than using performance, phenotype, and markers separately. The challenge is to determine what emphasis marker information should be given in the selection decision.

Molecular Markers

Until recently, researchers relied on information about how animals, plants, and their relatives perform to make observations about the genes they possess. Today, researchers can use *molecular markers* to find genes of interest that control how plants and animals perform. Some molecular markers are pieces of DNA that have no known function or impact on animal and plant performance. Other markers may involve the gene of interest itself.

Linked Markers

One type of molecular marker is called a *linked marker*. Using well-designed experiments, scientists can find molecular markers that are located very close to major genes of interest. The molecular marker is said to be linked to that gene. Linked markers are only near the gene of interest on the chromosome and are not part of the DNA of the gene of interest. Suppose that scientists are trying to locate a certain gene in an animal species. Choosing animals randomly from a population and studying them would give the scientists no clues about whether a marker is associated with the gene. However, if scientists studied the progeny (offspring) of the mating of male and female animals through many generations, they may determine the presence of a useful molecular marker.

Direct Markers

A second kind of molecular marker is one that is part of the gene of interest. *Direct markers* are easier to work with after they are found, but they often are more difficult to find than linked markers.

Genetic markers represent a powerful tool which plant and animal breeders can use to vastly enhance the potential of their R&D programs.

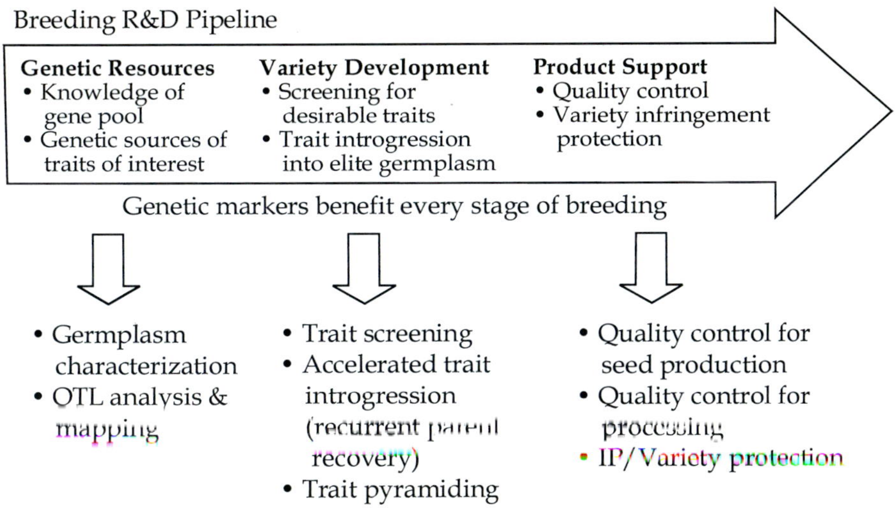

Definition of Appropriate Marker Technology

It is important to develop molecular marker technologies that can be used directly by plant breeders. Protocols must be at least as effective as conventional selection methods as well as being comparatively robust, inexpensive and simple since large number of individual plants require testing which potentially may consume time, money and labour.

There is a long list of molecular markers that have been developed so far. The most important among them are: RFLP, AFLP, SSRs, ISSRs, SNPs, ESTs etc.

Of these molecular marker systems, Simple Sequence Repeats (SSRs) or microsatellites, (the two terms are synonymous) are most suited to routine application in breeding programs. SSRs are short segments of repeated DNA sequences that are widely dispersed in the genome, and have been found to be highly variable between individuals. Analysis of SSRs utilizes the polymerase chain reaction (PCR) which requires only small quantities of DNA which in turn means only small tissue samples are required. PCR is relatively simple, quick and safe to use.

Applications in Plant Breeding

The most basic applications of the molecular marker techniques in marker assisted breeding include genetic diversity analysis, variety identification, isolation of markers tightly linked to specific genes, and marker assisted back-crossing.

Genotyping, Genetic Diversity and Seed Purity Analysis : Genotyping using DNA markers can be considered as the most reliable method for the identification of lines and varieties. Therefore, the DNA fingerprinting methods can be used to analyze the purity of seed lots. Genetic distance analysis can be a powerful tool for breeders to identify different heterotic groups and to increase the efficiency of finding crosses with good specific combinability (SCA). To determine the genetic distance between lines and groups of lines, the lines are fingerprinted and the marker-presence or absence is scored for each line. Based on the obtained score table, similarity indices can be calculated for all combinations of lines. Subsequently, the relatedness amongst the lines can be visualized using a dendrogram display or PCA plots.

Indirect selection : Indirect selection can be an advantageous method of selection in plant breeding. Especially for traits for which the phenotypic tests are unreliable, expensive or destructive to the plant (e.g. root parameters), markers can offer a solution. Before indirect selection can be applied, the genetic basis of the trait of interest needs to be elucidated and markers linked to the gene(s) of interest have to be identified. Once linked markers have been identified, the markers (if non-PCR like RFLP) can be converted into simple PCR assays (like STS), which allow screening of large numbers of plants for the trait of interest in a cost effective manner. A suitable linked DNA marker should allow the prediction of the phenotype in a range of the germplasm. The occurrence of multiple alleles in the germplasm for a desired locus may sometimes complicate the identification of markers with a good predictive value in the germplasm. There are different approaches for the monogenic and polygenic traits.

Monogenic traits : For the identification of markers linked with monogenic traits, different approaches can be followed. The preferable approaches are all based on screening a limited number of samples with a relatively large number of primer pairs. This way, many loci can be screened with a limited effort. The number of lanes per fingerprint can be limited by screening on set(s) of Near Isogenic Lines (NIL's), if these are available. Candidate markers that are identified in this way are then screened on a panel of phenotypically well characterized lines (for example male sterile and fertile lines, resistant and susceptible lines etc.) to confirm their linkage and to determine the predictive value of the markers. Another efficient approach consists of the 'Bulked Segregant Analysis' (B.S.A.) method (Michelmore *et al.*, 1991). For this type of screening, individuals from a segregating population are pooled on the basis of their phenotype (contrasting characters) , and the pools are then screened until a sufficient number of markers emerges. This method can be used for both dominant and recessive monogenic traits. For dominant genes, 'cis' markers (linked with the trait of interest) will emerge from the screening, whereas 'trans' markers (linked with the opposite allele) will be identified for recessive traits.

The B.S.A. approach may also be useful for the identification of linked markers for oligogenic traits.

Bulk Egregant Analysis (BSA)

Polygenic (quantitative) traits: The classical approach for the identification of loci involved in complex polygenic traits consists in the screening of a large number of individuals from a segregating population with a set of markers that are evenly distributed throughout the genome. Subsequently, statistical analysis is performed to identify regions in the genome that are involved in the trait. The laborious nature of this approach makes it unrealistic to screen sufficiently large populations to precisely locate the quantitative trait loci (QTLs). As a consequence, the QTLs cannot be localized precisely on the map and closely linked markers cannot be obtained, thereby preventing the broad scale application of indirect selection for quantitative traits.

For this reason, a new approach for the identification of QTLs markers, based on the B.S.A. principle, was investigated. With this goal, an oilseed rape F_2 population of » 2500 individuals segregating for two quantitative traits, glucosinolate and erucic acid contents, has been used. Based on the phenotypic scores, bulks were composed and approximately 2000 loci were screened on those bulks using the AFLP fingerprinting technique. Candidate markers that were identified on the bulk screening were then analyzed on » 200 randomly chosen individuals of the F_2 population. This screening demonstrated that the candidate markers identified using the B.S.A. screening approach were derived from three different loci involved in glucosinolate content and two QTLs involved in erucic acid content. The results have been confirmed by independent studies in which the same map positions have been identified to be involved in the respective traits (Toroser *et al.*, 1995; Jourdren *et al.*, 1996). The results demonstrate that a B.S.A. strategy may be useful even for the identification of markers for quantitative traits.

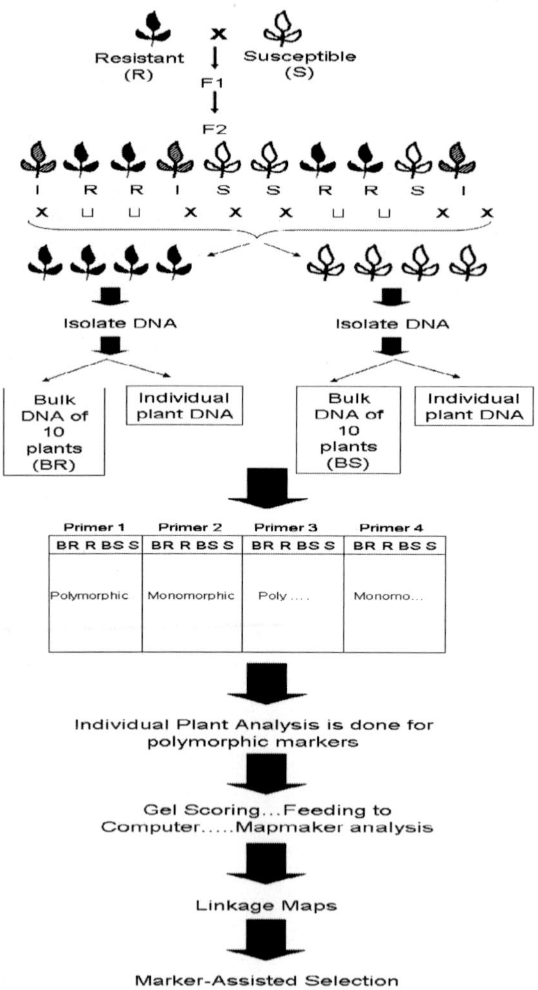

1. Make cross between parents differing for the target trait.
2. Self F_1 to get F_2 population segregating for the target trait. For example, parents differing for disease reaction as shown in the figure.
3. Plants with extreme phenotype (R and S) are selected and intermediates (I) are rejected.
4. Isolate DNA from each selected F_2 Plant and store at 4°C.
5. Bulk equal amount of DNA from 5-10 plants each from two extreme phenotypes. For example, BR= Bulk of 10 resistant plants and BS= Bulk of 10 susceptible plants. Remaining plants' DNA is stored for further use.
6. Run gel with multiple of subsets of BR-R-BS-S where R and S represent DNA from resistant and susceptible parents, respectively. Test 4-6 primers (SSRs/RAPDetc.) Look for the polymorphism and cosegregating marker. Select primers showing polymorphism along with cosegregation with the trait.
7. Go for the individual plant analysis on the gel with the set R-S-F_2 R plants-F_2 S plants
8. Score the gels in 0,1 or +, - way for the presence (+ or 1) or absence (0 or -) of bands in the parents and F_2 plants.
9. Go for linkage analysis using Map-Maker or any other appropriate software. Find out distances of various markers mapped on the genome.
10. The closest one may be used as marker in MAS.

The Benefits of Marker Technology

Application of genetic markers offers many clear advantages:

- *Speed*: Markers reduce the number of breeding cycles required to bring new lines to market.
- *Consistency*: Markers eliminate environmental variation from evaluation trials and also allow screening for resistance to pathogens without the biohazard of introducing the pathogen into the population.
- *Efficient*: Markers eliminate "also-rans" from a breeding program at a much earlier stage. A breeder can still work with the same number of progeny but early stage screening means that the progeny are already superior in many ways.
- *Effective*: Markers allow breeders to manage complex multigenic traits unlocking a whole new level of potential for their breeding program.

Overall genetic performance of a breeding population will follow a bell curve distribution.

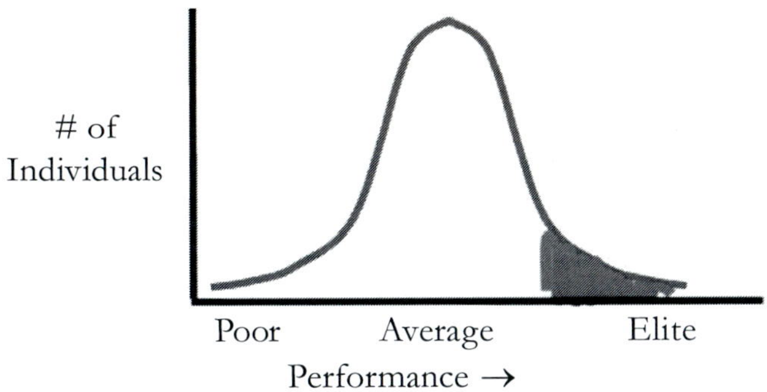

Markers allow breeders to find individuals in the right-hand tail of the curve more quickly, more consistently and at a more competitive cost

Case Study

With over a decade of experience in marker-assisted breeding projects, DNA LandMarks has had the opportunity to participate in some tremendous advances in our clients' breeding programs.

Case Study: Improving Yield in Popping Corn

- Morphology of popcorn limits yield
- Project focused on identifying and mapping "popping" QTL

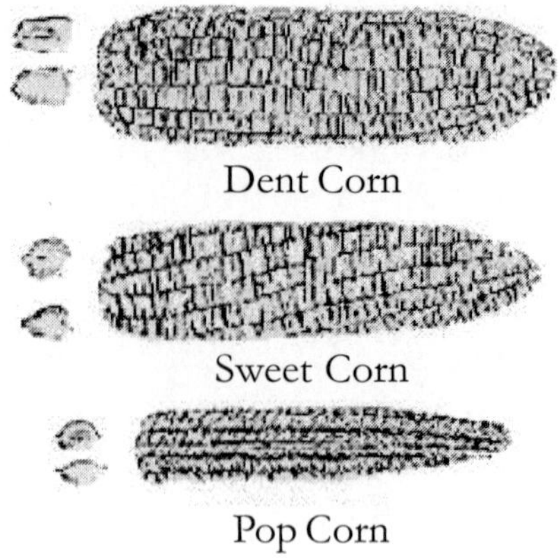

QTL Analysis

- Cross yellow dent variety with a popping variety
- 3 F2 populations (250 plants each)
- 15 traits assessed in F3 families segregating for popping QTL
- 12-13 QTL identified for various popping traits
- Traits tracked using linked SSR markers

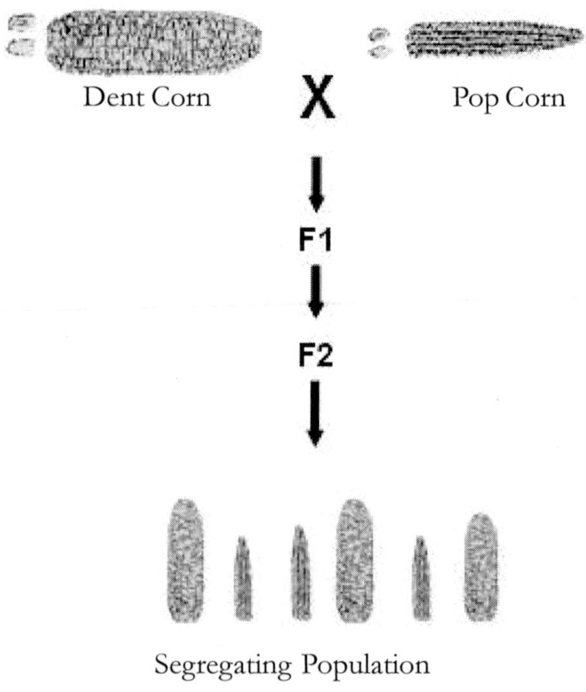

Once popping QTL had been identified and linked to SSRs, two populations were created using a popping donor parent and a yellow dent recurrent parent.

BC1: Two Populations

- Sample size: 13,442 (population 1) & 10,152 (population 2)
- Stepwise selection with 12-13 QTL-linked markers for heterozygosity
- Background selection with SSR markers for Recurrent Parent

BC2: Two Populations

- Sample size: 15,372 & 5,339
- Stepwise selection with 12-13 QTL-linked markers for heterozygosity
- Background selection with SSR markers for Recurrent Parent.

BC2S1

- Sample size: 1,139 & 976
- Selection with 12-13 QTL-linked markers to fix the donor alleles

BC2S2

- Sample size: 563 & 1,038
- Further selection with markers to fix the donor alleles
- IMP genotyping for Recurrent Parent - 87% similarity.

BC2S2 x Testers: Field Trials

Results

- 20-25% yield increase

Marker-Assisted Selection

Marker-assisted selection is the most widely used application of DNA markers. Once traits have been mapped and a closely linked marker has been found, it is possible to screen large numbers of samples for rapid identification of progeny that carry desirable characteristics.

Example: Resistant & susceptible lines in tomato

A marker link to the resistance gene scores at different positions for resistant and susceptible lines.

Using markers for trait selection has numerous advantages when compared to conventional plant breeding:

- *Speed* – DNA can be extracted from tissue from the first leaves or the cotyledons of a plant. Trait information can be discovered with markers prior to pollination allowing more informed crosses to be made.
- *Consistency* – Markers remove the impact of environmental variation that often complicates phenotypic evaluation.
- *Biosafety* – Using markers in screening for disease resistance means not having to introduce the pathogen into breeding populations. Particularly for livestock breeding this delivers a very important level of biosafety.
- *Efficiency* – Screening progeny early in the process allows a breeder to drop "also-rans" from the program more quickly. Most breeding programs that use markers still evaluate the same number of plants in the field however the level of genetic quality is vastly increased because of the early-stage screening that has been carried.
- *Complex traits* – Most multigenic traits are very difficult to manage through conventional plant breeding. The statistical chance of getting the required allele at each of a number of loci is very low. Markers allow you to skew the odds in your favour.

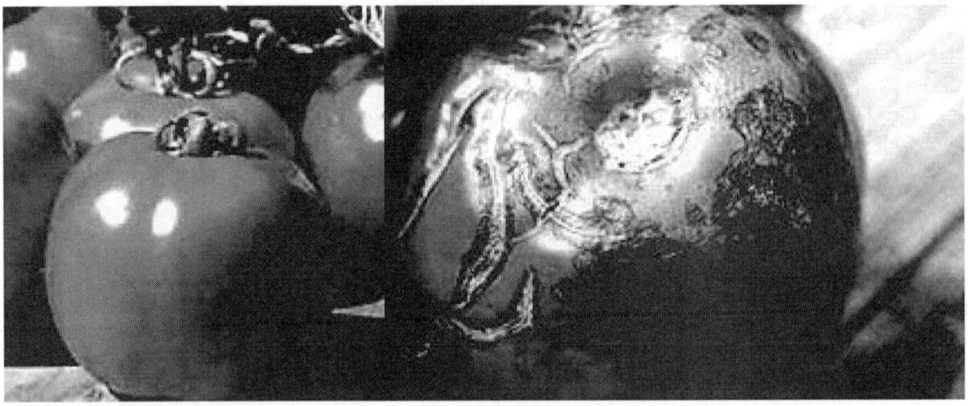

In most major crops, various linked markers exist to screen key traits. However if the trait has not been mapped to a marker, it is possible for DNA LandMarks to carry out the research necessary to find linked markers for your trait of interest (see Marker Development and Mapping).

Marker-Assisted Backcrossing

Backcrossing (or trait introgression) has long been a valuable strategy in plant breeding for a number of crops. The goal of backcrossing is to move a single trait of interest (e.g. disease resistance from a wild relative or a transgene from a donor line) into the genome of a commercially valuable variety without losing any part of the commercial variety's existing genome. The plant with the gene of interest is the donor parent, while the commercially viable variety is the recurrent parent.

Using DNA markers can accelerate a backcrossing program significantly. For example, most conventional corn backcrossing programs require 4-6 backcross generations before sufficient recurrent parent genome is recovered to release the line commercially.

Using markers allows a breeder to reach the same goal in 2 backcross generations thereby cutting 1-2 years of the product development cycle.

If **linkage drag** is a problem, markers that flank the gene of interest can be used to look for rare recombination events that break the linkage.

Marker-assisted backcrossing can be carried out with most marker types however IMP markers are particularly well suited to this task. Because each IMP reaction generates many, well distributed datapoints, it is possible to cover the genome by running only a handful (5-10) primers. IMP-assisted backcrossing can be carried out for a much lower cost compared to SSR markers.

How to Start Breeding Work to use Marker-Assisted Selection

1. Assess genetic variation for the target trait, particularly in landraces, elite breeding lines and other potential donors.
2. Identify molecular markers linked to the target trait.
3. Use the markers directly in the breeding programme to follow the introgression of desirable regions of the genome. This would enable us to breed directly for improved variety rather than depending on unreliable field screens and also to incorporate multiple resistance genes along with accumulating favourable alleles for other agronomically important traits (gene stacking).

Steps

Steps for MAS have been explained using an example in Pearl Millet for the transfer of drought tolerance and downy mildew resistance genes in parents of very famous hybrid HHB 67.

a. Creation of a molecular marker-based genetic linkage map of the crop genome

To do this the polymorphic DNA markers are used. Genetic mapping is based on the statistical likelihood that markers that are linked will be inherited together and this is used to order markers on the chromosome. The closer the markers, the more likely the linked markers will be inherited together. Mapping also indicates relative distances between markers and assigns them to linkage groups. For example, Pearl millet has seven linkage groups that correspond to its seven chromosomes ($2n = 2 \times 7 = 14$).

b. Search for associations with traits of interest

An individual's phenotype is a composite that depends on the segregation of several naturally occurring polymorphic genes. Areas or regions on the genome that affect a target trait are known as a quantitative trait loci or QTLs. To map QTLs for a particular trait, highly inbred homozygous parents are chosen that differ in their response to that trait. These are crossed to produce heterozygous F_1 seed. An F_1 plant is grown from a single seed, self-fertilised, and the resultant F_2 seed is sown to produce a segregating F_2 population. A tissue sample is taken from each F_2 plant for analysis using DNA markers (genotyping) and the individual segregating plants are self-fertilized and bulked to produce progenies. For each individual cross, a skeleton genetic map is developed with anchor loci that are polymorphic between the two parents of the cross. The bulk progenies are screened for target trait in wide range of environments.

C. Use of marker-assisted selection (MAS) into the ongoing breeding programme

Once the QTLs are mapped, MAS can be used to breed for resistance. Markers flanking the QTLs are used as tags to identify pieces of the donor parent linkage group containing the putative resistance QTLs. The F_1 is back-crossed several times to an agriculturally important variety. At each stage, molecular markers are used to identify individual plants that contain as much of the recurrent parent genome as possible, but that are heterozygous for the region carrying the resistance QTLs. Selected plants are then finally self-fertilized to obtain homozygous progeny. MAS has a number of important advantages in terms of transferring desirable genes, but the main one is that the breeding process is more rapid, selection being based on the test for the appropriate markers rather than on unreliable testing for phenotype in the field.

27

Mapping Populations

Genetic Mapping

Perhaps one of the most powerful applications of markers is genetic mapping. This technology allows researchers identify markers located within or very close to genes of interest. With this it is possible to trace these genes through generations of crosses.

If the trait in question is complex and multigenic, mapping will also be of tremendous help to dissect the various genes or loci that contribute to the overall characteristic. Determining the amount of impact each locus delivers is a process called Quantitative Trait Loci or QTL mapping.

Single Gene or QTL?

Determining whether a trait of interest is controlled by a single gene or by multiple loci is an important first step in genetic mapping. A general rule of thumb is that the stronger the heritability of the trait, the less complex (fewer loci) it encompasses. Another indication is the phenotypic data generated. Narrow distributions for specific phenotypes indicate single gene control whereas multigenic traits usually show a single, broader distribution of phenotypes.

In either case, the key to successful mapping is generating a population that shows good segregation for the trait(s) of interest.

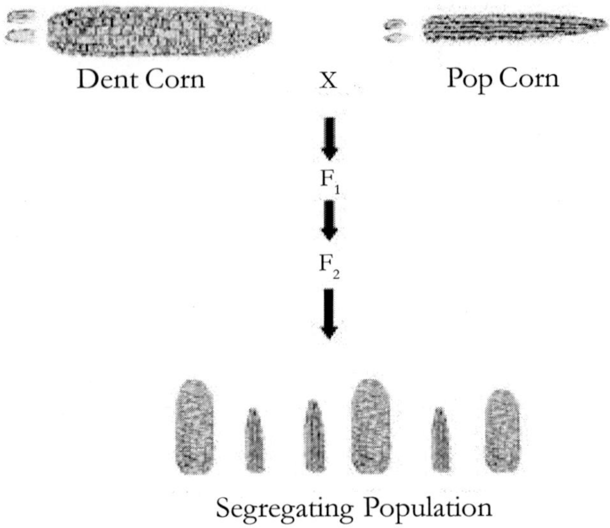

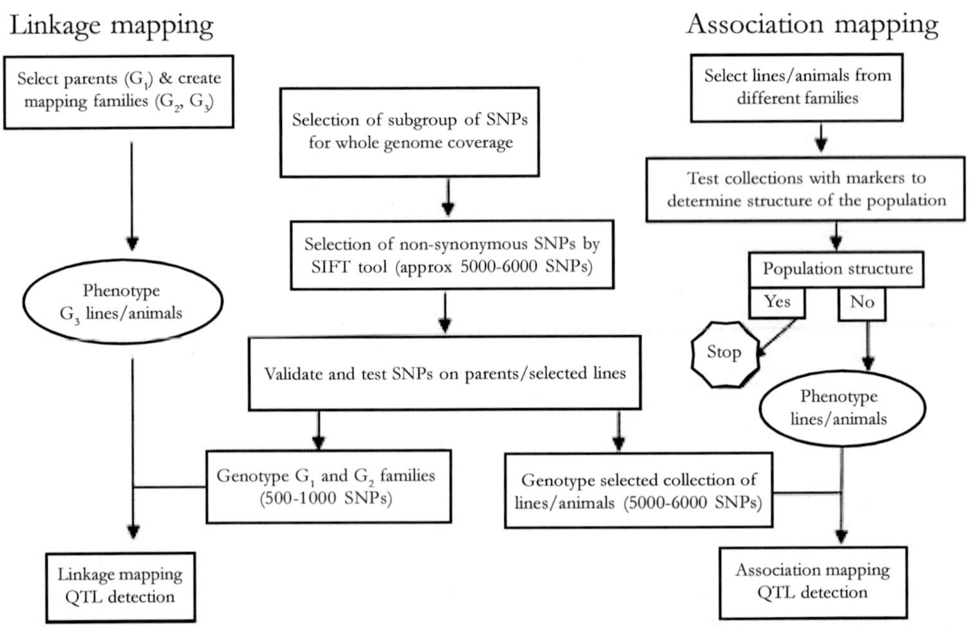

QTL Intervals usually covers large physical distance subsequent fine mapping with more markers and more lines/animals required to narrow down candidate genes

More precise location of candidate genes but need a non-structured population of lines/animals and more SNP markers

Association Mapping

In some cases, depending on the population being used, it might make more sense to take an Association Mapping approach. The following flowchart outlines the differences between linkage QTL analysis and association mapping:

In genetics and breeding, mapping population are the tools used to identify the genetic loci controlling measurable phenotypic traits. The development of molecular marker technology has caused renewed interest in genetic mapping. An appropriate mapping population, suitable marker system and the software for analyses of data are the key requirements for a molecular mapping and molecular breeding programme. Genetic map construction requires:

1. Appropriate mapping population(s)
2. Calculate pairwise recombination frequencies using these populations
3. Establish linkage groups and estimate map distances and
4. Determine map order

Mapping Populations
(F_2 *Population, RIL's, Back Cross Population, NIL's, Introgression lines*)
↓
Computer Software
[Linkage 1 (Suiter et al., 1983), GMendel (Echt et al., 1992), Mapmaker (Lander and Botstein, 1986; Lander et al., 1987), Mapmanager (Manly and Eliot, 1991) and Joinmap (Stam, 1993)]
↓
Recombination Frequency
↓
Linkage Map

Mapping Populations

A population used for gene mapping is commonly called a mapping population. Commonly used mapping populations are usually obtained from controlled crosses. Selection of a population for genome mapping involves choosing of parents and determining a mating scheme. Decisions on selection of parents, mating designs and the type of markers should depend upon the objectives of experiments.

Parents of mapping populations must have sufficient variation for the traits of interest at both the DNA sequence and the phenotype level. The variation at DNA level is essential to trace the recombination events. The more DNA sequence variation exists, the easier it is to find polymorphic informative makers. When the objective is to search for genes controlling a particular trait, genetic variation of trait between parents is important. If the parents are greatly different at phenotypic level for a trait, there is a reasonable chance that genetic variation exists between the parents, although uncontrolled environmental effects could create large phenotypic variation without any genetic basis for the effects. However, lack of phenotypic variation between parents does not mean that there is no genetic variation, as different sets of genes could result in same phenotype.

Selection of parents for developing mapping population : Selection of parents for developing mapping population is critical to successful map construction. Since a map's economic significance will depend upon marker-trait association, as many qualitatively inherited morphological traits as possible should be included in the genetic stocks chosen as parents for generating mapping population. Consideration must be given to the source of parents (adapted vs. exotic) used in developing mapping population. Chromosome pairing and recombination rates can be severely disturbed (suppressed) in wide crosses and generally yield greatly reduced linkage distances (Albine and Jones, 1987; Zamir and Tadmar, 1986). Wide crosses will usually provide segregating populations with a relatively large array of polymorphism when compared to progeny segregating in a narrow cross (adapted x adapted). To have significant value in crop improvement programme, a map made from a wide cross must be collinear (i.e. order of loci similar) with map constructed using adapted parents.

Types of mapping populations : Different types of mapping populations that are often used in linkage mapping are: (i) F_2 population; (ii) F_2 derived F_3 (F_2:F_3) populations; (iii) Backcrosses; (iv) Doubled Haploids (DHs); (v) Recombinant Inbred Lines (RILs); and (vi) Near-isogenic Lines (NILs). The characteristic features, merits and demerits of each of these populations are briefly presented below:

F_2 Population

- Produced by selfing or sib mating the individuals in segregating populations generated by crossing the selected parents.
- F_2 individuals are products of single meiotic cycle
- Ratio expected for dominant marker is 3:1 and for codominant marker is 1:2:1

Merits

- Best population for preliminary mapping
- Requires less time for development
- Can be developed with minimum efforts, when compared to other populations.

Demerits

- The major drawback to using F_2 or backcross populations is that the populations are not eternal. Therefore, your source of tissue to isolate DNA or protein will be exhausted at some point in time. You then would have to begin mapping again in another population.
- Linkage established using F_2 population is based on one cycle of meiosis
- F_2 populations are of limited use for fine mapping
- Quantitative traits cannot be precisely mapped using F_2 population as each individual is genetically different and cannot be evaluated in replicated trials over locations and years. Thus, the effect the G x E interaction on the expression of quantitative traits cannot be precisely estimated.
- Not a long-term population; impossible to construct exact replica or increase seed amount.

F_2 Derived F_3 ($F_{2:3}$) Population

- $F_{2:3}$ population is obtained by selfing the F_2 individuals for a single generation
- Suitable for specific situations like

  Mapping quantitative traits

  Mapping recessive genes

  The $F_{2:3}$ family can be used for reconstituting the genotype of respective F_2 plants, if needed, by pooling the DNA from plants in the family

Demerit : Like F_2 population, it is not 'immortal'.

Backcross Population

Backcross populations are generated by crossing the F_1 with either of the parents. Usually in genetic analysis, backcross with recessive parent (testcross) is used. With respect to molecular markers, the backcross with dominant parent (B_1) would segregate in a ratio 1:0 and 1:1 for dominant and codominant markers, respectively. However, backcross with recessive parent (B_2) or testcross would segregate in a ratio of 1:1 irrespective of the nature of marker. Like an F_2 population, the backcross populations require less time to be developed, but are not 'immortal'. However, the recombination information in case of backcrosses is based on only one parent (the F_1). The specific advantage of backcross populations is that, the populations can be further utilized for marker-assisted backcross breeding.

Doubled Haploids (DHs)

- Chromosome doubling of anther culture derived haploid plants from F_1 generates DHs. The suitability of doubled-haploid progenies for mapping project has been demonstrated in by Lefebvre et al. (1995) in pepper.
- DHs are also products of one meiotic cycle, and hence comparable to F_2 in terms of recombination information.
- The expected ratio for the marker is 1:1, irrespective of genetic nature of marker (whether dominant or codominant).

Merits

- DHs are permanent mapping population and hence can be replicated and evaluated over locations and years and maintained without any genotypic change
- Useful for mapping both qualitative and quantitative characters
- Instant production of homozygous lines, thus saving time

Demerits

- Recombination from the male side alone is accounted.
- Since it involves *in vitro* techniques, relatively more technical skills are required in comparison with the development of other mapping populations
- Often suitable culturing methods / haploid production methods are not available for number crops and different crops differ significantly for their tissue culture response. Further, anther culture induced variability should be taken care of.

Recombinant Inbred Lines (RILs)

The major drawback to using F_2 or backcross populations is that the populations are not eternal. Therefore, your source of tissue to isolate DNA or protein will be exhausted at some point in time. You then would have to begin mapping again in another population. Populations of recombinant inbred lines can be a powerful solution to this problem. Recombinant inbred lines are developed by single-seed selections from individual plants of an F_2 population. (Because of this procedure, these lines are also called F_2-derived lines). Single-seed descent is repeated for several generations. At this point, all of the seed from an individual plant is bulked. For example, a $F_{3:4}$ RI population underwent single-seed descent through the F_3 generation, and was bulked to develop the F_4. This population of seed can then be grown to obtain a large quantity of seed of each individual line. Importantly, each of the lines is fixed for many recombination events

- RILs are produced by continuous selfing or sib mating the progeny of individual members of an F_2 population until complete homozygous is achieved
- SSD method is best suited for developing RILs. Bulk method and pedigree methods without selection can also be used
- RILs also equalize marker types like DHs, the genetic segregation ratio for both dominant and co dominant marker would be 1:1
- RILs developed though brother-sister mating require more time than those developed through selfing. The number of inbred lines required is twice, in case they are developed through brother–sister mating compared to selfing particularly, when linkage is not very tight.

Merits

- Once homozygosity is achieved, RILs can be propagated indefinitely without further segregation
- Since RILs are immortal population, they can be replicated over locations and years and therefore are of immense value in mapping QTLs
- RILs being obtained after several cycles of meiosis, are very useful in identifying tightly linked makers
- RIL populations obtained by selfing have twice the amount of observed recombination between very closely linked markers as compared to population derived from a single cycle of meiosis.

Demerits

- Requires many seasons / generations to develop
- Developing RILs is relatively difficult in crops with high inbreeding depression.

Immortalized F_2 Population

Immortalized F_2 populations can be developed by paired crossing of the randomly chosen RILs derived from a cross in all possible combinations excluding reciprocals. The set of RILs used for crossing along with the F_1s produced, provide a true representation of all possible genotype combinations (including the heterozygotes) expected in the F_2 of the cross from which the RIls are derived. The RILs can be maintained by selfing and required quantity of F1 seed can be produced at will by fresh hybridization. This population therefore provides an opportunity to map heterotic QTLs and interaction effects from multilocation data.

Near-Isogenic Lines (NILs)

- NILs are generated either by repeated selfing or backcrossing the F_1 plants to the recurrent parents.
- NILs developed through backcrossing are similar to recurrent parent but for the gene of interest, while NILs developed though selfing are similar in pair but for the gene of interest (however, differ a lot with respect to the recurrent parent)
- Expected segregation ratio of the markers is 1:1 irrespective of the nature of marker.

Merits

- Like DHs and RILs, NILs are also 'immortal mapping population'
- Suitable population for tagging the trait, wherever such population is available
- NILs are quite useful in functional genomics.

Demerits

- Require many generations for development
- Directly useful only for molecular tagging of the gene concerned, but not for linkage mapping
- Linkage drag is a potential problem in constructing NILs, which has to be taken care of.

Besides these mapping populations, disomic chromosome substitution lines, addition lines can also be used for localizing genes and developmental mutants for characterizing sequence variations (ecotilling) using molecular markers.

Bulk Segregant Analysis

Often a plant breeder is not interested in developing a complete molecular map, but would rather find a marker or a few markers that are closely linked to a target trait to be improved. The identification of these markers is often achieved by a procedure called *Bulk Segregant Analysis*(BSA). The core of this procedure is the formation of a bulk sample of DNA for analysis by pooling DNA from individuals with similar phenotypes, may be from segregating population like F_2. For example, if we are interested in finding a molecular locus linked to a disease resistance locus, we would create two bulk DNA samples, one containing DNA from plants or lines that are resistant to the disease and a second bulk containing DNA from plants or lines that are susceptible to the disease. Each of these bulk DNA samples will contain a random sample of all the loci in the genome, except for those that are in the region of the gene upon which the bulking occurred. Therefore, any difference in DNA marker (RFLP or RAPD or SSR etc.) pattern between these two bulks should be linked to the locus upon which the bulk was developed. This is a powerful technique and has gained wide acceptance.

In BSA approach, any one of the above-mentioned populations (except NILs) may be used in gene tagging. Usually, the DNA isolated from 10 plants in each group is pooled to constitute resistant and susceptible bulks. The four samples i.e. resistant parent, susceptible parent, resistant bulk and susceptible bulk, are surveyed for polymorphism using polymorphic markers (see detailed illustration above and also elsewhere in this manual). A marker showing polymorphism between parents as well as bulks is considered putatively linked to the target trait, and is further used for mapping using individual F_2 plants. Conceptually, the genetic constitution of the two bulks is similar, but for the genomic region associated with the target trait. Hence, they serve the purpose of isogenic lines in principle.

It has been observed over experiments that when 10 plants are sampled in each group for constituting the bulk, the probability of a polymorphic marker (between parents as well as bulks) not being linked to the target trait is extremely low (10^{-19}). Hence, usually 10 plants are used for constituting the bulks. However, this number may vary depending upon the types of mapping populations used. In absence of isogenic lines, the BSA approach provides a very useful alternative for gene tagging (Michelmore *et al.*, 1991).

Combining Markers and Populations

The genetic segregation ratio at maker locus is jointly determined by the nature of marker (dominant / codominant) and types of mapping populations. Therefore, a thorough understanding of the nature of markers and mapping population is crucial for any mapping projects. Markers such RFLPs, microsatellites and CAPS etc. are codominant in nature, while AFLP, RAPD, ISSR are often scored as dominant markers. Mapping populations such as RILs and DHs equalize marker type because of fixation of parental alleles at marker locus in homozygous condition. These population result in 1: 1 segregation ratio at marker locus irrespective of genetic nature of markers, while an F_2 population segregates in 1: 2: 1 ratio for a codominant marker and in 3:1 ratio for dominant marker. Depending upon the segregation pattern, statistical analysis of marker data will vary.

Characterization of Mapping Populations

Precise molecular and phenotypic characterization of mapping population is vital for success of any mapping project. Since the molecular genotype of any individual is independent of environment, it is not influenced by G x E interaction. However, trait phenotype could be influenced by the environment, particularly in case of quantitative characters. Therefore, it becomes important to precisely estimate the trait value by evaluating the genotypes in multilocation testing over years using immortal mapping populations to have a valid marker-trait association.

28

QTL Mapping

QTL Traits

Polygenic inheritance, also known as quantitative or multifactorial inheritance refers to inheritance of a phenotypic characteristic (trait) that is attributable to two or more genes and their interaction with the environment. Unlike monogenic traits, polygenic traits do not follow patterns of Mendelian inheritance (qualitative traits). Instead, their phenotypes typically vary along a continuous gradient depicted by a bell curve. Though not necessarily genes themselves, quantitative trait loci (QTLs) are stretches of DNA that are closely linked to the genes that underlie the trait in question. QTLs can be molecularly identified (for example, with PCR or AFLP) to help map regions of the genome that contain genes involved in specifying a quantitative trait. This can be an early step in identifying and sequencing these genes.

Quantitative Trait Locus

'QTL' (Quantitative Trait Loci), a term first coined by Gelderman (1975). A QTL is defined as "a region of the genome that is associated with an effect on a quantitative trait". Conceptually, a QTL can be a single gene, or it may be a cluster of linked genes that affect the trait. Typically, QTLs underlie continuous traits (those traits that vary continuously, e.g. height) as opposed to discrete traits (traits that have two or several character values, e.g. eye colour in humans or smooth vs. wrinkled peas used by Mendel in his experiments).

Moreover, a single phenotypic trait is usually determined by many genes. Consequently, many QTLs are associated with a single trait.

A quantitative trait locus (QTL) is a region of DNA that is associated with a particular phenotypic trait - these QTLs are often found on different chromosomes. Knowing the number of QTLs that explains variation in the phenotypic trait tells us about the genetic architecture of a trait. It may tell us that plant height is controlled by many genes of small effect, or by a few genes of large effect.

Another use of QTLs is to identify candidate genes underlying a trait. Once a region of DNA is identified as contributing to a phenotype, it can be sequenced. The DNA sequence of any genes in this region can then be compared to a database of DNA for genes whose function is already known.

In a recent development, classical QTL analyses are combined with gene expression profiling i.e. by DNA microarrays. Such expression QTLs (e-QTLs) describe cis- and trans-controlling elements for the expression of often disease-associated genes. Observed epistatic effects have been found beneficial to identify the gene responsible by a cross-validation of genes within the interacting loci with metabolic pathway- and scientific literature databases.

QTL Mapping

The problem with analyzing polygenic traits like height is the fact that many genetic loci affect the trait. Hence, traditional techniques of mutational analysis cannot be solely employed to solve this question. The idea behind QTL Mapping is to determine the degree of association of a specific region on the genome to the inheritance of the trait of interest. If the association is high, there is a good probability that the region of genome we are looking at has something to do with expression of the trait.

Towards this goal, the first step in QTL Mapping is creation of a map of the genome using molecular markers like RAPD, RFLP, SNP etc. These markers serve as landmarks on the genome. The most ideal scenario is to have markers spread at regular intervals over the entire genome, but that is often not the case, Prior to the advent of molecular markers, visible polymorphisms and allozymes were used in this step. The idea is to look at the pattern of association of these markers with the inheritance of the trait. It must be noted that these molecular markers are not necessarily genes, and thus QTL Mapping ends up telling us the degree of association of a stretch of DNA with a trait rather than pointing directly at the gene responsible for that trait. Different statistical methods are used to ascertain whether the degree of association is significant or not.

QTL mapping is the statistical study of the alleles that occur in a locus and the phenotypes (physical forms or traits) that they produce. Because most traits of interest are governed by more than one gene, defining and studying the entire locus of genes related to a trait gives hope of understanding what effect the genotype of an individual might have in the real world.

Statistical analysis is required to demonstrate that different genes interact with one another and to determine whether they produce a significant effect on the phenotype. QTLs identify a particular region of the genome as containing a gene that is associated with the trait being assayed or measured. They are shown as intervals across a chromosome, where the probability of association is plotted for each marker used in the mapping experiment.

The QTL techniques were developed in the late 1980s and can be performed on inbred strains of any species.

To begin, a set of genetic markers must be developed for the species in question. A marker is an identifiable region of variable DNA. Biologists are interested in understanding the genetic basis of phenotypes (physical traits). The aim is to find a marker that is significantly more likely to co-occur with the trait than expected by chance, that is, a marker that has a statistical association with the trait. Ideally, they would be able to find the specific gene or genes in question, but this is a long and difficult undertaking. Instead, they can more readily find regions of DNA that are very close to the genes in question. When a QTL is found, it is often not the actual gene underlying the phenotypic trait, but rather a region of DNA that is closely linked with the gene.

For organisms whose genomes are known, one might now try to exclude genes in the identified region whose function is known with some certainty not to be connected with the trait in question. If the genome is not available, it may be an option to sequence the identified region and determine the putative functions of genes by their similarity to genes with known function, usually in other genomes.

Another interest of statistical geneticists using QTL mapping is to determine the complexity of the genetic architecture underlying a phenotypic trait. For example, they may be interested in knowing whether a phenotype is shaped by many independent loci, or by a few loci, and do those loci interact. This can provide information on how the phenotype may be evolving.

Principle of QTL Mapping

It is not difficult in populations of most crop plants to identify and map a good number of segregating markers (10 to 50) per chromosome. However, most of these markers would be in non-coding regions of the genome and might not affect the trait of interest directly; but, a few of these markers might be linked to genomic regions (QTLs) that do influence the trait of interest. Where such linkage occurs, the marker locus and the QTL will cosegregate. Therefore, the basic principle of determining whether a QTL is linked to a marker is to partition the mapping population into different genotypic classes based on genotypes at the marker locus, and the apply correlative statistics to determine whether the individuals of one genotype differ significantly with the individuals of other genotype with respect to the trait being measured. Situations where genes fail to segregate independently are said to display "linkage disequilibrium". QTL analysis, thus, depends on linkage disequilibrium.

With natural populations, consistent association between QTL and marker genotype will not usually exist, except in a very rare situation where the marker is completely linked to the QTL. Therefore, QTL analysis is usually undertaken in segregating mapping populations, such as F_2-derived populations, recombinant inbred lines (RILs), near-isogenic lines (NILs), doubled haploid lines (DHs), and backcross populations.

Objectives of QTL Mapping

The vast majority of molecular marker research in quantitative traits has been devoted to mapping QTL. These experiments basically have the following major objectives:

1. To identify the regions of the genome that affect the trait of interest
2. To analyze the effect of the QTL on the trait:
    - How much of the variation for the trait is caused by a specific region?
    - What is the gene action associated with the QTL (additive effect? dominant effect?)
    - Which allele is associated with the favorable effect?

Requirements for QTL Mapping

1. Measureable trait
2. Segregating Population
   a. F_2, F_3
   b. Recombinant Inbred Lines (RILs)
   c. Double Haploid Lines
3. Molecular Markers
   a. Restriction Fragment Length Polymorphism(RFLP)
   b. Amplified Fragment Length Polymorphism(AFLP)
   c. Cleaved Amplified Polymorphic Sequences(CAPS)
   d. Simple Sequence Length Polymorphism(SSLP)
4. Reliable phenotypic screening of mapping population
5. Appropriate statistical packages to analyze the genotypic information in combination with phenotypic information for QTL detection.

Flow Chart explaining QTL analysis

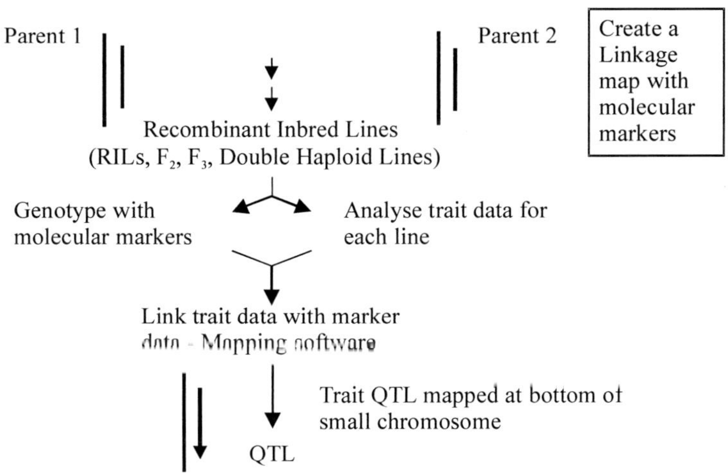

Types of Mapping Population

Different types of mapping populations that are often used in QTL mapping are: (i) F_2 population; (ii) F_2 derived F_3 (F_2:F_3) populations; (iii) Recombinant Inbred Lines (RILs); and (iv) Near-isogenic Lines (NILs). Random-mating populations are more difficult for QTL mapping, because linkage disequilibrium is a key to detecting QTLs with markers. It is essential to develop a suitable experimental mapping population using parental lines that are highly contrasting phenotypically for the target trait (ex., highly resistant and susceptible lines). Another important requirement is that these parental lines should be genetically divergent; this is important to enhance the possibility of identifying a large set of polymorphic markers that are well-distributed across the genome. To fulfill the second criterion, one may have to carry out molecular polymorphism survey across a set of potentially useful lines so as to identify the most suitable ones for generation of mapping population. The choice of a mapping population could vary based upon the objectives of the experiment, the timeframe as well as resources available for undertaking QTL analysis. But, the ability to detect QTLs or the information contained in F_2 or F_2-derived populations and RILs are relatively higher than others. The primary advantage of $F_{2:3}$ families is the ability to measure the effects of additive and dominance gene actions at specific loci. Because RILs are essentially homozygous, only additive gene action can be measured. The advantage, though, of the RILs is the ability to perform larger experiments at several locations and even in multiple years. For many crops, it is not possible to generate enough seed to perform a multi-location experiment with population of $F_{2:3}$ families. Modification of the genetic model is necessary to accommodate different types of populations.

The size of the mapping population for QTL analysis depends on several factors, including the type of mapping population employed for analysis, genetic nature of the target trait, objectives of the experiment, and the resources available for handling a sizable mapping population in terms of phenotyping and genotyping. While analysis of a large number of individuals (~500 or more) would enable detection of even QTLs having small effects on the target trait, from the practical point of view (MAS), the basic purpose of QTL mapping would be largely served if one can detect the QTLs with major effects. This would require, in general, a mapping population of a size of 200-300 individuals.

The characteristic features, of OTL mapping population can be described as below.

F_2 population

- Produced by selfing or sib mating the individuals in segregating populations generated by crossing the selected parents.
- F_2 individuals are products of single meiotic cycle
- Ratio expected for dominant marker is 3:1 and for codominant marker is 1:2:1

Merits

- Best population for preliminary mapping
- Requires less time for development
- Can be developed with minimum efforts, when compared to other populations.

Demerits

- The major drawback to using F_2 or backcross populations is that the populations are not eternal. Therefore, your source of tissue to isolate DNA or protein will be exhausted at some point in time. You then would have to begin mapping again in another population.
- Linkage established using F_2 population is based on one cycle of meiosis
- F_2 populations are of limited use for fine mapping
- Quantitative traits cannot be precisely mapped using F_2 population as each individual is genetically different and cannot be evaluated in replicated trials over locations and years. Thus, the effect the G x E interaction on the expression of quantitative traits cannot be precisely estimated.
- Not a long-term population; impossible to construct exact replica or increase seed amount.

F_2 derived F_3 ($F_{2:3}$) population:

- $F_{2:3}$ population is obtained by selfing the F_2 individuals for a single generation
- The $F_{2:3}$ family can be used for reconstituting the genotype of respective F_2 plants, if needed, by pooling the DNA from plants in the family.

Demerit: Like F_2 population, it is not 'immortal'

Recombinant Inbred Lines (RILs)

The major drawback to using F_2 or backcross populations is that the populations are not eternal. Therefore, your source of tissue to isolate DNA or protein will be exhausted at some point in time. You then would have to begin mapping again in another population. Populations of recombinant inbred lines can be a powerful solution to this problem. Recombinant inbred lines are developed by single-seed selections from individual plants of an F_2

population. (Because of this procedure, these lines are also called F_2-derived lines). Single-seed descent is repeated for several generations. At this point, all of the seed from an individual plant is bulked. For example, a $F_{3:4}$ RI population underwent single-seed descent through the F_3 generation, and was bulked to develop the F_4. This population of seed can then be grown to obtain a large quantity of seed of each individual line. Importantly, each of the lines is fixed for many recombination events.

- RILs are produced by continuous selfing or sib mating the progeny of individual members of an F_2 population until complete homozygous is achieved
- SSD method is best suited for developing RILs. Bulk method and pedigree methods without selection can also be used
- RILs also equalize marker types like DHs, the genetic segregation ratio for both dominant and co dominant marker would be 1:1
- RILs developed though brother-sister mating require more time than those developed through selfing. The number of inbred lines required is twice, in case they are developed through brother–sister mating compared to selfing particularly, when linkage is not very tight.

Merits

- Once homozygosity is achieved, RILs can be propagated indefinitely without further segregation
- Since RILs are immortal population, they can be replicated over locations and years and therefore are of immense value in mapping QTLs
- RILs being obtained after several cycles of meiosis, are very useful in identifying tightly linked makers
- RIL populations obtained by selfing have twice the amount of observed recombination between very closely linked markers as compared to population derived from a single cycle of meiosis.

Demerits

- Requires many seasons / generations to develop.

Developing RILs is relatively difficult in crops with high inbreeding depression.

Generating a Reasonably Saturated Linkage Map

By screening the mapping population using polymorphic molecular markers (popularly called as 'genotyping'), we can analyze the segregation patterns for each of the markers. The segregation patterns are usually in consonance with the type of mapping population used. The genotypic data is then analyzed using a statistical package such as MAPMAKER (Lander et al., 1987) or JOINMAP (Stam, 1993), for construction of a linkage map of the molecular markers analyzed in the study. Mapping means placing the markers in order, indicating the relative genetic distances between them, and assigning them to their linkage groups on the basis of recombination values from all pairwise combinations between the markers.

To perform a whole-genome QTL scan, it is desirable to have a saturated marker map. In such a map, markers are available for each chromosome from one end to the other, and adjacent markers are spaced sufficiently close that recombination events only rarely occur between them. For practical purposes, this is generally considered to be less than 10 recombinations per 100 meioses, or a map distance of less than 10 centiMorgans (cM). In the model plant Arabidopsis thaliana, which has a particularly small genome, this requires as few as 50 markers. Several-fold more markers are needed for plant genomes such as wheat and maize. In crops like maize, a broad 'rule-of-the-thumb' is to cover each of the chromosomal (bin) locations with at least one or two polymorphic molecular markers.

Phenotyping of Mapping Population and Sample Size

The target quantitative traits have to be measured as precisely as possible, and limited amounts of missing data can be tolerated. The power to resolve the QTL location is limited first by sample size, and then by genetic marker coverage of the genome. Often, the number of individuals in a sample might appear to be large, but missing data or skewed allele frequencies in the population cause the effective sample size to diminish, thus sacrificing statistical power. Sometimes, it may be necessary to sacrifice population size in favour of data quality, and this trade-off means that only major QTL (with relatively large effect) can be detected. Data is typically pooled over locations and replications to obtain a single quantitative trait value for the line. It is also preferable to measure the target trait(s) in experiments conducted in multiple (and appropriate) locations to have a better understanding of the QTL x environment interaction, if any.

Statistical Methods for QTL Mapping

The basic objective in QTL mapping studies is to detect QTL, while minimizing the occurrence of false positives (Type I errors, that is, declaring an association between a marker and QTL when in fact one does not exist). Tests for QTL/trait association are often performed by the following approaches:

a) Single Marker Approach

The single marker approach, sometimes referred to as the single factor analysis of variance (SF-ANOVA) or single point analysis, has been used extensively, especially with isozymes (Tanksley et al., 1982; Edwards et al., 1987). SF-ANOVA is done for each marker locus independent of information from other loci. F-tests provide evidence whether differences between marker locus genotype classes are significant or not.

Although computationally simple, this approach suffers from some major limitations: (i) the likelihood of QTL detection significantly decreases as the distance between the marker and QTL increases; (ii) the method cannot determine whether the markers are associated with one or more QTLs; (iii) the effects of QTL are likely to be underestimated because they are confounded with recombination frequencies.

When to use: For quick scanning of the entire genome (all chromosomes) to find best possible QTLs. Use single-marker analysis first to ensure your data file is clean; then move on more sophisticated analysis methods, such as Interval Mapping and Composite Interval Mapping.

How it works: Single-marker analysis is based on the idea that if there is an association between a marker genotype and trait value, it is likely that a QTL is close to that marker locus.

Comments: Single-marker analysis can be somewhat useful for a quick look at data, but it has been superseded by Interval Mapping and Composite Interval Mapping. IM and CIM are more thorough and accurate indicators of QTL.

b) Simple Interval Mapping (SIM)

Interval mapping (IM) is an extension of single-marker analysis. In singlemarker analysis, only one marker is used in QTL mapping but effects are underestimated and the QTL position cannot be determined. Interval mapping provides a systematic way to scan the whole genome for evidence of QTL. Lander Botstein (1989) developed the Interval Mapping (IM) approach in which one marker interval at a time is analyzed to construct a putative QTL by performing a likelihood ratio test (LRT) at every position in the interval. IM uses two observable flanking markers to construct an interval within which to search for QTL. A map function (either Haldane or Kosambi) is used to translate from recombination frequency to distance or vice visa. Then, a LOD score is calculated at each increment (walking step) in the interval. Finally, the LOD score profile is calculated for the whole genome. When a peak has exceeded the threshold value, we declare that a QTL have been found at that location.

SIM was first proposed by Lander and Botstein (1989) and it takes full advantage of a linkage map. The method evaluates the target association between the trait values and the genotype of a hypothetical QTL (target QTL) at multiple analysis points between pair of adjacent marker loci (the target interval). Presence of a putative QTL is estimated if the log of odds ratio (LOD) exceeds a critical threshold. Lander and Botstein (1989) developed formulae for calculating significance levels appropriate for interval mapping when the genome size, number of chromosomes, number of marker intervals, and the overall false positive rate desired are given. SIM has been the most widely approach as it can be easily accessed through statistical packages such as MAPMAKER/QTL. By using tightly linked markers for analysis, it is possible to compensate for recombination between markers and the QTL, thereby increasing the probability of statistically detecting the QTL, and providing an unbiased estimate of QTL effect. However, when multiple QTLs are segregating in a cross (which is usually the case), SIM fails to take into account genetic variance caused by other QTLs. In such a case, SIM suffers essentially from the same shortcomings of single marker analysis.

c) Composite Interval Mapping (CIM)

The approach of Interval Mapping (IM) considers one QTL at a time in the model for QTL mapping. Therefore, IM can bias identification and estimation of a QTL when multiple QTLs are located in the same linkage group. Jansen (1993) and Zeng (1993, 1994) independently proposed the idea of combining IM with multiple regression analysis in mapping. Zeng named this combination "composite interval mapping" (CIM). When testing for the putative QTL in an interval CIM uses other markers as covariates to control for other QTL and to reduce the residual variance such that the test can be improved. The model of CIM includes one QTL and markers.

What it is: Composite interval mapping (CIM) adds background loci to simple interval mapping (IM). CIM fits parameters for a target QTL in one interval while simultaneously fitting partial regression coefficients for "background markers" to account for variance caused by non-target QTL. "In theory, CIM gives more power and precision than simple IM because the effects of other QTL are not present as residual variance. Furthermore, CIM can remove the bias that would normally be caused by QTL that are linked to the position being tested."

Background markers are usually 20-40cM apart. CIM (Zeng, 1994) and MQM (multiple-QTL model or marker-QTL-marker analysis) developed by Jansen and Stam (1994) combine interval mapping for a single QTL in a given interval with multiple regression analysis on marker associated with other QTL. It considers a marker interval plus a few other well-chosen single markers in each analysis, so that n-1 tests for interval-QTL associations are performed on a chromosome with nmarkers. The advantages of CIM are as follows: (i) mapping of multiple QTLs can be accomplished by the search in one dimension; (ii) by using linked markers as cofactors,the test is not affected by QTL outside the region, thereby increasing the precision of QTLmapping; and (iii) By eliminating much of the genetic variance by other QTL, the residual variance is reduced, thereby increasing the power of detection of QTL. CIM is more powerful than SIM, but is yet to be used extensively in QTL mapping.

d) Multiple Interval Mapping (MIM)

Another recent and interesting development is multiple interval mapping (MIM). It uses multiple marker intervals simultaneously to fit multiple putative QTL directly in the model for mapping QTL. The MIM model is based on Cockerham's model for interpreting genetic parameters and the method of maximum likelihood for estimating genetic parameters. With the MIM approach, the precision and power of QTL mapping could be improved. Also, epistasis between QTL, genotypic values of individuals, and heritabilities of quantitative traits can be readily estimated and analyzed. Using the MIM model, a stepwise selection procedure with likelihood ratio test statistic as a criterion is proposed to identify QTL. This MIM method was applied to a mapping data set of radiata pine on three traits: brown cone number, tree diameter, and branch quality scores. Based on the MIM result, seven, six, and five QTL were detected for the three traits, respectively. The detected QTL individually contributed from approximately 1 to 27% of the total genetic variation. Significant epistasis between four pairs of QTL in two traits was detected, and the four pairs of QTL contributed approximately 10.38 and 14.14% of the total genetic variation. The asymptotic variances of QTL positions and effects were also provided to construct the confidence intervals. The estimated heritabilities were 0.5606, 0.5226, and 0.3630 for the three traits, respectively. With the estimated QTL effects and positions, the best strategy of marker-assisted selection for trait improvement for a specific purpose and requirement can be explored. MIM is the extension of interval mapping to multiple QTLs, just as multiple regression extends analysis of variance. MIM allows one to infer the location of QTLs to positions between markers, makes proper allowance for missing genotype data, and can allow interactions between QTLs.

Model selection is the principal problem in multiple QTL methods; the chief concern is the formation of appropriate criteria for comparing models. The simplest multiple QTL method, multiple regression, should be used more widely, although, like analysis of variance, it suffers in the presence of appreciable missing marker genotype data. A forward selection procedure using interval mapping (i.e., the calculation of conditional LOD curves) is appropriate in cases of QTLs that act additively, and makes proper allowance for missing genotype data. MIM is an improved method, that, although computationally intensive, can, in principle, map multiple QTLs and identify interactions between QTLs.

In determining whether a LOD score is sufficiently large for one to be confident of the presence of a QTL, consider the distribution of the LOD score under the null hypothesis of no segregating QTL. Adjustment must be made for the genomewide search for QTLs, so consider the distribution of the maximum LOD score genomewide. Permutation tests are valuable for determining significance landmarks for the LOD score; although computationally intensive, permutation tests allow for the observed phenotype distribution, marker density, and pattern of missing genotype data.

Once QTL are detected, the next step is to estimate the genotypic effect of the QTL and to localize the QTL to a precise genomic region. The interval mapping approach is superior to the ANOVA approach in terms of both localization and estimation of the effects of QTL. However, the reliability in terms of estimation of the QTL effect depends on the linkage between marker(s) and QTL, the number and type of progeny evaluated, and the heritability of the trait. From multiple regression analysis, one can also obtain an R^2 value which gives the percentage of the total genetic variance explained by all of the markers.

The R^2 value for the line is considered to be the amount of total genetic variation that is explained by the specific molecular marker. Recently developed statistical packages also offer the means to analyze the QTL x environment interactions.

Besides the QTL mapping methods described above, several other approaches are available, including Bayesian methods and the use of a genetic algorithm. The most commonly used statistical packages for QTL analysis are MAPMAKER/QTL, QTLCartographer, PLABQTL, QTL Mapper and Qgene. Most of these statistical packages yield essentially similar QTL locations and gene effects on a given data set, while there could be slight variation in the confidence intervals. Recent advances using nonparametric statistics or association-based approaches to identify QTLs and to calculate empirical critical (threshold) values for declaring significant QTLs will further refine QTL mapping.

Factors Affecting the Power of QTL Mapping

QTLs are statistically inferred from the data generated in an experiment. However, statistical influence does not always indicate biological significance due to multiple test problems associated with QTL mapping (Liu, 2002). The following factors affect the power of QTL mapping:

1. Number of genes controlling the target trait(s) and their genome positions
2. Distribution of genetic effects and existence of genetic interactions
3. Heritability of the trait
4. Number of genes segregating in a mapping population
5. Type and size of mapping population
6. Density and coverage of markers in the linkage map
7. Statistical methodology employed and significance level used for QTL mapping

Excellent discussions on the above factors and the possible means to improve the power of QTL mapping are available in literature (ex., Beavis, 1994; Kearsey and Farquhar,1998; Liu, 2002). Replicate progeny analysis, selective genotyping, sample pooling and sequential sampling are some of the suggested approaches for optimization of experimental designs, so as to enhance the power of QTL detection and estimation of QTL effects.

QTL Information – Utility and Prospects

Despite lack of precise information about the molecular nature of the QTL, introgression of QTLs into elite lines or germplasm, and marker-assisted selection (MAS) for QTLs in breeding could be undertaken in some crop plants such as maize, tomato and rice, with reasonable success. Plant breeders may not need to know the precise locations of the QTL,so long as the QTL has large effect, and can be introgressed using marker-assisted backcrossing. The methods available will enable them to pick such useful QTL, which could well have been missed by conventional phenotypic selection. Also, another important advantage of the markers is in the reduction of linkage drag during the introgression of QTL by backcrossing. At IARI, we have mapped and validated QTLs conferring resistance to downy mildews of maize (George *et al.*, 2003; Nair *et al.*, 2005) and have recently transferred two major QTLs for downy mildew resistance into CM139, an elite but downy mildew-susceptible inbred line.

There are still some important caveats regarding QTL analysis. Only the QTLs of largest effect, and those closest to a marker locus, will show statistically reliable associations. It may be difficult to estimate even the presence of QTL if they interact strongly in their effects. And the regions to which a QTL is localized can be quite large (several cM, where 1 cM can range from hundreds to thousands of kilobases). Such regions may contain many genes, and there is no guarantee that a QTL will correspond to only one gene. Thus, with QTLs in hand, much further work is necessary to truly dissect quantitative variation at the mechanistic level. Particularly important is fine-mapping or high-resolution mapping of the QTL, if the QTL information has to be effectively applied in basic/applied research. Once fine-mapped, QTLs can also serve as useful tools for comparative genomics, functional genomics and evolutionary studies.

While the methods for QTL analysis provide us with information about the putative locations and effects of QTLs influencing a quantitative trait, these do not tell us anything about the molecular nature of the QTL. In other words, are these QTL coding for specific enzymes involved in a particular pathway, do they act as regulators of gene expression, or are they non-coding regions that have some influence in the expression of the trait of interest? Identification of putative QTL locations and DNA markers linked to QTLs have opened up opportunities for isolation and molecular characterization of QTLs via map based cloning. During the last decade, considerable progress has been made in terms of positional cloning of several QTLs, including Brlx 9-2-5, fw2.2, Hd1, Hd6, FRI in tomato, rice and Arabidopsis (Frary *et al.*, 2000; Fridman *et al.*, 2000; Yano *et al.*, 2000; Takahashi *et al.*, 2000; Salvi *et al.*, 2002).

Recent advances in molecular marker technology, coupled with development of better theoretical models and high-throughput strategies, are expected to enable greater power and precision in detection of QTL and utilization of QTL information for crop improvement.

Merits of QTL Mapping

a. Identification of novel genes

b. Where mutant approaches fail to detect genes with phenotypic functions, QTL mapping can help

c. Good alternative when mutant screening is laborious and expensive e.g circadium rhythm screens

d. Can identify New functional alleles of known function genes

e. Natural variation studies provide insight into the origins of plant evolution.

(Demerits)

a. Mainly identifies loci with large effects

b. Less strong ones can be hard to pursue

c. No. of QTLs detected, their position and effects are subjected to statistical error

d. Small additive effects/epistatic loci are not detected and may require further analyses

e. Cloning can be challenging but not impossible

29

Southern Blotting

Introduction

Southern blotting was named after Edward M. Southern who developed this procedure at Edinburgh University in the 1970s. To oversimplify, DNA molecules are transferred from an agarose gel onto a membrane. Southern blotting is designed to locate a particular sequence of DNA within a complex mixture. For example, Southern Blotting could be used to locate a particular gene within an entire genome. The amount of DNA needed for this technique is dependent on the size and specific activity of the probe. Short probes tend to be more specific. Under optimal conditions, you can expect to detect 0.1 pg of the DNA for which you are probing.

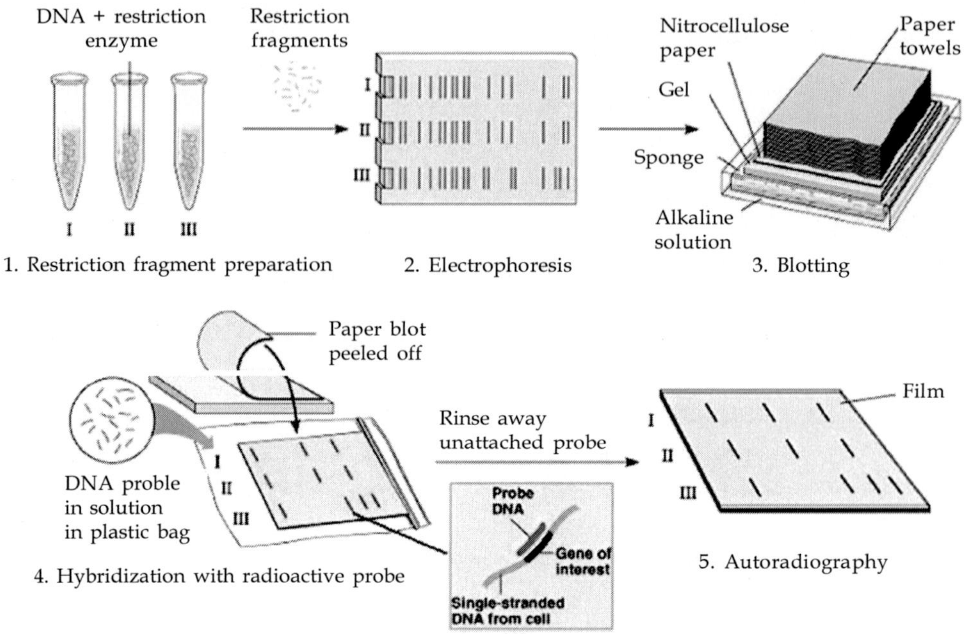

This diagram shows the basic steps involved in a Southern blot. Let's look at this technique in greater detail.

1. Digest the DNA with an appropriate restriction enzyme.
2. Run the digest on an agarose gel.
3. Denature the DNA (usually while it is still on the gel). For example, soak it in about 0.5M NaOH, which would separate double-stranded DNA into single-stranded DNA. Only ssDNA can transfer.

   A depurination step is optional. Fragments greater than 15 kb are hard to transfer to the blotting membrane. Depurination with HCl (about 0.2M HCl for 15 minutes) takes the purines out, cutting the DNA into smaller fragments. Be aware, however, that the procedure may also be hampered by fragments that are too small.

   Be sure to neutralize the acid after this step, or the base after the prior step if you don't depurinate.

4. Transfer the denatured DNA to the membrane. Traditionally, a nitrocellulose membrane is used, although nylon or a positively charged nylon membrane may be used. Nitrocellulose typically has a binding capacity of about 100μg/cm, while nylon has a binding capacity of about

500 µg/cm. Many scientists feel nylon is better since it binds more and is less fragile. Transfer is usually done by capillary action, which takes several hours. Capillary action transfer draws the buffer up by capillary action through the gel an into the membrane, which will bind ssDNA.

You may use a vacuum blot apparatus instead of capillary action. In this procedure, a vacuum sucks SSC through the membrane. This works similarly to capillary action, excepts more SSC goes through the gel and membrane, so it is faster (about an hour). (SSC provides the high salt level that you need to transfer DNA.)

After you transfer your DNA to the membrane, treat it with UV light. This cross links (via covalent bonds) the DNA to the membrane. (You can also bake nitrocellulose at about 80C for a couple of hours, but be aware that it is very combustible.)

5. Probe the membrane with labeled ssDNA. This is also known as hybridization.

   Whatever you call it, this process relies on the ssDNA hybridizing (annealing) to the DNA on the membrane due to the binding of complementary strands.

   Probing is often done with ^{32}P labeled ATP, biotin/streptavidin or a bioluminescent probe.

   A prehybridization step is required before hybridization to block non-specific sites, since you don't want your single-stranded probe binding just anywhere on the membrane. To hybridize, use the same buffer as for prehybridization, but add your specific probe.

6. Visualize your radioactively labeled target sequence. If you used a radiolabeled ^{32}P probe, then you would visualize by autoradiograph. Biotin/streptavidin detection is done by colorimetric methods, and bioluminescent visualization uses luminesence. ^{32}P labeled ATP Treat the dsDNA fragment that you are using as a probe with a limiting amount of Dnase, which causes double-stranded nicks in DNA. Add ^{32}P, dATP, and other dNTPs to DNA polymerase I, which has 5' to 3' polymerase activity and 5' to 3' exonuclease activity.

   Nick translation occurs and as the nick is translated down the DNA strand, the polymerase activity continues to nick while the exonuclease activity continues to fill in the nick. As this happens, ^{32}P becomes incorporated into, and thus labels, the DNA. Heat the DNA to make it single stranded, then immediately places it on ice to keep the two strands from reannealing to each other. (If the DNA is on ice, the DNA passes through the annealing temperature too quickly for the DNA to rehybridize into double-stranded DNA).

Prehybridization

To prehybridize, add non-specific ssDNA. Somicated salmon sperm DNA is commonly used. Add 20X SSC, Denhardt's solution (ficol and PVP, which are large molecules to take up space and generate more contact; and BSA, bovine serum albumen, a non-specific protein), SDS (sodium dodecyl sulfate), and formamide.

Altering the concentrations of formamide, SSC, and SDS affects "stringency," or specificity. If you have a higher stringency you should also have a higher degree of similarity between the probe and the target sequence.

Protocol

(Source: Laboratory protocols; CIMMYT Applied Molecular Genetics Laboratory)

The matrix we use is an MSI Magnagraph Nylon membrane, non-charged, 0.45μm pore size, 20 cm x 3 m rolls.

1. The best surface of a gel for regular contact with a membrane filter is that which was formed by the bottom of the gel mold. It is therefore advisable to flip the gel before constructing a blot and, preferably, before denaturation. Sandwich the gel between two thin acrylic plates, hold firmly at the corners, and flip it in one swift movement. Leave one of the plates under the gel to help in handling the gel in subsequent operations.
2. Denature gel for 30 min in 0.4 N NaOH, 0.6 M NaCl; treat each gel in about three times its volume of solution.
3. Transfer gel to another tray and neutralize for 30 min in 0.5 M Tris-7.5, 1.5 M NaCl; treat each gel in about three times its volume of solution.

Construction of Wet Blot Transfer System

4. Cut nylon membrane to the same dimensions as gel. Label (S&S marker pen) or nick the upper left corner of the membrane for later identification. Place in transfer buffer.
5. Place a plastic grid in a shallow tray to allow transfer buffer (25 mM NaPO4, pH 6.5) access to center of sponge.
6. Place a 6-8 cm thick, clean sponge on the center of the plastic grid; sponge surface should be equal to or greater than the gel to be blotted. Soak sponge thoroughly in transfer buffer.

7. Briefly, dip 1 sheet of blotting paper (extra thick) in transfer buffer and place on top of sponge.

   Note: Make sure there are **NO** air bubbles between blotting paper, gel, and membrane. Use transfer buffer between each layer and roll a glass pipette on the exposed surface to avoid bubble problems.

8. Place gel on blotting paper on sponge, open-side of wells facing down.

9. Place cut piece of matrix on gel, label-side down, to identify transfer side of matrix. Use a glass rod to smooth matrix on gel surface.

10. Place 1 sheet of wetted blotting paper on matrix.

11. Carefully place a 10 cm stack of paper towels on top of the blotting paper. A light weight can be placed on top, if used with a flat surface, to apply even pressure to blotting surface.

    Note: Paper towels do not need to cover entire area of gel. However, if they extend beyond the sides of the blotting paper, a piece of plastic (old X-ray film works well) or Saran Wrap should be placed between the two layers of blotting paper, isolating the paper towels from the lower blotting paper and buffer solution. This will avoid short-circuiting the transfer.

    Instead of paper towels, a second 6-8 cm sponge may be used on top. Wet the sponge with transfer buffer and wring out as much of the buffer as possible. Place on top of the blotting paper and place a light weight on top.

12. Add transfer buffer to tray, so that the buffer level remains high during blotting process.

13. Allow to transfer overnight (16-18 hours). It is a good idea to carefully remove the bottom layer of wet paper towels after the stack has absorbed 5-8 cm of transfer buffer.

    Note: If sponge is used, remove and wring out buffer after 4-5 hours of transfer.

14. Remove matrix and immediately place in 2X SSC. With a gloved hand, gently rub off any agar particles. Wash blot for 15 min, shaking in 2X SSC.

15. Air or drip-dry until moist but not wet (usually 2-5 min); do not allow to dry.

16. Place membrane on a moist filter paper and UV cross link in Stratagene UV Crosslinker using auto setting (120,000 μjoules/cm^2).
17. Bake at 95°C on or between clean filter paper for 1.5-2 h.
18. Briefly check transfer under UV light. If membrane was not previously labeled, label with a permanent marker pen or pencil on DNA bound side.
19. If blot is not going to be used for a week or more, store between clean filter paper in a sealed plastic bag in a cool, dry place (can be stored at 4°C).

Denaturation solution: 0.4 N NaOH, 0.6 M NaCl (1 liter/gel)

STOCK	1 liter	5 liters	10 liters	20 liters	40 liters
NaOH (MW=40.00)	16.0 g	80.0 g	160.0 g	320.0 g	640.0 g
NaCl (MW=58.44)	35.0 g	175.3 g	350.6 g	701.3 g	1402.6 g

Dissolve the NaCl first, then the NaOH to avoid precipitate formation.

Neutralization solution: 0.5 M Tris-7.5, 1.5 M NaCl (1 liter/gel)

STOCK	1 liter	5 liters	10 liters	20 liters	40 liters
Tris-HCl (MW=156.60)	63.5 g	317.5 g	630.5 g	1270.0 g	2540.0 g
Tris-base (MW=121.10)	11.8 g	59.0 g	118.0 g	236.0 g	472.0 g
NaCl (MW=58.44)	87.6 g	438.0 g	876.0 g	1752.0 g	3504.0 g
-OR-					
Tris-base (MW=121.10)	60.6 g	302.8 g	605.5 g	1211.0 g	2422.0 g
NaCl (MW=58.44)	87.7 g	438.3 g	876.6 g	1753.2 g	3506.4 g
Conc. HCl	25.0 ml	125.0 ml	250.0 ml	500.0 ml	1000.0 ml

Transfer buffer: 25 mM $NaPO_4$, pH 6.5 (5 liters/gel)

STOCK	1 liter	5 liters	10 liters	20 liters	40 liters
1 M NaP04 -6.5	25 ml	125 ml	250 ml	500 ml	1000 ml

2X SSC

STOCK	250 ml	500 ml	750 ml	1000 ml	2000 ml
25X SSC	20 ml	40 ml	60 ml	80 ml	160 ml

Dot Blot

A **Dot blot** (or **Slot blot**) is a technique in molecular biology used to detect biomolecules. It represents a simplification of the northern blot, Southern blot, or western blot methods. In a dot blot the biomolecules to be detected are not first separated by chromatography. Instead, a mixture containing the molecule to be detected is applied directly on a membrane as a dot. This is then followed by detection by either nucleotide probes (for a northern blot and Southern blot) or antibodies (for a western blot).

The technique offers significant savings in time, as chromatography or gel electrophoresis, and the complex blotting procedures for the gel are not required. However, it offers no information on the size of the target biomolecule. Furthermore, if two molecules of different sizes are detected, they will still appear as a single dot. Dot blots therefore can only confirm the presence or absence of a biomolecule or biomolecules which can be detected by the DNA probes or the antibody.

A radioactive sample can be hybridized to it allowing the researcher to detect variation between samples. The DNA is quantified and equal amounts are aliquoted into tubes in excess of the number of its targets in the samples, such as 10ug for a plasmid and 1ug for a PCR amplicon. These are denatured (NaOH and 95°C) and added to the wells where a vacuum sucks the water (with NaOH and NH_4OAc) from underneath the membrane (nylon or nitrocellulose).

Dot blot is a technique for detecting and identifying proteins, similar to the Western blot technique but differing in that protein samples are not separated electrophoretically but are spotted through circular templates directly onto the membrane or paper substrate. Antigens may be applied directly to nitrocellulose membrane as a discrete spot (dot) to give a simple and reliable assay.

30

Western Blotting

Introduction

The **western blot** (alternatively, **protein immunoblot**) is an analytical technique used to detect specific proteins in a given sample of tissue homogenate or extract. It uses gel electrophoresis to separate native or denatured proteins by the length of the polypeptide (denaturing conditions) or by the 3-D structure of the protein (native/non-denaturing conditions). The proteins are then transferred to a membrane (typically nitrocellulose or PVDF), where they are probed (detected) using antibodies specific to the target protein.

There are now many reagent companies that specialize in providing antibodies (both monoclonal and polyclonal antibodies) against tens of thousands of different proteins. Commercial antibodies can be expensive, although the unbound antibody can be reused between experiments. This method is used in the fields of molecular biology, biochemistry, immunogenetics and other molecular biology disciplines.

Other related techniques include using antibodies to detect proteins in tissues and cells by immunostaining and enzyme-linked immunosorbent assay (ELISA).

The method originated from the laboratory of George Stark at Stanford. The name **western blot** was given to the technique by W. Neal Burnette and is a play on the name Southern blot, a technique for DNA detection developed earlier by Edwin Southern. Detection of RNA is termed northern blotting and the detection of post-translational modification of protein is termed Eastern blotting.

Western blot analysis can detect **one** protein in a mixture of any number of proteins while giving you information about the size of the protein. It does not matter whether the protein has been synthesized in vivo or in vitro. This method is, however, dependent on the use of a high-quality antibody directed against a desired protein. So you must be able to produce at least a small portion of the protein from a cloned DNA fragment. You will use this antibody as a probe to detect the protein of interest.

Western blotting tells you how much protein has accumulated in cells. If you are interested in the rate of synthesis of a protein, Radio-Immune Precipitation (RIP) may be the best assay for you. Also, if a protein is degraded quickly, Western blotting won't detect it well; you'll need to use (RIP). See the section on RIP for more information, as well as a helpful comparative chart that illustrates the differences between these two techniques.

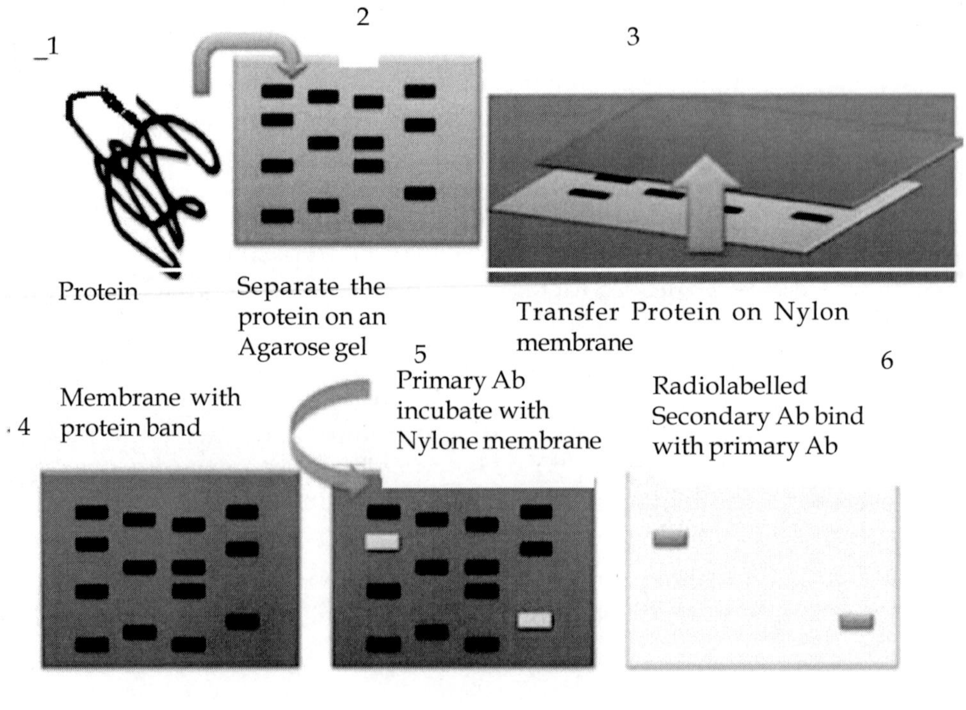

1. Separate the proteins using SDS-polyacrylamide gel electrophoresis (also known as SDS-PAGE). This separates the proteins by size.
2. Place a nitrocellulose membrane on the gel and, using electrophoresis, drive the protein (polypeptide) bands onto the nitrocellulose membrane. You want the negative charge to be on the side of the gel and the positive charge to be on the side of the nitrocellulose membrane to drive the negatively charged proteins over to the positively charged nitrocellulose membrane. This gives you a nitrocellulose membrane that is imprinted with the same protein bands as the gel.

   One thing to be aware of is that proteins bind better to nitrocellulose at a low pH. You may need to go through some trial-and-error to find the optimal pH. You also need to be sure there are no air bubbles between the nitrocellulose and the gel or your proteins will not transfer.
3. Incubate the nitrocellulose membrane with a primary antibody. sticks to your protein and forms an antibody-protein complex with the protein of interest.
4. Incubate the nitrocellulose membrane with a secondary antibody. This antibody should be an antibody-enzyme conjugate. The secondary antibody should be an antibody against the primary antibody. This means the secondary antibody will "stick" to the primary antibody, just like the primary antibody "stuck" to the protein. The conjugated enzyme is there to allow you to visualize all of this. It's kind of like a molecular flare stuck on the antibodies so you can visualize what s going on.
5. To actually see your enzyme in action, you'll need to incubate it in a reaction mix that is specific for your enzyme. If everything worked properly, you will see bands wherever there is a protein-primary antibody-secondary antibody-enzyme complex, or, in other words, wherever your protein is.
6. Put x-ray film on your gel to detect a flash of light, which is given off by the enzyme. The reaction usually runs out in about an hour.

Making a Primary Antibody

This description assumes you have available purified protein. Run the protein on an SDS-PAGE gel. Stain the gel with KCl. The KCl forms a precipitate with the SDS. Since the area with the protein has a low concentration of SDS, the area with the protein will not show a precipitate. This will allow you to see the protein band as a clear band against a milky white precipitate on the rest of the gel.

Carefully cut out the band and soak it in 1 mL PBS buffer. Crush it and make an emulsion with 1 mL Freund's Complete Adjuvant (which is an oily substance). The complete adjuvant contains microbacteria (an immune stimulant) to increase the immune response. Inject this subscapularly into a rabbit. This is your first inoculation. Only use the complete adjuvant for the first inoculation. **NEVER** inject a rabbit with complete adjuvant more than one time.

Rest the rabbit for one month, then repeat the process using an **incomplete** adjuvant. You can expect to see good antibody titers about 10 days after the second booster.

Bleed the rabbit. You can expect about 30 to 40 mL per bleeding, and about 50 percent of the volume is serum. Now you have rabbit antisera.

To get your primary antibody, dilute the rabbit antisera in blotto (aka Carnation Nonfat Dry Instant Milk) and apply it to your nitrocellulose blot. Make sure you dilute 1:500 to 1:100 in blotto; less dilution will give you background binding and really muddy up your results.

Secondary Antibody

This is much easier than the procedure for the primary antibody. Grab a catalogue and look for a goat-anti-rabbit antibody conjugated to horseradish peroxidase (HRP). The goat-anti-rabbit is your secondary antibody (the one that "sticks" to the primary antibody) and the HRP is the conjugated enzyme that will allow you to visualize your protein.

To separate the proteins through SDS-PAGE and detection followed by characterization of proteins through Western blotting.

Principle

Identification of protein separated by gel electrophoresis is limited by the small pore size of the gel, as the macromolecule probe for protein analysis cannot permeate the gel. This limitation is overcome by blotting the protein into an adsorbent porous membrane. The apparatus consists of a tank containing buffer, in which is located a cassette. Clamping the gel and the membrane tightly together, a current is applied from electrodes, and repeated on either side of the cassette to avoid heating effects. The proteins are separated according to their electrophoresis mobility and blotted onto the membrane identified, using suitable immunochemicals to locate the protein of interest. The individual techniques are explained below.

SDS-PAGE

Stage 1. Prepare a PAGE gel slab and fix to a vertical electrophoretic apparatus. Treat the sample with suitable buffer and load onto the gel slots.

Stage 2. Apply electric current. After a few minutes, proteins in the sample migrate according to their electrophoretic mobility in the stacking gel. The stacking gel has a polyacrylamide concentration resulting in higher pore size and a lower pH of 7. This enables the protein to concentrate into sharp bands due to isotachopharesis, or band-sharpening effect. At the end of the stacking gel, it meets the separating gel, which has a higher polyacrylamide concentration and higher pH. In the separating gel, the proteins travel according to their size.

Stage 3. When the dye front reaches the bottom of the separating gel, the proteins in the sample are resolved depending on their size. However the protein cannot be visualized directly. The gel needs to be stained with suitable stainer to visualize all the proteins. The identification of protein of interest can be done using a suitable probe and a developing system.

Western Blotting

Blotting is the transfer of resolved proteins from the gel to surface of suitable membrane. This is done commonly by electrophoresis (known as electroblotting). In this method, the transfer buffer has a low ionic strength which allows electrotransfer of proteins. Methane in the buffer increases binding of proteins to niotrocellulose and reduces gel smelling during transfer. The use of the membrane as a support for protein enables the case of manipulation efficient washing and faster reactions during the immunodetection, as proteins are more accessible for reaction.

(a) The membrane is in close contact with PAGE gel containing proteins. The proteins are electrotransferred to nitrocellulose membrane.

(b) At the end of electrotransfer, all proteins would have migrated to the NC membrane. The protein was transferred to the corresponding position on the membrane as on the gel. A mirror image of the gel was formed. However, the protein location and detection can only be assessed after immunodetection.

Applications

To characterize proteins and to identify specific antigens for antibodies.

Preparation of Reagents

1. *Blotting Buffer.* Add 25 mL of blotting buffer component A and 25 mL of component B to 150 mL of distilled water.
2. *Other Buffers.* Dilute the required amount of buffer concentrate to 1X concentration with water.
3. *Antibody.* Dilute primary Ab and label secondary HRP conjugate in an assay buffer.
4. *Substrate.* Dilute TMB/H_2O_2, 10X concentration 10 times with heat just before use.

Procedure

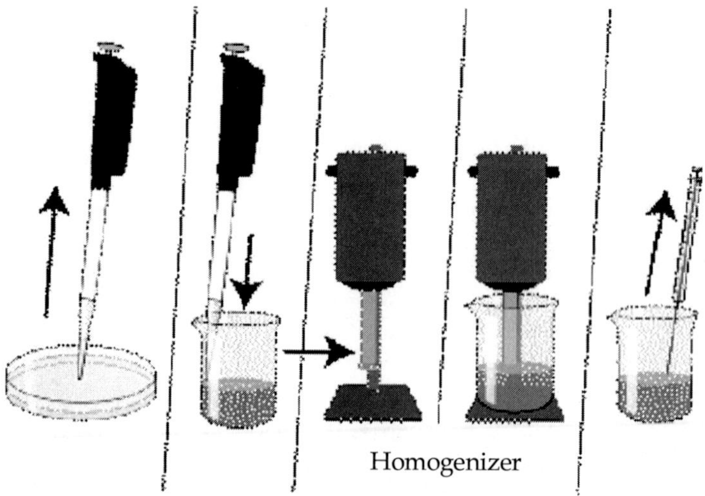

Tissue Preparation

Samples may be taken from whole tissue or from cell culture. In most cases, solid tissues are first broken down mechanically using a blender (for larger sample volumes), using a homogenizer (smaller volumes), or by sonication. Cells may also be broken open by one of the above mechanical methods. However, it should be noted that bacteria, virus or environmental samples can be the source of protein and thus Western blotting is not restricted to cellular studies only.

Assorted detergents, salts, and buffers may be employed to encourage lysis of cells and to solubilize proteins. Protease and phosphatase inhibitors are often added to prevent the digestion of the sample by its own enzymes. Tissue preparation is often done at cold temperatures to avoid protein denaturing.

A combination of biochemical and mechanical techniques - including various types of filtration and centrifugation - can be used to separate different cell compartments and organelles.

Gel Electrophoresis

The proteins of the sample are separated using gel electrophoresis. Separation of proteins may be by isoelectric point (pI), molecular weight, electric charge, or a combination of these factors. The nature of the separation depends on the treatment of the sample and the nature of the gel.

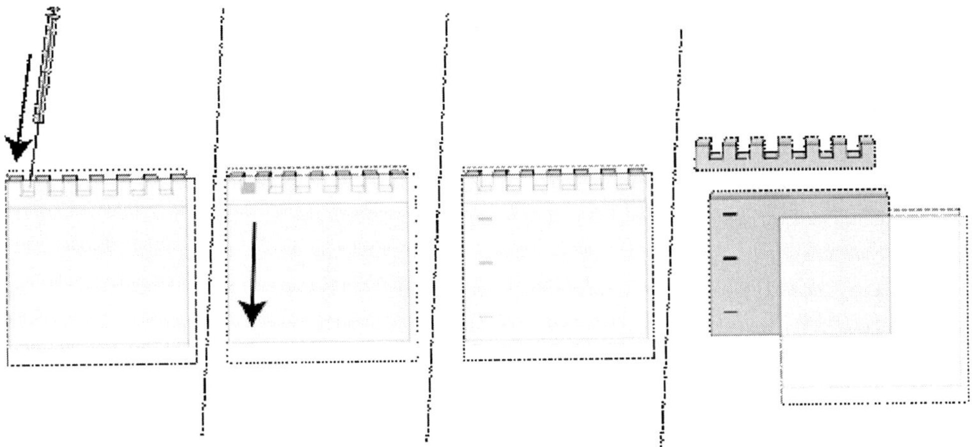

Gel Electrophoresis

By far the most common type of gel electrophoresis employs polyacrylamide gels and buffers loaded with sodium dodecyl sulfate (SDS). SDS-PAGE (SDS polyacrylamide gel electrophoresis) maintains polypeptides in a denatured state once they have been treated with strong reducing agents to remove secondary and tertiary structure (e.g. disulfide bonds [S-S] to sulfhydryl groups [SH and SH]) and thus allows separation of proteins by their molecular weight. Sampled proteins become covered in the negatively charged SDS and move to the positively charged electrode through the acrylamide mesh of the gel. Smaller proteins migrate faster through this mesh and the proteins are thus separated according to size (usually measured in kilodaltons, kDa). The concentration of acrylamide determines the resolution of the gel - the greater the acrylamide concentration the better the resolution of lower molecular weight proteins. The lower the acrylamide concentration the better the resolution of higher molecular weight proteins. Proteins travel only in one dimension along the gel for most blots.

Samples are loaded into *wells* in the gel. One lane is usually reserved for a *marker* or *ladder*, a commercially available mixture of proteins having defined molecular weights, typically stained so as to form visible, coloured bands. When voltage is applied along the gel, proteins migrate into it at different speeds. These different rates of advancement (different *electrophoretic mobilities*) separate into *bands* within each *lane*.

It is also possible to use a two-dimensional (2-D) gel which spreads the proteins from a single sample out in two dimensions. Proteins are separated according to isoelectric point (pH at which they have neutral net charge) in the first dimension, and according to their molecular weight in the second dimension.

Transfer

In order to make the proteins accessible to antibody detection, they are moved from within the gel onto a membrane made of *nitrocellulose or polyvinylidene difluoride (PVDF)*. The membrane is placed on top of the gel, and a stack of filter papers placed on top of that. The entire stack is placed in a buffer solution which moves up the paper by capillary action, bringing the proteins with it. Another method for transferring the proteins is called electroblotting and uses an electric current to pull proteins from the gel into the PVDF or nitrocellulose membrane. The proteins move from within the gel onto the membrane while maintaining the organization they had within the gel. As a result of this "blotting" process, the proteins are exposed on a thin surface layer for detection (see below). Both varieties of membrane are chosen for their non-specific protein binding properties (i.e. binds all proteins equally well). Protein binding is based upon hydrophobic interactions, as well as charged interactions between the membrane and protein. Nitrocellulose membranes are cheaper than PVDF, but are far more fragile and do not stand up well to repeated probings.

The uniformity and overall effectiveness of transfer of protein from the gel to the membrane can be checked by staining the membrane with Coomassie or Ponceau S dyes. Ponceau S is the more common of the two, due to Ponceau S's higher sensitivity and its water solubility makes it easier to subsequently destain and probe the membrane as described below.

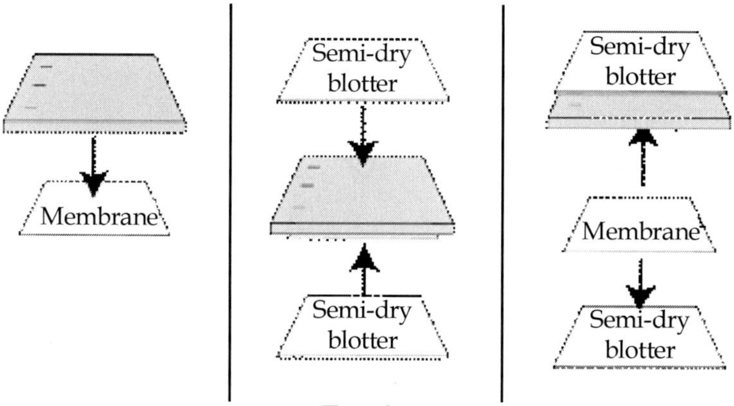

Transfer

Blocking

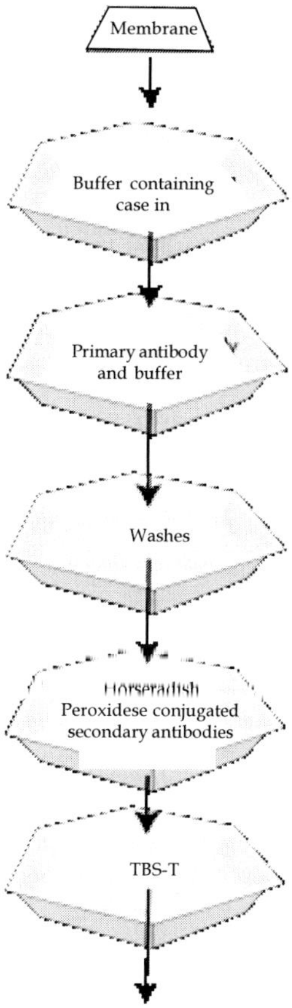

Blocking and Detection

Since the membrane has been chosen for its ability to bind protein, and both antibodies and the target are proteins, steps must be taken to prevent interactions between the membrane and the antibody used for detection of the target protein (since the antibody is a protein itself). Blocking of non-specific binding is achieved by placing the membrane in a dilute solution of protein - typically Bovine serum albumin (BSA) or non-fat dry milk (both are inexpensive), with a minute percentage of detergent such as Tween 20. The protein in the dilute solution attaches to the membrane in all places where the target proteins have not attached. Thus, when the antibody is added, there is no room on the membrane for it to attach other than on the binding sites of the specific target protein. This reduces "noise" in the final product of the Western blot, leading to clearer results, and eliminates false positives.

Detection

During the detection process the membrane is "probed" for the protein of interest with a modified antibody which is linked to a reporter enzyme, which when exposed to an appropriate substrate drives a colourimetric reaction and produces a colour. For a variety of reasons, this traditionally takes place in a two-step process, although there are now one-step detection methods available for certain applications.

Two Step

• *Primary Antibody*

Antibodies are generated when a host species or immune cell culture is exposed to the protein of interest (or a part thereof). Normally, this is part of the immune response, whereas here they are harvested and used as sensitive and specific detection tools that bind the protein directly.

After blocking, a dilute solution of primary antibody (generally between 0.5 and 5 micrograms/mL) is incubated with the membrane under gentle agitation. Typically, the solution is composed of buffered saline solution with a small percentage of detergent, and sometimes with powdered milk or BSA. The antibody solution and the membrane can be sealed and incubated together for anywhere from 30 minutes to overnight. It can also be incubated at different temperatures, with warmer temperatures being associated with more binding, both specific (to the target protein, the "signal") and non-specific ("noise").

• *Secondary Antibody*

After rinsing the membrane to remove unbound primary antibody, the membrane is exposed to another antibody, directed at a species-specific portion of the primary antibody. This is known as a secondary antibody, and due to its targeting properties, tends to be referred to as "anti-mouse," "anti-goat," etc. Antibodies come from animal sources (or animal sourced hybridoma cultures); an anti-mouse secondary will bind to just about any mouse-sourced primary antibody. This allows some cost savings by allowing an entire lab to share a single source of mass-produced antibody, and provides far more consistent results. The secondary antibody is usually linked to biotin or to a reporter enzyme such as alkaline phosphatase or horseradish peroxidase. This means that several secondary antibodies will bind to one primary antibody and enhance the signal.

Most commonly, a horseradish peroxidase-linked secondary is used in conjunction with a chemiluminescent agent, and the reaction product produces luminescence in proportion to the amount of protein. A sensitive sheet of photographic film is placed against the membrane, and exposure to the light from the reaction creates an image of the antibodies bound to the blot. A cheaper but less sensitive approach utilizes a 4-chloronaphthol stain with 1% hydrogen peroxide; reaction of peroxide radicals with 4-

chloronaphthol produces a dark brown stain that can be photographed without using specialized photographic film.

As with the ELISPOT and ELISA procedures, the enzyme can be provided with a substrate molecule that will be converted by the enzyme to a colored reaction product that will be visible on the membrane (see the figure below with blue bands).

Another method of secondary antibody detection utilizes a near-infrared (NIR) fluorophore-linked antibody. Light produced from the excitation of a fluorescent dye is static, making fluorescent detection a more precise and accurate measure of the difference in signal produced by labeled antibodies bound to proteins on a western blot. Proteins can be accurately quantified because the signal generated by the different amounts of proteins on the membranes is measured in a static state, as compared to chemiluminescence, in which light is measured in a dynamic state.

A third alternative is to use a radioactive label rather than an enzyme coupled to the secondary antibody, such as labeling an antibody-binding protein like *Staphylococcus* Protein A with a radioactive isotope of iodine. Since other methods are safer, quicker and cheaper this method is now rarely used.

One Step

Historically, the probing process was performed in two steps because of the relative ease of producing primary and secondary antibodies in separate processes. This gives researchers and corporations huge advantages in terms of flexibility, and adds an amplification step to the detection process. Given the advent of high-throughput protein analysis and lower limits of detection, however, there has been interest in developing one-step probing systems that would allow the process to occur faster and with less consumables. This requires a probe antibody which both recognizes the protein of interest and contains a detectable label, probes which are often available for known protein tags. The primary probe is incubated with the membrane in a manner similar to that for the primary antibody in a two-step process, and then is ready for direct detection after a series of wash steps.

Interpretation and Result

After the unbound probes are washed away, the western blot is ready for detection of the probes that are labeled and bound to the protein of interest. In practical terms, not all westerns reveal protein only at one band in a membrane. Size approximations are taken by comparing the stained bands to that of the marker or ladder loaded during electrophoresis. The process is repeated for a structural protein, such as actin or tubulin, that should not change between samples. The amount of target protein is indexed to the structural protein to control between groups. This practice ensures correction for the amount of total protein on the membrane in case of errors or incomplete transfers.

Colorimetric Detection

The colorimetric detection method depends on incubation of the western blot with a substrate that reacts with the reporter enzyme (such as peroxidase) that is bound to the secondary antibody. This converts the soluble dye into an insoluble form of a different color that precipitates next to the enzyme and thereby stains the membrane. Development of the blot is then stopped by washing away the soluble dye. Protein levels are evaluated through densitometry (how intense the stain is) or spectrophotometry.

Chemiluminescent Detection

Chemiluminescent detection methods depend on incubation of the western blot with a substrate that will luminesce when exposed to the reporter on the secondary antibody. The light is then detected by photographic film, and more recently by CCD cameras which captures a digital image of the western blot. The image is analysed by densitometry, which evaluates the relative amount of protein staining and quantifies the results in terms of optical density. Newer software allows further data analysis such as molecular weight analysis if appropriate standards are used.

Radioactive Detection

Radioactive labels do not require enzyme substrates, but rather allow the placement of medical X-ray film directly against the western blot which develops as it is exposed to the label and creates dark regions which correspond to the protein bands of interest (see image to the right). The importance of radioactive detections methods is declining, because it is very expensive, health and safety risks are high and ECL provides a useful alternative.

Fluorescent Detection

The fluorescently labeled probe is excited by light and the emission of the excitation is then detected by a photosensor such as CCD camera equipped with appropriate emission filters which captures a digital image of the western blot and allows further data analysis such as molecular weight analysis and a quantitative western blot analysis. Fluorescence is considered to be among the most sensitive detection methods for blotting analysis.

Secondary Probing

One major difference between nitrocellulose and PVDF membranes relates to the ability of each to support "stripping" antibodies off and reusing the membrane for subsequent antibody probes. While there are well-established protocols available for stripping nitrocellulose membranes, the sturdier PVDF allows for easier stripping, and for more reuse before background noise limits experiments. Another difference is that, unlike nitrocellulose, PVDF must be soaked in 95% ethanol, isopropanol or methanol before use. PVDF membranes also tend to be thicker and more resistant to damage during use.

2-D Gel Electrophoresis

Main article: Two-dimensional gel electrophoresis

2-dimensional SDS-PAGE uses the principles and techniques outlined above. 2-D SDS-PAGE, as the name suggests, involves the migration of polypeptides in 2 dimensions. For example, in the first dimension polypeptides are separated according to isoelectric point, while in the second dimension polypeptides are separated according to their molecular weight. The isoelectric point of a given protein is determined by the relative number of positively (e.g. lysine and arginine) and negatively (e.g. glutamate and aspartate) charged amino acids, with negatively charged amino acids contributing to a high isoelectric point and positively charged amino acids contributing to a low isoelectric point. Samples could also be separated first under nonreducing conditions using SDS-PAGE and under reducing conditions in the second dimension, which breaks apart disulfide bonds that hold subunits together. SDS-PAGE might also be coupled with urea-PAGE for a 2-dimensional gel.

In principle, this method allows for the separation of all cellular proteins on a single large gel. A major advantage of this method is that it often distinguishes between different isoforms of a particular protein - e.g. a protein that has been phosphorylated (by addition of a negatively charged group). Proteins that have been separated can be cut out of the gel and then analysed by mass spectrometry, which identifies the protein.

Far-Western Blotting

Far-western blotting is a molecular biological method which is based on the technique of western blotting. While usual western blotting uses an antibody to detect a protein of interest, far-western blotting uses a non-antibody protein, which can bind the protein of interest. Thus, whereas western blotting is used for the detection of certain proteins, far-western blotting is rather employed to detect protein:protein interactions.

Method

In conventional western blotting, gel electrophoresis is used to separate proteins from a sample; these proteins are then transferred to a membrane in a 'blotting' step. In a western blot, specific proteins are then identified using an antibody probe.

Far-western blotting employs non-antibody proteins to probe the protein(s) of interest on the blot. In this way, binding partners of the probe (or the blotted) protein may be identified. The probe protein is often produced in E coli using an expression cloning vector.

The probe protein can then be visualized through the usual methods— it may be radiolabelled; it may bear a specific affinity tag like His or FLAG for which antibodies exist; or there may be a protein specific antibody (to the probe protein).

31

Northern Blotting

Northern Blot

The **northern blot** is a technique used in molecular biology research to study gene expression by detection of RNA (or isolated mRNA) in a sample.

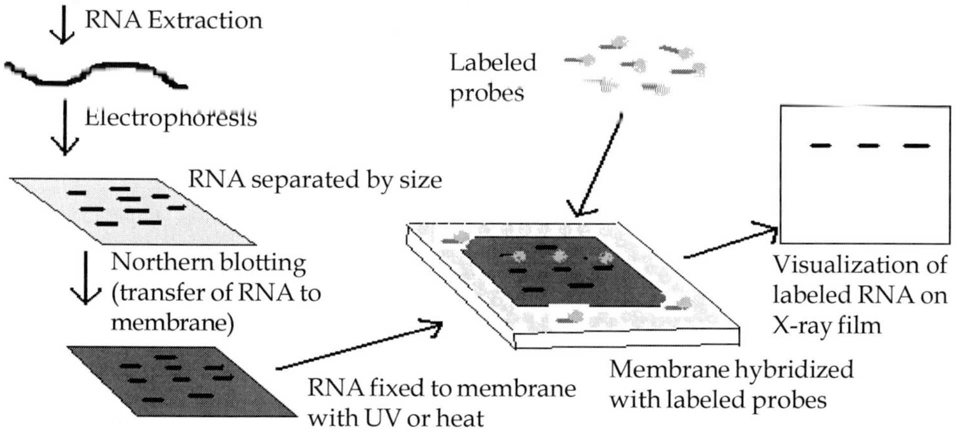

Flow diagram outlining the general procedure for RNA detection by northern blotting.

A northern blot is very similar to a Southern blot except that it is RNA rather than DNA which is extracted, run on a gel and transferred to a filter membrane. There are 3 types of RNA: tRNA (transfer RNA - active in assembly of polypeptide chains), rRNA (ribosomal RNA - part of the structure of ribosomes) and mRNA (messenger RNA - the product of DNA transcription and used for translation of a gene into a protein). It is mRNA which is isolated and hybridized in northern blots.

With northern blotting it is possible to observe cellular control over structure and function by determining the particular gene expression levels during differentiation, morphogenesis, as well as abnormal or diseased conditions. Northern blotting involves the use of electrophoresis to separate RNA samples by size, and detection with a hybridization probe complementary to part of or the entire target sequence. The term 'northern blot' actually refers specifically to the capillary transfer of RNA from the electrophoresis gel to the blotting membrane, however the entire process is commonly referred to as northern blotting. The northern blot technique was developed in 1977 by James Alwine, David Kemp, and George Stark at Stanford University. Northern blotting takes its name from its similarity to the first blotting technique, the Southern blot, named for biologist Edwin Southern. The major difference is that RNA, rather than DNA, is analyzed in the northern blot.

Procedure

A general blotting procedure starts with extraction of total RNA from a homogenized tissue sample. The mRNA can then be isolated through the use of oligo (dT) cellulose chromatography to maintain only those RNAs with a poly(A) tail. RNA samples are then separated by gel electrophoresis. Since the gels are fragile and the probes are unable to enter the matrix, the RNA samples, now separated by size, are transferred to a nylon membrane through a capillary or vacuum blotting system.

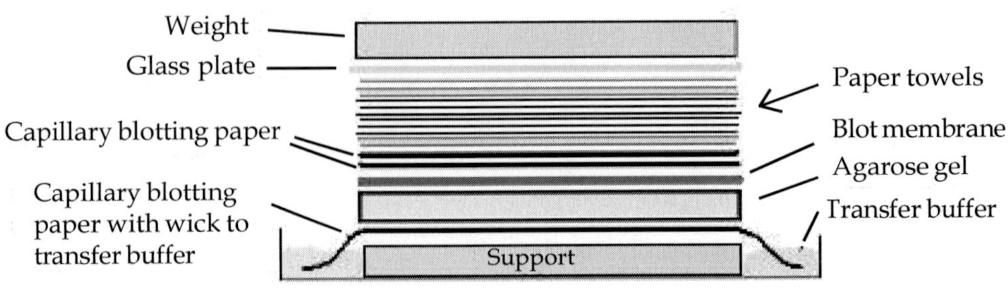

Capillary blotting system setup for the transfer of RNA from an electrophoresis gel to a blotting membrane.

A nylon membrane with a positive charge is the most effective for use in northern blotting since the negatively charged nucleic acids have a high affinity for them. The transfer buffer used for the blotting usually contains formamide because it lowers the annealing temperature of the probe-RNA interaction preventing RNA degradation by high temperatures. Once the RNA has been transferred to the membrane it is immobilized through covalent linkage to the membrane by UV light or heat. After a probe has been labeled, it is hybridized to the RNA on the membrane. Experimental conditions that can affect the efficiency and specificity of hybridization include ionic strength, viscosity, duplex length, mismatched base pairs, and base composition. The membrane is washed to ensure that the probe has bound specifically and to avoid background signals from arising. The hybrid signals are then detected by X-ray film and can be quantified by densitometry. To create controls for comparison in a northern blot, samples not displaying the gene product of interest can be used after determination by microarrays or RT-PCR.

Gels

RNA run on a formaldehyde agarose gel to highlight the 28S (top band) and 18S (lower band) ribosomal subunits.

The RNA samples are most commonly separated on agarose gels containing formaldehyde as a denaturing agent for the RNA to limit secondary structure. The gels can be stained with ethidium bromide (EtBr) and viewed under UV light to observe the quality and quantity of RNA before blotting. Polyacrylamide gel electrophoeresis with urea can also be used in RNA separation but it is most commonly used for fragmented or microRNAs. An RNA ladder is often run alongside the samples on an electrophoresis gel to observe the size of fragments obtained but in total RNA samples the ribosomal subunits can act as size markers. Since the large ribosomal subunit is 28S (approximately 5kb) and the small ribosomal subunit is 18S (approximately 2kb) two prominent bands will appear on the gel, the larger at close to twice the intensity of the smaller.

Probes

Probes for northern blotting are composed of nucleic acids with a complementary sequence to all or part of the RNA of interest, they can be DNA, RNA, or oligonucleotides with a minimum of 25 complementary bases to the target sequence. RNA probes (riboprobes) that are transcribed in vitro are able to withstand more rigorous washing steps preventing some of the background noise. Commonly cDNA is created with labelled primers for the RNA sequence of interest to act as the probe in the northern blot. The probes need to be labelled either with radioactive isotopes (^{32}P) or with chemiluminescence in which alkaline phosphatase or horseradish peroxidase breakdown chemiluminescent substrates producing a detectable emission of light. The chemiluminescent labelling can occur in a couple ways, the probe attached to the enzyme, or the probe labelled with a ligand (e.g. biotin) for which the antibody (e.g. avidin or streptavidin) is attached to the enzyme. X-ray film can detect both the radioactive and chemiluminescent signals and many researchers prefer the chemiluminescent signals because they are faster, more sensitive, and reduce the health hazards that go along with radioactive labels. The same membrane can be probed up to five times without a significant loss of the target RNA.

Applications

Northern blotting allows one to observe a particular gene's expression pattern between tissues, organs, developmental stages, environmental stress levels, pathogen infection, and over the course of treatment. The technique has been used to show overexpression of oncogenes and downregulation of tumor-suppressor genes in cancerous cells when compared to 'normal' tissue, as well as the gene expression in the rejection of transplanted organs. If an upregulated gene is observed by an abundance of mRNA on the northern blot the sample can then be sequenced to determine if the gene is known to researchers or if it is a novel finding. The expression patterns obtained under given conditions can provide insight into the function of that gene. Since the RNA is first separated by size, if only one probe type is used variance in the level of each band on the membrane can provide insight into the size of the product, suggesting alternative splice products of the same gene or repetitive sequence motifs. The variance in size of a gene product can also indicate deletions or errors in transcript processing, by altering the probe target used along the known sequence it is possible to determine which region of the RNA is missing.

BlotBase is an online database publishing northern blots. BlotBase has over 700 published northern blots of human and mouse samples, in over 650 genes across more than 25 different tissue types. Northern blots can be searched by a blot ID, paper reference, gene identifier, or by tissue. The results of a search provide the blot ID, species, tissue, gene, expression level, blot image (if available), and links to the publication that the work originated from. This new database provides sharing of information between members of the science community that was not previously seen in northern blotting as it was in sequence analysis, genome determination, protein structure, etc.

Advantages & Disadvantages

Analysis of gene expression can be done by several different methods including RT-PCR, RNase protection assays, microarrays, serial analysis of gene expression (SAGE), as well as northern blotting. Microarrays are quite commonly used and are usually consistent with data obtained from northern blots, however at times northern blotting is able to detect small changes in gene expression that microarrays cannot. The advantage that microarrays have over northern blots is that thousands of genes can be visualized at a time while northern blotting is usually looking at one or a small number of genes. A problem in northern blotting is often sample degradation by RNases (both endogenous to the sample and through environmental contamination) which can be avoided by proper sterilization of glassware and the use of RNase inhibitors such as DEPC (diethylpyrocarbonate). The chemicals used in most northern blots can be a risk to the researcher, since formaldehyde, radioactive material, ethidium bromide, DEPC, and UV light are all harmful under certain exposures. Compared to RT-PCR northern blotting has a low sensitivity but it also has a high specificity which is important to reduce false positive results. The advantages of using northern blotting include the detection of RNA size, the observation of alternate splice products, the use of probes with partial homology, the quality and quantity of RNA can be measured on the gel prior to blotting, and the membranes can be stored and reprobed for years after blotting.

Reverse Northern Blot

A variant of the procedure known as the reverse northern blot is occasionally used. In this procedure, the substrate nucleic acid (that is affixed to the membrane) is a collection of isolated DNA fragments, and the probe is RNA extracted from a tissue and radioactively labelled.

The use of DNA microarrays that have come into widespread use in the late 1990s and early 2000s is more akin to the reverse procedure, in that they involve the use of isolated DNA fragments affixed to a substrate, and hybridization with a probe made from cellular RNA. Thus the reverse procedure, though originally uncommon, enabled northern analysis to evolve into gene expression profiling, in which many (possibly all) of the genes in an organism may have their expression monitored.

Protocol

RNA Isolation

1. Harvest cells when they are about 80-85% confluent. Stationary phase cells have decreased RNA yield.
2. Wash cells with warm PBS and scrape them off in 7 ml cold GITC buffer.
3. Pour GITC/cell mix into (14x95 mm) polyallomer tubes with 4 ml CsCl buffer.
4. Balance and top off tubes with more GITC buffer.
5. Spin tubes on the SW40Ti rotor for 20 hr at 32,000 rpm (18°C).
6. Pour off liquid and keep tubes inverted because the RNA pellets on the bottom (DNA can be recovered from the CsCl layer).
7. Cut off the tube bottoms (keep inverted) and resuspend the pellet in 100μl 0.3 M NaOAc (pH 6) buffer. Transfer to a screw-cap 1.5 ml Eppi. Wash tube bottom twice more with 100μl buffer each time.
8. Add 750μl 100% EtOH and let precipitate overnight at -70°C.
9. Spin down at 4°C for 12 min, dry in the TC hood, and resuspend with DEPC water.

RNA Quantitation

1. Let resuspended RNA sit on ice for at least 30 min before quantitation.
2. Pipet up and down several times before taking 2μl into 998μl water.
3. Read at 260 nm.
4. RNA concentration (μg/μl) = OD x 500 x 0.04 μg/μl = OD x 20.

Formaldehyde Gel

1. Boil 2 g agarose in 144 ml water (check weight to account for evaporation)
2. Add 20 ml 10X MOPS and let cool on stir plate to about 56°C.
3. Add 36 ml 37% formaldehyde.
4. Pour into Hoefer appartatus when gel solution is below 50°C.
5. Running buffer: dilute 100 ml 10X MOPS to one liter.

Sample Preparation

1. To 0.5 ml Eppis, add:

   a) 15μg RNA (approxiamately 4.5μl)

b) 14.5μl sample buffer

c) Adjust with DEPC water so that final volumes are equal

d) Heat for 5 min at 68°C, then chill on ice for 5 min.

e) Add 2μl loading buffer and 1μl 0.4 mg/ml EtBr.

2. For the ladder lanes, use 3μg of 1 kb ladder and treat the same way as with the samples.

Running the Gel

1. For the Hoefer, run between 70-85 V for about 3 hr or to a predetermined length.
2. Recirculate the buffer with a peristaltic pump.
3. As the gel is running, cut membrane, paper towels, etc.
4. At the end, take a picture (f8, 4s) of the gel using a fluorescent ruler.

Transferring

1. Fill reservoir with 10X SSPE. Lay a piece of chromatography paper across a glass plate. Allow the SSPE to wick up the paper (or just soak the paper wet).
2. Remove all bubbles between the paper and the plate by rolling a pipet across.
3. Lay gel top-side down and remove bubbles between gel and paper.
4. Surround the gel with Parafilm strips.
5. Soak the MagnaGraph membrane in ddH_2O for 3 min and lay it on top of the gel.
6. Place 2 pieces of chromatography paper on top of membrane.
7. Stack at least 3 inches of cut-to-size brown paper towels on top of paper.
8. Place an oversized piece of plastic or glass on top of the towels. Weigh the whole set up down with about 400 gm (like water in tupperware).
9. Allow the RNA to transfer overnight.

Fixing

1. Bake membrane in 80°C vacuum oven for 20 min.
2. UV crosslink (on optimal setting of Fisher Crosslinker) the membrane (face side up).
3. Store between 2 pieces of chromatography paper in a plastic bag.

Labelling the Probe

1. Make a 10 ng/µl dilution of the probe.
2. Add 10µl (100 ng) of probe to a 0.5 ml screw-cap eppendorf tube. Adjust volume to 28µl with water. (*i.e.*, 7µl of probe needs 21µl H_2O.)
3. Boil for 5 min and then chill on ice for 5 min.
4. Briefly spin up/down in the microfuge.
5. Following the protocol from the Amersham Multiprime Kit (RPN.1601):
   a. Add 10µl of Buffer (solution 1, blue cap vial)
   b. Add 5µl of Primer (solution 2, black cap vial)
   c. Add 5µl of [□-^{32}P]dCTP (3000 Ci/mmol)
   d. Add 2µl of Enzyme (solution 3, Klenow fragment) Final volume at this point should be 50µl.
   e. Incubate for 1 hr at 37°C or overnight at room temperature.

Prehybing the blot

6. For hybridization in a tube, about 3-5 ml of buffer is sufficient. If a bag is used, it's about 0.2 ml of buffer per square cm of membrane. It is critical that the membrane be wet thoroughly.
7. Boil 200µg/ml of sheared salmon sperm DNA in a 50 ml conical tube for 5 min and chill on ice for 5 min. (If it's a total of 10 ml, use 2 mg salmon sperm.)
8. Add the hybridization buffer (5 ml) and the formamide (5 ml) to the salmon sperm.
9. Put the membrane in a tube and wet it with this mixture.
10. Incubate at 42°C for at least 30 min.

Hybridization

11. Check the incorporation of [32P] dCTP into the probe first.
    a. Spot 1 µl of reaction mixture onto the center of a small piece of DE81 paper that is placed into the bottom of a 7 ml scintillation vial. Do this twice (i.e., into two vials. One will be washed; the other will be the control).
    b. Let the spot dry (about 2 min).
    c. Fill the wash vial with 5 ml of 0.5 M Na_2HPO_4 and let it sit for 3 min. Then aspirate off the wash. Do this twice more.

d. Add 5 ml of scintillation fluid and count.

e. % incorporation = (cpm of washed/cpm of unwashed) x 100. Anything over 10% is usually usable. Using the kit, it's usually 40%.

12. The specific activity (S.A.) of the probe is also important. S.A. = [(cpm of washed) x (volume of labelling reaction)]/ µg amount of DNA. If S.A. is over $1x10^8$, the probe is OK.

13. Boil the probe for 5 min and chill it on ice for 5 min.

14. Up/down spin the tube to bring down the condensation on the lid.

15. Add it to the tube or bag.

16. Let it hybridize overnight (at least 16 hr) at 42°C.

Washing the blot

17. Add about 20 ml 2X SSPE/0.1% SDS to the tube. Mix and pour out. Be careful of loose drops.

18. Fill tube ≈ 1/3 -1/2 full with 2X SSPE/0.1% SDS and incubate at 56°C for 30 min.

19. Repeat with 1X SSPE/0.1% SDS and 0.1X SSPE/0.1% SDS until background counts are below 1000 cpm. Keep washing with 0.1X SSPE/ 0.1% SDS if the counts are high. Washes can go longer than 20 min. The stringency of the washes depends on the nature of the probe.

20. Wrap the blot with Saran wrap making sure there are no wrinkles.

21. Expose to autoradiography film overnight or to a Molecular Imager screen for 1 hr. Adjust exposure time accordingly. The time on the screen is roughly one-tenth to what it would be on film.

Solutions

GITC Buffer (200 ml)

1. 94.53 g Guanidine isothiocyanate (4 M final)
2. 1.67 ml 3 M NaOAc, pH 6
3. Bring volume to 200 ml
4. Add 1.67 ml β-mercaptoethanol
5. Store a 4°C and away from light

CsCl Buffer (100 ml)

1. 95.97 g CsCl (5.7 M final)
2. 0.83 ml 3 M NaOAc, pH 6
3. Bring volume to 100 ml
4. Sterile filter with 0.8μm
   (This is optional)

10X MOPS

0.4 M 3-[N-Morpholino] propanesulfonic acid, pH 7.0

0.1 M Sodium acetate

10 mM EDTA (pH 8.0)

Prepare in DEPC water. OK if solution turns yellow.

Autoclave to sterilize.

DEPC Water

Add DEPC to ddH_2O at 1:1000 dilution

Let it sit overnight at room temp (destroys RNase)

Autoclave the next day (destroys DEPC)

RNA Sample Buffer (7.25 ml)

0.5 ml 10X MOPS

1.75 ml 37% Formaldehyde

5.0 ml Formamide

Keep away from light!

Store at -20°C

RNA Loading Buffer

50% Glycerol

1 mM EDTA

0.25% Bromophenol blue

0.25% Xylene cyanol

Use DEPC water, keep at 4°C

2X Slaman Hybe

2% SDS

2 M NaCl

Will not be in solution at RT

Keep at 4°C

20X SSPE

3.6 M NaCl

0.2 M NaH_2PO_4-H_2O

0.02 M EDTA

pH to 7.7

32

Eastern Blotting

Eastern blotting is a technique to detect protein post translational modifications (PTM) and is an extension of the biochemical technique of western blotting. There are multiple versions of the method, but in general proteins or lipids are blotted from one or two-dimensional SDS-PAGE gel on to a PVDF or nitrocellulose membrane are analyzed for post-translational protein modifications using probes that may detect lipids, carbohydrate, phosphorylation or any other protein modification. Eastern blotting should be used to refer to methods that detect their targets through specific interaction of the PTM and the probe, distinguishing them from a standard Far-western blot. The most common application is lectin blotting, however, a variety of other probes can be used for detecting different PTMs.

History and Multiple Definitions

Definition of the term Eastern blotting is somewhat confused due to multiple sets of authors dubbing a new method as Eastern blotting, or a derivative therof. All of the definitions are a derivative of the technique of Western blotting developed by Towbin in 1979. The current definitions are summarized below in order of the first use of the name, however all are based on some earlier works. In some cases, the technique had been in practice for some time before the introduction of the term.

- (1996) Eastern-Western blot was first used by Bogdanov et al. The method involved blotting of phospholipids on PVDF membrane, and probing with antibodies that recognized carbohydrate epitopes. This method is based on earlier work by Taki et al in 1994, which they originally dubbed TLC blotting, and was based on a similar method introduced by Towbin in 1984.
- (2000) Far-Eastern blotting seems to have been first named in 2000 by Ishikawa & Taki. The method is described more fully in the article on Far-Eastern blotting, but is based on antibody or lectin staining of lipids transferred to PVDF membranes.
- (2002) Eastern blot has also been used to describe an immunoblot performed on proteins blotted to a PVDF membrane from a PAGE gel run with opposite polarity. Since this is essentially a Western blot, the charge reversal was used to dub this method an Eastern blot.
- (2005) Eastern blot has been used to describe a blot of proteins on PVDF membrane where the probe is an aptamer rather than an antibody. This could be seen as similar to a Southern blot, however the interaction is between a DNA molecule(the aptamer) and a protein, rather than two DNA molecules.
- (2006) Eastern blotting was described again as a new technique for detecting glycoconjugates generated by blotting BSA onto PVDF membranes, followed by periodate treatment. The oxidized protein is then treated with a complex mixture, generating a new conjugate on the membrane. The membrane is then probed with antibodies for epitopes of interest.
- (2009) Eastern blotting has most recently been re-dubbed by Thomas et al. as a technique which probes proteins or lipids blotted to PVDF membrane with lectins and chemical stains. These authors distinguish the method from the Far-eastern blot named by Taki et al. in that they use lectin probes and other staining reagents.

It is not clear why so many different groups have re-named the technique. Perhaps the commonly known Southern, Northern, and Western blots leave the appearance of a vacuum for an Eastern that many would like to fill. Since these techniques are all quite similar (most detect carbohydrates) it is surprising that different groups have not recognized the previous invocation of the term

Eastern blot. Regardless, it seems that each of these schools of thought will continue using their own version of the technique.

The principles used for Eastern blotting or glycans can be traced back to the use of lectins to detect protein glycosylation. The earliest example is Tanner and Anstee in 1976, where lectins were used to detect glycosylated proteins isolated from human erythrocytes. The specific detection of glycosylation through blotting is usually referred to as lectin blotting. A summary of more recent improvements of the protocol has been provided by H. Freeze.

Though a commonly employed set of techniques, the term Eastern blotting is used only rarely in primary literature, and has appeared more in recent publications, sometimes referring to significantly different versions of the experiment. One application of the technique includes detection of protein modifications in two bacterial species Ehrlichia- E. muris and IOE. Cholera toxin B subunit (which binds to gangliosides), Concanavalin A (which detects mannose-containing glycans) and nitrophospho molybdate-methyl green (which detects phosphoproteins) were used to detect protein modifications. The technique showed that the antigenic proteins of the non-virulent E.muris is more post-translationally modified than the highly virulent IOE.

Significance

Most proteins that are translated from mRNA undergo modifications before becoming functional in cells. The modifications collectively, are known as post-translational modifications (PTMs). The nascent or folded proteins, which are stable under physiological conditions, are then subjected to a battery of specific enzyme-catalyzed modifications on the side chains or backbones.

Post-translational protein modifications can include: *acetylation, acylation (myristoylation, palmitoylation), alkylation, arginylation, biotinylation, formylation, geranylgeranylation, glutamylation, glycosylation, glycylation, hydroxylation, isoprenylation, lipoylation, methylation, nitroalkylation, phosphopantetheinylation, phosphorylation, prenylation, selenation, S-nitrosylation, sulfation, transglutamination and ubiquitination (sumoylation).*[14][15]

Post-translational modifications occurring at the N-terminus of the amino acid chain play an important role in translocation across biological membranes. These include secretory proteins in prokaryotes and eukaryotes and also proteins that are intended to be incorporated in various cellular and organelle membranes such as lysosomes, chloroplast, mitochondria and plasma membrane. Expression of posttranslated proteins is important in several diseases.

Far-Eastern Blotting

Far-Eastern blotting is a technique developed in the 1990s by T. Taki and colleagues at the Cellular Technology Institute of Otsuka Pharmaceutical Co., Japan for the analysis of lipids separated by high-performance thin layer chromatography (HPTLC). The lipids are transferred from the HPTLC plate to a PVDF membrane for further analysis, for example by enzymatic or ligand binding assays or mass spectrometry.

Cholesterol, glycerophospholipids and sphingolipids are major constituents of the cell membrane and in certain cases function as second messengers in cell proliferation, apoptosis and cell adhesion in inflammation and tumor metastasis. Far-eastern blotting was established as a method for transferring lipids from an HPTLC plate to a polyvinyledene difluoride (PVDF) membrane within a minute. Applications of this with other methods have been studied. Far-eastern blotting allows for the following techniques:

- Purification of glycosphingolipids and phospholipids.
- Structural analysis of lipids in conjunction with direct mass spectrometry.
- Binding study using various ligands such as antibodies, lectins, bacterium, viruses, and toxins, and
- Enzyme reaction on membranes.

Not only analysis of lipids but also metabolites of drugs and natural compounds from plants, and environmental hormones are possible by this method.

33

DNA Sequencing and Designing of Primers

DNA Sequencing

DNA sequencing is the determination of the precise sequence of nucleotides in a sample of DNA. Sequencing can be done on a short fragment of DNA to cover the entire genome. The maps thus constructed using the nucleotide sequences in the correct order is called as sequence map.

In 1977, twenty-four years after the discovery of the structure of DNA, two separate methods for sequencing DNA were developed: the **chain termination method** (Sanger *et al.*, 1977) and the **chemical degradation method** (Maxam and Gilbert, 1977). Both methods were equally popular to begin with, but, for many reasons, the chain termination method is the method more commonly used today. In chemical degradation method, chemicals specific to each base remove the last base, which has been labelled. The method uses toxic chemicals which required high care while handling.

In chain termination method, DNA polymerase extends a DNA strand in which dideoxynucleotide becomes incorporated terminating the chain extension (phosphodiester formation) at the 3′OH site making an array of fragments differing in length. This method is based on the principle that single-stranded DNA molecules that differ in length by just a single nucleotide can be separated from one another using polyacrylamide gel electrophoresis.

The Chain Termination Method

The technique uses specifically synthesized nucleotide molecules called dideoxynucleotides or ddNTPs. The ddNTPs differ from their conventional counterparts – the dNTPs- by having an H atom instead of the OH- at the 3′ carbon atom in the nucleotide skeleton. The absence of the hydroxyl group prevents the ddNTPs in extending the DNA chain, since they are not capable of forming the phosphodiester bonds. This is because dideoxynucleotides are missing a special group of molecules, called a 3'-hydroxyl group, needed to form a connection with the next nucleotide (Figure below). This method is otherwise called as 'dideoxy method'.

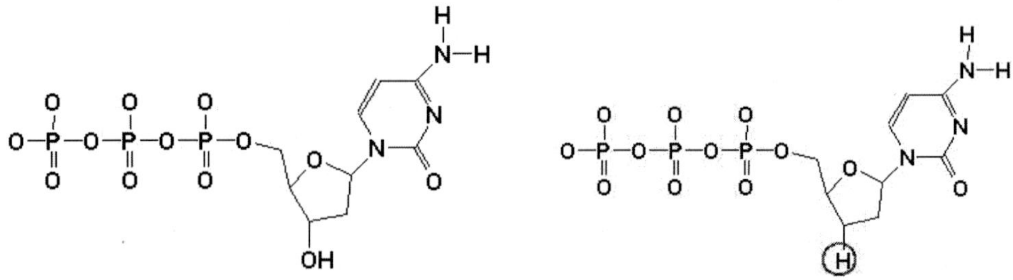

Deoxycytosine (dCTP) Dideoxycytosine (ddCTP)

Structures of deoxycytosine tri phosphate (dCTP) and dideoxycytosine tri phosphate (ddCTP)

DNA sequencing reactions are just like the PCR reactions for replicating DNA. The reaction mix includes the template DNA, free nucleotides, an enzyme (usually a variant of *Taq* polymerase) and a 'primer' - a small piece of single-stranded DNA about 20-30 nucleotides long that can hybridize to one strand of the template DNA.

There are two types of sequencing approaches followed.

a) Manual sequencing
b) Automated sequencing

Manual Sequencing

The DNA to be sequenced, called the template DNA, is first prepared as a single stranded DNA. Then, a short oligonucleotide is annealed, or joined, to the same position on each template strand. The oligonucleotide acts as a primer for the synthesis o f a new DNA strand that will be complimentary to the template DNA. This technique requires that four nucleotide-specific reactions—one each for G, A, C, and T—be performed on four identical samples of DNA independently in four different tubes. The four sequencing reactions require the addition of all the components necessary to synthesize and label new DNA, including:

- A DNA template;
- A primer tagged with a mildly radioactive molecule or a light-emitting chemical;
- DNA polymerase—an enzyme that drives the synthesis of DNA;
- Four deoxynucleotides (G, A, C, T); and
- One dideoxynucleotide, either ddG, ddA, ddC, or ddT.

After the first deoxynucleotide is added to the growing complementary sequence, DNA polymerase moves a long the template and continues to add base after base. The strand synthesis reaction continues until a dideoxynucleotide is added, blocking further elongation. Only a small amount of a dideoxynucleotide is added to each reaction, allowing different reactions to proceed for various lengths of time, until by chance, DNA polymerase inserts a dideoxynucleotide, terminating the reaction. Therefore, the result is a set of new chains, all of different lengths.

Usually the reaction mix is supplied with a copious amount of normal dNTPs, and limited amount of special ddNTPs. For example, in the case of DNA extension in the presence of ddTTP, each of the time when a 'TTP' is required to make the new strand, the enzyme will get a good one most of the time. After adding a T, the enzyme will go ahead and add more nucleotides. However, in 5% of the time, the enzyme will get a dideoxy-T, and that strand can never again be elongated. It eventually breaks away from the enzyme, as a dead end product.

Sooner or later all of the copies will get terminated by a T, but each time the enzyme makes a new strand, the place it gets stopped will be random. In millions of starts, there will be strands stopping at every possible T along the way. All of the strands started at one exact position, and all of them end with a T. There are billions of them ... many millions at each possible T position. To find out where all the T's are in our newly synthesized strand, all we have to do is find out the sizes of all the terminated products!

Sanger Sequencing

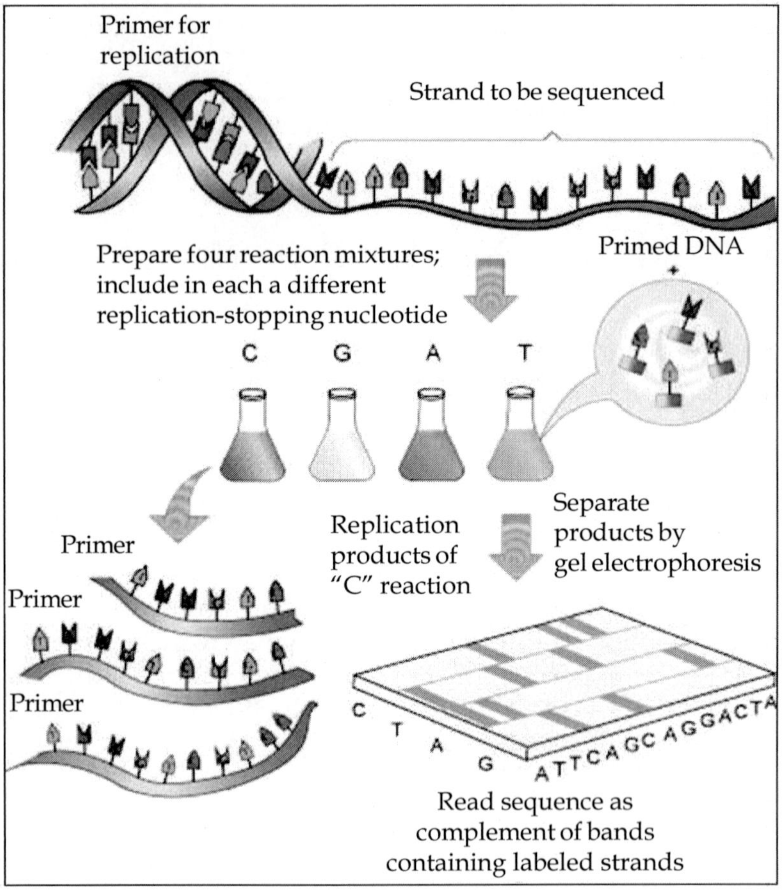

Source : http--www_ncbl_nlm_nih_gov.Class-NAWBIS.Modules-Genomes2Other-images.slide 13 gif.mht

To read the newly generated sequence for all the four nucleotides, the four reactions are run side-by-side on a polyacrylamide sequencing gel. The family of molecules generated in the presence of ddATP is loaded into one lane of the gel and the other three families, generated with dCTP, ddGTP, and ddTTP, are loaded into three adjacent lanes. After electrophoresis, the DNA sequence can be read directly from the positions of the bands in the gel.

Now by putting all four deoxynucleotides into the picture, by comparing their relative positions, one can determine the order of their occurrence in the DNA strand. The bases are read from bottom to top, starting with the smallest fragment first (figure ahead).

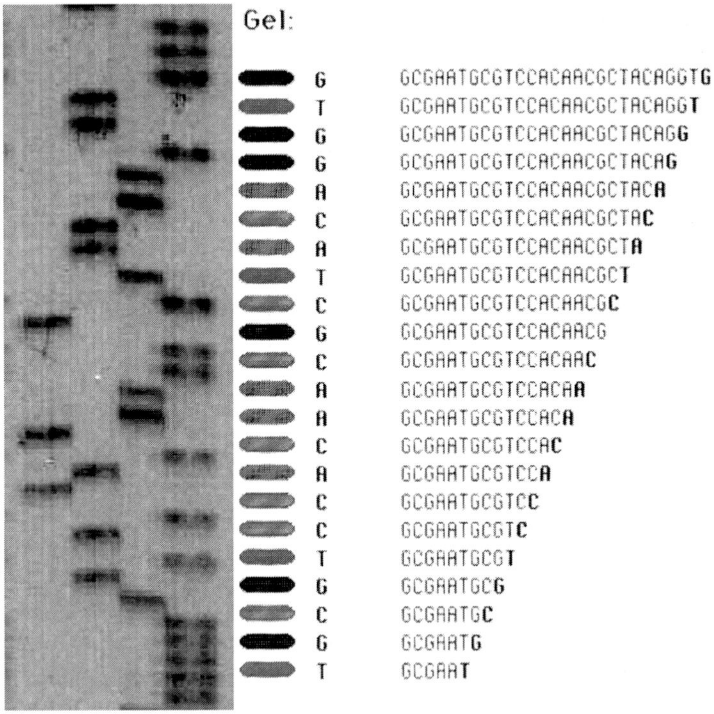

Separating gel in manual sequencing and interpretation of the bands

Automated Sequencing

Automatic sequencing is done with help of sequencing machines which are controlled by computers. Here, unlike in the manual method, the entire reaction is done in single tube, and all the ddNTPs are added together in the reaction. The ddNTPs are prepared especially so that they can emit a characteristic colour individually when transilluminated with laser beam. A gel separates the resulting fragments by size and we can 'read' the sequence from it, bottom to top.

These are used in a large-scale sequencing lab, where a machine is used to run the gels and to monitor the different colours as they come off the bottom. It's called an automated DNA sequencer. There's an ultraviolet laser built into the machine that shoots through the gel near the bottom and scans side to side, checking for bands of fluorescent colours to pass through its beam. There might be as many as 96 'lanes' of samples running in one gel. Figure 3 shows how a real fragment of a gel looks like. The four colours red, green, blue and yellow each represent one of the four nucleotides. The actual gel image, if you could get a monitor large enough to see it all at this magnification, would be perhaps 3 or 4 meters long and 30 or 40 cm wide.

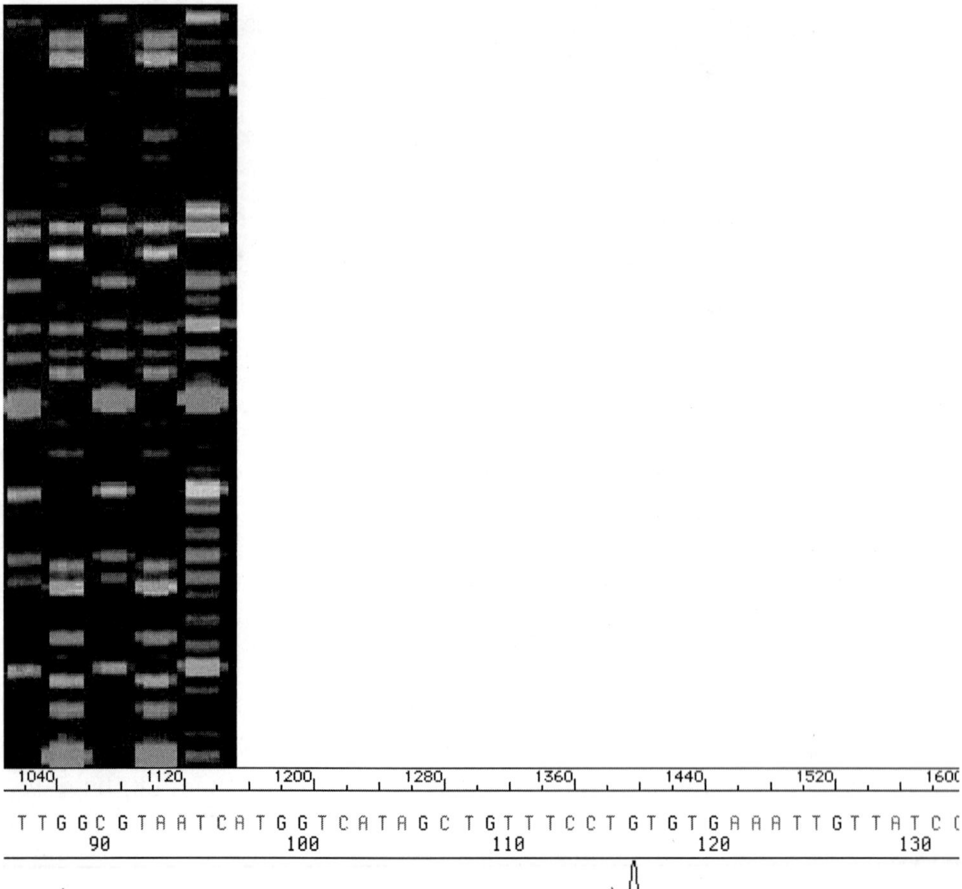

Computer simulated gel scan and electrophorgram in automated sequencing

The 'Scan' of each gel lane is done by the computer through the fluorescent signal generated by the laser scanner, and the computer process the signal to generate an electrophorogram (figure above). This is a plot of the colours detected in one 'lane' of a gel (one sample), scanned from smallest fragments to largest. The computer can interprets the colours by printing the nucleotide sequence across the top of the plot. This is just a fragment of the entire file, which would span around 700 or so nucleotides of accurate sequence. The sequencer also gives the operator a text file containing just the nucleotide sequence, without the colour traces. Recent advancements uses multigel, multidye scanning for accurate sequencing as shown in figure below.

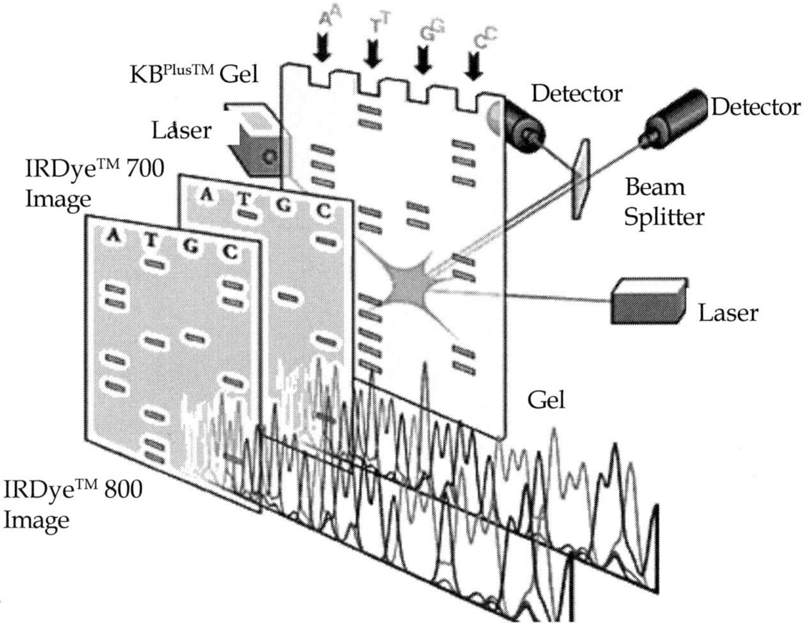

Advanced methods in automated sequencing

Whole Genome Sequencing

The prevailing method of determining the sequence of a long DNA segment is the shotgun sequencing approach, in which a random sampling of short fragment sequences is acquired and then assembled by a computer program to infer the sampled segment's sequence. In the early 1980s, such segments were typically on the order of 5,000 to 10,000 base pairs (5 to 10 kbp). By 1990, this method was sequencing segments on the order o f 40 kbp, and by 1995, the entire 1 ,800-kbp *H. Influenzae* bacterium had been sequenced.

Shotgun Sequencing: Assembly of Random Sequence Fragments

In shot gun method, to sequence a bacterial artificial chromosome (BAC) of typically 200-300kbp long, first millions of copies of it are constructed and chop them all up randomly. The fragments are then inserted into plasmids and each one are grown in lots in bacteria and sequence the insert.

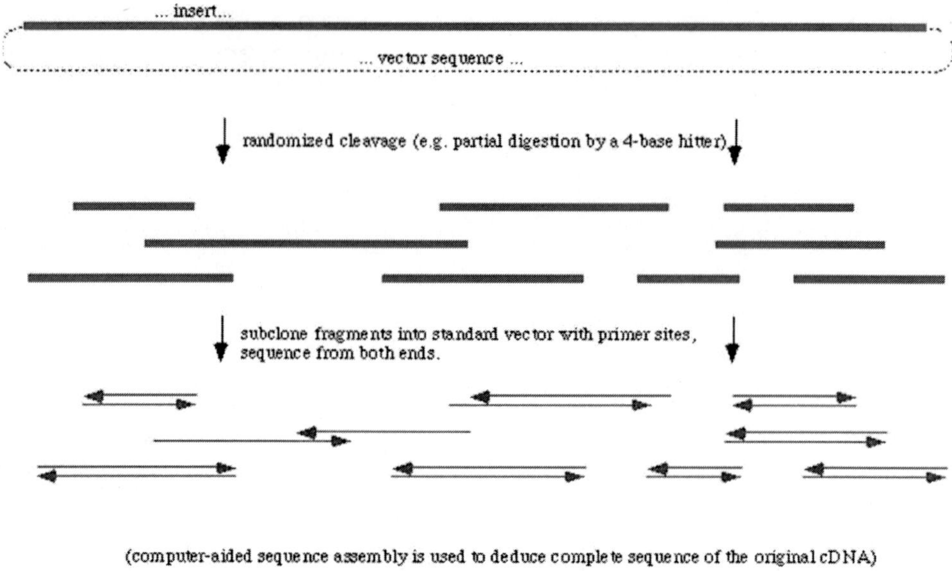

Cartoon depicting shotgun sequencing

The method used are,

1. The sample DNA is randomly fractured either using sound (sonication) or passing it through a nozzle under pressure (nebulation), which produces a uniformly random partitioning of each copy of the source strand into a collection of DNA fragments.
2. To remove fragments that are too large or too small, this pool of fragments is size selected, typically using size separation under gel electrophoresis and then simply excising a band of the gel containing the desired size. With care, this procedure produces a normally distributed collection of fragment sizes with a 10% variance.
3. The size-selected fragments are then inserted into the DNA of a genetically engineered bacterial virus (phage), called a vector. Usually, at most one fragment is inserted at a predetermined point, called the cloning site, in the vector. Typically, the number of vectors where more than one fragment gets inserted is less than 1%, but can be as low as 0.01% for some meticulously executed protocols. The fragments at this point are often called inserts and the collection of inserts is a library.

4. A bacterium is then infected with a single vector, which reproduces to produce a bacterial colony containing millions of copies of the vector and its associated insert. The procedure thus has effectively cloned a pure sample of the given insert. This procedure repeats simultaneously for as many inserts as desired for sequencing in the final step.
5. By design, the vector t hen permits a sequencing reaction to be performed, starting just to the left or right of a source fragment's insertion point. The sequencing reaction produces a read of the first 300 to 900 bases of one end of the insert.
6. When all the fragments are sequenced, eventually the entire sequence is reconstructed to the sequence of the original BAC based on the overlapping fragments (figure above).

Advantages

- Low quantities of template DNA required (10-100 ng per reaction)
- High reproducibility
- Fragments studied are of known identity
- Amenable to automation

Disadvantages

- Laborious and technically demanding.

Primer Designing

Primers are short strands of DNA that are capable of annealing to the target DNA in a predictable location and on a predictable strand. They must be capable of extension by *Taq* DNA Polymerase. Primers are extensively used in modern molecular techniques for in vitro multiplication or amplification of the DNA strands. Most common uses of primers are in Polymerase chain reaction (PCR) and in DNA sequencing.

Perhaps the most critical parameter for successful PCR is the design of Primers. All things being equal, a poorly designed primer can result in a PCR reaction that will not work. The primer sequence determines several things such as the length of the product, its melting temperature and ultimately the yield. A poorly designed primer can result in little or no product due to non-specific amplification and/or primer-dimer formation, which can become competitive enough to suppress product formation. This a p p lication note is provided to give rules that should be taken into account when designing primers for PCR. More comprehensive coverage of this subject can be found elsewhere (Dieffenbach *et al.*, 1995).

Primer Selection

Several variables must be taken into account when designing PCR Primers. Among the most critical are:

- Primer length
- Melting Temperature (*T*m)
- Specificity
- Complementary Primer Sequences
- G/C content and Polypyrimidine (T, C) or polypurine (A, G) stretches
- 3′-end Sequence

Each of these critical elements will be discussed in turn.

Primer Length

Since both specificity and the temperature and time of annealing are at least partly dependent on primer length, this parameter is critical for successful PCR. In general, oligonucleotides between 18 and 24 bases are extremely sequence specific, provided that the annealing temperature is optimal. Primer length is also proportional to annealing efficiency: in general, the longer the primer, more inefficient the annealing. With fewer templates primed at each step, this can result in a significant decrease in amp lified product. The primers should not b e too short, however, unless the application specifically calls for it. As discussed below, the goal should be to design a primer with an annealing temperature of at least 50°C.

The relationship between annealing temperature and melting temperature is one of the "Black Boxes" of PCR. A general rule-of-thumb is to use an annealing temperature that is 5°C lower than the melting temperature. Thus, when aiming for an annealing temperature of at least 50°C, this corresponds to a primer with a calculated melting temperature (*T*m) ~55°C. Often, the annealing temperature determined in this fashion will not be optimal and empirical experiments will have to be performed to determine the optimal temperature. This is most easily accomplished using a gradient thermal cycler like Eppendorf's Mastercycler® gradient.

Melting Temperature (*T*m)

It is important to keep in mind that there are two primers added to a PCR reaction. Both of the oligonucleotide primers should be designed such that they have similar melting temperatures. If primers are mismatched in terms of *T*m, amplification will be less efficient or may not work at all since the primer with the higher *Tm* will mis-prime at lower temperatures and the primer with the lower *Tm* may not work at higher temperatures.

The melting temperatures of oligos are most accurately calculated using nearest neighbour thermodynamic calculations with the formula:

$$Tm\text{primer} = \Delta H\ [\Delta S + R \ln (c/4)] - 273.15°C + 16.6 \log 10\ [K+]$$

where, H is the enthalpy and S is the entropy for helix formation, *R* is the molar gas constant and c is the concentration of primer. This is most easily accomplished using any of a number of primer design software packages on the market (Sharrocks, 1994). Fortunately, a good working approximation of this value (generally valid for oligos in the 18–24 base range) can be calculated using the formula:

$$Tm = 2(A+T) + 4(G+C).$$

In addition to calculating the melting temperatures of the primers, care must be taken to ensure that the melting temperature of the product is low enough to ensure 100% melting at 92°C. This parameter will help ensure a more efficient PCR, but isnot always necessary for successful PCR. In general, products between 100–600 base pairs are efficiently amplified in many PCR reactions. If there is doubt, the product *Tm* can be calculated using the formula:

$$Tm = 81.5 + 16.6\ (\log10[K+] + 0.41\ (\%G+C) - 675/\text{length}.$$

Under standard PCR conditions of 50 mM KCL (Sharrocks, 1994), this reduces to, $Tm = 59.9 + 0.41\ (\%G+C) - 675/\text{length}$.

Specificity

As mentioned above, primer specificity is at least partly dependent on primer length. It is evident that there are many more unique 24 base oligos than there are 15 base pair oligos. That being said, primers must be chosen so that they have a unique sequence within the template DNA that is to be amplified. A primer designed with a highly repetitive sequence will result in a smear when amplifying genomic DNA.

However, the same primer may give a single band if a single clone from a genomic library is amplified. Because *Taq* Polymerase is active over a broad range of temperatures, primer extension will occur at the lower temperatures of annealing. If the temperature is too low, non-specific priming may occur which can be extended by the polymerase if there is a short homology at the 3' end. In general, a melting temperature of 55°C–72°C gives the best results (Note that this corresponds to a primer length of 18–24 bases using Wallace's rule above).

Complementary Primer Sequences

Primers need to be designed with absolutely no intra-primer homology beyond 3 base pairs. If a primer has such a region of self-homology, "snap back", partially double- tranded structures, can occur which will interfere with annealing to the template.

Another related danger is inter-primer homology. Partial homology in the middle regions of two primers can interfere with hybridization. If the homology should occur at the 3' end of either primer, Primer dimer formation will occur which, more often than not, will p revent the formation of the desired product via competition.

G/C Content and Polypyrimidine (T, C) or Polypurine (A, G) Stretches

The base composition of primers should be between 45% and 55% GC. The primer sequence must be chosen such that there is no PolyG or PolyC stretches that can promote non-specific annealing. Poly A and Poly T stretches are also to be avoided as these will "breath" and open up stretches of the primer-template complex. This can lower the efficiency of amplification. Polypyrimidine (T, C) and polypurine (A, G) stretches should also be avoided. Ideally the primer will have a near random mix of nucleotides, a 50% GC content and be ~20 bases long. This will put the *T*m in the range of 56°C - 62°C (Dieffenbach *et al.*, 1995).

3'-end Sequence

It is well established that the 3' terminal position in PCR primers is essential for the control of mis-priming (Kwok *et al.*, 1990). We have already explored the problem of primer homologies occurring at these regions. Another variable to look at is the inclusion of a G or C residue a t the 3 ' end of p rimers. This "GC Clamp" helps to ensure correct binding at the 3' end due to the stronger hydrogen bonding of G/C residues. It also helps to improve the efficiency of the reaction by minimizing any "breathing" that might occur.

Designing a Sequencing Primer

Generally, the basics of designing sequence primer are very similar to that of the PCR primers. However, following aspects are to be taken into consideration while synthesising a sequence primer.

a. Design Primers only from Accurate Sequence Data

Automated sequencing (and in fact any sequencing) has a finite probablility of producing errors. Sequence obtained too far away from the primer must be considered questionable. Select a region for primer placement where the possibility of sequence error is low.

b. Restrict Search to Regions that Best Reflect the Goals

We may be interested in maximizing the sequence data obtained, or may only need to examine the sequence at a very specific location in the template. Such needs dictate very different primer placements.

i. Maximize sequence obtained while minimizing the potential for errors : Generally, you should design the primer as far to the 3' as you can manage as long as you have confidence in the accuracy of the sequence from which the primer is drawn. Primers on opposite strands should be placed in staggered fashion as much as possible.

ii. Targeted sequencing of a specific region : Position the primer so the desired sequence falls in the most accurate region of the chromatogram. Sequence data is often most accurate about 80-150 nucleotides away from the primer. Do not count on seeing good sequence less than 50 nucleotides away from the primer or more than 300 nucleotides away (although we often get sequence starting immediately after the primer, and we often return 700 nucleotides of accurate sequence).

c. Locate Candidate Primers

Identify potential sequencing primers that produce stable base pairing with the template DNA under conditions appropriate for cycle sequencing. It is **strongly** suggested that you use a computer at this step.

Suggested Primer Characteristics

i. Length should be between 18 and 30 nucleotides, with optimal being 20-25 nucleotides. (Although there are some successes with primers longer than 30 and shorter than 18).

ii. G-C content of 40-60% is desirable.

iii. The T*m* should be between 55°C and 75°C.

d. Discard Candidate Primers that Show Undesirable Self-hybridization

Primers that can self-hybridize will be unavailable for hybridization to the template. Generally avoid primers that can form 4 or more consecutive bonds with itself, or 8 or more bonds total. Example of a marginally problematic primer:

```
        5'-ACGATTCATCGGACAAAGC-3'
               |||| ||||
3'-CGAAACAGGCTACTTAGCA-5'
```

This oligo forms a substantially stable dimer with itself, with four consecutive bonds at two places and a total of eight inter-strand bonds. Primers with 3' ends hybridizing even transiently will become extended due to polymerase action, thus ruining the primer and generating false bands. Be more stringent in avoiding 3' dimers. For example, the following primer self-dimerizes with perfect 3' hybridization on itself:

```
5'-CGATAGTGGGATCTAGATCCC-3'
          ||||||||||||||
        3'-CCCTAGATCTAGGGTGATACG-5'
```

The above oligo is pretty bad, and almost guaranteed to cause problems. Note that the polymersase will extend the 3' end during the sequencing reaction, giving very strong sequence ACTATGC. These bands will appear at the start of your 'real' data as immense p eaks, o ccluding the correct s equence. Most p rimer d esign p rograms will correctly spot such self-dimerizing primers, and will warn to avoid them.

It may be noted that no computer program or rule-of-thumb assessment can accurately predict either success or failure of a primer. A primer that seems marginal may perform well, while another that appears to be flawless may not work at all.

e. Verify the Site-specificity of the Primer

Perform a sequence homology search (e.g. dot-plot homology comparison) through all known template sequence to check for alternative priming sites. Discard any primers that display 'significant' tendency to bind to such sites. Avoid primers where alternative sites are present with,

1. More than 90% homology to the primary site or
2. More than 7 consecutive homologous nucleotides at the 3' end or
3. Abundance greater than 5-fold higher than the intended priming site.

f. Choosing Among Candidate Primers

If there are several candidate primers, select one or a few that are more A-T rich at the 3 ' end. T hese tend t o b e s lightly m ore s p ecific in a ction, a ccording t o some investigators. Also use of more than one primer can maximise the likelihood of success.

If there are no candidate primers that survived the criteria above, then relax the stringency of the selection requirements. Ultimately, the test of a good primer is only in its use, and cannot be accurately predicted by these simplistic rules-of-thumb.

34

Gene Transfer to Plant

The transfer foreign gene to the plant cell is a very crucial method to production of transgenic plant. So improvement of plant quality requires to transfer our interest gene or we can say useful for production of our interest product. To achieve genetic transformation in plants, we need the construction of a vector (genetic vehicle) which transports the genes of interest, flanked by the necessary controlling sequences i.e. promoter and terminator, and deliver the genes into the host plant. The two kinds of gene transfer methods in plants are.

1. Vector-mediated or Indirect Gene Transfer

Among the various vectors used in plant transformation, the Ti plasmid of ***Agrobacterium tumefaciens*** has been widely used. This bacteria is known as "natural genetic engineer" of plants because these bacteria have natural ability to transfer T-DNA of their plasmids into plant genome upon infection of cells at the wound site and cause an unorganized growth of a cell mass known as crown gall. Ti plasmids are used as gene vectors for delivering useful foreign genes into target plant cells and tissues. The foreign gene is cloned in the T-DNA region of Ti-plasmid in place of unwanted sequences.

To transform plants, leaf discs (in case of dicots) or embryogenic callus (in case of monocots) are collected and infected with ***Agrobacterium*** carrying recombinant disarmed Ti-plasmid vector. The infected tissue is then cultured (co-cultivation) on shoot regeneration medium for 2-3 days during which time the transfer of T-DNA along with foreign genes takes place. After this, the transformed tissues (leaf discs/calli) are transferred onto selection cum plant regeneration medium supplemented with usually lethal concentration of an antibiotic to selectively eliminate non-transformed tissues. After 3-5 weeks, the regenerated shoots (from leaf discs) are transferred to root-inducing medium, and after another 3-4 weeks, complete plants are transferred to soil following the hardening (acclimatization) of regenerated plants. The molecular techniques like PCR and southern hybridization are used to detect the presence of foreign genes in the transgenic plants.

2. Vectorless or Direct Gene Transfer

In the direct gene transfer methods, now are day many methods were developed to the foreign gene of interest is delivered into the host plant cell without the help of a vector. The methods used for direct gene transfer in plants.

Microinjection

where the DNA is directly injected into plant protoplasts or cells (specifically into the nucleus or cytoplasm) using fine tipped (0.5 - 1.0 micrometerdiameter) glass needle or micropipette. This method of gene transfer is used to introduce DNA into large cells such as oocytes, eggs, and the cells of early embryo.

The process is observed and controlled under the microscope. The DNA is then integrated into the plant genome - probably during the cell's own DNA repair processes.

The advantage of microinjection is that with this method, genetic modification in theory no longer requires a **marker gene**.

- The target gene, which confers a new trait, is introduced directly into a single cell.
- The cells transformed in this way are easy to identify if a dye is injected along with the DNA.
- If the process works, it will no longer be necessary to select the transformed cells using antibiotic resistance or herbicide resistance markers.

So far this method has been used primarily with animal cells. One of the ways in which plant cells differ from animal cells is that they have a stable cell wall. To use microinjection with plants the method has to be adapted to their special characteristics.

The most important difference is the cell wall and the internal cell pressure of up to 30 bar. If the kind of cannula used for animal cells is used on plant cells it causes a drastic loss of pressure and the cell dies. To solve the problem, even thinner cannula tips are being developed and used in combination with a finely adjustable pressure generator. In this way the cannula enters the plant cell without destroying it.

Another approach to improving microinjection as a transformation method for plant cells focuses on the efficiency with which the injected genes are integrated into the plant genome. One way of improving this is to inject special proteins along with the target gene sequences which support the stable integration of 'foreign' genes into the plant genome and are produced by *Agrobacterium tumefaciens* . This soil bacterium is naturally capable of inserting its own genes into the genome of plants.

Chemically Stimulated DNA Uptake by Protoplasts

Direct DNA uptake by protoplasts can be stimulated by chemicals like polyethylene glycol (PEG) and dextran sulphate The technique is so efficient that virtually every protoplast system has proven transformable.

PEG is also used to stimulate, the uptake of liposomes ,and to improve the efficiency, of electroporation. PEG at high concentration (15-25%) will precipitate ionic macromolecules such as DNA and stimulate their uptake by endocytosis without any gross damage to protoplasts. This is followed by cell wall formation and initiation of cell division. These cells can now be plated at low density on selection medium.

However, other plant systems (rice, maize, etc.) were also successfully used later. In these methods, PEG was used in combination with pure Ti plasmid, or calcium phosphate precipitated Ti plasmid mixed with a carrier DNA.

Transformation frequencies up to 1 in 100 have been achieved by this method. Nevertheless, there are serious problems in using this method for getting transgenic plants and all these problems relate to plant regeneration from protoplasts.

Electroporation

This method is based on the use of short electrical impulses of high field strength. a pulse of high voltage applied to protoplasts/cells/ tissues to make transient (temporary) pores in the plasma membrane which facilitates the uptake of foreign DNA. If the DNA is in direct contact with the membrane.

In view of this, for delivery of DNA to protoplasts, electroporation is one of the several routine techniques for efficient transformation. However, since regeneration from protoplasts is not always possible, cultured cells or tissue explants are often used.

Consequently, it is important to test whether electroporation could transfer genes into walled cells. The electroporation pulse is generated by discharging a capacitor across the electrodes in a specially designed electroporation chamber.Either a high voltage (1.5 kV) rectangular wave pulse of short duration or a low voltage (350V) pulse of long duration is used.

Using electroporation method, successful transfer of genes was achieved with the protoplasts of tobacco, petunia, maize, rice, wheat and sorghum. In most of these cases cat gene associated with a suitable promoter sequence was transferred.

Transformation frequencies can be further improved by

(i) using field strength of 1.25kV/cm,

(ii) adding PEG after adding DNA,

(iii) heat shocking protoplasts at 45°C for 5 minutes before adding DNA and by using linear instead of circular DNA.

Particle Gun/Particle Bombardment / Microprojectiles / Biolistics

In recent years, it has been shown that DNA delivery to plant cells is also possible, when heavy microparticles (tungsten. or gold) coated with the DNA of interest are accelerated to a very high initial velocity (1,400 ft per, sec). These microprojectiles, normally 1-3pm in diameter, are carried by a 'macroprojectile' or the 'bullet' and are accelerated into living plant cells (target cells can be pollen, cultured cells, cells in differentiated tissues and meristems) so that they can penetrate cell walls of intact tissue.

The acceleration is achieved either by an explosive charge (cordite explosion) or by using shock waves initiated by a high voltage electric discharge. The design of two particle guns used for acceleration of microprojectiles.

Transformed plants using the above technique have been obtained in many cases including soybean, tobacco, maize, rice, wheat, etc. Transient expression of genes transferred in cells by this method has also been observed in onion, maize, rice and wheat.

There is no other gene transfer approach, which has met with so much of enthusiasm. Consequently considerable investment has been made in experimentation and manpower for development of this technique.

Liposome Mediated Gene Transfer or Lipofection

Liposomes are small, circular lipid molecules with an aqueous interior that can carry nucleic acids. Liposomes encapsulate the DNA fragments and then adhere to the cell membranes and fuse with them to transfer DNA fragments. Thus, the DNA enters the cell and then to the nucleus. Lipofection is a very efficient technique used to transfer genes in bacterial, animal and plant cells.

The technique, offers following advantages:

- Protection of DNAIRNA from nuclease digestion,
- Low cell toxicity,
- Stability and storage of nucleic acids due to encapsulation in liposomes,
- High degree of reproducibility and
- Applicability to a wide range of cell types. In this technique, DNA enters the protoplasts due to endocytosis of liposomes, involving the following steps:
    (i) Adhesion of the liposomes to the protoplast surface,
    (ii) Fusion of liposomes at the site of adhesion and
    (iii) Release of plasmids inside the cell. The technique has been successfully used to deliver DNA into the protoplasts of a number of plant species (e.g. tobacco, petunia, carrot, etc.).

Transformation by Ultrasonication for Gene Transfer

In wheat, tobacco and sugarbeet, explants after being cultured for a few days, were sonicated with plasmid DNA (carrying marker genes like cat, npt II and gus). When sonicated calli were transferred to selective medium, shoots were obtained, although all control calli (not sonicated with plasmid DNA) died. In tobacco transgenic plants were obtained at a frequency of 22 per cent. This work was conducted at the Biotechnology Research Center at Beijing, China.

Gene Transformation using Pollen OR Pollen Tube

There has been a hope that DNA can be taken up by the germinating pollen and can either integrate into sperm nuclei or reach the zygote through the pollen tube pathway. Both these approaches have been tried and interesting phenotypic alterations suggesting gene transfer have been obtained.

In no case, however, unequivocal proof of gene transfer has been available. In a number of experiments, when marker genes were used for transfer, only negative results were obtained.

Several problems exist in this method and these include the presence of cell wall, nucleases, heterochromatic state of acceptor DNA, callose plugs in pollen tube, etc. Transgenic plants have never been recovered using this approach and this method, though very attractive, seems to have little potential for gene transfer.

Calcium Phosphate Precipitation Method for Gene Transfer

Foreign DNA can also be carried with the Ca ++ ions, to be released inside the cell due to the precipitation of calcium in the form of calcium phosphate. In the past, this method was considered to be very important for gene transfer in plants.

Incubation of Dry Seeds, Embroys, Tissues OR Cells in DNA

Incubation of dry seeds, embryos, tissues or cells in known DNA (viral or non viral having defined marker genes) has been tried in many cases and expression of defined genes has been witnessed. However, in no case proof of integrative transformation could be available. In all these cases, plant cell walls not only work as efficient barriers, but are also efficient traps for DNA molecules. It would be very surprising if DNA can cross cell walls efficiently without permeabilizing them either by PEG, or by electroporation or by anher device.

Selection of Transformed Cells from Untransformed Cells

The selection of transformed plant cells from untransformed cells is an important step in the plant genetic engineering. For this, a marker gene (e.g. for antibiotic resistance) is introduced into the plant along with the transgene followed by the selection of an appropriate selection medium (containing the antibiotic). The segregation and stability of the transgene integration and expression in the subsequent generations can be studied by genetic and molecular analyses (Northern, Southern, Western blot, PCR).

Vector-Mediated or Indirect Gene Transfer

Vectors for Gene Transfer The common features of vectors used for transformation of eukaryotic cells. Most vectors carry marker genes, which allow recognition of transformed cells (other cells die due to the action of an antibiotic or herbicide) and are described as selectable markers. Among these marker genes, the most common selectable marker is npt II, providing kanamycin resistance.

Other common features of suitable transformation vector include the following:

(i) multiple unique restriction sites (a synthetic polylinker);

(ii) bacterial origins of replication (*e.g.* ColE1).

The vectors having these properties may not necessarily have features, which facilitate their transfer to plant cells or integration into the plant nuclear genome.

Therefore, *Agrobacterium* Ti plasmid is preferred over all other vectors, because of wide host range of this bacterial system and the capacity to transfer genes due to the presence of T - DNA border sequences.

Virus Vectors

Many viruses or their isolated genomes are capable of infecting intact plant tissue. This made them suitable for use as plant transformation vectors. But since viral genomes can not integrate with plant genome, they can be used only for a study of transient expression of transferred genes.

All the three kind of viruses including *caulimoviruses* (double stranded DNA), *gemini* viruses (single stranded DNA) and tobacco mosaic virus (RNA) are capable of delivering genes into intact plant tissues, where they are expressed.

However, virus vectors can not be used for the production of stably transformed plants, popularly described as transgenic plants.

Selectable Marker Genes in Vectors Used for Gene Transfer

Selectable marker gene	Substrate used for selection
	I. Antibiotics
1. Neomycin phosphotransferase (npt II)	G418, kanamycin, neomycin
2. Hygromycin phosphotransferase (hpt)	Hygromycin B
3. Dihydrofoate reductase (dhfr)	Methotrexate trimethoprim
4. Gene ble (enzyme not known)	Bleomycin
5. Gentamycin acetyltransferase	Gentamycin
6. Streptomycin phosphotransferase	Streptomycin
	II. Herbicides
7. Mutant form of acetolactate synthase (als)	Chlorosulfuron imidazolinones
8. Bromoxynil nitrilase	Bromoxynil
9. Phosphinothricin acetyl transferase (bar)	L - phosphinothricin (PPT; also known as bialaphos)
10. S. enolpyruvylshikimate -3–phosphate (EPSP) synthase (aro A)	Glyphosate (Roundup)

Ti Plasmid (Tumor Inducing Plasmid)

- Ti plasmid is considered as the most efficient Plant vector. It was first time discovered in 1983.
- This plasmid is found in *Agrobacteriaum tumefaciens*.
- This organism is a typical plant pathogen that enters the dicotyledonous plants from their root system and damages them.
- Agrobacterium enter through the fresh wound and attaches itself to the wall of an intact cell. Later on it transfer a short piece of Ti plasmid into the nucleus of plant cell. This DNA piece is known as T-DNA.
- T-DNA carries few genes which are expressed within the plant Construction of Ti plasmid :-

(1) One gene codes for an enzyme which is responsible for the synthesis of opine from amino acids.

- Opines are not found normally in plants and not metabolized by plants, but they can be used as a nutrient by Agrobacteriaum.
- Total 24 types of opines are detected in nature the particular opine produced by plant depends on the strain of bacterium.

- Some strains produce a gene responsible for the synthesis of "Nopaline" and others have the gene for the synthesis of "Octopine". These genes are respectively known as Nos & OCS.

(2) The other gene found in T-DNA is responsible for the induction of disorganized growth of infected plant cells that forms callus, Gail or tumor. It is called one gene.

- It is also observed that certain Agrobacteriaum strains contain genes for plant growth hormones Auxin & cytokinine instead of onc gene, such hormones also contribute to accelerate plant growth.

• Transformation of Ti plasmid

- Inside the plant cell; the T-DNA does not remain independent but it becomes integrated into plant chromosome.
- The integration occurs due to the presence of two repeated sequences of 25 base pairs which are located dear the ends of T-DNA.
- Once Ti plasmid is nicked, T-DNA is released from and it travels into plant cell.
- T-DNA becomes circular in its intermediate form, and ultimately it is inserted into plant chromosome DNA.
- Genes that remain on Ti plasmid include those for attachment of bacterium to plant cell wall, those for transfer of T-DNA and those for uptake and catabolism of opine.
- The gene which is responsible for the transfer of T-DNA lies on Ti plasmid and it is called vir-Region.
- This transformation is irreversible and callus cells can continuously produce opine once they are infected (One type of opine) is also present on Ti plasmid and it is called noc gene.

Onc: Gene for tumor induction

Vir: Gene for the transfer of (Oncogene) T-DNA (Virulence)

nos: Nopaline synthesis

noc:Nopaline catabolism

Fig. Show the mechanism of Ti-plasmid transfer bacterialcell to plant cell

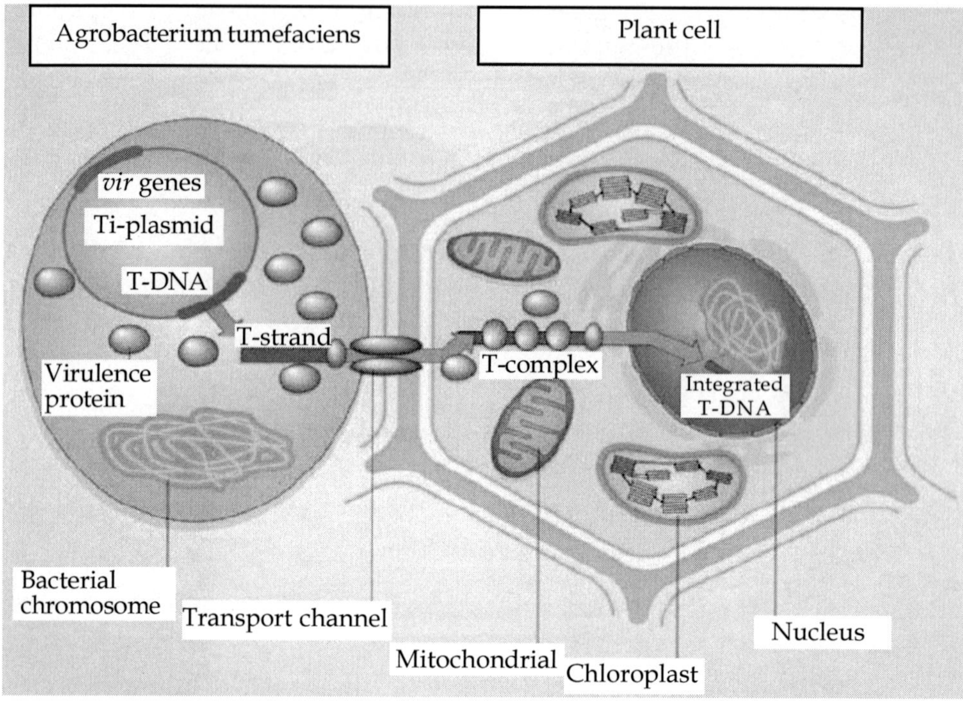

- *Ti Plasmid as a Vector*
  - Usefulness of Ti plasmid as a vector during plant genetic manipulation is due to many features of Ti-plasmid these features include...

(1) The genes in T-DNA are eukaryotic in nature. Even though these genes are of bacterial origin, they are transcribed by plants RNA polymerase. These genes contain introns and they are removed correctly during m-RNA splicing.

(2) Only some part of T-DNA is required if one wishes to use it as a vector.

- Opine production genes and tumor formation genes can be replaced by foreign DNA.
- Thus nos/ocs & onc genes can be deleted.
- If nos or ocs genes are removed, their promoters are retained to ensure the gene expression of foreign DNA.
- Such experiments have been successfully carried out for the expression of kanamycin & methotrexate resistance in plants.
- Phaseolin is synthesized only in developing seeds. Therefore it seems that their promoter is regulated one.

Schematic represent of transfer of gene using Ti- plasmid

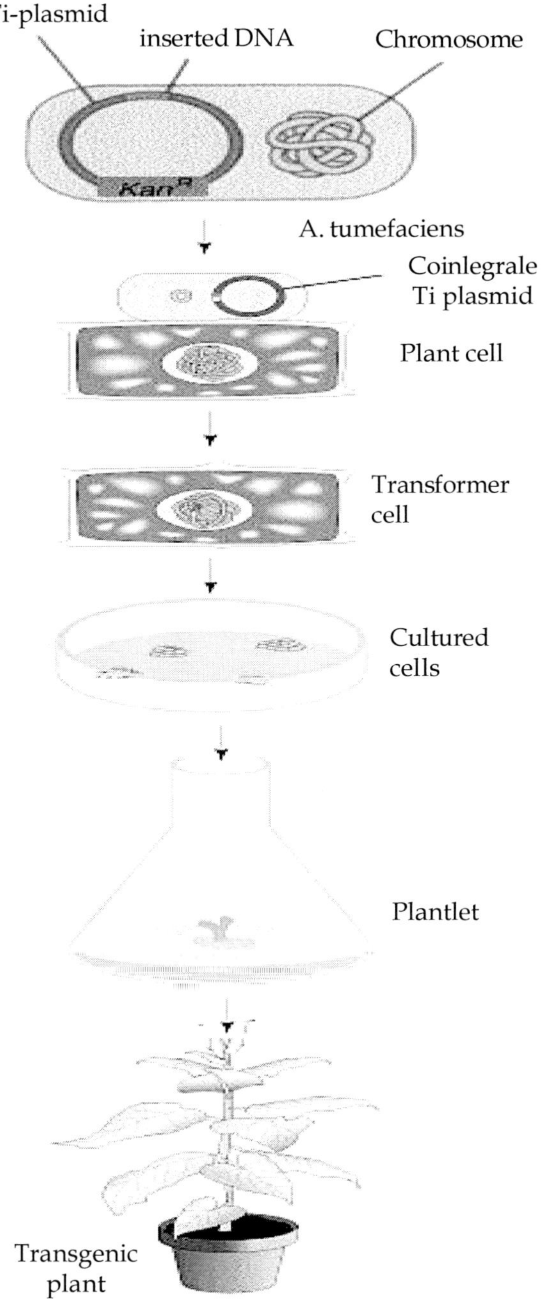

• Direct Transformation

- In this technique, plant protoplasts are prepared by treating plant cells with poly ethylene Glycol (PEG).
- Protoplasts can take up DNA from their surrounding medium.
- This uptakes DNA can be stably integrated into plant chromosomal DNA.
- This technique has been used to transform plant cells with a gene for kanamycin resistance, linked to a strong promoter.
- Plants regenerated from this plant were resistant to kanamycin, and this resistance was inheritable.
- The limitation of direct transformation is that the uptake DNA is randomly integrated in plant genome.
- As along as one can select the desired plant among all undesirables, this method should not present a great problem.

Preparation of Shoot Apex Explant

1. Cotton variety Coker 312 was used in the transformation experiments.
2. Seeds of Coker 312 were treated with 20% hydrogen peroxide for 2 hours and rinsed three times with double-distilled water. The seeds were then placed on a rotor shaker at 100 rpm overnight. After removing the seed coat, seeds were germinated in MS basal medium
3. Shoot apices were cultured on MS medium with 0.1mg/L Kinetin for 3 days.

Agrobacterium Strain and Plasmid

Agrobacterium strain LBA4404 harboring a 'super-binary' vector pTOK233 was used to develop the optimized transformation protocol. This strain has been successfully used in transformation of rice The T-DNA of pTOK233 contains a hygromycin-resistance gene (*HPT*), a kanamycin-resistance gene (*NPTII*), and a *GUS* gene which has an intron in the N- terminal region of the coding sequence and which is fused to the CaMV35S promoter

The intron –gus gene expresses GUS activity in plant cells, but not in cells of *A. tumefaciens Agrobacterium* strain EHA 105 harboring both NPT II and *bar* genes was used to transfer a herbicide resistance trait into cotton. The *bar* gene was originally cloned from the bacterium *Streptomycin hygroscopius*. It encodes for phosphinothricin acetyltransferase (*PAT*) (Thompson *et al.* 1987) that detoxifies phosphinothricin or glufosinate, the active ingredient of the herbicides Liberty and Basta (DeBlock *et al.* 1987). Therefore, plants expressing the *bar* gene are tolerant to herbicides Liberty and Basta.

Pretreatment of Shoot Apex

The shoot apical meristem (SAM) is a population of cells located at the tip of the shoot axis. The shoot apex is divided into three layers, To obtain germline transformation, the shoot apices were wounded in the middle tip by using a scalpel to expose layer III cell before co-culturing with *Agrobacterium*.

Agrobacterium Co-cultivation and Transgenic Plants Regeneration

1. The *Agrobacterium* strains were cultured in LB medium (contains 10g/L Bacto Tryptone, Bacto, 5g/L Yeast extract and 10g/L NaCl).
2. Twenty ml of LB medium plus antibiotics (50mg/L kanamycin and 50 mg/L hygromycin for strain LBA 4404 or kanamycin 50mg/L for strain EHA 105) was inoculated with *Agrobacterium* and incubated in a 100ml Erlenmeyer flask overnight (about 17 hours) on a shaker set for 180 to 220 rpm at 28°C.
3. Then 2ml of the overnight culture was withdrawn and used to inoculate 50ml of LB medium without antibiotics.
4. Acetosyringone was added to the culture at a final concentration of 100 μM. After incubation for 3 to 4 hours at 28°C with shaking, those cultures were diluted with additional LB medium (containing 100 μM acetosyringone) to a concentration (OD_{600} 0.6) for transformation. Equal numbers of shoot apices were randomly distributed to two independent treatments, one with *Agrobacterium* co-cultivation and one without *Agrobacterium* co-cultivation
5. Shoot apices were inoculated by placing one drop of *Agrobacterium* solution onto each shoot apex in co-culture medium (MS + 100 μM acetosyringone) and incubating at 28 °C under dark conditions for approximately 1 to 4 days.
6. After co-cultivation, explants were washed three times with sterile distilled water. Cleaned apices were blotted dry using a sterile paper towel and cultured on the selection medium consisting of MS with 400 mg/L timentin and 50 ml/L kanamycin.
7. Shoot apices not inoculated with *Agrobacterium* were plated on the selection medium as a negative control. Timentin was included in the

selection medium to suppress the *Agrobacterium* growth. The Petri dishes were incubated at a temperature of 28 °C under an 18 hours photoperiod and sub-cultured every 3 weeks. The process was repeated until controls, not co-cultivated with *Agrobacterium*, were totally dead.

8. After this period the surviving shoot apices were transferred to an MS medium without kanamycin to root the plants. Rooted plants were then transferred to soiland grown to maturity in a greenhouse.

β-Glucuronidase (GUS) Histochemical Analysis

The histochemical assay for GUS gene expression was performed by established methods (Jefferson, 1987; Kosugi *et al.*, 1990). Following co-cultivation, apices were harvested for GUS staining.

1. The apices were incubated overnight in a solution containing 25 mg/l X gluc, 10 mM EDTA, 100 mM NaH_2PO_4, 0.1% Triton X-100 and 50% methanol, pH 8.0) at 37°C.
2. The number of apices that stained with blue spots was recorded. Young leaves of putative transgenic plants were also collected for GUS staining to confirm the transformation event.

Kanamycin and Glufosinate Leaf Test

In the putative transgenic plants, expression of the transgene (*NPT II*) or *bar* gene was analyzed by first establishing the lowest concentration of Kanamycin or glufosinate that would kill untransformed plants. Leaves of control plants were painted with a cotton swab when they had two totally opened true leaves using 0, 0.1, 1, 2, or 3% (W/V) of kanamycin or 0, 0.1, 0.2, 0.3, 0.4, and 0.5 ml/L Liberty. The lowest level (2%) of kanamycin that caused damage to the controls was used to evaluate for resistance to kanamycin in the greenhouse. The lowest level (0.3 ml/L) of Liberty was used to evaluate for resistance to glufosinate. Plants were evaluated for resistance 7 days after leaf application of kanamycin or Liberty.

Polymerase Chain Reaction Analysis

1. DNA was isolated from young leaves of putative transgenic plants using the DNAeasy Plant Mini Kit (Qiagen, Santa Clarita, CA).
2. The DNA samples were tested for the presence of the T-DNA region using a pair of *nptII* specific primers (upstream 5'-AGAACTCGTCAAGAAGGCGA-3' and downstream 5'CTGAATGAACTGCAGGACGA-3') to amplify the 700 bp *nptII* fragments.

3. Regenerated plants transformed by EHA101 were screened for the presence of the *bar* gene by PCR using the *bar* gene specific primers (upstream 5′- CATCGTCAACCACTACATCGAG-3′ and downstream 5′- CAGCTGCCAGAAACCCACGTCA-3′).
4. The 25uL amplification mixture contained 2.5uL 10X PCR II buffer (50mM Tris (PH 8.3); 500 mM KCl);1.5 mM MgCl2; 1.0 mM dNTP mix (Pharmacia Biotech); 0.2uM primer; 0.5 unit of AmpliTaq DNA polymerase (Promega); and 20 ng of genomic DNA as template.
5. DNA was amplified in a Perkin Elmer Geneamp PCR System 9600, programmed for a first denaturation step of 2 minutes at 94°C followed by 45 cycles of 94°C for 1 minute, 35°C for 1 minute, and 72°C for 2 minutes.
6. After the completion of 45 cycles, a final extension at 72°C was carried out for 5 minutes. The completed reactions were then held at 4°C until electrophoresis was done.
7. PCR products were separated by loading 12uL of each sample and 2uL of loading buffer type II on a 1.2 % agarose gel prepared with 1.0X TBE buffer. The sample were subject to electrophoresis at 90-100V for 4 hours in 1.0X TBE buffer. The gel was stained with ethidium bromide and visualized under UV light.

Southern Blot Analysis

DNA was isolated from young leaves of putative transgenic plants using the DNAeasy Plant Mini Kit (Qiagen, Santa Clarita, CA) and completely digested with *HindIII*. Based on the construct of the plasmid, Hind III digested genomic DNA will result in a 3.1 Kb fragment in LBA 4404 transformed plants and a 1.8 Kb fragment in EHA 105 transformed plants. Twenty μg of genomic DNA was digested with *Hind* III overnight in a 37°C water bath. The digested DNA fragments were electrophoresed on an 0.8% agarose gel in 0.5x Tris-Borate-EDTA (TBE) buffer, and transferred to a nylon membrane by the alkaline transfer method (Reed and Mann, 1985). The [^{32}P]- labeled probes for LBA 4404 transformed plants were made from a 0.5-kb polymerase chain reaction (PCR) product (Primer: 5'-CTG TAG AAA CCC CAA CCC GTG-3' and 5'-CAT TAC GCT GCG ATG GAT CCC-3') containing the *GUS* coding region. The probes for EHA 105 transformed plants were made from a 430 bp PCR product (Primer: 5′- CAT CGTCAACCACTACATCGAG-3′ and 5′- CAGCTGCCAGAAACCCACGTCA-3′). The band was excised from agarose gel and purified using a Pre A gene Kit (Bio-Rad, Hercules, CA). The probe was then labeled with^{32}P-dCTP.

Transfer of Plants from Tissue Culture

1. As soon as possible remove the transgenic plants from the antibiotic containing media plates and transfer the plants to soil
2. Gently remove the plants from the agar surface. It might help to first loosen the agar surrounding the plants using forceps and then give the plant a gentle tug and (Note that in lower concentrations of phytagar; *e.g.*, ~0.8%, the roots tend to penetrate the agar whereas in higher concentrations; e.g., ~1.0% the roots tend to grow more on the surface).
3. If the objective is to grow numerous independent transgenic plants in very close proximity, then transfer single individual plants in an individual cell in one of the cell packs (these are like what you buy bedding plants from a nursery in; they come in all sizes: 6 pack 8 packs etc).
4. Cover the newly transplanted plants with plastic wrap. Leave this on for several days until visible signs of growth from the transplants.
5. To prevent cross contamination of seed between adjacent plants, tie each plant to a hyacinth stake once large enough.

II. Solutions, media and other stuff

A. YEP 1L

Yeast Extract	10g
Peptone	10g
NaCl	5g
H_2O to	1L
Agar	15g
Cool to	~ 50°C

Add appropriate quantity of antibiotic.

B. Antibiotics

Prepare each needed antibiotic by dissolving it in the appropriate solvent (the following are all water soluble except):

Antibiotic Stock conc For 10ml of stock Final conc. in media

1. Gen 50mg/ml 0.5g 50μg/ml
2. Kanamycin 50mg/ml 0.5g 50μg/ml
3. Hygo 50mg/ml 0.5g 15-50μg/ml
4. Rif 50mg/ml 0.5g 50μg/ml (in DMSO)

5. Strep 300mg/ml 3g 10µg/ml *E. coli* 300µg/ml *Agro*
6. Spectinomycin 100mg/ml 1g 100µg/ml *Agro*
7. Timenton 300mg/ml 3g 300µg/ml Arabidopsis
   Filter-sterilize a/b using syringe and 0.2µ filter unit
   Add a/b to media after it has cooled to ~50°C.
   Store a/b at –20°C.

C. 0.1% phytagar
1. Phytagar 0.1g
2. dH_2O 100ml
3. A/C

D. 5% sucrose / 0.05% Silwet-L77
1. dH_20 250ml
2. Sucrose 12.5g
3. Silwet-L77 125µL Silwet-L77

35

Fluorescent In Situ Hybridization (FISH)

What is FISH ?

Fluorescent in situ hybridization is a technique in which single-stranded nucleic acids (usually DNA, but RNA may also be used) are permitted to interact so that complexes, or hybrids, are formed by molecules with sufficiently similar, complementary sequences. Through nucleic acid hybridization, the degree of sequence identity can be determined, and specific sequences can be detected and located on a given chromosome. Fluorescent in situ hybridization (FISH) is a powerful technique for detecting RNA or DNA sequences in cells, tissues, and tumors. FISH provides a unique link among the studies of cell biology, cytogenetics, and molecular genetics.

Principle of FISH

FISH involves the preparation of short sequences of single-stranded DNA, called probes, which are complementary to the DNA sequences the researchers wish to paint and examine. These probes hybridize, or bind, to the complementary DNA and, because they are labeled with fluorescent tags, allow researchers to see the location of those sequences of DNA. Unlike most other techniques used to study chromosomes, which require that the cells be actively dividing, FISH can also be performed on non dividing cells, making it a highly versatile procedure.

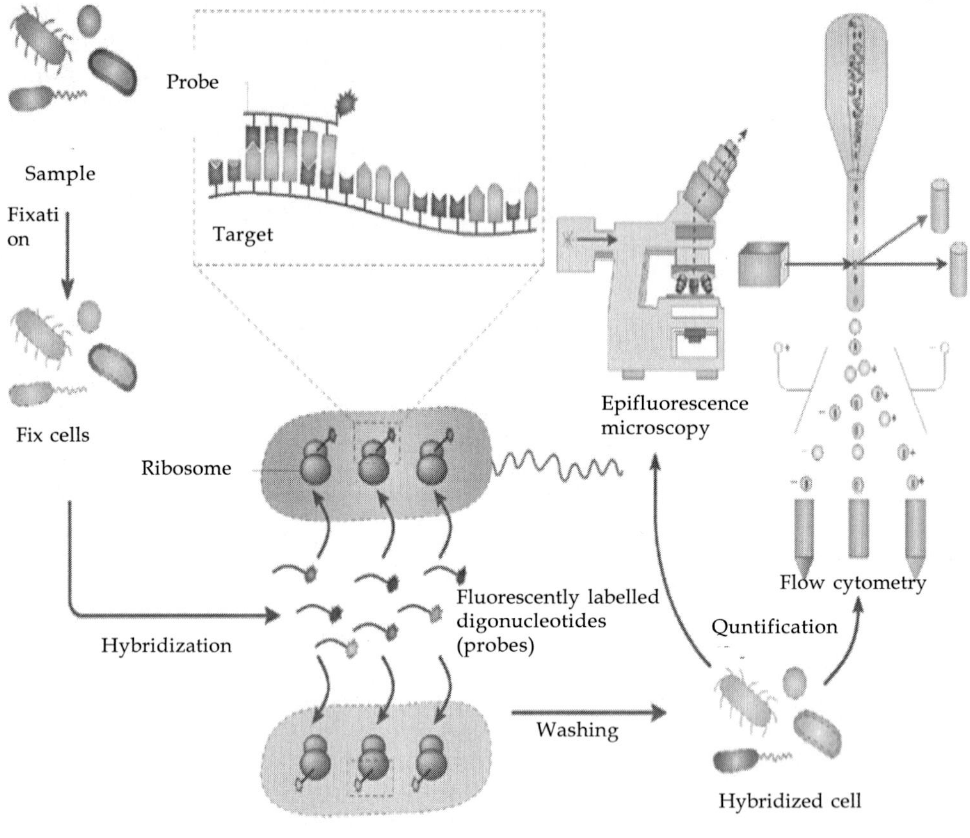

The method comprises of three basic steps:

1. Fixation of a specimen on a microscope slide,
2. Hybridization of labeled probe to homologous fragments of genomic DNA,
3. Enzymatic detection of the tagged target hybrids.

While probe sequences were initially detected with isotopic reagents, non isotopic hybridization has become increasingly popular, with fluorescent hybridization now a common choice. Protocols involving non isotopic probes are considerably faster, with greater signal resolution, and provide options to visualize different targets simultaneously by combining various detection methods.

Fluorescence Microscope

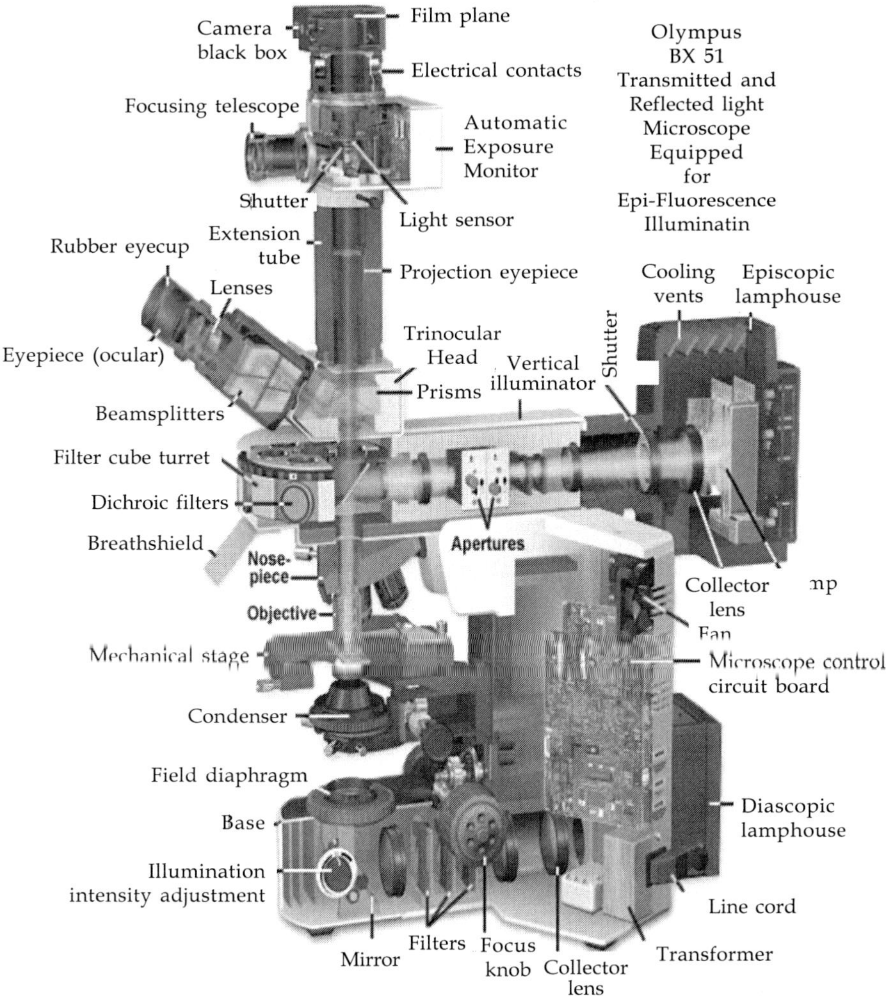

The detection of sequences on the target chromosomes is performed indirectly, commonly with biotinylated or digoxigenin-labeled probes detected via a fluorochrome-conjugated detection reagent, such as an antibody conjugated with fluorescein. As a result, the direct visualization of the relative position of the probes is possible. Increasingly, nucleic acid probes labeled directly with fluorochromes are used for the detection of large target sequences. This method takes less time and results in lower background; however, lower signal intensity is generated. Higher sensitivity can be obtained by building layers of detection reagents, resulting in amplification of the signal. Using such means, it is possible to detect single-copy sequences on chromosome with probes shorter than 0.8 kb.

Probes can vary in length from a few base pairs for synthetic oligonucleotides to larger than one Mbp. Probes of different types can be used to detect distinct DNA types. PCR-amplified repeated DNA sequences, oligonucleotides specific for repeat elements, or cloned repeat elements can be used to detect clusters of repetitive DNA in heterochromatin blocks or centromeric regions of individual chromosomes. These are useful in determining aberrations in the number of chromosomes present in a cell. In contrast, for detecting single locus targets, cDNAs or pieces of cloned genomic DNA, from 100 bp to 1 Mbp in size, can be used.

To detect specific chromosomes or chromosomal regions, chromosome-specific DNA libraries can be used as probes to delineate individual chromosomes from the full chromosomal complement. Specific probes have been commercially available for each of the human chromosomes since 1991.

Any given tissue or cell source, such as sections of frozen tumors, imprinted cells, cultured cells, or embedded sections, may be hybridized. The DNA probes are hybridized to chromosomes from dividing (metaphase) or non-dividing (interphase) cells.

The observation of the hybridized sequences is done using epifluorescence microscopy. White light from a source lamp is filtered so that only the relevant wavelengths for excitation of the fluorescent molecules reach the sample. The light emitted by fluorochromes is generally of larger wavelengths, which allows the distinction between excitation and emission light by means of a second optical filter. Therefore, it is possible to see bright colored signals on a dark background. It is also possible to distinguish between several excitation and emission bands, thus between several fluorochromes, which allows the observation of many different probes on the same target.

Application of FISH

Scientists use three different types of FISH probes, each of which has a different application: Locus specific probes hybridize to a particular region of a chromosome. This type of probe is useful when scientists have isolated a small portion of a gene and wish to determine which chromosome that gene is located on. They prepare a probe from the piece of the gene and observe which chromosome the probe hybridizes to.

Alphoid or centromeric repeat probes are generated from repetitive sequences found at the centromeres of chromosomes. Because each chromosome can be painted in a different color, researchers use this technique to determine whether an individual has the correct number of chromosomes or, for example, whether a person has an extra copy of a chromosome.

Whole chromosome probes are actually collections of smaller probes, each of which hybridizes to a different sequence along the length of the same chromosome. Using these libraries of probes, scientists

are able to paint an entire chromosome and generate a spectral karyotype. This full color image of the chromosomes allows scientists to distinguish between the chromosomes based on their colors, rather than based on their dark and light banding patterns, viewed in black and white through traditional karyotyping. Whole chromosome probes are particularly useful for examining chromosomal abnormalities, for example, when a piece of one chromosome is attached to the end of another chromosome.

FISH has a large number of applications in molecular biology and medical science, including gene mapping, diagnosis of chromosomal abnormalities, and studies of cellular structure and function. Chromosomes in three-dimensionally preserved nuclei can be "painted" using FISH. In clinical research, FISH can be used for prenatal diagnosis of inherited chromosomal aberrations, postnatal diagnosis of carriers of genetic disease, diagnosis of infectious disease, viral and bacterial disease, tumor cytogenetic diagnosis, and detection of aberrant gene expression. In laboratory research, FISH can be used for mapping chromosomal genes, to study the evolution of genomes (Zoo FISH), analyzing nuclear organization, visualization of chromosomal territories and chromatin in interphase cells, to analyze dynamic nuclear processes, somatic hybrid cells, replication, chromosome sorting, and to study tumor biology. It can also be used in developmental biology to study the temporal expression of genes during differentiation and development. Recently, high resolution FISH has become a popular method for ordering genes or DNA markers within chromosomal regions of interest. Medical applications

Often parents of children with a developmental delay want to know more about their child's conditions before choosing to have another child. These concerns can be addressed by analysis of the parents' and child's DNA. In cases where the child's developmental delay is not understood, the cause of it can be determined using FISH and cytogenetic techniques. Examples of diseases that are diagnosed using FISH include Prader-Willi syndrome, Angelman syndrome, 22q13 deletion syndrome, chronic myelogenous leukemia, acute lymphoblastic leukemia, Cri-du-chat, Velocardiofacial syndrome, and Down syndrome.

In medicine, FISH can be used to form a diagnosis, to evaluate prognosis, or to evaluate remission of a disease, such as cancer. Treatment can then be specifically tailored. A traditional exam involving metaphase chromosome analysis is often unable to identify features that distinguish one disease from another, due to subtle chromosomal features; FISH can elucidate these differences. FISH can also be used to detect diseased cells more easily than standard Cytogenetic methods, which require dividing cells and requires labor and time-intensive manual preparation and analysis of the slides by a technologist. FISH, on the other hand, does not require living cells and can be quantified automatically, a computer counts the fluorescent dots present. However, a trained technologist is required to distinguish subtle differences in banding patterns on bent and twisted metaphase chromosomes.

Species Identification

FISH is often used in clinical studies. If a patient is infected with a suspected pathogen, bacteria, from the patient's tissues or fluids, are typically grown on agar to determine the identity of the pathogen. Many bacteria, however, even well-known species, do not grow well under laboratory conditions. FISH can be used to detect directly the presence of the suspect on small samples of patient's tissue.

FISH can also be to used compare the genomes of two biological species, to deduce evolutionary relationships. A similar hybridization technique is called a zoo blot. Bacterial FISH probes are often primers for the 16s rRNA region.

FISH is widely used in the field of microbial ecology, to identify microorganisms. Biofilms, for example, are composed of complex (often) multi-species bacterial organizations. Preparing DNA probes for one species and performing FISH with this probe allows one to visualize the distribution of this specific species within the biofilm. Preparing probes (in two different colors) for two species allows to visualize/study co-localization of these two species in the biofilm, and can be useful in determining the fine architecture of the biofilm.

Microfluidic chip that automates the interphase FISH procedure. The microchip shown requires only minutes of setup time by technician, as opposed to the hours or days of labour needed to perform FISH with conventional equipment.

Although interphase fluorescence in situ hybridization (FISH) is a sensitive diagnostic tool used for the detection of chromosomal abnormalities on cell-by-cell basis, the cost-per-test and the technical complexity of current FISH protocols has inhibited its widespread utilization. Lab-on-a-chip or microfluidic devices, incorporate networks of microchannels that can miniaturize, integrate and automate conventional analytical techniques onto chip-style platforms. Since microchannels permit sophisticated levels of fluid control (down to picolitres), these devices can reduce analysis times, lower reagent consumption, and minimize human intervention.

Currently, FISH has been performed on glass microfluidic platforms that standardize much of the protocol offering repeatable results that are accurate, cost-effective and easier to obtain in a clinical setting.

Compared to conventional FISH methods, these first implementations of on-chip FISH provide a 10-fold higher throughput and a 10-fold reduction in the cost of testing, enabling the simultaneous assessment of several chromosomal abnormalities or patients.It is increasingly essential that diagnostic tests determine the type and extent of chromosomal abnormalities for more informed diagnosis and for appropriate choice of treatment strategies. Since the on-chip FISH technique is 10-20 times more cost-effective than conventional methods, and can be fully integrated and automated, this technology will make widespread genetic testing of patients more accessible in a clinical setting.

Virtual Karyotype

Virtual karyotyping is another cost-effective, clinically available alternative to FISH panels uses thousands to millions of probes on a single array to detect copy number changes, genome-wide, at unprecedented resolution.

Protocol for Fluorescence *in situ* Hybridization (FISH)

FISH is a very widely used technique on not only cytogenetic studies, but also other biological fields. It include metaphase & interphase FISH. This protocol will be divided into three parts: Probe labeling, Hybridization, and Washing.

Probe Labeling

Several methods are used to do probe labeling: Nick translation, Random Priming, and PCR.

Probes Labeling for FISH by Nick Translation

Materials

Nick Translation Kit

Components:

10X dNTP Mix

0.2 mM each dCTP, dGTP, dTTP

0.1 mM dATP

0.1 mM biotin-14-dATP

500 mM Tris-HCl (pH 7.8)

50 mM MgCl2

100 mM beta-mercaptoethanol

100μg/ml nuclease-free BSA

10X Enzyme Mix

0.5 U/μl DNA Polymerase I

0.007 U/μl DNase I

50 mM Tris-HCl (pH 7.5)

5 mM magnesium chloride

0.1 mM phenylmethylsulfonyl fluoride

50% (v/v) glycerol

100μg/ml nuclease-free BSA

Stop Buffer 0.5 M EDTA (pH 8.0)

Distilled H_2O

Procedures

1. Place a 0.6mL microcentrifuge tube on ice and allow the tube to cool.
2. Pipet the following components to the tube:

Volume	Reagent
5μl	10X dNTP mix (minus dATP)
5μl	10X dATP + Biotin-dATP
x μl	1μg DNA (YAC, BAC)
35– x μl	dH_2O
45μl	Total volume

3. Mix the tube briefly. Add 5uL Pol 10X Enzyme Mix. Mix thoroughly but gently. Centrifuge briefly in a centrifuge to bring liquid to the bottom of the tube.
4. Incubate at 15°C for 1-2 hours in a PCR machine.
5. At 1 hour, stop the reaction by placing the tubes in -20°C.
6. Check the size of the labeled probes (2 µl) by gel electrophoresis in 0.7~2% agarose gel, looking for the peak size between 50-500 bp (or 100-300 bp) DNA fragments.
7. If the size range is larger than this, add a further 5mL enzyme mix, place at 15°C for a further 30-60 min, and run another on a gel to test the size.
8. Stop the reaction by either adding the Stop Buffer or heating in a 75°C water bath for 10 minutes (or incubate at 75°C in a PCR machine).
9. Chill on ice.

Precipitating the biotin-labeled probe

10. Combine the following in a 1.5mL microcentrifuge tube: 48µl Biotin-labeled DNA 1µl glycogen or 50µg salmon sperm DNA
11. Then, add 51µl dH_2O to make up to 100µl. Add 0.1 volume 3M Sodium Acetate solution (pH 5.6) into the DNA sample; e.g., 100µl DNA solution + 10µl 3M Sodium Acetate solution.
12. Add 2.5X volume (250µl) cold absolute ethanol to the mixture, mix well.
13. Incubate the mixture at –20°C for at least 30-60 minutes.
14. Centrifuge the mixture for 20 minutes at highest speed at 4°C
15. Discard the supernatant, then vacuum dry the DNA for about 10 minutes.
16. Store the DNA probe in dry form at –20°C until use.
17. Add x µl of dH_2O (usually 10-20 µl) to the precipitated DNA (according to the size of the pellet). To give a final concentration of 50ng / µl. Allow the DNA to dissolve at RT for 1-2 h or at 4°C overnight with occasional mixing. Purified, labeled probes are stable for several years stored at -20°C.

Probes Labeling for FISH by Random Priming Method

Materials

BioPrime DNA Labeling System (Cat#: 18094-011)

Components

2.5X Random Primers Solution

[125 mM Tris-HCl (pH 6.8), 12.5 mM MgCl2, 25 mM 2-mercaptoethanol, 750 μg/ml oligodeoxyribonucleotide primers (random octamers)]

10X dNTP Mixture

[1 mM biotin-14-dCTP, 1 mM dCTP, 2 mM dATP, 2 mM dGTP, 2 mM dTTP in 10 mM Tris-HCl (pH 7.5), 1 mM Na2EDTA]

Klenow Fragment (Large Fragment of DNA Polymerase I)

[40 U/□l Klenow Fragment in 50 mM Potassium Phosphate (pH 7.0), 100 mM KCl,

1 mM DTT, 50% Glycerol]

Stop Buffer: [0.5 M Na2EDTA (pH 8.0)]

Distilled Water

Procedures

1. Dissolve 100ng DNA in 5-20ul of dilute buffer in a microcentrifuge tube. On ice, add 20μl 2.5X Random Primers Solution.
2. Denature by heating for 5 min in a boiling water bath; immediately cool on ice. (The amount of template per reaction has been varied from 25-500 ng with satisfactory results).

   Perform the following additions on ice: 5ul 10X dNTP Mixture

   Distilled Water to a total volume of 49ul
3. Mix briefly.
4. Add 1ul Klenow Fragment. Mix gently but thoroughly. Centrifuge 15-30 sec.
5. Incubate at 37°C for 60 min.
6. Add 5ul Stop Buffer.

Hybridization

First, a probe is constructed. The probe must be large enough to hybridize specifically with its target but not so large as to impede the hybridization process. The probe is tagged directly with fluorophores, with targets for antibodies or with biotin. Tagging can be done in various ways, such as nick translation, or PCR using tagged nucleotides.

Then, an interphase or metaphase chromosome preparation is produced. The chromosomes are firmly attached to a substrate, usually glass. Repetitive DNA sequences must be blocked by adding short fragments of DNA to the sample. The probe is then applied to the chromosome DNA and incubated for approximately 12 hours while hybridizing. Several wash steps remove all unhybridized or partially-hybridized probes. The results are then visualized and quantified using a microscope that is capable of exciting the dye and recording images.

If the fluorescent signal is weak, amplification of the signal may be necessary in order to exceed the detection threshold of the microscope. Fluorescent signal strength depends on many factors such as probe labeling efficiency, the type of probe, and the type of dye. Fluorescently-tagged antibodies or streptavidin are bound to the dye molecule. These secondary components are selected so that they have a strong signal.

FISH experiments designed to detect or localize gene expression within cells and tissues rely on the use of a reporter gene, such as one expressing green fluorescent protein, to provide the fluorescence signal.

Materials

Slide Pretreatment

1. (Optional: not recommended for metaphase FISH): PK stock solution: 5 mg proteinase K (Boehringer, Mannheim, Germany), 50μl 1M Tris-HCl (pH 7.5), 20μl 0.5M EDTA (pH 7.0), 2μl *5M* NaCl, make up to 1 mL in filtered double-distilled water; make fresh as required.
2. 20X standard saline citrate (SSC) stock solution: 3.0*M* NaCl, 0.3M Na-citrate; set up with double-distilled water, adjust to pH 7.0, autoclave, and store at room temperature.
3. RNase stock solution: 10μg/μl of RNase type A (Boehringer); set up with filtered double-distilled water; aliquot and store at -20°C.
4. RNase solution: per slide 200μl 2X SSC plus 10μl of RNase stock solution are necessary; make fresh as required.
5. (Optional: not recommended for metaphase FISH): Pepsin stock solution 10% (w/v): dissolve 100 mg pepsin (Serva, Heidelberg, Germany) in 1 mL of filtered double-distilled water at 37°C; aliquot and store at -20°C.

6. (Optional: not recommended for metaphase FISH): Pepsin buffer: Add 1 mL of 1*M* HCl to 99 mL of distilled water and incubate at 37°C for about 20 min; then add 50µl of the pepsin stock solution 10% (wlv) and leave the coplin jar at 37°C; make fresh as required.
7. (Optional: not recommended for metaphase FISH): 1X PBS/ $MgCl_2$: 5% (v/v) 1M $MgCl_2$ in 1X PBS. (2.5mL 1M $MgCl_2$ in 47.5mL 1X PBS)
8. (Optional: not recommended for metaphase FISH): Formalin buffer: 3% (v/v) of acid-free formaldehyde (37%; Roth) in 1X PBS; make fresh as required.

Fluorescence *in situ* Hybridization (FISH)

Slide Denaturation

1. Denaturation buffer (preferred): 70% (v/v) deionized formamide, 10% (v/v) filtered double-distilled water, 10% (v/v) 20X SSC, 10% (v/v) phosphate buffer; make fresh as required. OR:Denature solution: 70% (v/v) formamide, 2X SSC (pH7.0), 0.1mM EDTA, pH7.0 Add 175mL formamide, 25mL 20X SSC (pH7.0), 50µl 0.5M EDTA, pH7.0 and 50mL purified H_2O to make 250mL solution and mix thoroughly. Verify that the pH is 7.0-7.5 by measuring the pH at ambient temperature. Between use, store covered at 4°C. Discard after 7 days.
2. Deionized formamide: Add 5 g of ion exchanger Amberlite MB1 (Serva) to 100 mL of formamide (Merck, Darmstadt, Mannheim, Germany) stir for 2 h (room temperature) and filter twice through Whatmann no. 1 filter paper. Aliquot and store at -20°C.
3. Phosphate buffer: prepare 0.5M Na_2HPO_4 and 0.5M NaH_2PO_4, mix these two solutions (1: 1) to get pH 7.0, then aliquot and store at -20°C.

Probe Denaturation

1. Hybridization buffer: Dissolve 2 g dextran sulfate in 10 mL 50% deionized formamide/2X SSC/50 m*M* phosphate buffer for 3 h at 70°C. Aliquot and store at -20°C.

   OR: Hybridization solution: MM2.1: 5.5mL formamide
1g Dextran sulfate
0.5mL 20X SSC

Heat to 70°C for several hours to dissolve the dextran sulfate, then cool and adjust to pH 7.0 and add water to volume of 7mL.

1.2.3. Posthybridization and Detection Washing

1. Washing solution I: 50% (v/v) formamide (Merck), 10% (v/v) 20X SSC, 40% (v/v) distilled water; make fresh as required.
2. Washing solution 2 (WS-2) (4X SSC/0.05% Tween 20).
3. Washing solution 3 (WS-3) (4X SSC).
4. Blocking solution: 1-3% (w/v) BSA in 4X SSC, 0.05% (v/v) Triton X-100 (make up fresh).
5. PN buffer: 0.1M NaH_2PO_4/0.1M Na_2HPO_4 M, pH 8.0; 0.1% NP-40.
6. PNM buffer:

   Add 5% (w/v) non fat dry milk to PN buffer plus 0.02% (w/v) Na-azide, incubate at 37°C overnight. It will look terrible. Centrifuge the solution for 5 minutes at 1000g. Transfer the supernatant to a clean tube and store at 4°C.
7. FITC-Avidin (2µg/2µl): Add 398mL PNM buffer to make up 5µg/mL.

   OR

   Solution 1: FITC-avidin (CAMON Vector Laboratories)/4X SSC/0.2 %Tween/5% BSA (1: 300 both Sigma, St. Louis, MO); make fresh as required.
8. Anti-Avidin (2µg/2µl): Add 398mL PNM buffer to make up 5µg/mL.

   OR

   Solution ll: Biotinylated antiavidin (CAMON Vector Laboratories)/Anti-digoxigenin- rhodan-dne (Boehringer Mannheim, Germany)/4X SSC/ 0.2%Tween/5% BSA (1: 20: 100); make fresh as required.
9. Antifade solution

   100mg p-phenylenediamine dihydrochloride in 10mL PBS. Adjust to pH 8.0 with 0.5M carbonate-bicarbonate buffer (0.42g $NaHCO_3$ in 10mL dH_2O, adjust pH to 9.0 with NaOH). Add to 90mL with glycerol. Filter with 0.22µ membrane to remove undissolved particals. If necessary, add 0.5-1?g/mL DAPI.

   OR

   DAPI-solution: Dissolve 5µl of DAPI (4,6-diamidino-2 phenylindol. 2HCl stock- solution; Serva) in 100 mL 4X SSC/0.2% Tween; make fresh as required.

   OR

   70% (v/v) deionized formamide, 10% (v/v) filtered double-distilled water, 10% (v/v) 20X SSC, 10% (v/v) phosphate buffer, make fresh as required.

Procedures for Fluorescence *in situ* Hybridization (FISH)

Pre-treatment of Slides

1. (Optional) If the slide is not dry enough, dehydrate the slide by immersing the slide into 100% ethanol for 1 min. Air dry.
2. Pretreat slide with 200μl diluted RNase solution (0.1μg/μl) for about 30 to 60 minutes at 37°C.
3. Wash slide with 2X SSC for 5 minutes (with agitation).
4. Pepsin treatment: (Optional, if the chromosome targets are bone marrow smear, bone marrow progenitors from methyl cellulose-grown colony assays, tumour preparations)
   i. Add 200μl diluted (with 0.01M HCl) pepsin to the slide
   ii. Incubate slides at 37°C for 5-10 minutes.
   iii. Wash 2X with 1X PBS for 5 min at RT with shaking.

   Note: over-digestion can also cause problems (loss of cells from the slide), so only use when absolutely necessary.
5. Place slides in PBS/50mM $MgCl_2$ for 5 min.
6. (Optional) Fix in PBS/50mM $MgCl_2$/1% formaldehyde for 10 min.
7. Wash in PBS for 5 min (with agitation).
8. Dehydrate the slide with 70%, 90%, and 100% ethanol for 1-2 min each and allow to air dry. Slides can be stored desiccated at 4°C for up to one month before use.

   Pre-hybridization of DNA probes
9. During incubation step 1, prepare hybridization mixture as follows:
   i. Add 2μl (100ng) diluted Biotin-labeled DNA + 1μl (2.5 μg) Cot-1 DNA + 7μl MM2.1 (warmed to RT).
   ii. Denature the hybridization mix at 75°C for 5-7 minutes.
   iii. Place the probe on ice for 3-5 mins. (very **IMPORTANT**)
   iv. Transfer the denatured probe to 37°C for 15 min – 2 h for pre-hybridization.

   Denature metaphase chromosome slides
10. Incubate slides in denaturing solution (in water-bath) for **2** minutes at 75°C.

11. Wash slides in **cold** 2X SSC, followed by two changes of 2X SSC.
12. Dehydrate the slide through a **cold** alcohol series (70%, 90%, and 100% ethanol for 1 min each).
13. Air dry the slides and place on a hot plate at ~42°C.
14. Apply 10µl of denatured probe mix to the slide.
15. Immediately apply a coverslip and seal with rubber cement. Keep the slide in a moist chamber at 37°C overnight – four days for hybridization.

    Washing slides

16. Place the wash tanks containing WS-1 in a 45°C water bath for at least 30 minutes prior to use.
17. Remove rubber cement by using forceps. Coverslips can then be removed either by soaking in 2X SSC or gently tipping them off into the glass disposal bin (never pull them off).
18. Wash the slides 3 times with WS-1 at 45°C for 5–10 min each (usu. 5min).
19. Wash the slides 3 times with WS-2 at RT for 2 min each.
20. a) Blocking treatment with 1-3% BSA in 4X SSC for 20 min at RT.
21. b) Wash the slides 3 times with WS-2 at RT for 2 min each.
22. Wash the slides with WS-3 for 2 min at RT.

    Signal enhancement

23. Add 40uL Avidin-FITC (5 µg/mL Avidin in PNM buffer) onto slides and cover with coverslip. Keep the slides in a moist chamber in dark for 20 min at RT.
24. Wash the slide as steps 19, 21 (do not perform step 18 & 20). Then, proceed from 24 to 26 steps or directly jump to steps 27-29.
25. (Optional): Add 40uL anti-Avidin (5ug/mL anti-Avidin in PNM buffer) onto slides and cover with coverslip. Keep the slides in a moist chamber in dark for 20 min at RT.
26. Wash the slides as steps 23.
27. Repeat steps 22-23 for one additional Avidin-FITC treatment.

Visualizing the hybridization

28. Dehydrate the slides with 70%, 90%, and 100% ethanol for 1-2 min each and allow to air dry.
29. Apply 40uL of DAPI II counterstain and a coverslip to hybridization location.
30. Tore in dark if not use, otherwise, examine the slide at once under fluorescence microscope.

Source : http://www.hku.hk/oncology/lcg/ProtocolforFISH.htm

36

Genotypic Barcoding

The biological diversity of each country is a valuable and vulnerable natural resource. Sampling, identifying, and studying biological specimens are among the first steps toward protecting and benefiting from biodiversity. Knowledge about biodiversity and the ability to identify organisms that comes with it are global public goods. That is, this knowledge can create benefits for all countries - benefits that no one country can create by itself.

DNA Barcoding: Documenting Biodiversity with a Gene Sequence

"DNA barcoding" is an exciting new tool for taxonomic research. The DNA barcode is a very short, standardized DNA sequence in a well-known gene. It provides a way to identify the species to which a plant, animal or fungus belongs. The Consortium for the Barcode of Life (CBOL) is promoting international partnerships that will enable people in all countries to better understand and protect their biodiversity.

DNA barcoding is a taxonomic method that uses a short genetic marker in an organism's mitochondrial DNA to identify it as belonging to a particular species. It is based on a relatively simple concept: most eukaryote cells contain mitochondria and mitochondrial DNA (mtDNA) has a relatively fast mutation rate, which results in significant variance in mtDNA sequences between species and, in principle, a comparatively small variance within species. However, because all mtDNA genes are maternally inherited, any occurrences of hybridization, male-killing microoroganisms, cytoplasmic incompatibility-inducing symbionts (e.g., *Wolbachia*), horizontal gene transfer (such as via cellular symbionts), or other "reticulate" evolutionary phenomena in a lineage can lead to misleading results (i.e., it is possible for two different species to share mtDNA, or for one species to have more than one mtDNA sequence exhibited among different individuals). A 648-bp region of the mitochondrial cytochrome c oxidase subunit I (COI) gene was initially proposed as a potential 'barcode'. "There are a thousand species of plants within a long stone's throw. Nearly every one of them is a described species with a proper scientific name, a handle that you can plug into Google and come up with something. Nearly all of these species or their near relatives have been studied, sampled, thought about and are in the literature. And I cannot identify a single species. Imagine ... if you could walk up to any plant anywhere – seedling, sapling, 40 m tree, grass, root, pressed leaf, or fallen log – and know in a few seconds its scientific name ... That capacity would transform far more than the science of plant biology, the conservation of plants, and the superficial ways we currently make use of the incredible diversity of form, physiology, genetics and chemistry of plants. It would be to plants what the printing press was to stories, education, science, law, medicine and communication." A rapid and accurate method is now being developed for the quick identification of plant species based on extracting DNA from a tiny tissue sample of a leaf, flower or fruit. Appropriately called 'DNA barcoding', referring to the coded labels on many products, DNA barcodes are a short sequence of DNA between 400 and 800 base pairs long that can be easily extracted and characterized for all species on the planet. These genetic barcodes could then be accessed through a digital library and used to identify unknown plants in the field, garden or market. Once fully developed, DNA barcoding has the potential to completely revolutionize our knowledge of plant diversity and our relationship to Nature. People will be able to quickly and cheaply recognize known species and retrieve information about them. Barcoding could be a vital new tool for appreciating and managing the Earth's immense and changing biodiversity. The use of short DNA sequences for biological identifications was first proposed by Paul Herbert and colleagues several years ago at the University of Guelph. Their aim was quick and reliable identification of species across all forms of life including animals, plants and micro-organisms. Although the value of such approaches has long been accepted for identifying micro-organisms where morphological data is lacking, so far this concept has been applied most successfully to animals. Plants have been notably absent in the early stages of barcoding – despite the formation of the Consortium for the Barcode of Life to drive the process forward and build the digital library. Despite the delayed start, DNA barcodes have now been successfully achieved for flowering plants. A successful DNA barcode must be a) short enough to be quickly sequenced, b) easily identified in all plant species, and c) variable enough to provide a unique sequence for each species. The cytochrome c oxidase (CO1) sequence, which is widely used in animal barcoding, is not suitable for plants because of a much slower rate of COI gene evolution in higher plants than in animals. Instead we have identified two other regions: the nuclear Internal Transcribed Spacer (ITS) and the plastid trnH-psbA intergenic spacer. ITS is the one of the most commonly sequenced locus used in plant phylo-genetic investigations at the species level and shows high levels of interspecific divergence. The trnH-psbA spacer, though short in length (~450 base pairs), is the most variable plastid region in angiosperms and is easily amplified across a broad range of land plants. Together, these two DNA regions may unlock the key to DNA barcoding in plants. The millions of plant specimens in herbaria provide a documented source for building the library of plant barcodes. Problem Advances in genotyping assays have resulted in commercially available, cost- effective whole-genome analyses. Once relegated to major genomics centers, genotyping capabilities are now spreading to smaller academic and industrial labs worldwide. With vast numbers of microbial sequences published in on- line databases, there is increasing interest in not only detecting the presence of pathogens, but also determining their relative virulence. Whereby microarrays are used to detect a pattern, or signature, based on the presence or absence of multiple genomic sequences. Commercially available software is used to import and align tens of sequences, output as a text file. It is highly time consuming to visually search for informative sections in these alignment _les, as they are often thousands to tens of thousands of characters long. Manually developing a phylogenetic tree based on the classification calls of individual SNPs is very inefficient and prone to errors. A computer program designed to choose SNPs from alignment files would greatly simplify the genotyping assay development process. Several classification problems will be used to test the program. The genes responsible for the virulence of other bacteria such as *E. coli* and *Salmonella spp*. have been more thoroughly studied and will be ideal for follow-up work. A novel barcode for *H. inuenzae*, an important pathogen in chronic inammatory respiratory diseases such as Asthma, COPD, and emphysema, will be developed based on tens of 16S RNA sequences.

How is DNA Barcoding Done?

The DNA barcode of an unidentified specimen can be read using standard gene sequencing techniques. DNA barcoding includes three types of activities:

- Working with organisms: Collecting, identifying, and preserving voucher specimens in secure repositories
- Laboratory procedures: Sampling and processing tissue from specimens to obtain DNA barcode gene sequences
- Managing data: Sharing the DNA barcode sequence and data about its voucher specimen in a public database

The team behind the discovery found that DNA sequences of the gene 'matK' differ among plant species, but are nearly identical in plants of the same species. This means that the matK gene can provide scientists with an easy way of distinguishing between different plants, even closely related species that may look the same to the human eye.

The researchers made this discovery by analysing the DNA from different plant species. They found that when one plant species was closely related to another, differences were usually detected in the matK DNA.

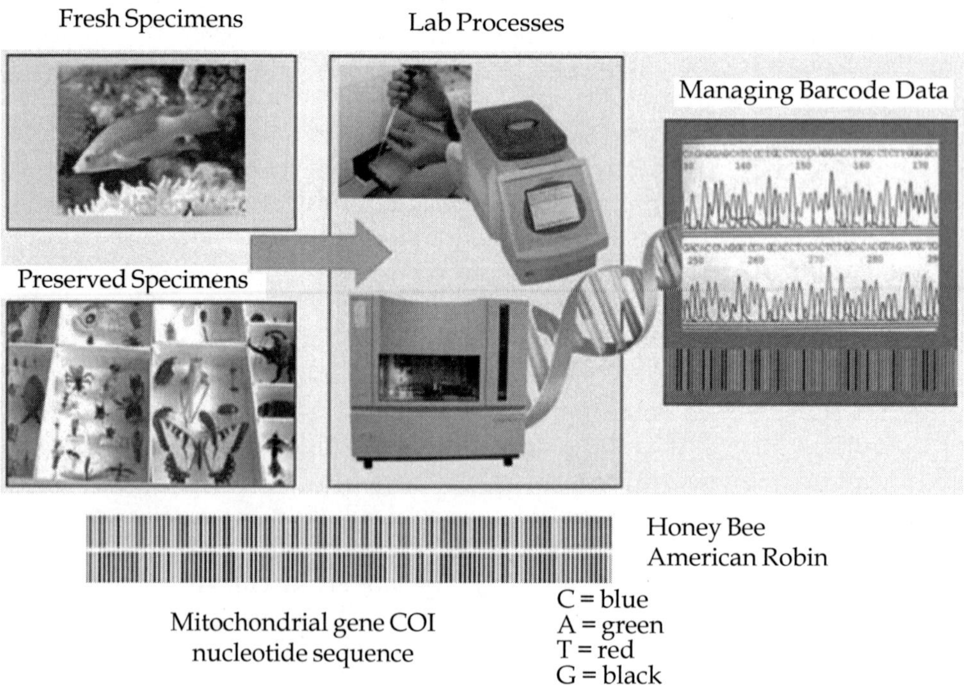

Uses and Global Benefits of DNA Barcoding

DNA barcoding can help to achieve many of the Millennium Development Goals (MDGs) and reach the objectives of the Convention on Biological Diversity.

Controlling Agricultural Pests

Pest damage to agriculture costs farmers the equivalent of billions of US dollars each year. DNA barcoding can identify pests in any life stage, making it easier to control them. The global Tephritid Barcoding Initiative will contribute to the management of fruitflies and will provide border inspectors with tools to identify and stop fruitflies at the border. Ensuring pest-free trade will gaurantee better access to global markets.

Identifying Disease Vectors: Combating Diseases

Many of the most serious human and animal infectious diseases such as malaria are transmitted through vector species. DNA barcoding enables non-taxonomists to identify these vectors, thereby helping to understand and curb disease-carrying pests and pathogens. A global Mosquito Barcoding Initiative is building a reference barcode library that can assist public health officials to control this important disease vector more effectively and with less reliance on insecticides. Protecting Endangered Species Hunting for bushmeat has reduced primate populations by 90% in some parts of Africa. Law enforcement can use DNA barcoding to distinguish bushmeat in local markets obtained from endangered species, thereby contributing to the conservation of biological diversity. A DNA barcoding initiative for conservation has recently been launched, with the goal of creating a barcode library for critically endangered species. The Consortium for the Barcode of Life has organized the Mosquito Barcoding Initiative (MBI), dedicated to developing a system for the unambiguous identification of more than 85% of all known mosquitoes. MBI will become an invaluable resource for preventing the spread of disease bearing mosquitoes. In addition, more accurate identification using barcodes can minimize the use of insecticides by eradication programs.

Monitoring Water Quality

Drinking water is rapidly becoming a precious resource. The health of lakes, rivers and streams is often measured by studying the organisms living in them. DNA barcoding is being used to document these indicator species that can otherwise be difficult to identify. Environmental agencies can use barcoding to improve their water quality assessments, and to create and enforce better policies that ensure sustainable supply of safe drinking water.

Environmental Sustainability

Sustaining Natural Resources over-harvesting of natural resources such as fish and hardwood trees is leading to species depletion, extinction and the economic collapse of industries that rely on them. Natural resource managers and regulators can monitor the illegal trade of processed products using barcoding. Reference barcode libraries are being constructed for fish (FISH-BOL) and hardwood trees (TreeBOL), with the goal of improving the management and conservation of these natural resources. Barcoding can improve monitoring of population size and growth because barcode data can be used to identify larvae, seeds, and juveniles. Barcoding also enables resource managers and government regulators to monitor how much of each species is being used because processed proucts such as fish filets and lumber can be identified using barcode data. Enforcement officials can also use barcoding to identify and prosecute illegal harvesting. Data on population growth and use form the basis of wiser policies for sustainable

Benefits to Country Gain from DNA Barcoding

- The ability to identify specimens quickly and cheaply,
- Better ability to control the movement of species across national borders,
- Opportunities for training of local students and researchers,
- Involvement of local researchers in global networks and biodiversity initiatives, and
- Opportunities to improve the national research infrastructure of specimen collections, molecular biology labs, and biodiversity databases.

Why to Go with Barcoding ?

1. Works with fragments. Barcoding can identify a species from bits and pieces. When established, barcoding will quickly identify undesirable animal or plant material in processed foodstuffs and detect commercial products derived from regulated species. Barcoding will help reconstruct food cycles by identifying fragments in stomachs and assist plant science by identifying roots sampled from soil layers.

2. Works for all stages of life. Barcoding can identify a species in its many forms, from eggs and seed, through larvae and seedlings, to adults and flowers.

3. Unmasks look-alikes. Barcoding can distinguish among species that look alike, uncovering dangerous organisms masquerading as harmless ones and enabling a more accurate view of biodiversity.

4. Written as a sequence of four discrete nucleotides - CATG – along a uniform locality on genomes, a barcode of life provides a digital identifying feature, supplementing the more analog gradations of words, shapes and colors. A library of digital barcodes will provide an unambiguous reference that will facilitate identifying species invading and retreating across the globe and through centuries.

5. The bewildering diversity of about 2 million species already known confines even an expert to morphological identification of only a small part of the plant and animal kingdoms. Foreseeing millions more species to go, scientists can equip themselves with barcoding to speed identification of known organisms and facilitate rapid recognition of new species.

6. Democratizes access. A standardized library of barcodes will empower many more people to call by name the species around them. It will make possible identification of species whether abundant or rare, native or invasive, engendering appreciation of biodiversity locally and globally.

7. Opens the way for an electronic handheld field guide, the Life Barcoder. Barcoding links biological identification to advancing frontiers in DNA sequencing, miniaturization in electronics, and computerized information storage. Integrating those links will lead to portable desktop devices and ultimately to hand-held barcoders. Imagine the promise of a schoolchild with a barcoder in hand learning to read wild biodiversity, the power granted to a field ecologist surveying with a barcoder and global positioning system, or the security imparted by a port inspector with a barcoder linked to a central computer!

8. Sprouts new leaves on the tree of life. Since Darwin, biologists seeking a natural system of classification have drawn genealogical trees to represent evolutionary history. Barcoding the similarities and differences among the nearly 2 million species already named will provide a wealth of genetic detail, helping to draw the tree of life on Earth. Barcoding newly discovered species will help show where they belong among known species, sprouting new leaves on the tree of life.

9. Demonstrates value of collections. Compiling the library of barcodes begins with the multimillions of specimens in museums, herbaria, zoos and gardens, and other biological repositories. The spotlight that barcoding shines on these institutions and their collections will strengthen their ongoing efforts to preserve Earth's biodiversity.

10. Demonstrates value of collections. Compiling the library of barcodes begins with the multimillions of specimens in museums, herbaria, zoos and gardens, and other biological repositories. The spotlight that barcoding shines on these institutions and their collections will strengthen their ongoing efforts to preserve Earth's biodiversity.

37

mtDNA

In cell biology, a **mitochondrion** is a membrane-enclosed organelle found in most eukaryotic cells These organelles range from 0.5–10 micrometers (μm) in diameter. Mitochondria are sometimes described as "cellular power plants" because they generate most of the cell's supply energy in form of adenosine triphosphate (ATP), used as a source of chemical energy. In addition to supplying cellular energy, mitochondria are involved in a range of other processes, such as signaling, cellular differentiation, cell death, as well as the control of the cell cycle and cell growth. Mitochondria have been implicated in several human diseases, including mitochondrial disorders and cardiac dysfunction, and may play a role in the aging process.

Several characteristics make mitochondria unique. The number of mitochondria in a cell varies widely by organism and tissue type. Many cells have only a single mitochondrion, whereas others can contain several thousand mitochondria. The organelle is composed of compartments that carry out specialized functions. These compartments or regions include the outer membrane, the intermembrane space, the inner membrane, and the cristae and matrix. Mitochondrial proteins vary depending on the tissue and the species. In humans, 615 distinct types of proteins have been identified from cardiac mitochondria; whereas in Murinae (rats), 940 proteins encoded by distinct genes have been reported. The number of proteins encoded in plant mitochondrial DNA is probably not much higher than the number encoded in mammalian mitochondria.Experiments have shown that most of the protein synthesis in isolated maize mitochondria can be accounted for by some 18-20 polypeptidesThe mitochondrial proteome is thought to be dynamically regulated. Although most of a cell's DNA is contained in the cell nucleus, the mitochondrion has its own independent genome. Further, its DNA shows substantial similarity to bacterial genomes.

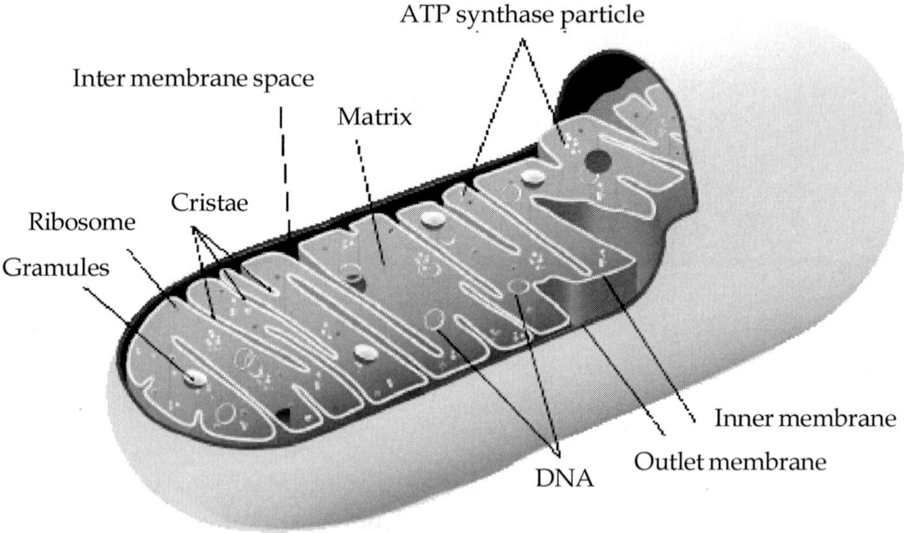

Mitochondrial DNA (mtDNA)

Mitochondrial DNA (mtDNA) is the DNA located in organelles called mitochondria. Most other DNA present in eukaryotic organisms is found in the cell nucleus. Mitochondrial DNA was discovered by Margit M. K. Nass and Sylvan Nass by electron microscopy as DNAase-sensitive thread inside mitochondria, and by Ellen Haslbrunner, Hans Tuppy and Gottfried Schatz by biochemical assays on highly purified mitochondrial fractions. Higher plant mtDNA can be circular or linear and varies from 200 kbp (in brassicas) up to greater than 2500 kbp (in muskmelon).

The genome of plant mitochon dria is thus very large and may also be divided between one or more DNA molecules.

Nuclear and mitochondrial DNA are thought to be of separate evolutionary origin, with the mtDNA being derived from the circular genomes of the bacteria that were engulfed by the early ancestors of today's eukaryotic cells. Each mitochondrion is estimated to contain 2-10 mtDNA copies. In the cells of extant organisms, the vast majority of the proteins present in the mitochondria (numbering approximately 1500 different types in mammals) are coded for by nuclear DNA, but the genes for some of them, if not most, are thought to have originally been of bacterial origin, having since been transferred to the eukaryotic nucleus during evolution. In most multicellular organisms, mtDNA is inherited from the mother (maternally inherited). Mechanisms for this include simple dilution (an egg contains 100,000 to 1,000,000 mtDNA molecules, whereas a sperm contains only 100 to 1000), degradation of sperm mtDNA in the fertilized egg, and, at least in a few organisms, failure of sperm mtDNA to enter the egg. Whatever the mechanism, this single parent (uniparental) pattern of mtDNA inheritance is found in most animals, most plants and in fungi as well. mtDNA is particularly susceptible to reactive oxygen species generated by the respiratory chain due to its close proximity. Though mtDNA is packaged by proteins and harbors significant DNA repair capacity, these protective functions are less robust than those operating on nuclear DNA and therefore thought to contribute to enhanced susceptibility of mtDNA to oxidative damage. Mutations in mtDNA can in some cases cause maternally inherited diseases and some evidence suggests that they might be major contributors to the aging process and age-associated pathologies.

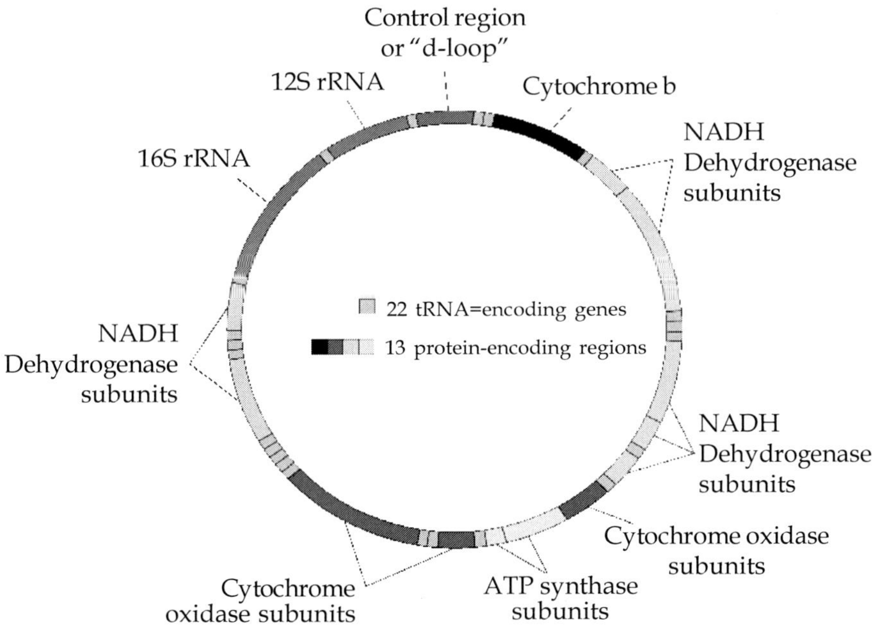

Use in Identification

Mitochondria are structures within cells that convert the energy from food into a form that cells can use. Although most DNA is packaged in chromosomes within the nucleus, mitochondria also have a small amount of their own DNA. This genetic material is known as mitochondrial DNA or mtDNA. In humans, mitochondrial DNA spans about 16,500 DNA building blocks (base pairs), representing a fraction of the total DNA in cells.

Unlike nuclear DNA, which is inherited from both parents and in which genes are rearranged in the process of recombination, there is usually no change in mtDNA from parent to offspring. Although mtDNA also recombines, it does so with copies of itself within the same mitochondrion. Because of this and because the mutation rate of animal mtDNA is higher than that of nuclear DNA, mtDNA is a powerful tool for tracking ancestry through females (matrilineage) and has been used in this role to track the ancestry of many species back hundreds of generations.

Human mtDNA can also be used to identify individuals. Forensic laboratories occasionally use mtDNA comparison to identify human remains, and especially to identify older unidentified skeletal remains. Although unlike nuclear DNA mtDNA is not specific to one individual, it can be used in combination with other evidence (anthropological evidence, circumstantial evidence, and the like) to establish identification. mtDNA is also used to exclude possible matches between missing persons and unidentified remains Many researchers believe that mtDNA is better suited to identification of older skeletal remains than nuclear DNA because the greater number of copies of mtDNA per cell increases the chance of obtaining a useful sample, and because a match with a living relative is possible even if numerous maternal generations separate the two. American outlaw Jesse James's remains were identified using a comparison between mtDNA extracted from his remains and the mtDNA of the son of the female-line great-granddaughter of his sister Similarly, the remains of Alexandra Feodorovna (Alix of Hesse), last Empress of Russia, and her children were identified by comparison of their mitochondrial DNA with that of Prince Philip, Duke of Edinburgh, whose maternal grandmother was Alexandra's sister Victoria of Hesse.

The low effective population size and rapid mutation rate (in animals) makes mtDNA useful for assessing genetic relationships of individuals or groups within a species and also for identifying and quantifying the phylogeny (evolutionary relationships; see phylogenetics) among different species, provided they are not too distantly related. To do this, biologists determine and then compare the mtDNA sequences from different individuals or species.

Data from the comparisons is used to construct a network of relationships among the sequences, which provides an estimate of the relationships among the individuals or species from which the mtDNAs were taken. This approach has limits that are imposed by the rate of mtDNA sequence change. In animals, the rapid rate of change makes mtDNA most useful for comparisons of individuals within species and for comparisons of species that are closely or moderately-closely related, among which the number of sequence differences can be easily counted. As the species become more distantly related, the number of sequence differences becomes very large; changes begin to accumulate on changes until an accurate count becomes impossible.

Land plants exhibit a significant evolutionary plasticity in their mitochondrial DNA (mtDNA), which contrasts with the more conservative evolution of their chloroplast genomes. Frequent genomic rearrangements, the incorporation of foreign DNA from the nuclear and chloroplast genomes, an ongoing transfer of genes to the nucleus in recent evolutionary times and the disruption of gene continuity in introns or exons are the hallmarks of plant mtDNA, at least in flowering plants. Peculiarities of gene expression, most notably RNA editing and trans-splicing, are significantly more pronounced in land plant mitochondria than in chloroplasts. At the same time, mtDNA is generally the most slowly evolving of the three plant cell genomes on the sequence level, with unique exceptions in only some plant lineages. The slow sequence evolution and a variable occurrence of introns in plant mtDNA provide an attractive reservoir of phylogenetic information to trace the phylogeny of older land plant clades, which is as yet not fully resolved. This review attempts to summarize the unique aspects of land plant mitochondrial evolution from a phylogenetic perspective.

Materials and Methods

Solutions

Homogenization Buffer A:	50 mM Tris-HCl (pH 8.0), 1.3 M NaCl, 25 mM EDTA(pH 8.0), 0.2% BSA. Add 0.05% cysteine 56 mM β-mercaptoethanol immediately before use.
Homogenization Buffer B:	100 mM Tris-HCl (pH 8.0), 2.6 M NaCl, 50 mM EDTA (pH 8.0), 0.4% BSA. Add 0.1% cysteine and 56 mM β-mercaptoethanol immediately before use.

Lysis Buffer:	25 mM Tris-HCl (pH 8.0), 20 mM EDTA (pH 8.0), 0.5% SDS
	Proteinase K: 10 mg/mL
	Ammonium acetate: 2 M
	TE: 10 mM Tris-HCl (pH 8.0), 1 mM EDTA
	TE-saturated phenol/chloroform (50:50)
	Water-saturated phenol/chloroform/isoamyl alcohol (50:49:1)
RNAse A:	10 mg/mL
RNA Extraction Buffer:	4 M guanidine thiocyanate, 25 mM sodium citrate (pH 7.0), 0.5% sarcosyl. Add 0.1% β-mercaptoethanol immediately before use.
LiCl:	4 M

Procedure

Plant Material

Preparation of Mitochondria

The crucial step is to isolate intact organelles and lyse the nuclei in a medium of high ionic strength. This step was adapted from a previous protocol for the preparation of chloroplasts (Bookjans et al., 1984). Plastids are eliminated by low speed centrifugations. All steps must be carried out at 4°C either in the cold room or on ice.

Depending on the tissue, two procedures were used to prepare the extract:

(i) Portions (1-15 g) of different fresh potato tissues (leaves, flowers, flower buds) were homogenized in a Waring blender, for a few seconds at moderate speed, using 5-75 mL of cold Homogenization Buffer A in a cold room.

(ii) Small potato tubers (20 g) were passed through a Moulinex juice extractor (the latter was used to avoid a large amount of buffer and to increase the yield of extraction), the juice being directly collected in an equal volume of cold Homogenization Buffer B (20 g of potato tubers give about 10 mL of juice). This second method can also be applied to other tissues such as embryos or hypocotyls from different plant species.

- After filtration through a nylon filter (100 ?m mesh), the suspension was centrifuged at 500 *g* for 10 min to eliminate large cell fragments.

- To eliminate plastids, the supernatant was centrifuged twice at 2600 g for 15 and 10 min, respectively. 67b *Scotti et al.*
- To sediment mitochondria, the supernatant was centrifuged at 14,500 g for 15 min. Supernatant containing lysed nuclei components was discarded and the mitochondrial pellet, carefully resuspended in 10 to 30 mL of cold Homogenization Buffer A, was centrifuged at 14,500 g for 15 min. This mitochondrial pellet can be stored at -80°C after freezing in liquid nitrogen or directly used for mtDNA or mtRNA isolation.

Mitochondrial DNA Isolation

To isolate mtDNA, the mitochondria were lysed with a detergent and the resulting mtDNA was dissociated from the proteins by a proteolytic agent.

i. The mitochondrial pellet was resuspended carefully in 0.5-1 mL of Lysis Buffer.

ii. Proteinase K was added to a final concentration of 50 □g/mL and the suspension was incubated at 37°C for 1 h with gentle shaking to avoid mechanical breaking of DNA.

iii. After the addition of 0.1 volume of 2 M ammonium acetate, the nucleic acids were extracted with an equal volume of TE-saturated phenol/ chloroform (50:50).

iv. Nucleic acids were precipitated by the addition of 2 volumes of 100% ethanol and overnight incubation at -20°C.

v. After centrifugation at 18,000 *g* for 20 min, the pellet was re-dissolved in 10 to 50 microliter of water.

At this step, although RNA was still present, the DNA was suitable for restriction enzyme analysis or PCR experiment. If necessary, RNAse A treatment (Maréchal Drouard and Guillemaut, 1995) followed by phenol extraction and ethanol precipitation can be performed.

Mitochondrial RNA Isolation

To isolate mtRNA, the mitochondria were lysed and RNA was extracted using a guanidine thiocyanate medium (Chomeczynski and Sacchi, 1987).

i. The mitochondrial pellet was resuspended in 0.4-0.8 mL of RNA Extraction Buffer and incubated at 65°C for 1-2 min.

ii. After the addition of 0.1 volume of 2 M sodium acetate pH 4.0, the suspension was extracted with water-equilibrated phenol/

chloroform:isoamyl alcohol (50:49:1), precipitated with 1 volume of isopropanol at -20°C for 1 h and centrifuged at 18,000 *g* for 20 min.

iii. The pellet was resuspended in 0.2-0.4 mL of 4 M LiCl and centrifuged again at 18,000 g for 20 min.

iv. After centrifugation, the pellet was dissolved in 0.3 mL of RNA Extraction Buffer by heating at 65°C for 1-2 min. Nucleic acids were precipitated with 1 volume of isopropanol at -20°C for 1 h, then centrifuged at 18000 g for 20 min.

v. The mtRNA pellet was washed in 75% ethanol twice, air dried and dissolved in 10-20 micro liter of water.

At this step the mtRNA was of sufficient purity to perform northern and RTPCR analyses.

Plant Mitochondrial DNA and RNA Preparation 67°C Molecular Analysis

The purity of mtDNA was evaluated by restriction enzyme digestions followed by Southern hybridizations. The purity of mtRNA was checked by northern analysis. Highly purified mtDNA prepared from potato tubers according to Kemble (1987) and chloroplast DNA (cpDNA) purified from potato leaves according to Bookjans *et al.* (1984) were used as controls for Southern experiments. mtDNA (1 microgram) was digested with *Eco*R I, electrophoresed on an 0.8% agarose gel and blotted on positively charged Hybond N+ membrane under alkaline conditions according to the manufacturer (Amersham Pharmacia Biotech). mtRNA (1 □g) and total RNA (10 microgram) were electrophoresed on a 1.2% denaturing formaldehyde-agarose gel and blotted onto a positively charged Hybond N+ membrane. Probes described by Cardi *et al.* (1999) were used for both Southern and northern analyses.

Second Method for Mitochondrial DNA Isolation

Solutions

(do not autoclave, store all solutions at 4°C)

Grinding Buffer (GB)

Need 4 liters (Mix at least 2 hour)

0.35 M Sorbitol 255.08 g

50 mM Tris pH 7.6 200.00 ml (1.0M Stock)

5 mM EDTA 40.00 ml (0.5M Stock)

0.2% BSA 8.0 g

1.0% PVP 40.0 g

0.025% Spermine 1.0 g

0.025% Spermidine 1.0 g

0.125% B-ME 5.0 ml (Add day of extraction)

Wash Buffer (WB)

Need 1 liter

0.35 M Sorbitol 63.77 g

50 mM Tris pH 7.6 50.00 ml

0.1% BSA 1.0 g

Wash Buffer + EDTA (WBE)

Need 2 liters

0.35M Sorbitol 127.54 g

50 mM Tris pH 7.6 100.00 ml

20 mM EDTA 80.00 ml

Lysis Buffer (LB)

Need 500 ml

1.0% Sarkosyl 50.00 ml (10% Stock)

50 mM Tris pH 8.0 25.00 ml

25 mM EDTA 50.00 ml Proteinase K 10mg/ml (Add directly to lysis reaction)

Other Solutions Needed

25:24:1 Phenol:Chloroform:Isoamyl Alcohol

TE (10mM Tris, 1mM EDTA (pH 8.0)

Percoll gradients

5X Gradient Buffer (1.5M Sucrose, 50mM MOPS, 10mM EDTA); Percoll (Sigma P-4937)

18% = 6 ml 5x, 5.5 ml Percoll, 18.5 ml ddH20

23% = 6 ml 5x, 7 ml Percoll, 17 ml dd H20

40% = 6 ml 5x, 12 ml Percoll, 12 ml ddH20

Materials Needed

Waring blender

Large flasks and funnels

Cheesecloth/Miracloth (or 48uM nylon membrane)

Soft paintbrush

Waterbath

Corex tubes

Ultra-centrifuge tubes

GSA Rotor

SS-34 Rotor

Centrifuge tubes

Procedure

1. Grind in mortar and pestle or Waring blender with 5-7 volumes buffer A per g tissue. Use MCE at 350 l/L, and if necessary, with 5 ml 1 M DIECA/L.
2. Squeeze through cheese cloth, two layers of Miracloth.
3. 10 min at 1000 g
4. Decant supernatant and centrifuge 10 min at 15,900 g.
5. Resuspend each pellet in a few drops of buffer G with paint brush; combine; bring to about 10 ml/50 g, 15 ml/75 g.
6. 10 min at 1000 g; pour off most; swirl pellet to remove fluffy layer; combine.
7. Bring supernatant to 10 mM $MgCl_2$ (100 l 1M/10 ml). Bring to 20 g DNase/ml (100 l 2mg/ml/10 ml).
8. 60 min. 4°C.
9. Underlay shelf buffer, 20 ml/10-15 ml; always use 20 ml or more.
10. 20 min at 12000 g.
11. Resuspend in small volume shelf buffer with brush; bring to about 10 ml/50-100 g.
12. 10 min at 15900 g.
13. Resuspend pellets in NN (lysis) buffer (4-5 ml/50-75 g).

14. Add SDS to 0.5% (250 l of 10%/5 ml NN). Swirl thoroughly.
15. Add proteinase K to 100 g/ml (25 l of 20 mg/ml/5 ml NN). Swirl gently.
16. 60 min. 37°C.
17. Add equal volume of 3:1 water-saturated phenol, chloroform-isoamyl alcohol mixture. Emulsify ca. 5 min.
18. 10 min at 7000 g.
19. Collect supernatant; repeat 17 and 18: 3 total extractions.
20. Final supernatant; add 0.1 volume 8 M Ammonium acetate; then add 2 volumes of absolute ethanol.
21. 60 min, -80°C; 10 min at 8000-9000 g; drain; add equal volume 70% ethanol; let sit 10 min; 10 min at 8000-9000 g; drain dry. Vacuum dry pellet, 30 min. Two small corex tubes are better than one 30 ml Corex.
22. Add 100-500 l 0.1X NTE, 10 l RNase mixture. Typically use 500 l per 50 g tissue.
23. Hydrate 30 min., 37°C.

38

Micro RNA

In genetics, **microRNAs (miRNA)** are single-stranded RNA molecules of 21-23 nucleotides in length, which regulate gene expression. miRNAs are encoded by genes from whose DNA they are transcribed but miRNAs are not translated into protein (i.e. they are non-coding RNAs); instead each primary transcript (a *pri-miRNA*) is processed into a short stem-loop structure called a *pre-miRNA* and finally into a functional miRNA. Mature miRNA molecules are partially complementary to one or more messenger RNA (mRNA) molecules, and their main function is to down-regulate gene expression. They were first described in 1993 by Lee and colleagues in the Victor Ambros lab .

MicroRNAs can be encoded by independent genes, but also be processed (via the enzyme Dicer) from a variety of different RNA species, including introns, 3' UTRs of mRNAs, long noncoding RNAs, snoRNAs and transposons.

The genes encoding miRNAs are much longer than the processed mature miRNA molecule; miRNAs are first transcribed as primary transcripts or pri-miRNA with a cap and poly-A tail and processed to short, 70-nucleotide stem-loop structures known as pre-miRNA in the cell nucleus. These pre-miRNAs are then processed to mature miRNAs in the cytoplasm by interaction with the endonuclease Dicer, which also initiates the formation of the RNA-induced silencing complex (RISC). This complex is responsible for the gene silencing observed due to miRNA expression and RNA interference. The pathway in plants varies slightly due to their lack of Drosha homologs; instead, Dicer homologs alone effect several processing steps. The pathway is also different for miRNAs derived from intronic stem-loops; these are processed by Dicer but not by Drosha. Either the sense strand or antisense strand of DNA can function as templates to give rise to miRNA.

Efficient processing of pri-miRNA by Drosha requires the presence of extended single-stranded RNA on both 3'- and 5'-ends of hairpin molecule. These ssRNA motifs could be of different composition while their length is of high importance if processing is to take place at all. The Drosha complex cleaves the RNA molecule ~22 nucleotides away from the terminal loop. Most pre-miRNAs don't have a perfect double-stranded RNA (dsRNA) structure topped by a terminal loop. There are few possible explanations for such selectivity. One could be that dsRNAs longer than 21 base pairs activate interferon response and anti-viral machinery in the cell. Another plausible explanation could be that the thermodynamic profile of pre-miRNA determines which strand will be incorporated into Dicer complex. Indeed, clear similarities between pri-miRNAs encoded in respective (5'- or 3'-) strands have been demonstrated.

When Dicer cleaves the pre-miRNA stem-loop, two complementary short RNA molecules are formed, but only one is integrated into the RISC complex. This strand is known as the *guide strand* and is selected by the argonaute protein, the catalytically active RNase in the RISC complex, on the basis of the stability of the 5' end. The remaining strand, known as the *miRNA**, *anti-guide* or *passenger strand*, is degraded as a RISC complex substrate. After integration into the active RISC complex, miRNAs base pair with their complementary mRNA molecules and inhibit translation or sometimes induce mRNA degradation by argonaute proteins, the catalytically active members of the RISC complex. It is as yet unclear how the activated RISC complex locates the mRNA targets in the cell, though it has been shown that the process is not coupled to ongoing protein translation from the mRNA.

Cellular Functions

The function of miRNAs appears to be in gene regulation. For that purpose, a miRNA is complementary to a part of one or more messenger RNAs (mRNAs). Animal miRNAs are usually complementary to a site in the 3' UTR whereas plant miRNAs are usually complementary to coding regions of mRNAs. Perfect or near perfect base pairing with the target RNA promotes cleavage of the RNA. This is the primary mode of plant microRNAs.

MicroRNAs that are partially complementary to the target can also speed up deadenylation, causing mRNAs to be degraded sooner. For partially complementary microRNA to recognise their targets, the nucleotides 2–7 of the miRNA ('seed region'), still have to be perfectly complementary. miRNAs occasionally also causes DNA methylation of promoter sites and therefore affecting the expression of targeted genes. miRNAs function in association with a complement of proteins collectively termed the miRNP. Human miRNPs contain eIF2C2 (also known as Argonaute 2), DDX20, GEMIN4 and microRNA.

Gene Activation

dsRNA can also activate gene expression, a mechanism that has been termed "small RNA-induced gene activation" or RNAa. dsRNAs targeting gene promoters can induce potent transcriptional activation of associated genes. This was demonstrated in human cells using synthetic dsRNAs termed small activating RNAs (saRNAs), but has also been demonstrated for endogenous microRNA.

Genomics of MicroRNA

It was initially thought that miRNA genes were located in intergenic regions, however, later study showed that several miRNA genes were located within introns of either protein-coding or noncoding genes, or in intergenic regions, while only a few were located in exons of noncoding RNAs or UTR of protein coding genes. Recently, it has shown that miRNA genes overlap with the protein-coding region of the genes of a multigene family.

miRNA and Disease

Just as miRNA is involved in the normal functioning of eukaryotic cells, so has dysregulation of miRNA been associated with disease. Disease association in turn has led to increased funding opportunities for academic research and financial incentives for development and commercialization of miRNA-based diagnostics and therapeutics. After early commercialization aimed at academic research support was established, the initial research focus based on products and services requested was on cancer and neuroscience research. During 2007, interests indicated by product and services requested broadened to include cardiac research, virology, cell biology in general and plant biology.

Isolation of RNA

There are two main methods for isolating RNA: phenol-based extraction and silica matrix or glass fiber filter (GFF)-based binding. Phenol-based reagents contain a combination of denaturants and RNase inhibitors for cell and tissue disruption and subsequent separation of RNA from contaminants.

Phenol-based isolation procedures can recover RNA species in the 10-200-nucleotide range (e.g., the miRNAs, 5S rRNA, 5.8S rRNA, and U1 snRNA). If a sample of "total" RNA was purified by the popular silica matrix column or GFF procedure (i.e. Qiagen RNEasy columns), it will be significantly depleted in small RNAs. Extraction procedures like Trizol/TriReagent, however will purify all RNAs, large and small, and are the recommend methods for isolating total RNA from biological samples that will contain miRNAs/siRNAs.

This protocol utilizes the powerful guanidine isothiocyanate–phenol:chloroform extraction method which allows the rapid isolation of RNA. The conditions for extraction enable the partitioning of proteins and DNA into the organic layer of the biphasic solution interface, while retaining RNA in the upper aqueous layer. The aqueous phase is removed to a second tube, and RNA is precipitated with an equal volume of isopropanol. High yields of pure, undegraded total RNA can be recovered from even small quantities of tissue or cells. Large numbers of samples may be performed simultaneously, because of the simplicity of the technique.

Materials and Reagents

1. Denaturing solution
   - 4 M guanidine isothiocyanate (GITC)
   - 0.02 M sodium citrate
   - 0.5% sarcosyl
2. Solution D: Solution D is prepared by adding 0.72μl of β–ME for every 100μl of denaturing solution
3. 2 M Sodium acetate (pH 4.0)
4. Chloroform:isoamyl alcohol
5. Isopropanol
6. Phenol (equilibrated to pH 5.3–5.7 with 0.1 M succinic acid)
7. β-Mercaptoethanol (β-ME) (14.4 M)
8. Glycogen (2 mg/ml)
9. Diethylpyrocarbonate (DEPC)
10. DEPC-treated water

11. 75% (v/v) ethanol (prepared using DEPC-treated water)
12. Dounce, micro-Dounce, or mechanical microhomogenizer (for tissue samples only)
13. 1× PPBS (for collecting adherent cells only)
14. 1× Trypsin and EDTA solution (for collecting adherent cells only)
   - 0.05% trypsin
   - 0.53 mM EDTA

Procedure

1. Collect sample
2. Flash-freeze sample in liquid nitrogen, weigh sample
3. Use 100μl of solution D for every 10 mg or 1 × 10^6 cells.
4. Homogenize sample in solution D
5. The volume of solution D = Volume D
6. Add 1/10 volume D of 2 M sodium acetate
7. Add a volume equal to volume D of phenol (equilibrated to pH 5.3–5.7 with succinic acid)
8. Add 1/5 volume D of chloroform:isoamyl alcohol; Cap and vortex thoroughly
9. Spin the sample in a microcentrifuge at 13–14k rpm for 5 minutes
10. Transfer aqueous layer to a tube containing a volume equal to volume D of isopropanol. For samples less than 10 mg or 1 × 10^6cells, mix 1μl of glycogen with the RNA solution before adding the isopropanol. Spin the sample in a microcentrifuge for 30 minutes for maximal recovery of the RNA.
11. Spin the sample in a microcentrifuge at 13–14k rpm for 5 minutes
12. Remove supernatant completely
13. Wash pellet in 2× volume D of 75% ethanol–25% DEPC-treated water
14. Remove supernatant completely
15. Dry for 5 minutes or less
16. Resuspend pellet in a volume equal to 1/2 to 1/10th of volume D in DEPC-treated water

Spectrophotometric Quantification of RNA

1. Zero the spectrophotometer at 260 nm with DEPC-treated water or 5 mM Tris-HCl (pH 7.5), or TE buffer. Note If the DEPC-treated water has a pH <7, the quantitation should be performed in 5 mM Tris-HCl (pH 7.5). The low pH will alter the OD measurements between 260 and 280 nm, indicating a low purity.
2. If using a 500-μl cuvette, place 5μl of the RNA solution into 495μl of the diluent. Place a piece of Parafilm® over the top of the cuvette and mix the sample well. The conversion factor for RNA is 0.040μg/μl per OD260 unit. Take the spectrophotometric reading. For a reading of 0.10, calculate the concentration as follows:

0.10	× 500/5	× 0.04 μg/μl	= 0.4μg/μl
spec. reading at A260	Dilution factor	Conversion factor A260	Final concentration

3. Calculate the yield of RNA by multiplying the volume in microliters by the concentration. For example, in the sample above a volume of 100μl results in a yield of 40μg.

Rezero the spectrophotometer with desired solution at 280 nm. Calculate the purity of the RNA by measuring the OD at 280 nm. The ratio of the 260 nm measurement to the 280 nm measurement indicates purity. Ratios of 1.8 to 2.0 are very pure. Lower ratios indicate possible protein contamination, or low pH in the solution used as a diluent for the spectrophotometric readings.

Procedural Outline

1. Oligo Design
2. ImpA Synthesis
3. Synthesizing Oligos
4. Oligonucleotide Adenylation
5. Small RNA Purification
6. 3' Adaptor Ligation
7. 5' Adaptor Ligation
8. RT-PCR
9. Concatamerization
10. Cloning into Topo vector
11. Screen and Sequence

RNA Cloning Method Flowchart

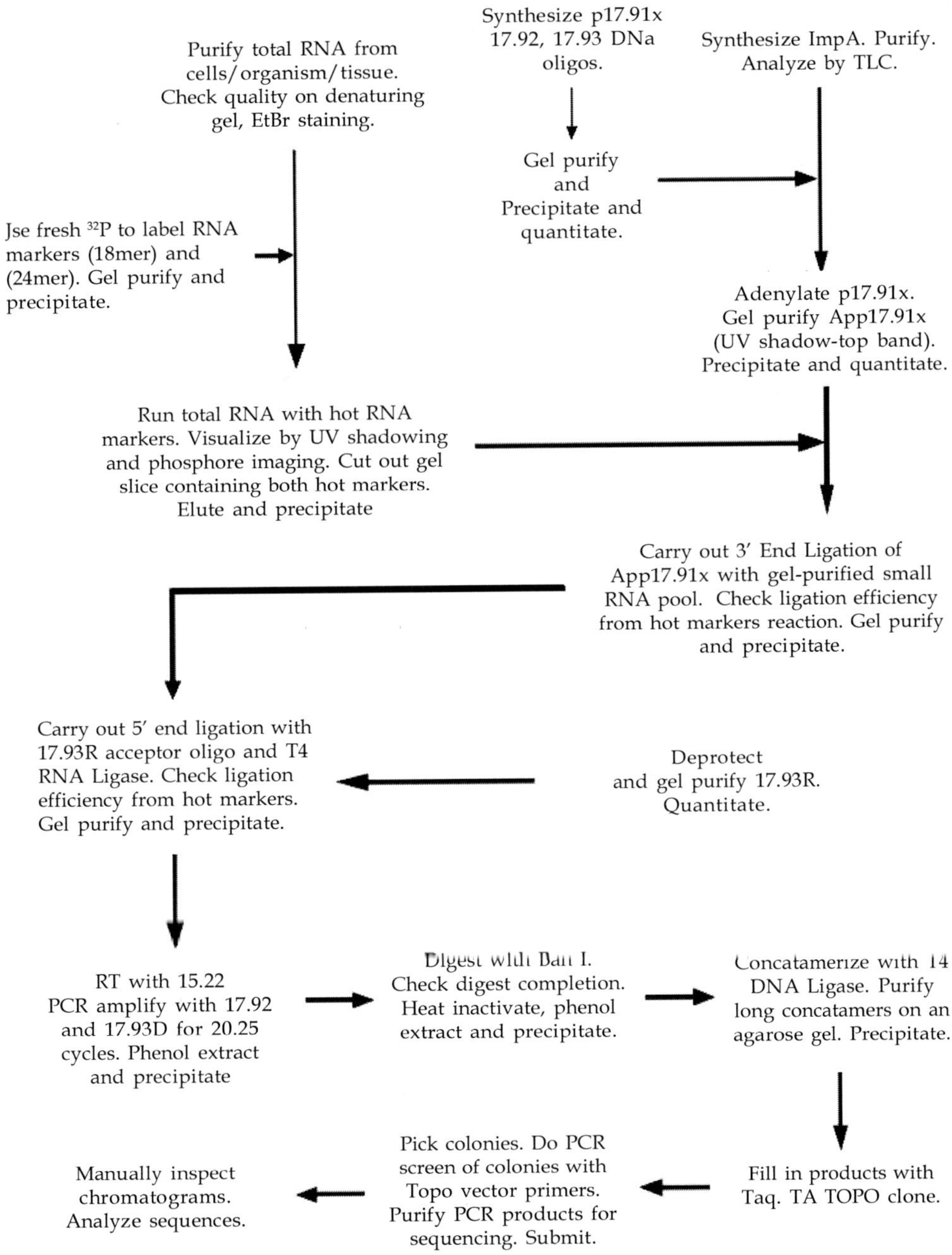

- **Oligos Design Oligos for RNA Cloning** [Underline is Ban I site (G¯GYRCC)]

  3' End Donor Oligo (requires 5' Phosphorylation and 3' Modification)

  pCTGTAGGCACCATCAAx (x: DMT-O-C3 -CPG)

  5' End Acceptor Oligo R : ATCGTaggcacctgaaa (RNA/DNA version, lowercase RNA)

  3' RT Primer Oligo ATTGATGGTGCCTAC

  5' PCR oligo D : ATCGTAGGCACCTGAAA (DNA version, calcTm = 42oC)

  3' PCR oligo ATTGATGGTGCCTACAG (calcTm = 40oC)

  24mer and 18mer marker RNAs with restriction sites:

  T7 template for 44.12R (italicized sequence is T7 promoter) TCACTATTGTTGAGAACGTTGGCC*TATAGTGAGTCGTATTACGC*

  RNA Marker transcribed off 44.12D (with Acl I site) GGCCAA¯CGUUCUCAACAAUAGUGA

  RNA marker purchased from Dharmacon (contains BamH I site) AGCGUGUAGG¯GAUCCAAA

Synthesizing ImpA

Prepare following separate mixes in clean beakers that are rinsed in acetonitrile to dry out the glass. Reagents for ImpA synthesis can be purchased from Sigma Chemical. Buy grades that are at least 99% pure.

Mixture A

0.5 mmol AMP (5' p) free acid, 174mg

15 ml Dimethylformamide

Mixture B

1 mmol Triphenylphosphine , 262 mg,

1 mmol 2,2'-dipyridyldisulfide , 220 mg,

2.5 mmol Imidazole , 170 mg

0.90 ml Triethylamine , d=0.726

15 ml Dimethylformamide

Add mixture A slowly into mixture B while stirring with vigour (clean, dry stir bar). The mixtures initially are not very soluble, but the reaction begins to turn yellow-green, solids dissolve, some oil appears on bottom, and this also eventually dissolves in.

Stir for 1-1.5 hr at room temp with a cover over beaker or reaction flask.

Purification

To precipitate and purify ImpA from reaction, add dropwise above reaction to vigorously stirred anhydrous solution containing: 9 mmol NaClO4 (FW 122.4), 1.1g 225 ml Acetone 115 ml Ether (anhydrous ethyl ether)

The ImpA precipitates out as puffy aggregates. Let this settle, and then remove solvent phase down to about 60 ml. Save some of this solvent for analysis. Transfer precipitate to Corex tubes, rinse remaining chunks with acetone. Centrifuge at 5000 rpm (3000 g on ss34 rotor) for 10 min, pour off acetone, wash and spin down twice again in acetone. Do one final wash with just ether, and spin down at 3000 g for 20 min. Removing all the ether will be difficult. Dry overnight in a vaccuum vessel between 22.5°C and 45°C, scrape out white chunks with a clean spatula into a scintillation vial with cap for dry-storage at –20°C (stable for a couple weeks).

Analysis

Note weight of recovered white chunks. I obtained ~80 mg of ImpA product. Pre-soak Cellulose-F TLC (JT Baker) in 1:10 saturated (NH4)2SO4. Let air-dry or blow dry.Prepare much-scaled-down mini-samples of mixture A and B (stoichiometries no longer matter here, just need sample mixtures to compare before and after states of reaction). Dissolve AMP (mixture A) and ImpA in water instead of dimethylformamide for this analysis (should dissolve better).

Dot following samples with cappillary tubes: AMP (mixture A), ImpA, Mix B, solvent from purification, and AMP again. Use 80% EtOH as solvent, run up about 9 cm. The AMP should run the slowest, ImpA should run out higher, Mix B and leftover solvent from purification should run with the EtOH solvent front. Visualize with hand-held UV lamp.

Synthesizing Oligos

All regular oligos were synthesized by standard protocols for DNA synthesis on the Expedite 8900 DNA synthesizer (Applied Biosystems) For making p17.91x, first purchase 3'-Spacer C3 CPG and chemical phosphorylating agent. Buy either the 4-pack 1mmole pre-packed columns, or buy the 0.1g powder and pack your own 4 columns.

Prime the machine twice, then carry out a standard synthesis program of the 17.91 sequence for DNA synthesis at 1 mmole scale. Note the trityl readings. Once done, remove the columns and set aside. Load on the phosphorylation reagent. Dissolve into 3 mL acetonitrile. Load this into a clean amber bottle on the machine.

Before loading the columns back on, prime all reagents for each column (1 and 2). Be careful with the phosphorylating reagent because the small volume can be quickly primed away and wasted. Now configure the machine for RNA synthesis. This tells the machine to take longer coupling times. Program in the phosphorylating reagent as the last coupling, and monitor efficiency with the last trityl reading.

Deprotect all oligos by standard ammonia incubation, followed by butanol precipitation. Only oligos used in reverse transcription or T7 RNA transcription should be gel-purified. p17.91x does not need to be gel purified for the next step.

Adenylation – Synthesis of App 17.91x

Set up reaction in distilled H_2O (i.e. in 500 mL reaction):

50 mM ImpA (FW 423) 420 mL dH_2O with 9 mg

25 mM MgCl2 7 mL from 2M solutionn

0.2 mM p17.91x 80 mL of 1.3 mM solution

Incubate at 50°C for 3 hrs. Gel purify on a denaturing 20% polyacrylamide gel. Be careful not to overload the gel, or else it gets hard to visualize the App17.91x, which should run about 1 nt slower than the p17.91x (top band is App17.91x).

For gel purification, pour 1.5 mm thick gel (prep gel thickness) using a 6-lane comb (lane width 23 mm) and load about 25 nmoles per lane. Cut out gel piece, elute overnight in 0.3 M NaCl, precipitate with 2x volume EtOH at -20°C for more than 1hr. Spin at full speed for >15 minutes, and resuspend pellet in dH_2O (want to make final stocks of 200 mM App17.91x).

Purifying 18-26mers from Total RNA

Pour a 15% 1.5 mm denaturing polyacrylamide gel with wide wells (23mm). Prerun to warm up gel. Make sure the lane is quite flat for nice loading and resolution of markers.Prepare an aliquot of total RNA (50-500μg), adding trace but very high specific activity radiolabeled marker RNA and 1X volume of 8M Urea, 0.5 mM EDTA Loading Dye. Heat for 5 min in 80°C heatblock and load entire volume in one lane. Electrophorese until the BB dye reaches the bottom. Expose gel, cut out gel slice that includes both top and bottom hot markers.

Elute ON in 0.3M NaCl, precipitate in 2X volume EtOH (>2 hrs) with glycogen (1µg/ml). Spin down (full speed, 30 min) and resuspend in 10-20µl dH_2O.

3' Adaptor Ligation and Purification

Prepare 5X T4 RNA Ligase Buffer.

250 mM Hepes pH 8.3

50 mM $MgCl_2$

16.5 mM DTT

50 ug/ml BSA

41.5% glycerol

Use RNase-free reagents and techniques. Store buffer at –20°C.

Set a 3' Adaptor Ligation Reaction. Incubate at Room Temp for 2-6 hrs.

2 mL 5x Ligation Buffer

2 mL 200 mM App17.91x

1 mL T4 RNA Ligase

5 mL small RNAs with Hot Markers

Stop reaction with 10 mL 2X Loading Dye.

Prepare a 10% (0.5 mm) denaturing polyacrylamide gel. Prerun, then load into 2-4 lanes (spread out the reaction to prevent overloading). Run gel until good deal separation of BB and XC (about 3-4 inches).

Separate one of the plates, keeping gel on other plate, cover with saran wrap. Expose on a phosphor plate, locate ligated bands (higher mobility), cut out gel fragments and transfer into siliconized tubes. You should expect to only cut out bands between 35-43 nt long. There shouldn't be too many artifacts, but you may see a 53mer artifact (18+18+17) and <16mer artifact (degradation or minor circularization of 18mer). Elute, EtOH precipitate with glycogen , spin down. Resuspend all pellets together into 10-20 mL dH_2O.

5' Adaptor Ligation and Purification

Set a 5' Adaptor Ligation Reaction. Incubate at Room Temp for 2-6 hrs.

2 mL 5x Ligation Buffer

2 mL 200 mM 17.93R

1 mL 4 mM ATP

1 mL T4 RNA Ligase

5 mL small RNAs from 3' Adaptor ligation reaction

Stop reaction with 10 mL 2X Urea Loading Dye. Prepare gel and purify 5' adaptor ligation products in the same way as for the 3' ligation products. For band identification, just use the freshly kinased 10 bp ladder as a reference for size.You should cut out the 52-60mer bands, and you may see 35-43mers remaining from ligaton. Resuspend pellets in a total 10-20 mL dH_2O.

RT-PCR of small RNAs Ligated with Adaptors

Using siliconized tubes, set up a reverse transcription reaction:

5 mL of ligated RNAs

3 mL 100 mM 15.22

5 mL dH_2O

5 mL 5X First Strand Buffer (from Invitrogen)

7 mL 10X dNTP's

3 mL 100mM DTT

3 mL SuperScript II RT (200U/mL) final

Incubate at 48°C for 1 hour, then add 1 mL 0.1M EDTA and 3.8 mL 1 M KOH. Incubate at 90°C for 10 minutes to degrade all the RNA. Do all steps in parallel with the (-) RT control. Neutralize the reaction by adding 4 mL 1M HCl-Tris pH 1, and 1 mL 0.2M $MgCl_2$. Use the entire RT(+) reaction for six 100μL PCR amplifications.

Set up 100 mL PCR reaction from the RT(+) [six rxns] and RT(-) [one rxn] samples. Use fresh tubes. Use up all of the RT reaction – it does not store well.

5 mL of RT reaction

10 mL 10X PCR Buffer

10 mL 10X dNTPs

1 mL 100 mM 17.92

1 mL 100 mM 17.93D

2 mL Taq Polymerase

71 mL dH_2O

Analyze reactions with a 15% denaturing polyacrylamide gel. Take 3 mL from each RT-PCR reaction, add loading dye, heat well before loading, and load onto a pre-run midi-thickness gel.

Run using the 10bp ladder to follow bands. Do not use EtBr for staining, because the sensitivity is very weak for these small DNAs. Use the SYBR Gold stain from Molecular Dynamics. You should see a good smear in the size range of small RNAs ligated with linkers. Use filter tips.

Two times phenol extract. Two times chloroform extract. Add NaCl to make 0.3M/EtOH precipitate (glycogen optional). Spin down pellet and resuspend each RT(+) reaction in 40 mL.

Concatamerization

Set up a Ban I digest of PCR products

40 mL of PCR products

30 mL NEBuffer 4

10 mL Ban I 20U/mL ® 0.67 U/ìL final

220 mL dH_2O 4 hrs incubation at 37°C

Check 15 mL from digestion on a 15% denaturing polyacrylamide gel. Use 1 mL from the PCR and the 10 bp ladder as markers, then stain the gel with SYBR Gold. Two times phenol extract.

Two times chloroform extract. Add NaCl to make 0.3M and EtOH precipitate (glycogen optional). Spin down pellet.

Take the entire pellet from a single Ban I digestion and add the following for concatamerization:

8 μL dH_2O

1 μL 10X T4 Ligase Buffer (USB or NEB brand is fine)

1 μL T4 DNA Ligase

Incubate at room temp for 30 min.

Take a mini-gel casting tray for agarose and rinse thoroughly. Prepare a 2% GTG Nusieve agarose gel with 1X TAE, pre-stained with EtBr. Load entire concatamerization reaction with glycerol loading dye into a lane, run with the 100bp marker (MBI Fermentas). Run a short time, when the ladder can be visualized. Using the low energy, high wavelength setting on transilluminator, locate smear corresponding

>300 bp concatamers and cut out with a clean razor blade. Add 10 volumes gel melting solution (20 mM TrisHCl pH 8, 1 mM EDTA pH 8) and melt for 5 minutes at 65°C. You may need to distribute this to a couple of siliconized tubes.

Add an equal volume of phenol, vortex for 20 seconds, chill on ice for 5 min, then spin at 5000g for 10 min (4°C). Remove aqueous phase, reextract with a 1:1 phenol, chloroform mix, and reextract again finally with just chloroform. Add 0.06 volume of 5M NaCl and 2.5 volume EtOH and precipitate at –20°C with glycogen for >2 hrs.

Cloning into TOPO Vector

Spin down pellet and resuspend in the following *Taq* Fill-In Reaction

11.5 mL dH_2O

1.5 mL 10x PCR Buffer

1.5 mL dNTPs

0.5 mL Taq

Heat to 72°C for 5 min. Have the TOPO TA cloning kit reaction tube set up. Freeze after taking 5 mL from the Fill-In Reaction and adding to 1 mL salt solution and 1 mL TA Cloning vector from TOPO cloning kit. Use all of Topo reaction for the transformation, add 500 µl SOC, let the cells grow for only 45 min (no longer), and plate out as 75 mL, 150 mL, 300µL (remaining) on LB Amp S-Gal plates. Grow ON at 37°C.

Screen and Sequence

Pick white colonies, and restreak on a master plate. Let this master plate grow ON. Screen only the white colonies by PCR in microplate format – 30µL reactions enough.

3 mL 10X PCR Buffer

3 mL 10X dNTPs

0.2 mL 100 mM M13F

0.2 mL 100 mM M13R

0.5 mL Taq Polymerase

23 mL dH_2O

Pick a colonies from master plates with a pipette tip, swish around in a PCR reaction well. Check on 2% agarose gel, look for inserts greater than 220 bp (expect 500-800 bp inserts). You can now either purify remaining PCRs and submit directly, or regrow colonies to extract plasmids. Submit to commercial sequencing facility, using M13F or M13R as sequencing primers.

39

Graphical Approach to Calculate Molecular Weights of DNA/Protein Fragments

After the DNA is run on the gel, the polymorphic bands or the bands cosegregating with a target trait are tagged. The bands identification is either done by calculating their relative mobilities (Rm / Rf values in case of isoenzymes) or by calculating their molecular weights (in case of proteins and DNA).

There are several methods to calculate molecular weights of DNA bands on the gel. Graphical approach is one of them. The approaches include (1) use of computer software to generate molecular weights and (2) running unknown DNA fragments with the fragments of known molecular weights (standard markers available from private companies, see examples in the figure). An example is given here (see figure) to illustrate the method to calculate molecular weights of unknown DNA fragments. This method is usually called as Graphical Approach.

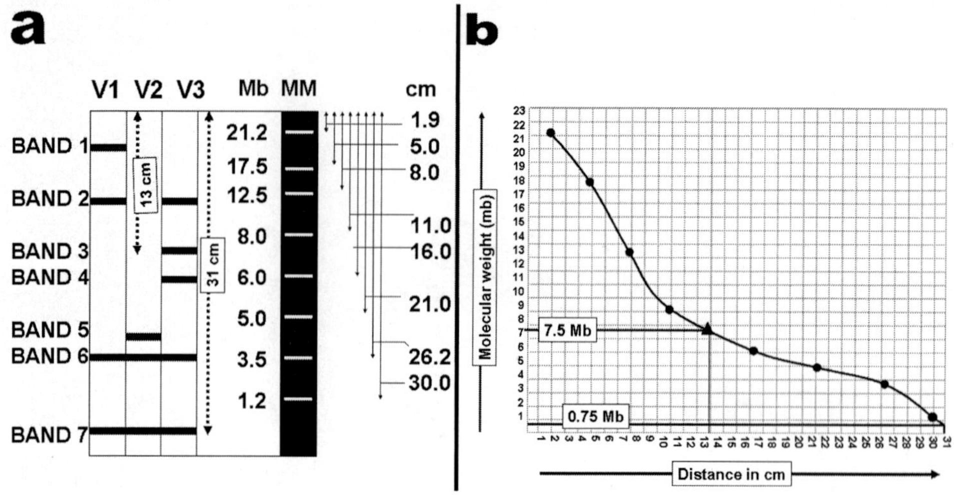

Legends Used in Above Figure

V1 = Variety 1

V2 = Variety 2

V3 = Variety 3

Mb = Molecular weight of DNA

MM = Standard molecular marker (here hypothetical figure has been used)

cm = Distance of band in cm measured from center of the well to center of the band (thickness).

BAND1, BAND 2 ...... BAND 7 = are DNA bands arranged as per their relative mobilities (higher molecular weight bands near to the well with lower mobility and lower molecular weight bands near to the bottom edge of the gel with higher mobility).

Procedure

Step 1: Measure the distance of each band (unknown / test) from the center of the loading well to the center of the band using an ordinary scale (Figure a above).

Step 2: Plot the graph of the bands obtained by standard molecular weight marker (MM), putting distance on x-axis and molecular weight on y-axis (Figure b above).

Step 3: The graph so obtained is not a linear graph, but follows a logarithmic trend and forms the shape of a parabola. (Figure b above).

Step 4: The molecular weight of unknown DNA fragment is now calculated by placing the distance value of that band on the x-axis, extending it upward to find out joining point on the standard curve and then expending the point to obtain MW towards left hand side (y-axis).

40

Calculation of Similarity Index Values and Construction of Dendrogram Using NTSYSpc 2.0

NTSYSpc 2.0 software is used for various types of analysis including calculation of similarity index values and construction of dendrogram. Construction of dendrogram using gel data in the form of bands. This software accepts zero-one data.

STEP 1:

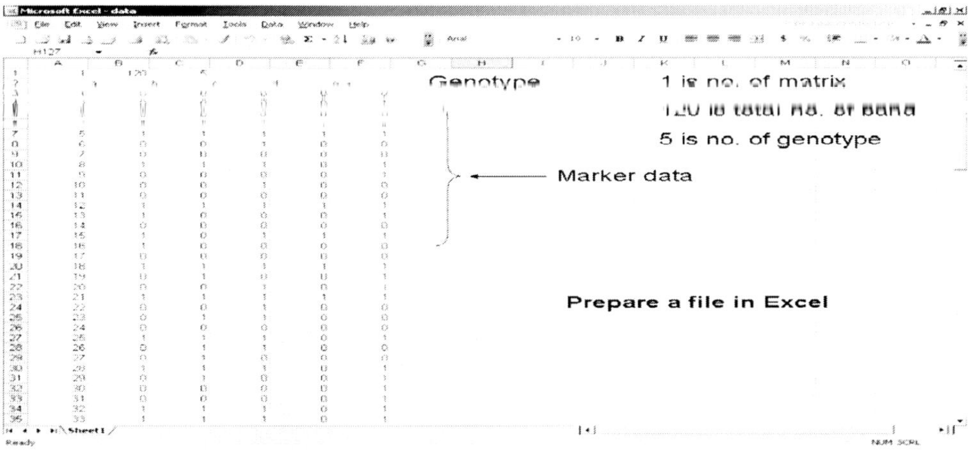

STEP 2:

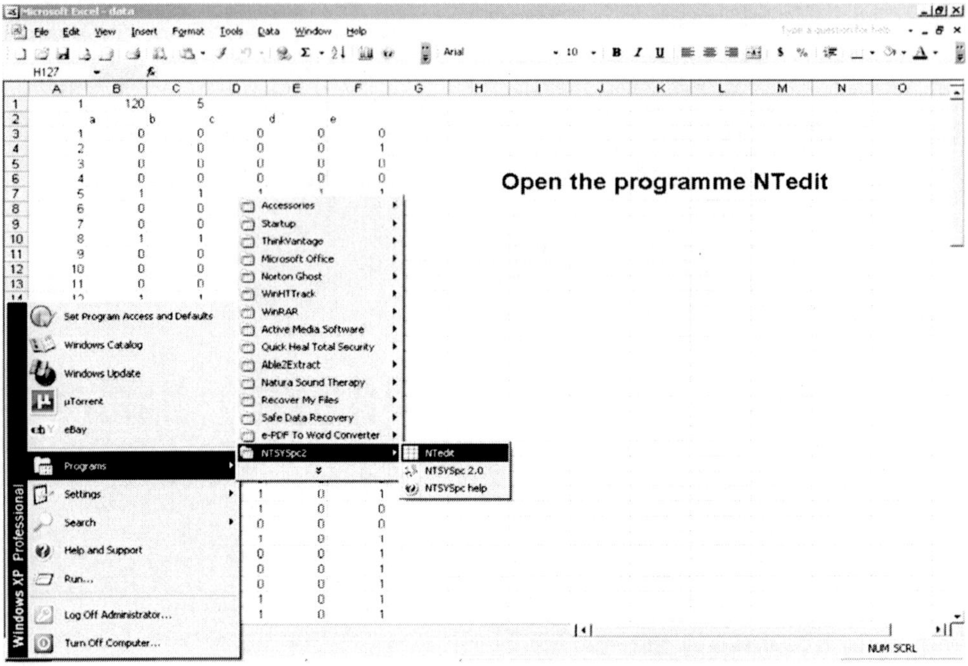

STEP 3:

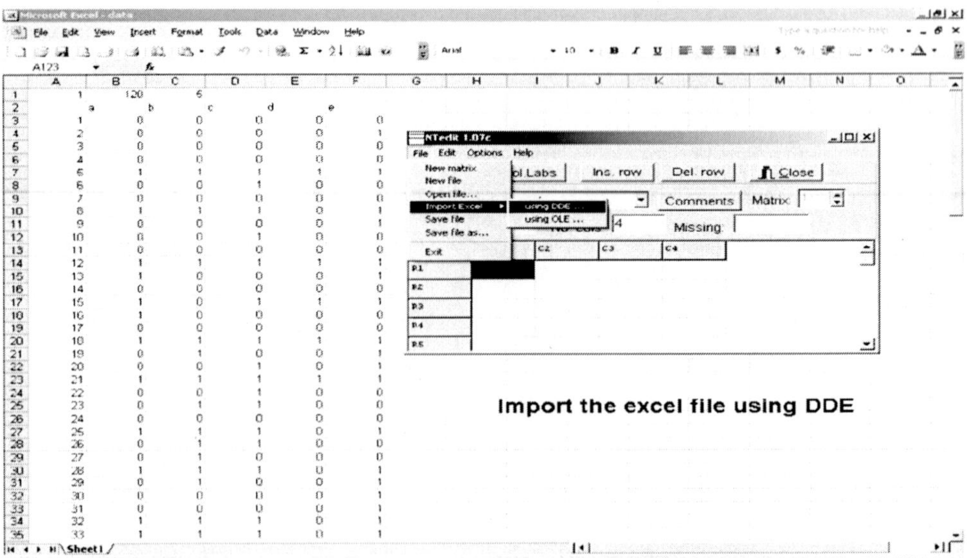

STEP 4:

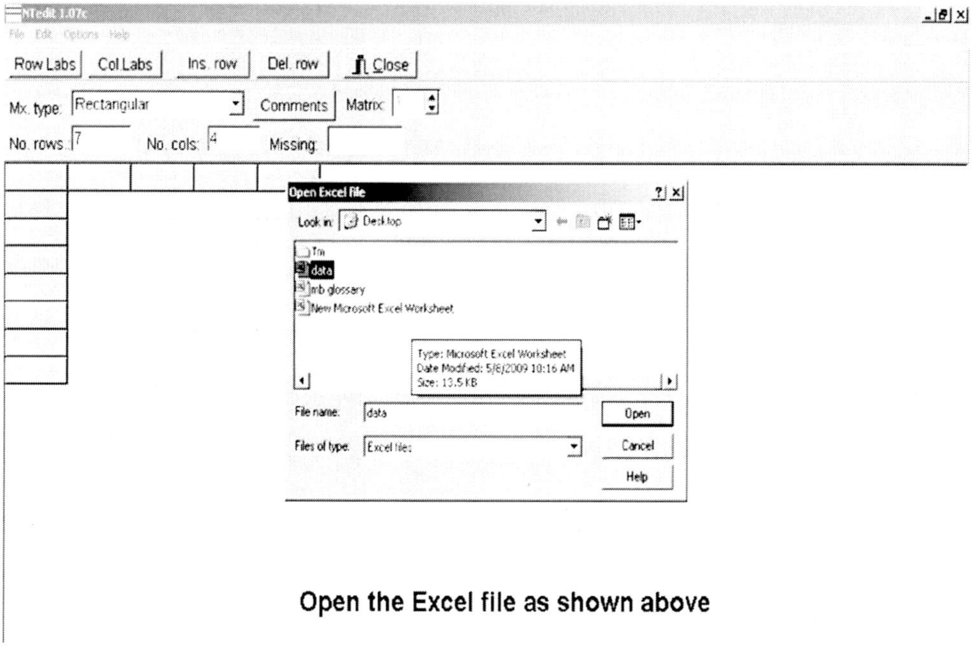
NTedit 1.07c
File Edit Options Help
Row Labs
Col Labs
Ins. row
Del. row
Close
Mx. type: Rectangular
Comments
Matrix
No. rows: 7
No. cols: 4
Missing:
Open Excel file
Look in: Desktop
Tm
data
mb glossary
New Microsoft Excel Worksheet
Type: Microsoft Excel Worksheet
Date Modified: 5/8/2009 10:16 AM
Size: 13.5 KB
File name: data
Open
Files of type: Excel files
Cancel
Help
Open the Excel file as shown above

STEP 5:

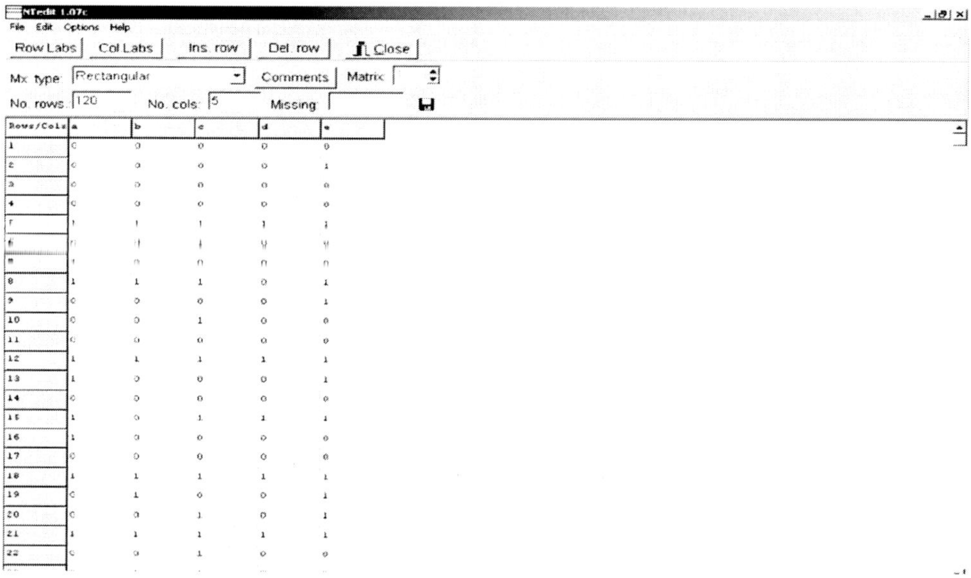
NTedit 1.07c
File Edit Options Help
Row Labs
Col Labs
Ins. row
Del. row
Close
Mx. type: Rectangular
Comments
Matrix
No. rows: 120
No. cols: 5
Missing:
Rows/Cols
a
b
c
d
e

STEP 6:

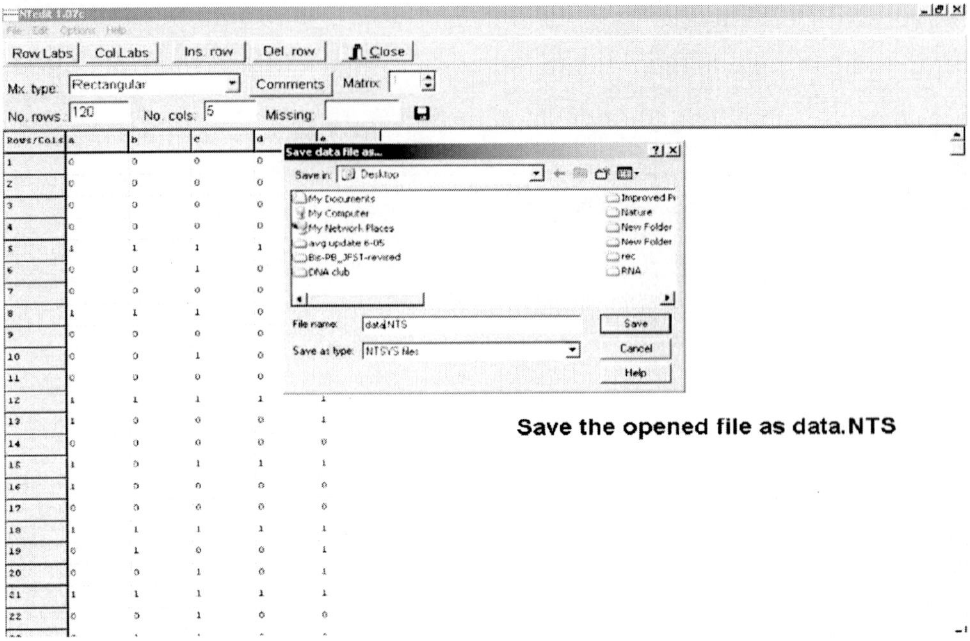

Save the opened file as data.NTS

STEP 7:

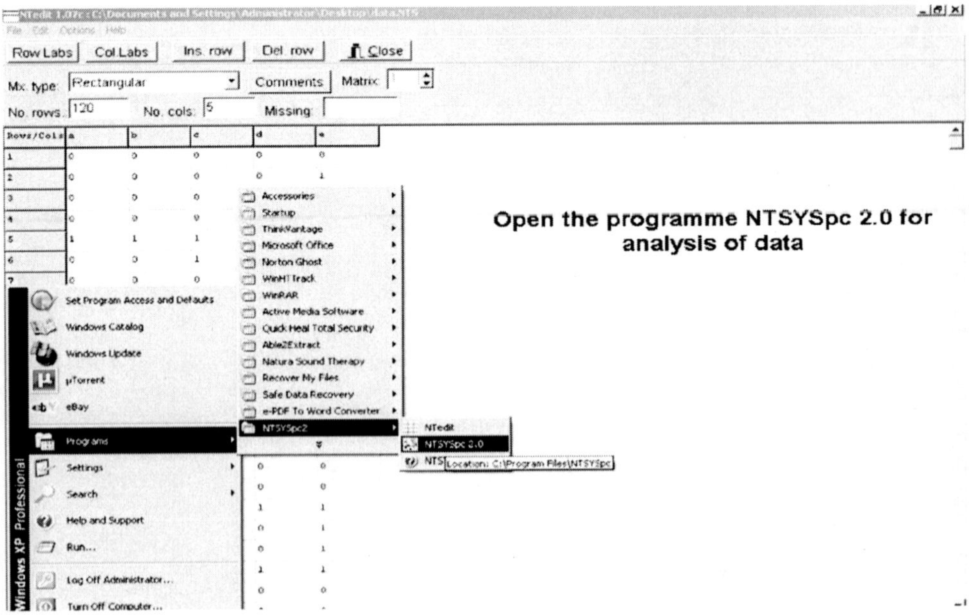

Open the programme NTSYSpc 2.0 for analysis of data

STEP 8:

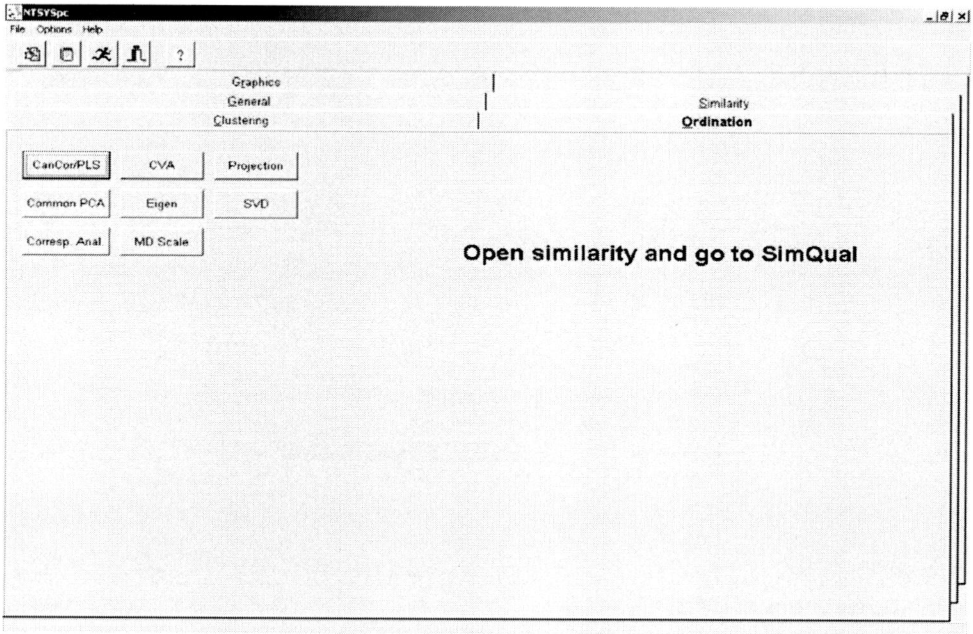
NTSYSpc
File Options Help
Graphics
General
Clustering
Similarity
Ordination
CanCor/PLS
CVA
Projection
Common PCA
Eigen
SVD
Corresp. Anal.
MD Scale
Open similarity and go to SimQual

STEP 9:

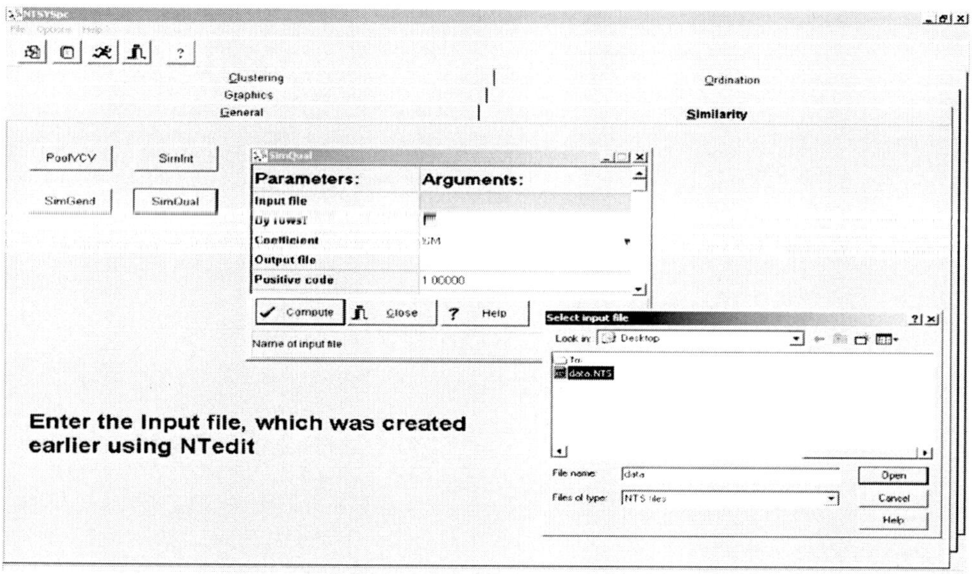
Clustering
Graphics
General
Ordination
Similarity
PoolVCV
SimInt
SimGend
SimQual
Parameters:
Arguments:
Input file
Coefficient
Output file
Positive code
Compute
Close
Help
Name of input file
Select input file
Look in: Desktop
data.NTS
File name:
Files of type:
Open
Cancel
Help
Enter the input file, which was created earlier using NTedit

STEP 10:

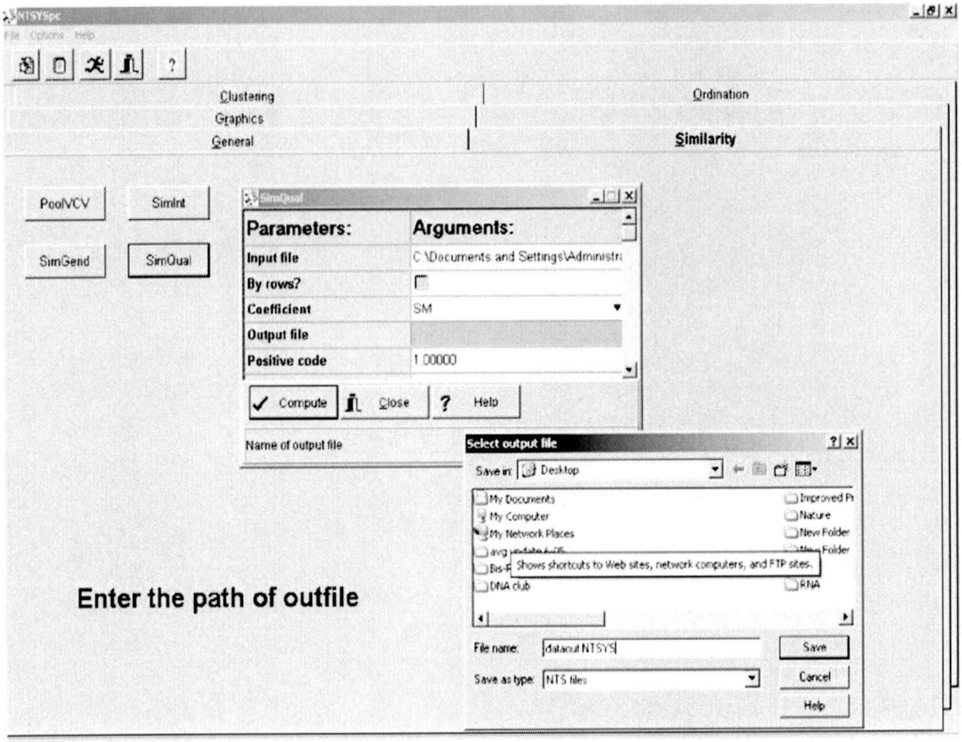

STEP 11:

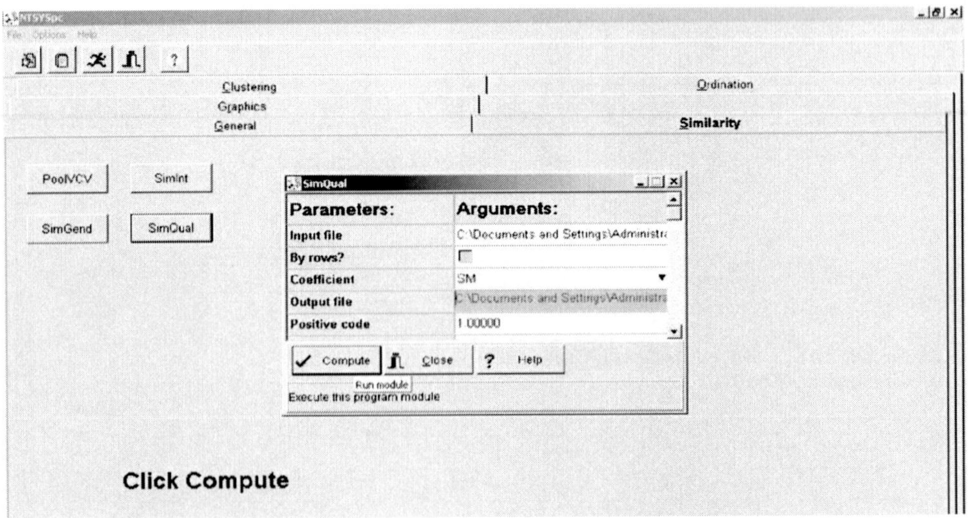

STEP 12:

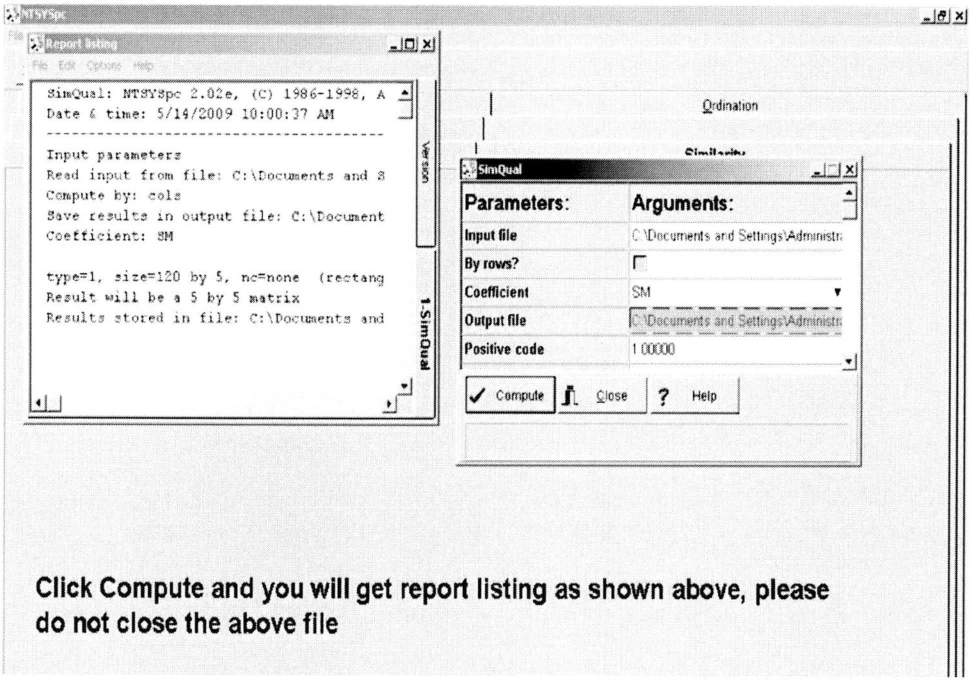

STEP 13 :

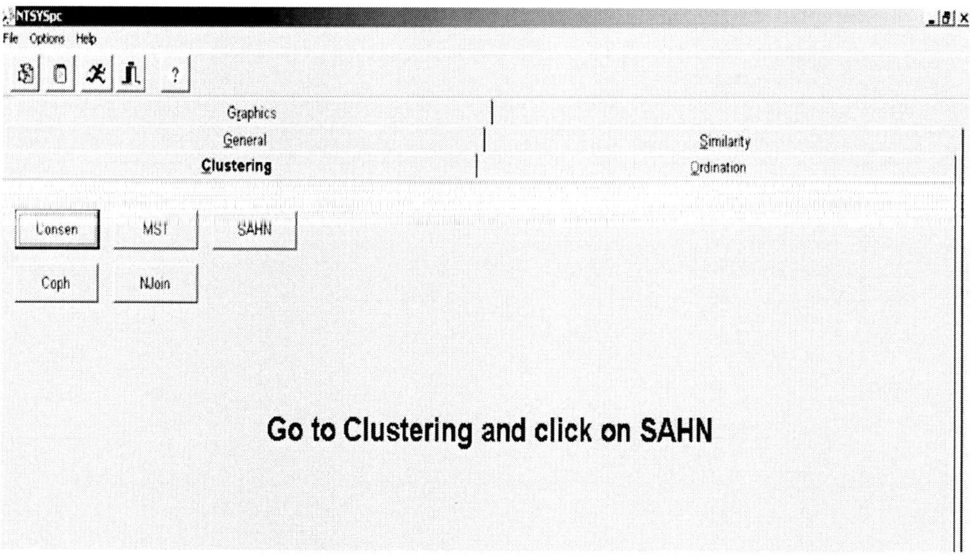

STEP 14:

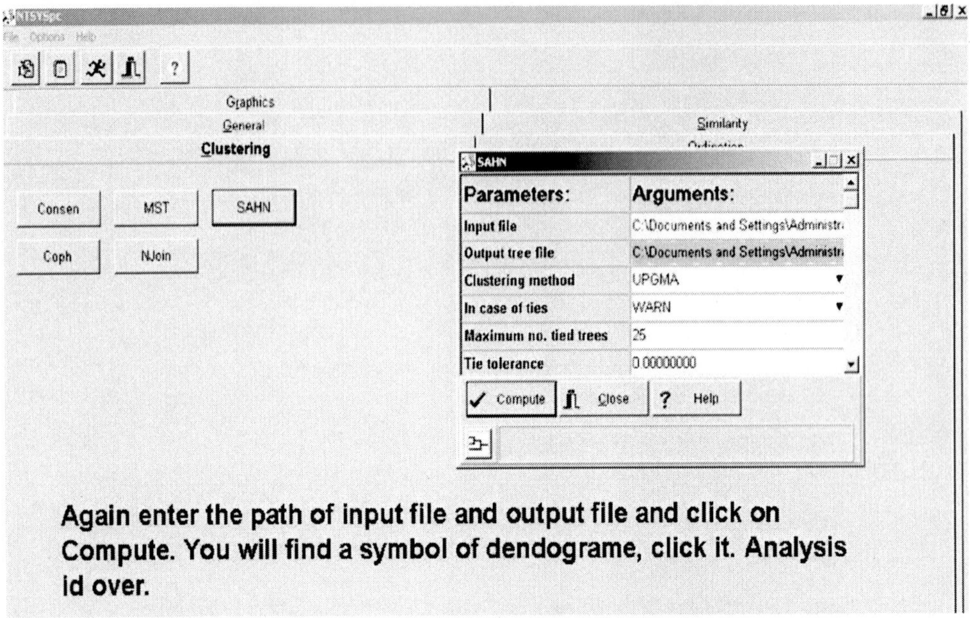

STEP 15:

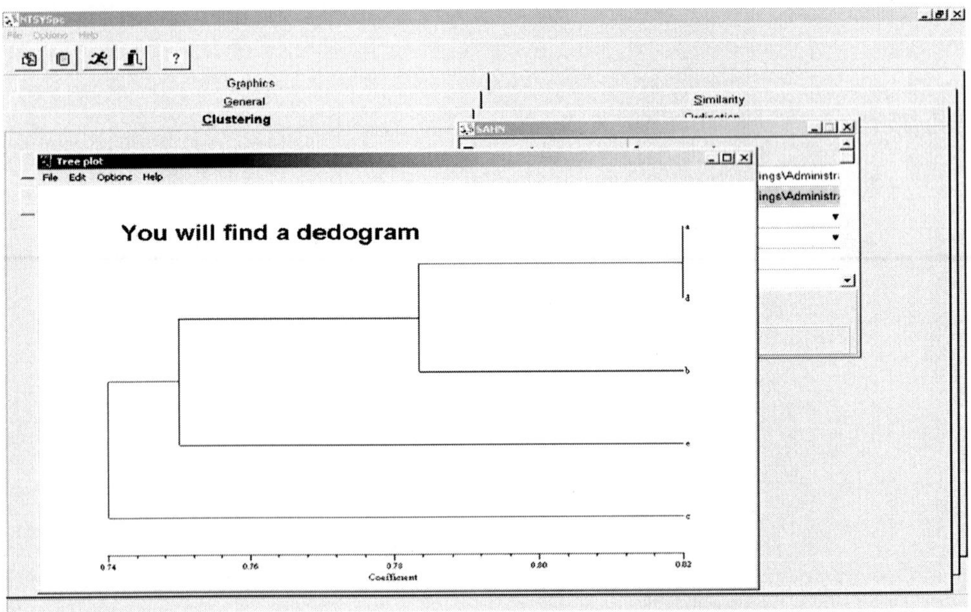

Appendix-I
Commonly Used Abbreviations

A	adenine or adenosine; one-letter code for alanine
A260	absorbance at 260nm
ACES	N-(2-acetamido)-2-aminoethanesulfonic acid
Ad-2	adenovirus-2
ADA	N-(2-acetamido)-2-iminodiacetic acid
ADP	adenosine 5'-diphosphate
AEC	3-amino-9-ethylcarbazole
AMP	adenosine monophosphate
AP	alkaline phosphatase
ATP	adenosine 5'-triphosphate
BAP	bacterial alkaline phosphatase
BCIP	5-bromo-4-chloro-3-indolyl phosphate
BES	N,N-bis (2-hydroxyethyl)-2-aminoethanesulfonic acid

BICINE	N,N-bis(2-hydroxyethyl)glycine
Bio-dNTP	biotin-deoxynucleoside triphosphate
bp	base pair
Bq	Becquerel
BSA	bovine serum albumin
B/W	blue/white cloning
C	cytosine or cytidine; one-letter code for cysteine
CA	casamino-acids
cDNA	complementary deoxyribonucleic acid
Ci	Curie
CIAP	calf intestinal alkaline phosphatase
cpm	counts per minute
CTP	cytidine 5'-triphosphate
Da	Dalton
DAB	3,3'-diaminobenzidine tetrahydrochloride
dAMP	deoxyadenosine monophosphate
dATP	deoxyadenosine triphosphate
dCTP	deoxycytidine triphosphate
ddATP	dideoxyadenosine triphosphate
ddCTP	dideoxycytidine triphosphate
ddGTP	dideoxyguanosine triphosphate
ddNTP	dideoxythymidine triphosphate
DE-81	Whatman® diethylaminoethyl cellulose paper
DEPC	diethyl pyrocarbonate
DIG	digoxigenin
dGTP	deoxyguanosine triphosphate
DMSO	dimethyl sulfoxide
DNA	deoxyribonucleic acid

DNase	deoxyribonuclease
dNTP	deoxynucleoside triphosphate
dpm	disintegrations per minute
ds	double-stranded
DTE	dithioerythritol
DTT	dithiothreitol
dTTP	deoxythymidine triphosphate
dUTP	deoxyuridine triphosphate
EDTA	ethylenediaminetetraacetic acid
ELISA	enzyme-liked immunosorbent assay
EMBL	European Molecular Biology Laboratory
ENDO	endodeoxyribonuclease assay
exo	exonuclease
EXO	5' and 3'- exodeoxyribonuclease assay
FOA (5-FOA)	5-fluoroorotic acid
G	guanine or guanosine; one-letter code for glycine
Gal	D-galactose
GUS	beta-D-glucuronidase
HC	high concentration
HEPES	N-(2-hydroxyethyl)piperazine-N'-(2-ethanesulfonic acid)
HRP	horseradish peroxidase
HPLC	high-performance liquid chromatography
IEF	isoelectric focussing
IPTG	isopropyl-beta-D-thiogalactopyranoside
kb	kilobase
kDa	kiloDalton
LB	Luria Bertani media
LC	low concentration

LO	labeled oligonucleotide
MCS	multiple cloning site
MES	2-(N-morpholino)ethanesulfonic acid
M-MuLV	Moloney Murine Leukemia Virus
MOPS	3-(N-morpholino)propanesulfonic acid
mRNA	messenger ribonucleic acid
MSPU	mini power supply unit
MW	molecular weight
NAD	nicotinamide adenine dinucleotide
NADH	nicotinamide adenine dinucleotide, reduced form
NADP	nicotinamide adenine dinucleotide phosphate
NADPH	nicotinamide adenine dinucleotide phosphate, reduced form
NBT	nitro blue tetrazolium
nd	not determined
NR	not recommended
NP-40	Nonidet P-40 (detergent)
nt	nucleotide
NTP	nucleoside triphosphate
oligo(A)	oligoadenylic acid
oligo(dT)	oligodeoxythymidylic acid
OMP	orotidine monophosphate
OPD	1,2-phenylenediamine; ortho-phenylenediamine
Pi	inorganic phosphate
PAGE	polyacrylamide-gel electrophoresis
PBS	phosphate-buffered saline
PCR	polymerase chain reaction
PEI	polyethylenimine
PEG	polyethylene glycol

PIPES	piperazine-N,N′-bis(2-ethanesulfonic acid)
PNK	polynucleotide kinase
pNPP	4-nitrophenyl phosphate; para-nitrophenyl phosphate
poly(A)	polyadenylic acid
poly(A)$^+$	polyadenylated (mRNA)
poly(dA-dT)	poly (deoxyadenylic acid - deoxythymidylic acid)
poly(dT)	polydeoxythymidylic acid
PSU	power supply unit
QCA	quality control assay
RE	restriction enzyme
RNA	ribonucleic acid
RNase	ribonuclease
R-M	restriction-modification
rRNA	ribosomal ribonucleic acid
RT	reverse transcriptase
SAM	S-adenosylmethionine
SDS	sodium dodecyl sulfate
ss	single-stranded
SSC	sodium chloride/sodium citrate (buffer)
SSPE	sodium chloride/sodium phosphate/EDTA (buffer)
T	thymine or thymidine; one-letter code for threonine
TAE	Tris/acetate/EDTA (buffer)
Taq	*Thermus aquaticus*
TBE	Tris/borate/EDTA (buffer)
TdT	terminal deoxynucleotidyl transferase
TE	Tris/EDTA (buffer)
TEMED	N,N,N′,N′-tetramethylethylenediamine
TES	N-tris(hydroxymethyl)methyl-2-aminoethanesulfonic acid

TLC	thin layer chromatography
T_m	melting temperature
TMB	3,3', 5,5'-tetramethylbenzidine
TRICINE	N-tris(hydroxymethyl)methylglycine
Tris	tris(hydroxymethyl)aminomethane
tRNA	transfer ribonucleic acid
TPE	Tris/phosphate/EDTA (buffer)
TX-100	triton X-100
U	uracil or uridine
u	unit
UTP	uridine 5'-triphosphate
UV	ultraviolet
v/v	volume/volume
w/v	weight/volume
X-Gal	5-bromo-4-chloro-3-indolyl-beta-D-galactopyranoside
X-Gluc	5-bromo-4-chloro-3-indolyl-beta-D-glucuronic acid

Appendix-II
Equations for Calculating Amounts and Concentrations of Reagents

Mole (mol)

1 mole of any substance contains 6.02×10^{23} units. The number of moles in a given weight of substance can be calculated using following equation.

$$mol = \frac{wt\,(g)}{mol\,wt}$$

The weight of a given mount of moles can be calculated using equation given below

$$wt\ (g) = mol \times mol\ wt\ (g/mol)$$

Equivalent Weight (equiv wt)

This is the weight of an acid or a base containing 1 mole of replaceable H^+ or OH^-. Gram equivalent weight is used when equivalent weight is expressed in grams, as calculated in equation given below

$$equiv\ wt = mol\ wt/n$$

Equivalent (equiv)

One equivalent of a compound contain 1 g-equiv wt of the compound, as calculated with following equation

$$\text{equiv} = \frac{\text{wt (g)}}{\text{equiv wt}} \text{ or equiv} = \text{mol} \times n$$

Molarity (M)

This defines the number of moles of a substance per liter (L) of solution, as calculated with following.

$$M = \frac{\text{mol}}{L} \text{ or } M = \text{mol} \times n$$

1M = 1GMW/L
(g/mole) x (moles/L) = g/L
(g/mole) x (moles/L) x L = g

Example

How many grams of H_3BO_3 (Boric acid, MW=61.83g/M) are required to make 2 liters of a 2M solution?

61.83g/M x 2M/L x 2L = Zg
Z = 247.32g

Example

What is the molarity of a solution that has 60g of NaOH (sodium hydroxide, MW=40g/M) in a total volume of 2 liters?

40g/M x XM/L x 2L = 60g
XM/L = 60g ¸ (40g/M x 2L)
XM/L =(60/80)M/L
X = 0.75M/L = 0.75M

Example

How many liters of 1M NaOH can be made with 80g of substance?

40g/M x 1M/L x XL = 80g
XL = 80g ¸ (40g/M x 1M/L)
XL = (80/40)L
X = 2L

MOLALITY

Moles of solute/Kg of solvent, this is a W/W expression. Molality is not a practical expression of concentration for routine laboratory use.

Normality (N)

"Normality is the number of "equivalent weights" of the solute per litre of solution".

- Biologists are most likely to encounter normality in reference to acids and bases.
- **For an acid,** 1 equivalent weight is equal to the number of grams of that acid that react to yield. 1 mole of H+ ions i.e. 1 mole of reactive group.
- **For bases,** 1 equivalent weight is equal to the number of grams of that base that supplies 1 mole of OH^- i.e.1 mole of reactive group.

NaOH	$Na^+ + OH^-$
H_2SO_4	$SO_4^{-2} + 2H^+$

1 equivalent weight of NaOH is the number of grams that will produce 1 mole of OH^- ions, so equivalent weight of NaOH is 40 gms (equal to molecular weight.).

1 mole of H_2SO_4 dissociates to form 2 molecules of H^+ ions. Therefore molecular weight of H_2SO_4 is 98.1 gm. & equivalent weight = 98.1 / 2 = 49 gms

$$N = \frac{\text{equiv}}{L} \text{ or } N = \frac{\text{wt}}{\text{mol wt} \times L}$$

Example

NaCl has a valence of +1 in solution; for every molecule of NaCl, there is one molecule of Na^+ that can replace a molecule of hydrogen ions. For H_2SO_4, there is the equivalent of two molecules of H^+ for every molecule of H_2SO_4. Therefore the gram equivalent weight of H_2SO_4 is equal to the gram molecular weight (98) divided by 2; 98/2 = 49.

Calculating Concentrations Based on Weight Percent (weight/weight) or % (w/w)

The percent weight of a pure analyte in a sample is calculated using following equation (e.g the weight in g of an analyte in a 100g sample).

$$\% \text{ (w/w)} = \frac{\text{g of analyte in sample}}{\text{g of sample}} \times 100\%$$

Percent (Weight/Volume) or %(w/v)

The percent weight of a pure analyte in a solution is calculated using equation (e.g. the weight in g of an analyte in 100ml sample).

Part dilutions : "Part solutions tell us how many parts of each component to mix together". The parts may have any units but must be the same for all the components of the mixture.

For example : A Solution that is 3:2:1 ethylene: chloroform: isoamylalcohol is:

1 parts ethylene, 2 parts Chloroform, 1 part isoamylalcohol.

Parts Per Million and Parts Per Billion

"Parts per million" (ppm) is the number of parts of solute per 1 million parts of total solution. Any units may be used but must be the same for the solute and the total Solution.

"Parts per billion" (ppb) is the number of parts of solute per billion parts of solution. Concentration is most often expressed in terms of ppm (or ppb) in environmental applications.

For example: an environmental scientist might speak of a pollutant in a lake as being present at a concentration of 5 ppm.

For any solute,

$$1\,\text{ppm in water} = \frac{1\,\mu\text{g}}{\text{ml}}$$

Also,

$$1\,\text{ppb in water} = \frac{1\,ng}{\text{ml}}$$

Gram Atomic Weight

The weight of a mole of a given element is equal to its atomic weight in grams or its gram atomic weight e.g. one mole of the element carbon weights 12.0 gms.

Compounds are composed of atoms of two or more elements that are bounded together. A mole of a compound contains 6.023 x 10^{23} molecules of that compound.

Gram Molecular Weight (MW) or Gram Formula Weight (FW)

Weight in grams of 1 mole of the compound. The FW is calculated by adding the atomic weights of the atoms that make up the compound.

For example: The gram molecular weight of Na_2SO_4 Sodium Sulphate is 142.04 gm

2	Sodium atoms	2 x 22.99	=	45.98 g
1	Sulfur atom	1 x 32.06	=	32.06 g
4	Oxygen atoms	4 x 16.60	=	64.00.g

				142.04 Gms.

Normality

"Normality is the number of "equivalent weights" of the solute per litre of solution".

- Biologists are most likely to encounter normality in reference to acids and bases.
- **For an acid,** 1 equivalent weight is equal to the number of grams of that acid that react to yield. 1 mole of H+ ions i.e. 1 mole of reactive group.
- **For bases,** 1 equivalent weight is equal to the number of grams of that base that supplies 1 mole of OH^- i.e.1 mole of reactive group.

  NaOH $Na^+ + OH^-$

  H_2SO_4 $SO_4^{-2} + 2H^+$

  1 equivalent weight of NaOH is the number of grams that will produce 1 mole of OH^- ions, so equivalent weight of NaOH is 40 gms (equal to molecular weight.).

  1 mole of H_2SO_4 dissociates to form 2 molecules of H^+ ions. Therefore molecular weight of H_2SO_4 is 98.1 gm. & equivalent weight = 98.1 / 2 = 49 gms

Parts Per Million (Weight/Weight), or ppm (w/w)

The grams of a pure analyte in 10^6 g of sample is calculated using equation 10.

$$\%\,(w/w) = \frac{\text{g of analyte in sample}}{\text{g of sample}} \times 10^6 \text{ ppm}$$

Part Per Million (Weight/Volume) or ppm (w/v)

The grams of pure analyte in 10^6 mL of sample is calculated using equation.

$$\%\,(w/v) = \frac{\text{g of analyte in sample}}{\text{mL of sample}} \times 10^6 \text{ ppm}$$

Density (p) and Specific Gravity (sp gr)

Density is the mass per volume ration of a substance and is calculated using equation.

$$p = \frac{wt}{vol}$$

Specific gravity is the density of a fluid or other material relative to that of H_2O as expressed in equation.

$$\text{sp gr} = \frac{p\,\text{sample}}{p\,H_2O} = p_{\text{angle}} \text{ at } 4^\circ\text{C}$$

Equations for Calculating the Weight or Volume of a Reagent Needed to Prepare a Given Solution

To calculate the number of gram of a substance to create a pre-assigned molar concentration use equation

$$\text{wt (g)} = \text{mol wt (g)} \times M \times L$$

When correcting for % purity use equation.

$$\text{wt (g)} = \text{mol wt (g)} \times M \times L \times \frac{100}{\text{purity of reagent (\%)}}$$

For % and ppm concentration use equation

$$\text{wt (g)} = \frac{\%\,(w/v) \times \text{mL of solution}}{100\%}$$

$$\text{wt (g)} = \frac{\text{ppm (w/v)} \times \text{mL of solution}}{10^{-6}\,\text{ppm}}$$

Dilution Factor in Concentration and Volume Calculations

To calculate the volume of a stock reagent necessary to add to a known volume of diluent to yield a desired concentration use equation.

$$V_1 = \frac{C_2 V_2}{C_1}$$

where V_1 is the volume of the stock reagent which, when diluted to a final volume V_2, yields a desired concentration C_2; C_1 is the concentration of the stock reagent. The units of C_1 and C_2 must be identical, and those of V_1 and V_2 must also be identical.

The densities of liquids

	Density [g/ml]	Additional information
Acetic acid glacial	1.049	T_b=118°C
Acetic acid 36%	1.049	
Acetone	0.788	T_b=56.5°C
AcOK, 5.0M	1.218	
1.0M	1.046	
AcONa, 3.0M	1.121	
2.5M	1.147	
$AcONH_4$, 10M	1.073	
Acrilamide:BIS, 40%	1.044	
Ammonium hydroxide, NH_4OH, 28%	0.89	
Blocking reag., 10%	1.038	
Butanol (n)	0.81	T_b=117-118°C
$CaCl_2$, 1M	1.086	
Chlorophorm	1.48	T_b=61-62°C
Denhardt sol., 50X	1.004	
DMFA	0.945	
DMSO	1.100	
EDTA, 0.5M	1.096	
Ethanol, anhydr.	0.79	
96%	0.80	
70%	0.868	T_b=78.5°C
FGRB, 50X	1.104	

Contd...

Formaldehyde, 37%	1.09	
Formamid	1.133	
Glucose, 10% (w/v)	1.040	
2M	1.13	
Glycerol	1.25	
GTGB, 20X	1.094	
HCl, 37% (12.07M)	1.19	
Isoamil alchogol	0.813	T_b=132°C
Isobutanol	0.871	T_b=108°C
Isopropanol	0.785	T_b=82.5°C
KCl, 0.25M	1.044	
KOH, 5.0M	1.217	
LB	1.024	
LiCl, 9M	1.194	
8M	1.172	
Mercaptoethanol (b)	1.114	14.192M, Mw 78.13
$MgCl_2$, 1M	1.073	
$MgSO_4$, 1M	1.112	
$MnCl_2$, 1M	1.099	
Nitric acid, HNO_3 71%	1.42	
NaCl, 5M	1.186	
NaOH, 50%	1.53	
NaOH, 10M	1.329	
$NaxHPO_4$, 1M, pH~7	1.104	
Nonident P-40	1.064	
Octane	0.71	T_b=125.6°C
PEG 6000, 20%	1.041	
40%	1.075	
PEG/NaCl, (20%/2.5M)	1.145	
PEG/TAE, (15%/1X)	1.032	
Phosphoric acid H_3PO_4, 85%	1.70	
Phenol	1.054	T_b=182°C, T_m=43°C

Contd...

Sulfuric acid H_2SO_4, 96%	1.84
Sarcosile, 10% (w/v)	1.017
30% (w/v)	1.027
SDS, 10%	1.018
SSC, 20X	1.163
30X	1.245
SSPE, 20X	1.136
TAE, 50X	1.088
TBE, 10X	1.070
Tris.Cl, 1M, pH 7-8	1.041
2M, pH7.9	1.10
TritonX-100, 10%	1.016
TritonX-100, 100%	1.06
Tween 20, 100%	1.108 (1.101)
$ZnCl_2$, 1M	1.111

Buffer Preparation

Preparation of 0.1 M Sodium and Potassium Acetate buffers.

Solution A: 11.55ml glacial acetic acid/liter (0.2M).

Solution B: 0.2M Sodium/Potassium Acetate.

Desired pH	Solution A (ml)	Solution B (ml)	5 ml	
			Solution A (ml)	Solution B (ml)
3.6	46.3ml	3.7ml	4.63	0.37
3.8	44.0ml	6.0ml	4.4	0.6
4.0	41.0ml	9.0ml	4.1	0.9
4.2	36.8ml	13.2ml	3.68	1.32
4.4	30.5ml	19.5ml	3.05	1.95
4.6	25.5ml	24.5ml	2.55	2.45
4.8	20.0ml	30.0ml	2	3
5.0	14.8ml	35.2ml	1.48	3.52
5.2	10.5ml	39.5ml	1.05	3.95
5.4	8.8ml	41.2ml	0.88	4.12
5.6	4.8ml	45.2ml	0.48	4.52

Preparation of 1 M Potassium Phosphate buffer.

Potassium Phosphate/Monobasic, KH_2PO_4, Mw = 136.1g/M; 1M = 136.1g/L

Desired pH	1M K_2HP_4	1M KH_2P_4	5 ml	
			1M K_2HP_4	1M KH_2P_4
5.8	8.5ml	91.5ml	0.43	4.58
6.0	13.2ml	86.8ml	0.66	4.34
6.2	19.2ml	80.8ml	0.96	4.04
6.4	27.8ml	72.2ml	1.39	3.61
6.6	38.1ml	61.9ml	1.91	3.1
6.8	49.7ml	50.3ml	2.49	2.52
7.0	61.5ml	38.5ml	3.08	1.93
7.2	71.7ml	28.3ml	3.59	1.42
7.4	80.2ml	19.8ml	4.01	0.99
7.6	86.6ml	13.4ml	4.33	0.67
7.8	90.8ml	9.2ml	4.54	0.46
8.0	94.0ml	6.0ml	4.7	0.3

Preparation of 1 M Sodium Phosphate buffer.

NaH_2PO_4

Sodium Phosphate/Monobasic, NaH_2PO_4, Mw=120.0g/M;

Na_2HPO_4

Sodium Phosphate/Dibasic, Na_2HPO_4, Mw=142.0g/M;

Desired pH	1M Na_2HP_4	1M NaH_2P_4	5ml	
			1M Na_2HP_4	1M NaH_2P_4
5.8	7.9ml	92.1ml	0.4	4.61
6.0	12.0ml	88.0ml	0.6	4.4
6.2	17.8ml	82.2ml	0.89	4.11
6.4	25.5ml	74.5ml	1.28	3.73
6.6	35.2ml	64.8ml	1.76	3.24
6.8	46.3ml	53.7ml	2.32	2.69
7.0	57.7ml	42.3ml	2.89	2.12
7.2	68.4ml	31.6ml	3.42	1.58
7.4	77.4ml	22.6ml	3.87	1.13
7.6	84.5ml	15.5ml	4.23	0.78
7.8	89.6ml	10.4ml	4.48	0.52
8.0	93.2ml	6.8ml	4.66	0.34

Solution Conversions

1) Percent solution to molarity
   M = (% X 10)/Molecular wt.

2) Percent solution to normality
   N = (% X 10)/Equivalent wt

3) Molarity to Normality
   M = Normality/Valence

4) Normality to Molarity
   N = Molarity X Valence

Unit Conversion

Ratios and proportion can be used to convert among other units of measure; i.e. English to Metric, Metric to English. With enough time and determination it is possible for the student to derive all the necessary conversion factors in a manner similar to that used to convert among the different temperature scales. The information listed below can be used to develop factors for converting among commonly used units of measure.

Length	1 meter (m) = 39.37 inches
	2.54 centimeters (cm) = 1 inch
	30.48 cm = 1 foot
	1.61 kilometers (km) = 1 mile
Mass	1 kilogram (kg) = 2.205 pounds
	453.6 grams (g) = 1 pound
	28.35 g = 1 ounce
Volume	1 liter (L) = 1.057 quarts
	29.57 milliliters (ml) = 1 fluid ounce
	3.78 L = 1 gallon (U.S.)

In addition to being used to convert among units of measure, factors can be developed to relate or convert among other types of units or to simplify calculations.

Example

Convert amount of nitrogen assayed in a blood test to amount of urea present. This needs to be done since some tests assay for nitrogen not urea. The test assumes that all the nitrogen detected is in the form of urea.

Molecular Weight (MW) of nitrogen = 14

Urea consists of two atoms of nitrogen, one carbon four hydrogen. Thus, the molecular weight of urea is 60.

Factor : 60 parts urea/28 pts N = X pts urea/1 pt N

Parts urea nitrogen x 2.14 = parts urea

Example

Determine how much NaCl is needed to yield a given concentration of Cl^-?

MW Na = 23 MW Cl = 35.5

58.5/35.5 = 1.65 = X parts NaCl/1 pt Cl

To yield 100 mg Cl, 165 mg NaCl is needed.

Converting Among Metric Units

The use of the base 10 system makes converting among different metric units easy. In general, subtract the exponent of the desired unit from the exponent of the given unit to determine the exponent for the conversion factor.

Example

Convert 7.5 milligrams (mg) to nanograms (ng). First convert each unit to the base unit. In this example convert mg and ng to grams.

$1 \text{ mg} = 10^{-3}\text{g}$
$1 \text{ ng} = 10^{-9}\text{g}$

Subtract Exponents

$-3 - (-9) = 6$
$7.5\text{mg} = (7.5 \times 10^{6}) = 7.5\times10^{6}\text{ng}$

Alternate Method

$(10^{-3}\text{g/mg}) \text{ ¸ } (10^{-9}\text{g/ng}) = (10^{-3}/10^{-9})\text{ng/mg} = 10^{6}\text{ng/mg}$
$7.5\text{mg} \times 10^{6}\text{ng/mg} = 7.5\times10^{6}\text{ng}$

Correction Factor

At times it is necessary to perform a test or assay with other than the specified amount of material. In general correction for variation in procedure quantities follows the general rule:

Correction Factor = amount specified/amount used

Example

Need 5 ml of urine for sodium determination; received only 3.5 ml

Take 3 ml and dilute to 5 ml with H_2O

5/3 = 1.67

1.67 x Test result = corrected answer

Test result 40 mEq Na

40 x 1.67 = 66.8 mEq Na in original specimen

Converting Molarity to Percent Solution

MW x M = g/L Therefore M = (g/L)/MW

M = (g/L)/MW = (g/100ml x 10)/MW

M= ($\%^{w/v}$ x 10)/MW

Example

Convert a 30$\%^{w/v}$ solution of NaCl (MW = 58.5) to a molar solution.

M = (% x 10)/MW
M = (30 x 10)/58.5
M = 5.13

Therefore 30% solution of NaCl is a 5.13M solution.

Example

Convert a 6M solution of NaOH (MW = 40g) to a $\%^{w/v}$ solution.

Rearrange the formula from above: M x MW = (% x 10)

(M x MW)/10 = %
6 x 40/10 = 24%
A 6M solution of NaOH is equivalent to a 24$\%^{w/v}$ solution.

Converting A Normal Solution to Percent Solution

g/EqW x N = g/L
N = (g/L)¸ EqW
N = (% x 10)¸ EqW

Example

Convert a 3N H_2SO_4 (MW = 98, % assay = 96, SG = 1.84) solution to a $\%^{w/v}$ solution.

N = (% x 10)/EqW

% = N x EqW/10

% = 3x49/10

% = 14.7 $\%^{w/v}$

Note : That for this problem the information on % assay and SG was not needed since the original normality concentration accounted for this.

Conversion of mg/dl to mEq/L

The old unit for reporting out the amount of sodium, potassium or chloride ions in body fluids was mg/dl. The new standard for reporting is mEq/L (milliequivalents/L or "millinormals"/L). Depending on the procedure used to measure these ions in solution, it may be necessary to convert from the old to the new units. It may also be helpful to some physicians to have the values converted to the older, more familiar units of concentration.

mEq/L = mgEqW/L (milligram equivalent weight/L)

mEq/L = (mg/L) ¸ EqW = (mg/dl x 10)/EqW.

Example

Convert 300mg/dl of Cl^- to mEq/L of Cl^-.

mEq/L = (300 x 10)/35.5 = 3000/35.5

mEq/L = 84.5

Therefore 300mg/dl of Cl^- is equivalent to 84.5mEq/L.

Example

Convert 150 mEq/L NaCl to mg/dl of NaCl.

mEq/L = (mg/dl x 10)/EqW

mg/dl = (mEq/L x Eq)/10

mg/dl = (150 x 58)/10

870 mg/dl NaCl

Conversion of Complex Metric Units

Conversion among complex metric units of concentration is an extension of conversion among metric units of measure. The complex unit of concentration, for example mg/dl is broken into its component parts, mg and dl. Each of these is then converted to the corresponding desired unit of measure, for example Tg and ml.

Convert mg/dl to Tg/ml

Tg = 10^{-3}mg

mg x 10^{3} = Tg

ml = 10^{-2}dl

dl x 10^{2} = ml

mg/dl x $10^{3}/10^{2}$ = mg/dl x 10^{1} = Tg/ml

Example

Convert 21g/L to mg/Tl.

g x 10^{3} = mg

L x 10^{6} = Tl

21 x $10^{3}/10^{6}$ = 21 x 10^{-3}

21g/L = 2.1 x 10^{-2} = 0.021mg/Tl

Example

Convert 316 Tg/ml to mg/dl.

Tg x 10^{-3} = mg

ml x 10^{-2} = dl

316Tg/ml x $10^{-3}/10^{-2}$ = 316 x 10 $^{-1}$ = 31.6mg/dl

1. If you have 20 grams of a substance in 100ml then how many grams in 20 ml?

   Known ratio 20g/100ml

   Unknown ratio Xg/20ml

   20g/100ml = Xg/20ml

   (100ml) x (Xg) = 20gx20ml

   Xg = (20gx20ml)/100ml

   Xg = (400g/ml)/100ml

   X = 4g

   Therefore 20g/100ml = 4g/20ml

2. How would you set up a 1:1000 dilution from undiluted serum?

   Answer : Format : 1 volume undiluted serum + 999 volumes of diluent = 1000 volumes or 1:1000 dilution

Obviously, if one actually had to prepare this dilution, 1000 ml of a 1:1000 dilution would be too great a volume to be practical. It can be decreased while maintaining the 1:1000 dilution as follows:

1 + 999 = 1000 or 1:1000

X 0.1 ml

0.1 ml + 99.9 = ml of 1:1000 (100 ml is still too large a quantity for practicality)

or

1 + 999

X 0.01 ml

0.01 ml serum + 9.99 ml saline = 10 ml of 1:1000

This is a practical volume of serum with which to work. Plan to measure small quantities such as 0.01 ml of undiluted serum with a 0.01 ml serologic pipet to insure accuracy. **Note:** When the decimal point is moved to the left on the left side of the formula (0.1 to 0.01), it must also be moved to the left on the right side (99.0 to 9.99) in order to maintain the 1:1000 dilution. The decimal must be moved the same distance on both sides.

3. Prepare a 1:80 dilution from an undiluted serum.

   Answer : Format 1 + 79 = 1:80

   1 volume of undiluted serum + 79 volumes of saline = 80 volumes of a 1:80 dilution. Since 80 ml is too great a volume with which to work, use 0.1 ml undiluted serum + 7.9 ml

   saline = 8.0 ml of a 1:80 dilution.

4. Prepare a 1:8 dilution directly from 1:1 or undiluted serum.

   Answer : 1 volume of serum + 7 volumes of diluent = 8 volumes

   or

   1 volume of serum is diluted to 8 volumes

   or the serum is diluted 1:8

   The above example indicates how a 1:8 dilution is prepared. Now, how can a 1:8 dilution be made if there is only 0.1 ml of undiluted serum with which to make that dilution?

   First : multiply each side of the basic formula (1 and 7) by 0.1 ml serum.

   1+ 7 = 8

   X 0.1 ml

   0.1 ml serum + 0.7 ml diluent = 0.8 ml total volume

Thus, by adding 0.1 ml of undiluted serum to 0.7 ml saline, 0.8 ml of a 1:8 dilution has been prepared.

5. Prepare a dilution of 1:64 using 0.5 ml of serum.

   Answer : 1 volume of serum + 63 volumes of diluent = 64 volumes or 1:64

   Multiply each side of the formula (1 and 63) by 0.5 ml serum

   1 + 63 = 64 or 1:64

   X 0.5 ml

   0.5 ml serum + 31.5 ml diluent = 32 ml of a 1:64 dilution

   Note that one can also prepare a dilution of 1:512 on paper similarly:

   1 volume of serum + 511 volumes of saline = 512 volumes of 1:512 dilution

   OR

   if 0.1 ml of serum is used,

   0.1 ml serum + 51.1 ml saline = 51.2 ml of a 1:512 dilution.

6. Prepare a twofold dilution - any volume

   When a twofold dilution was described above in STEP 7, the final volume in each tube was 1.0 ml; however, any final volume can be prepared. If one used 0.2 ml of each serum dilution instead of 0.1 ml, the final volume in each tube would be 0.2 ml. This is how a twofold dilution with a final volume of 0.2 ml would be prepared:

   1 + 1 = 2 (1:2) twofold

   X 0.2 ml

   0.2 ml serum + 0.2 ml saline = 0.4 ml

   If each side is multiplied (1 and 1) by 0.2 ml, a result of 0.2 ml serum + 0.2 ml of diluent is obtained, which equals 0.4 ml. the result is 0.4 ml of a twofold dilution. Now if one wishes to set up a twofold dilution using 0.2 ml of each serum dilution, after mixing, 0.2 ml must be removed, leaving 0.2 ml of the dilution in the tube. In this way, all test tubes have a final volume of 0.2 ml. **Note :** That to set up a twofold dilution using 0.2 ml of serum in tube 2, a total volume of 0.4 ml undiluted serum is needed (0.2 ml for tube 1 and 0.2 ml for tube 2). The less serum required, the better, since remaining serum may be used for other tests, if requested.

7. Prepare a two fold dilution-minimum serum volume.

   To economize without adversely affecting accuracy, perform steps 1-7 by using 0.1 ml of serum in tubes 1 and 2, stipulating the format on paper, and diagrammatically prepare the twofold dilution to determine how one would actually perform this procedure using this quantity of serum.

   1 + 1 = 2 (1:2)

   X 0.1 ml

   0.1 ml of serum + 0.1 ml saline = 0.2 ml of a 1:2 dilution

8. You have a frozen 1:10 dilution of serum and you need to prepare from this a 1:35 dilution for an antibody test. Show how you would do this.

   Answer :

   Step 1. Find the working dilution to be prepared: $\frac{1:35}{1:10} = 1:3.5$

   A dilution of 1:35 is 3.5 times more dilute than the 1:10. Note that a 1:3.5 of a 1:10 gives a final dilution of a 1:35.
   Step 2. Write the format: 1:3.5 = 1 + 2.5 = 3.5 ml of a 1:3.5 dilution

   Step 3. Write the directions: Add 1 ml of a 1:10 dilution to 2.5 ml saline, to prepare a 1:3.5 dilution.

9. You have just begun your Senior Honor's Thesis. Your advisor asks you to make a series of stock solutions. She also explains that you may use a pH meter to adjust the pH. Calculate how much of the solid reagent you would add to make the following:

   a. 500 ml 1 M Tris, pH 8.0 (MW=121.4 g/mole)
   b. 1.0 L 5 M NaCl (MW= 58.44 g/mole)
   c. 10 ml 100 mg/ml ampicillin (MW=371.4 g/mole)
   d. 500 ml 1 M $MgCl_2$ (**Note**: $MgCl_2$ is sold by the chemical company as co-crystallized with H_2O. Thus, the MW of $MgCl_2$ $6H_2O$ is 203.30 g/mole)

   Answers :

   a $\frac{\text{1 mole of Tris}}{\text{L}} \times \frac{\text{121.4 g}}{\text{mole}} \times 0.5\ \text{L} = 60.7\ \text{g of Tris}$

Dissolve 60.7 g of Tris in about 300 ml of H_2O. Once dissolved adjust the pH to 8.0, and then make up to 500 ml mark with H_2O in a graduated cylinder.

b $$\frac{5 \text{ moles of NaCl}}{\text{L}} \times \frac{58.44 \text{ g}}{\text{mole}} \times 1 \text{ L} = 292.2 \text{ g of NaCl}$$

Dissolve 292.2 g of NaCl in about 0.7 L of H_2O. Once dissolved, dilute to 1 L with H_2O in a graduated cylinder.

c $$\frac{100 \text{ mg of ampicillin}}{\text{ml}} \times 10 \text{ ml} = 1 \text{ g of ampicillin}$$

Dissolve 1 g of ampicillin in about 5 ml of H_2O. Once dissolved, dilute to 10 ml with H_2O in a graduated cylinder.

d $$\frac{1 \text{ mole of } MgCl_2}{\text{L}} \times \frac{203.3 \text{ g}}{\text{mole}} \times 0.5 \text{ L} = 101.65 \text{ g of } MgCl_2$$

Dissolve 101.65 g of $MgCl_2$ in about 250 ml of H_2O. Once dissolved, dilute to 500 ml with H_2O in a graduated cylinder.

10. For your first experiment, you need to make a solution that is commonly called TE and stands for Tris-EDTA. It is comprised of 10 mM Tris, 1 mM EDTA, pH 8.0. Calculate and describe how you would make 100 ml of TE using the stock solutions (stock EDTA: 0.5 M EDTA pH 8.0) or that you have already made above in

Answer :

For dilutions, use the formula $C_1V_1=C_2V_2$.

This solution has 2 components:

Tris-dilute 1M to 10 mM $C_1V_1=C_2V_2$

$1000 \text{ mM} \times V_1 = 10 \text{ mM} \times 100 \text{ ml}$

$$V_1 = \frac{10 \text{ mM} \times 100 \text{ ml}}{1000 \text{ mM}}$$

$= 1$ ml of 1 M Tris

EDTA-dilute 0.5 M to 1 mM $C_1V_1=C_2V_2$

$500 \text{ mM} \times V_1 = 1 \text{ mM} \times 100 \text{ ml}$

$$V_2 = \frac{10 \text{ mM} \times 100 \text{ ml}}{500 \text{ mM}}$$

= 0.2 ml or 200 mL of 0.5 M EDTA

Using a 100 ml graduated cylinder, add 1 ml of 1M Tris and 200 mL of 0.5 M EDTA. Add water to dilute to a final volume of 100 ml.

11. Your advisor asks you to make two concentrated stock solutions for the whole laboratory to use. She explains that it is common to make a concentrated solution to dilute for use. Thus, a 10x solution refers to one that is 10-fold concentrated. Calculate how you would make 1.0 L of each of the following:

    a. 10x TBE (0.9 M Tris, 0.9 M Boric Acid, 20 mM EDTA)
    Tris MW=121.4 g/mole
    Boric Acid MW=61.84 g/mole
    EDTA MW=292.2 g/mole

    b. 50x TAE (2 M Tris, 2 M Acetic acid, 0.5 M EDTA)
    Tris MW=121.4 g/mole
    Concentrated Acetic acid (glacial) comes as a solution and is 17.4 M
    EDTA MW=292.2 g/mole

Answers :

a

Tris

$$\frac{0.9 \text{ moles of Tris}}{\text{L}} \times \frac{121.4 \text{ g}}{\text{mole}} \times 1.0 \text{ L} = 109.3 \text{ g of Tris}$$

Boric Acid

$$\frac{0.9 \text{ moles of Boric Acid}}{\text{L}} \times \frac{61.84 \text{ g}}{\text{mole}} \times 1.0 \text{ L} = 55.6 \text{ g of Boric Acid}$$

EDTA

$$\frac{20 \text{ mM of EDTA}}{\text{L}} \times \frac{292.2 \text{ g}}{\text{mole}} \times 1.0 \text{ L} \times \frac{1 \text{ mole}}{1000 \text{ mmoles}} = 5.84 \text{ g of EDTA}$$

Dissolve 109.3 g of Tris, 55.6 g of Boric Acid and 5.84 g of EDTA in about 0.5 L of H_2O. Once dissolved, dilute the stock solution to 1.0 L with H_2O in a graduated cylinder to form 10x TBE.

b

Tris

$$\frac{2 \text{ moles of Tris}}{L} \times \frac{121.4 \text{ g}}{\text{mole}} \times 1.0 \text{ L} = 242.8 \text{ g of Tris}$$

EDTA

$$\frac{0.5 \text{ moles of EDTA}}{L} \times \frac{292.2 \text{ g}}{\text{mole}} \times 1.0 \text{ L} = 146.1 \text{ g of EDTA}$$

Acetic Acid 17.4M to 2M $C_1V_1 = C_2V_2$

$17.4 \text{ M} \times V_1 = 2 \text{ M} \times 1.0 \text{ L}$

$$V_1 = \frac{2 \text{ M} \times 1.0 \text{ L}}{17.4 \text{ M}}$$

=0.115 L or 115 ml of 17.4 M Acetic

Acid

Dissolve 242.8 g of Tris, 146.1 g of EDTA and also dilute 115 ml of concentrated acetic acid in about 0.5 L of H_2O. Once the solids are dissolved, dilute the stock solution to 1.0 L with H_2O in a graduated cylinder to make 50x TAE.

12. Our polyacrylamide gel electrophoresis experiment requires 1.0 L of 0.5x TBE. Calculate and describe how you would make this solution using your stock solutions.

Answer :

TBE-dilute 10x to 0.5x $C_1V_1 = C_2V_2$

$(10x) \times V_1 = (0.5x) \times 1.0 \text{ L}$

$$V_1 = \frac{(0.5x) \times 1.0 \text{ L}}{(10x)}$$

= 0.05 L or 50 ml of 10x TBE

Dilute 50 ml of 10x TBE to 1.0 L with H_2O in a graduated cylinder to make 0.5x TBE. (Also, you could say 50 ml of 10x TBE plus 950 ml of H_2O.)

13. Your agarose gel electrophoresis experiment requires 500 ml of 1x TAE. Calculate and describe how you would make this solution using your stock solutions.

Answer :

TAE-dilute 50x to 1x $C_1V_1 = C_2V_2$

$(50x) \times V_1 = (1x) \times 500$ ml

$$V_1 = \frac{(1x) \times 500 \text{ ml}}{(50x)}$$

= 10 ml of 50x TAE

Dilute 10 ml of 50x TAE to 500 ml with H_2O in a graduated cylinder to make 1x TAE. (Also, you could say 10 ml of 50x TAE and 490 ml of H_2O).

14. Your experiment requires 100 ml 20 mM Tris, pH 8.0 which also contains 0.5 mM EDTA. Calculate and describe how you would make this solution using a stock of 1 M Tris, pH 8.0 and 0.5M EDTA, pH 8.0.

Answer :

20 mM Tris from 1M:

$C_1V_1 = C_2V_2$ 1 M=1000 mM

$(1000 \text{ mM}) \times V_1 = (20 \text{ mM}) \times 100$ ml

$$V_1 = \frac{(20 \text{ mM}) \times 100 \text{ mL}}{(1000\text{mM})}$$

= 2.0 ml of 1 M Tris

0.5 mM EDTA from 0.5 M

$C_1V_1 = C_2V_2$ 0.5 M=500 mM

$(500 \text{ mM}) \times V_1 = (0.5 \text{ mM}) \times 100$ ml

$$V_1 = \frac{(0.5 \text{ mM}) \times 100 \text{ mL}}{(500 \text{ mM})}$$

= 0.1 ml or 100 ml of 0.5 M EDTA

Add 2.0 ml of 1 M Tris and 0.1 ml of 0.5 M EDTA in a graduated cylinder and fill it up with H_2O to 100 ml.

15. Your thesis advisor asks you to do your first restriction digest. This is a molecular biology technique that uses enzymes to cut DNA at specific sites. You are excited to hear that there are no solutions to be made up because your advisor has purchased enzymes from the Promega company and the concentrated stocks of reaction buffer including 10x buffer and 100x BSA (bovine serum albumen) come with the enzyme. The enzyme you need is EcoRI and it has a stock concentration of 1 U/ml. Your DNA concentration is 5 mg/ml. You need to set up a 50 ml reaction which incorporates 1 mg DNA, 1x buffer, 1x BSA, and 0.5 U EcoRI. Calculate and describe how much of each stock solution, DNA, enzyme and water you would add to make up your restriction digest.

Answer :

1. 50 µl Reaction	Reaction Mixture
1 mg DNA	0.2 ml 5 mg/ml DNA
1x Buffer	5.0 ml 10x Buffer
1x BSA	0.5 ml 100x BSA
0.5 U EcoR I (U=units)	0.5 ml 1 U/ml EcoR I
	43.8 ml H_2O

	50.0 ml total volume

DNA

$$1\ \mu g \times \frac{1\ \mu l}{5\ \mu g} = 0.2\ \mu l \text{ of } 5\ \mu g/\mu l \text{ DNA}$$

Buffer

$$C_1V_1 = C_2V_2$$

$$(10x) \times V_1 = (1x) \times 50\ \mu l$$

$$V_1 = \frac{(1x) \times 50\ \mu l}{(10x)}$$

$$= 5\ \mu l \text{ of } 10x \text{ Buffer}$$

BSA

$C_1V_1 = C_2V_2$

$(100x) \times V_1 = (1x) \times 50$ ml

$$V_1 = \frac{(1x) \times 50 \text{ ml}}{(100x)}$$

= 0.5 ml of 100x BSA

EcoR I Enzyme

$$0.5 \text{ U} \times \frac{1 \ \mu l}{1 \text{ U}} = 0.5 \ \mu l \text{ of } 0.5 \text{ U}/\mu l \ \textit{EcoR} \text{ I}$$

16. After your reaction has been running for 1 hour, you need to check the progress of the restriction digest by agarose gel electrophoresis. You need to make 100 ml of a 1% solution of agarose in 1x TAE. You remember that you made up a 50x TAE stock solution about a month ago. Calculate and describe how you would make your agarose gel.

Answer :

100 ml of 1% Agarose, 1x TAE

Agarose 1% = 1 g/100 ml = 1 g of Agarose

TAE

$C_1V_1 = C_2V_2$

$(50x) \times V_1 = (1x) \times 100$ ml

$$V_1 = \frac{(1x) \times 100\text{ml}}{(50x)}$$

= 2 ml of 50x TAE

Combine 2.0 ml of 50x TAE and 1 g of Agarose. Once dissolved, dilute to 100 ml with H_2O.

17. Given a 1:3 dilution, prepare a 1:60 dilution.

Answer :

Step 1. Working dilution:

$$\frac{1:60}{1:3} = 1:20$$

A dilution of 1:60 is 20 times more dilute than the 1:3. Note that a 1:20 of a 1:3 gives a final dilution of 1:60.

Step 2. Format: 1:20 = 1 + 19.0 = 20 ml of a 1:20

Step 3. Directions

Add 1 ml of a 1:3 to 19 ml of saline, to prepare a 1:60 dilution or 0.1 ml of a 1:3 to 1:9 ml of saline = 2.0 ml final volume.

18. Given a 1:60 dilution, prepare a 1:100 dilution.

Answer :

Step 1. 1:100
------- = 1:1.66
1:60

Step 2. 1:1.66 = 1 ml + 0.66 ml of diluent = 1.66 ml

Step 3. Add 1 ml of the 1:60 dilution to 0.66 ml of saline to prepare a 1:100 dilution.

Answer the following, using the approach demonstrated previously

19. Given a 1:5 dilution, prepare a 1:50 on paper and show your mathematical calculations:

Step 1. Working dilution:

Step 2. Format:

Step 3. Directions:

20. Given a 1:100 dilution, prepare a 1:10,000 dilution on paper and show your mathematical calculations:

Step 1. Working dilution:

Step 2. Format:

Step 3. Directions

21. Given a 1:20 dilution, prepare a 1:65 dilution on paper and show your mathematical calculations:

Step 1. Working dilution:

Step 2. Format:

Step 3. Directions:

Answers to Problems 19-21

19. Step 1. Working dilution: 1:50
----- = 1:10
1.5

Step 2. Format: 1:10 = 1 + 9 ml = 10 ml of a 1:10 dilution

Step 3. Directions: Add 1 ml of the 1:5 dilution to 9 ml saline to prepare a 1:50 dilution.

20. Step 1. Working dilution: 1:10,000
---------- = 1:100
1:100

Step 2. Format: 1:100 = 1 ml + 99 diluent = 100 mil of 1:100 dilution

Step 3. Directions: Add 1 ml of the 1:100 to 99 ml of diluent to prepare a 1:10,000 dilution.

21. Step 1. Working dilution: 1:65
-------- = 1:3.25
1:20

Step 2. Format: 1:3.25 = 1 ml + 2.25 ml of diluent = 1:3.25

Step 3. Directions: Add 1 ml of the 1:20 dilution to 2.25 ml of saline to prepare a 1:65 dilution.

In order to demonstrate your problem-solving expertise, go back and see how you solve problem 10 if you had only 0.2 ml serum. Show your mathematical calculations.

Format : 1:10 dilution (1:50 is a 1:10 dilution of a 1:5)

1 + 9 = 10 ml of a 1:10

X 0.2 ml

0.2 ml + 1.8 ml = 2.0 ml of a 1:10

Directions: Add 0.2 ml of the 1:5 to 1:8 saline to prepare 2.0 ml of a 1:50.

For electrophoresis Ethidium bromide, $C_{21}H_{20}N_3Br$; Mw=394.3g/M; (store at NT in the dark);

	Conc.	5ml	10ml	50ml	5ml
EtBr	10mg/ml	50mg	100mg	500mg	0.05ml

□□□

Appendix-III
Preparation of Reagents

Preparation of reagents with appropriate and precise molarity / concentrations is utmost important in Molecular Biology Experiments. A few examples have been given below to prepare some of the chemical constituents:

Example: How much Tris you would add in one liter of water to prepare 90mM solution?

Tris:		
Molecular Wt.	=	121.14
1 M	=	121.14 g / 1000 ml
or 1 mM	–	121.14 mg / 1000 ml
or 90 mM	–	121.14 x 90 / 1000 ml
	=	121.14 x 90 x 2 (for 2 liter)
	=	21805.2 mg/2 liter
	=	21.8051 g/2 liter
	=	10.9026 g/l
Boric Acid		
Mol. Wt.	=	61.83
	=	61.83 x 90 x 2 (for 2 liter)
	=	11.1294 g/2 l
	=	5.5647 g/ l

EDTA

Mol. Wt.	=	372.24
1M	=	372.24 g/1000 ml
1mM	=	0.37224 mg/1000ml
2.5 mM	=	0.37224 x 2.5

DNA Migration in Agarose and Polyacrylamide Gels

Recommended Gel Percentages for Separation of Linear DNA

Agarose gel, %	**Range of separation, bp**	**Polyacrylamide gel, %**	**Range of separation, bp**
0.5	1,000-30,000	3.5	100-1,000
0.7	800-12,000	5.0	80-500
1.0	500-10,000	8.0	60-400
1.2	400-7,000	12.0	40-200
1.4	200-4,000	20.0	5-100
2.0	50-2,000		

Migration Rates of the Marker Dyes through Polyacrylamide Gels (1)

Polyacrylamide gel, %	**Bromophenol blue***	**Xylene cyanol FF***
Non-denaturing gels		
3.5	100 bp	460 bp
5.0	65 bp	260 bp
8.0	45 bp	160 bp
12.0	20 bp	70 bp
15.0	15 bp	60 bp
20.0	12 bp	45 bp
Denaturing gels		
5.0	35 bases	130 bases
6.0	29 bases	106 bases
8.0	26 bases	76 bases
10.0	12 bases	55 bases
20.0	8 bases	28 bases

DNA Size Migration with Sample Loading Dyes

Agarose concentration, %	Xylene cyanol FF	Bromophenol blue	Orange G
0.7-1.7	~4000bp	~300bp	~50bp
2.5-3.0	~800bp	~100bp	~30bp

Common Conversions of Nucleic Acids

Molar	**Conversions**
1 μg of 1000 bp DNA	= 1.52 pmol
1 μg of pUC18/19 DNA (2686 bp)	= 0.57 pmol
1 μg of pBR322 DNA (4361 bp)	= 0.35 pmol
1 μg of SV40 DNA (5243 bp)	= 0.29 pmol
1 μg of PhiX174 DNA (5386 bp)	= 0.28 pmol
1 μg of M13mp18/19 DNA (7250 bp)	= 0.21 pmol
1 μg of lambda phage DNA (48502 bp)	= 0.03 pmol
1 pmol of 1000 bp DNA	= 0.66 μg
1 pmol of pUC18/19 DNA (2686 bp)	= 1.77 μg
1 pmol of pBR322 DNA (4361 bp)	= 2.88 μg
1 pmol of SV40 DNA (5243 bp)	= 3.46 μg
1 pmol of PhiX174 DNA (5386 bp)	= 3.54 μg
1 pmol of M13mp18/19 DNA (7250 bp)	= 4.78 μg
1 pmol of lambda phage DNA (48502 bp)	= 32.01 μg
Spectrophotometric	**Conversions**
1 A260 of dsDNA = 50 μg/ml	= 0.15 mM (in nucleotides)
1 A260 of ssDNA = 33 μg/ml	= 0.1 mM (in nucleotides)
1 A260 of ssRNA = 40 μg/ml	= 0.12 mM (in nucleotides)
1 mM (in nucleotides) of dsDNA	= 6.7 A260 units
1 mM (in nucleotides) of ssDNA	= 10.0 A260 units
1 mM (in nucleotides) of ssRNA	= 8.3 A260 units
The average MW of a deoxyribonucleotide base	= 333 Daltons
The average MW of a ribonucleotide base	= 340 Daltons

Estimation of Ends (3' or 5') Concentration

Circular **DNA**

pmol ends = pmol DNA x number of cuts x 2

Linear **DNA**

pmol ends = pmol DNA x (number of cuts x 2 + 2)

1 μg of 1000 bp DNA	=	3.04 pmol ends
1 μg of linear pUC18/19 DNA	=	1.14 pmol ends
1 μg of linear pBR322 DNA	=	0.7 pmol ends
1 μg of linear SV40 DNA	=	0.58 pmol ends
1 μg of linear PhiX174 DNA	=	0.56 pmol ends
1 μg of linear M13mp18/19 DNA	=	0.42 pmol ends
1 μg of lambda phage DNA	=	0.06 pmol ends

DNA/Protein Conversions

1 kb of DNA = 333 amino acids approx. is 3.7 x 10E4 Da

10 kDa protein approx. is 270 bp DNA

30 kDa protein approx. is 810 bp DNA

50 kDa protein approx. is 1.32 kb DNA

100 kDa protein approx. is 2.7 kb DNA

OD_{260} Units of Nucleic Acid to Concentration

1 OD_{260} Unit = 50μg/ml for dsDNA

1 OD_{260} Unit = 40μg/ml ssRNA

1 OD_{260} Unit = 35μg/ml ssDNA

1 OD_{260} Unit = 20μg/ml for single-stranded oligo

Protein Conversions

Protein Molar Conversion

$$\frac{109\ \mu g}{g} \times \frac{mole}{1012\ pmol}$$

Micrograms Protein
= Protein Size (kDa) x Picomoles protein x

$$\text{Picomoles Protein} = \frac{\mu\text{g Protein}}{\text{Protein Size (kDa)}} \times \frac{\text{kg}}{109\ \mu\text{g}} \times \frac{1012\ \text{pmol}}{\text{mole}}$$

$$\text{Protein Size (kDa)} = \frac{\mu\text{g Protein}}{\text{Picomoles Protein}} \times \frac{\text{kg}}{109\ \mu\text{g}} \times \frac{1012\ \text{pmol}}{\text{mole}}$$

Formulas for DNA Molar Conversions

For dsDNA

To convert pmol to μg: pmol × N × 660 pg/pmol × 1 μg/106pg = μg

To convert μg to pmol μg × 106pg/1 μg × pmol/660pg × 1/N = pmol

For ssDNA

To convert pmol to μg pmol × N × 330 pg/pmol × 1μg/106pg = μg

To convert μg to pmol μg × 106pg/1 μg × pmol/330pg × 1/N = pmol

Protein Molar Conversions

100 pmol of 100 kDa protein	=	10μg
100 pmol of 50 kDa protein5	=	μg
100 pmol of 10 kDa protein1	=	μg
100 pmol of 1 kDa protein	=	100ng

Protein/DNA Conversions

1kbDNA	=	333 amino acids = 37 kDa protein
270b DNA	=	10 kDa protein
810 bDNA	=	30 kDa protein
1.35 kbDNA	=	50 kDa protein
2.7 kb DNA	=	100 kDa protein
average MW of an amino acid	=	110 daltons

Dalton (Da) is an alternate name for the atomic mass unit, and kilodalton (kDa) is 1,000 daltons. Thus a protein with a mass of 64 kDa has a molecular weight of 64,000 grams per mole.

Agarose Gel (%): Resolution of Linear DNA

Recommended % Agarose	Optimum Resolution for Linear DNA (Size of fragments in nucleotides;bp)
0.5	1,000-30,000
0.7	800-12,000
1.0	500-10,000
1.2	400-7,000
1.5	200-3,000
2.0	50-2,000

Polyacrylamide Gel (%): Resolution of Protein

Recommended % Acryl amide	Protein Size Range
8	40-200 kDa
10	21-100 kDa
12	10-40 kDa

Length/M.W. of Common Nucleic Acids

Nucleic Acid	Number of Nucleotides	Molecular Weight
lambda DNA	48,502 (dsDNA)	3.2 × 107
pBR322DNA	4,361 (dsDNA)	2.8 × 106
28S rRNA	4,800	1.6 × 106
23S rRNA (E.coli)	2,900	1.0 × 106
18S rRNA	1,900	6.5 × 105
16S rRNA E.coli)	1,500	5.1 × 105
5S rRNA (E.coli)	120	4.1 × 104
tRNA (E.coli)	75	2.5 × 104

*Molecular weights based on actual sequence.

Standards

1. Average MW of dsDNA base pair 600.
2. Average MW of ssDNA base 330.
3. Average MW of RNA base 340.

Concentrations of Acids and Bases

Substance	Formula	MW	Moles/liter*	Grams/liter	% by weight	Specific gravity	ml/liter to prepare 1 M solution
Acetic acid, glacial	CH_3COOH	60.05	17.4	1045	99.5	1.05	57.5
Acetic acid	CH_3COOH	60.05	6.27	376	36	1.045	159.5
Formic acid	$HCOOH$	46.02	23.4	1080	90	1.20	42.7
Hydrochloric acid	HCl	36.5	11.6	424	36	1.18	86.2
Nitric acid	HNO_3	63.02	15.99	1008	71	1.42	62.5
Perchloric acid	$HClO_4$	100.5	11.65	1172	70	1.67	85.8
Phosphoric acid	H_3PO_4	97.9	14.8	1445	85	1.70	67.7
Sulfuric acid	H_2SO_4	98.1	18.0	1766	96	1.84	55.6
Ammonium hydroxide	NH_4OH	35.0	14.8	251	28	0.898	67.6
Potassium hydroxide	KOH	56.1	13.5	757	50	1.52	74.1
Sodium hydroxide	$NaOH$	40.0	19.1	763	50	1.53	52.4

Common Conversions of Oligonucleotides

Molecular Weight MW = 333 x N MW - molecular weight,

Da Concentration of Oligonucleotides A_{260} - absorbance at 260nm

C (μM or pmol/μl) = A_{260} / (0.01 x N) N - number of bases

C (ng/ml) = (A_{260} x MW) / (0.01 x N)

Melting Temperature of Duplex DNA and Oligonucleotides

For Duplex Oligonucleotide shorter than 25 bp, "The Wallace Rule" (1)

Tm (in °C) = 2 (A+T) + 4 (C+G), where

(A+T) - the sum of the A and T residues in the oligonucleotide,

(C+G) - the sum of G and C residues in the oligonucleotide.

Presence of m5C in oligonucleotide increases melting temperature of duplex.

Presence of m4C or m6A decreases melting temperature (2).

For Duplex DNA, <100 bp long (3)

Tm (in °C) = 81.5°C+16.6(log10[Na+])+0.41(%[G+C])-675/n-1.0m, where

n - number of bases in the oligonucleotide

m - the percentage of base-pair mismatches

Conversion Formula

C = A / e x 10E3, where

C mM concentration of compounds

A -observed absorbance at lambdamax (nm)

e - molar absorption coefficient (M-1 x cm-1)

Commonly used Media, Stock Solutions and Buffers

Growth Media	Per liter	Final 1X concentration:
LB Medium (Adjust pH to 7.0)		
Tryptone	10g	1.0% (w/v)
Yeast extract	5g	0.5% (w/v)
NaCl	10g	1.0% (w/v)
H_2O	to 1 liter	

Contd...

Low Salt LB Medium

Tryptone	10g	1.0% (w/v)
Yeast extract	5g	0.5% (w/v)
NaCl	5g	0.5% (w/v)
H_2O	to 1 liter	
Adjust pH to 7.0		

Terrific Broth Medium

Tryptone	12g	1.2% (w/v)
Yeast extract	24g	2.4% (w/v)
Glycerol	4 ml	0.4% (w/v)
Add H_2O	to 900 ml	

Autoclave, cool to 60°C or less before adding 100ml of filter sterilized 10XTB phosphate (0.17M KH_2 PO_4, 0.72M K_2HPO_4).

SOB Medium

Tryptone	20g	2.0% (w/v)
Yeast extract	5g	0.5% (w/v)
NaCl	0.5g	0.05% (w/v)
250 mM KCl	10 ml	2.5 mM
H_2O	to 900 ml	

Adjust pH to 7.0 and add H_2O to 990 ml.
Autoclave, cool to room temperature and add 10ml of sterile solution of 1M $MgCl$ before use.

M9 Minimal Medium

5X M9 salts	200 ml	10 mM
Sterile H_2O	to 1 liter	
1M $MgSO_4$	2 ml	1 mM
20% glucose	20 ml	2.0% (w/v)
1M $CaCl_2$	0.1 ml	0.1 mM

5X M9 Salts

$Na_2HPO_4 \cdot 7H_2O$	64g	47.8 mM
KH_2PO_4	15g	22mM
NaCl	2.5g	8.6 mM
NH_4Cl	5g	18.7 mM

Additives:

Antibiotics (if required)

Ampicillin	to 50 µg/ml
Chloramphenicol	to 20 µg/ml
Kanamycin	to 30 µg/ml
Tetracycline	to 12 µg/ml

Contd...

Additives:

Media containing agar or agarose		
Agar (for plates)	15g	1.5% (w/v)
Agar(for top agar)	7g	0.7% (w/v)
Agarose(for plates)	15g	1.5% (w/v)
Agarose (for top agarose)	7g	0.7% (w/v)

Stock Solutions

10 M Ammonium Acetate

Ammonium acetate	385.4g
H_2O	to 500 ml

100X Denhardt Solution

Ficoll 400	10g
Polyvinylpyrrolidone	10g
Bovine serum albumin	10g
H_2O	to 500 ml

Filter sterilize and store at -20°C in 25ml aliquots

10 mg/ml Ethidium Bromide

Ethidium bromide	0.2g
H_2O	to 20 ml

Mix well and store at 4°C in dark.

1 M $CaCl_2$

$CaCl_2x2H_2O$	147g
H_2O	to 1 liter

0.5M EDTA (ethylenediamine tetraacetic acid) (pH 8.0)

$Na_2EDTAx2H_2O$	186.1g
H_2O	to 700 ml

Adjust pH to 8.0 with 10M NaOH (~50ml)

H_2O	to 1 liter

1 M Dithiothreitol (DTT)

DTT	15.45g
H_2O	to 100 ml

Store at -20°C

Caution : Ethidium bromide is a mutagen and must be handled carefully.

1 M KCl

KCl	74.6g
H_2O	to 1 liter

1 M $MgSO_4$

$MgSO_4x7H_2O$	24.6g
H_2O	to 100 ml

10 M NaOH

NaOH	400g
H_2O	to 1 liter

1 M $MgCl_2$

$MgCl_2x6H_2O$	20.3g
H_2O	to 100 ml

5 M NaCl

NaCl	292g
H_2O	to 1 liter

1 M Tris-HCl [tris(hydroxymethyl)aminomethane]

Tris base	121g
H_2O	to 800 ml

Adjust to desired pH with concentrated HCl. Mix and add H_2O to 1 liter

3M Sodium Acetate (pH 5.2 and 7.0)

Sodiumacetate.$3H_2O$	408.1g
H_2O	to 800 ml

Adjust the pH to 5.2 with glacial acetic acid or adjust the pH to 7.0 with dilute acetic acid.

H_2O	to 1 liter

Buffers

Buffers	Per liter	Final 1X concentration:
10X Stock Phosphate-buffered Saline (PBS		
NaCl	80g	137 mM
KCl	2g	2.7 mM
Na_2HPO_4	14.4g	100 mM
KH_2PO_4	2.4g	2 mM
H_2O	to 800 ml	
Adjust pH 7.4 with HCl		
H_2O	to 1 liter	
20X SSC		
NaCl	175.3g	150 mM
$Na_3citratexH_2O$	88.2g	15 mM
H_2O	to 800 ml	
Adjust pH to 7.0 with 1 M HCl		
H_2O	to 1 liter	
20X SSPE		
NaCl	175.3g	150 mM
$NaH_2PO_4xH_2O$	27.6g	10 mM
Na_2EDTA	7.4g	10 mM
H_2O	to 800 ml	
Adjust pH to 7.4 with 10 M NaOH		
H_2O	to 1 liter	
5X Tris-glycine Electrophoresis Buffer		
Tris base	15.1g	25 mM
Glycine	72.0g	192 mM
H_2O	to 1 liter	
The pH of diluted solution is 8.3		
5X Tris-glycine-SDS Electrophoresis Buffer		
Tris base	15.1g	25 mM
Glycine	72.0g	192 mM
SDS	5.0g	0.1% (w/v)
H_2O	to 1 liter	
The pH of diluted solution is 8.3		
5X Tris-tricine-SDS Electrophoresis Buffer		
Tris base	121.1g	0.1M
Tricine	179.2g	0.1M
SDS	10.0g	0.1% (w/v)
H_2O	to 1 liter	
The pH of diluted solution is 8.3.		

Contd...

50X TAE (Tris-acetate-EDTA) Electrophoresis Buffer

Tris base	242g	40 mM
Glacial acetic acid	57.1 ml	20 mM
0.5M EDTA (pH 8.0)	100 ml	1 mM
H_2O	to 1 liter	

The pH of diluted solution is ~8.5

10X TBE (Tris-borate-EDTA) Electrophoresis Buffer

Tris base	108g	90 mM
Boric acid	55g	90 mM
0.5M EDTA(pH 8.0)	40 ml	1 mM
H_2O	to 1 liter	

10X TPE (Tris-phosphate-EDTA) Electrophoresis Buffer

Tris base	108g	90 mM
Phosphoric acid (85%)	15.5 ml	23 mM
0.5M EDTA(pH 8.0)	40 ml	1 mM
H_2O	to 1 liter	

TE (Tris-EDTA) Buffer, pH 7.4, 7.6 or 8.0

1 M Tris, pH 7.4, 7.6, 8.0	10 ml	10 mM
0.5M EDTA(pH 8.0)	2 ml	1 mM
H_2O	to 1 liter	

TE

1x; pH7.4; 7.6; 8.0; (store at 4°C).

	Conc.	Stock	5ml	10ml	50ml	400ml	5ml
Tris Cl	10mM	1M	50µl	100µl	500µl	4ml	0.05ml
EDTA	1mM	0.5M	10µl	20µl	100µl	800µl	0.01ml
H_2O		mQ	4.94ml	9.88ml	49.4ml	395ml	4.94ml

10x

	Conc.	Stock	1ml	5ml	5ml
Tris Cl	0.1M	1M	0.1ml	0.5ml	0.5ml
EDTA	10mM	0.5M	20µl	0.1ml	0.1ml
H_2O	mQ		880µl	4.4ml	4.4ml

Genome Comparisons

Virus	Length, nt	Approx. MW, Da
Bacteriophage PhiX174	5,380	3.5 x 10E6
Bacteriophage Lambda	48,502	3.1 x 10E7
Human Immunodeficiency Virus 1	9,181	3.1 x 10E6
Rous Sarcoma Virus	9,392	3.2 x 10E6
SARS Coronavirus	29,751	1.0 x 10E7
Simian Virus 40 (SV 40)	5,224	3.4 x 10E6
Vaccinia Virus	191,737	1.2 x 10E8
Variolla (Small pox) Virus	185,578	1.2 x 10E8
Archaebacteria	**Length of DNA, bp**	**Approx. MW, Da**
Halobacterium sp. NRC-1	2.01 x 10E6	1.31 x 10E9
Methanosarcina acetivorans C2A	5.75 x 10E6	3.74 x 10E9
Methanococcus jannaschii DSM2661	1.66 x 10E6	1.08 x 10E9
Methanosarcina mazei Go1	4.10 x 10E6	2.66 x 10E9
Methanobacterium thermoautotrophicum delta H	1.75 x 10E6	1.14 x 10E9
Pyrococcus horikoshii OT3	1.74 x 10E6	1.13 x 10E9
Pyrococcus abysii GE5	1.76 x 10E6	1.14 x 10E9
Pyrococcus furiosus DSM3638	1.91 x 10E6	1.24 x 10E9
Sulfolobus solfataricus P2	2.99 x 10E6	1.94 x 10E9
Thermoplasma acidophilum	1.56 x 10E6	1.01 x 10E9
Bacteria	**Length of DNA, bp**	**Approx. MW, Da**
Agrobacterium tumefaciens C58-Cereon	4.91 x 10E6	3.19 x 10E9
Aquifex aeolicus VF5	1.55 x 10E6	1.00 x 10E9
Bacillus subtilis 168	4.20 x 10E6	2.73 x 10E9
Bacillus halodurans	4.20 x 10E6	2.73 x 10E9
Bordetella pertussis Tohama I NCTC-13251	4.07 x 10E6	2.65 x 10E9
Chlamydophila pneumoniae CWL029	1.23 x 10E6	8.00 x 10E8

Deinococcus radiodurans R1	3.28 x 10E6	2.13 x 10E9
Escherichia coli K12	4.64 x 10E6	3.02 x 10E9
Escherichia coli 0157:H7 EDL933	4.10 x 10E6	2.66 x 10E9
Haemophilus influenzae KW20	1.83 x 10E6	1.19 x 10E9
Helicobacter pylori 26695	1.66 x 10E6	1.08 x 10E9
Lactococcus lactis IL1403	2.36 x 10E6	1.53 x 10E9
Mycobacterium leprae TN	3.27 x 10E6	2.12 x 10E9
Mycobacterium tuberculosis H37Rv	4.45 x 10E6	2.89 x 10E9
Mycoplasma genitalium G037	5.80 x 10E5	3.77 x 10E8
Mycoplasma pneumoniae M129	8.16 x 10E5	5.30 x 10E8
Neisseria meningitidis Z2491	2.18 x 10E6	1.42 x 10E9
Salmonella typhimurium LT2 SGSC1412	4.86 x 10E6	3.16 x 10E9
Staphylococcus aureus MW2	2.82 x 10E6	1.83 x 10E9
Streptomyces coelicolor A3(2)	8.67 x 10E6	5.64 x 10E9
Streptococcus pneumoniae R6	2.04 x 10E6	1.33 x 10E9
Streptococcus pyogenes SF370 (M1)	1.85 x 10E6	1.20 x 10E9
Pseudomonas aeruginosa	6.26 x 10E6	4.07 x 10E9
Vibrio cholerae N16961	4.00 x 10E6	2.60 x 10E9
Eukaryotes	**Length of DNA, bp**	**Approx. MW, Da**
Anopheles gambiae PEST (malaria mosquito)	2.78 x 10E8	1.80 x 10E11
Arabidopsis thaliana (flowering plant)	1.15 x 10E8	7.47 x 10E10
Caenorhabditis briggsae (soil-dwelling nematode)	1.04 x 10E8	6.76 x 10E10
Caenorhabditis elegans (round worm)	1.21 x 10E7	7.86 x 10E9
Ciona intestinalis (ascidian tadpole)	1.16 x 10E8	7.54 x 10E10
Drosophila melanogaster (fruit fly)	1.37 x 10E8	8.90 x 10E10
Guillardia theta (chromophyte algae)	5.51 x 10E5	3.58 x 10E8
Homo sapiens (human)	3,15 x 10E9	2.07 x 10E12
Mus musculus (mouse)	~3,00 x 10E9	~1.95 x 10E12

Neurospora crassa OR74A (filamentous fungus)	4.30 x 10E7	2.80 x 10E10
Saccharomyces cerevisiae S288C (budding yeast)	1.21 x 10E7	7.86 x 10E9
Schizosaccharomyces pombe (fission yeast)	1.40 x 10E7	9.10 x 10E9
Plasmodium falciparum 3D7 (human malaria parasite)	2.29 x 10E7	1.49 x 10E10
Oryza sativa japonica (rice)	4.20 x 10E8	2.73 x 10E11

Common Stock Solutions

10 M Ammonium Acetate

To prepare a 10 M solution in 100 ml, dissolve 77 g of ammonium acetate in 70 ml of H_2O at room temperature. To prepare a 5 M solution in 100 ml, dissolve 38.5 g in 70 ml of H_2O. Adjust the volume to 100 ml with H_2O. Sterilize the solution by passing it through a 0.22 μm filter. Store the solution in tightly sealed bottles at 4°C or at room temperature. Ammonium acetate decomposes in hot H_2O and solutions containing it should not be autoclaved.

Ampicillin

Prepare a stock of 100 mg/ml in water. Sterilize by filtration. Store at -20°C but avoid repeated freeze/thaw cycles. Use at a final concentration of 100 μg/ml.

Cresol Red Loading Dye - 2.5X - for PCR Reactions

1M sucrose, 0.02% cresol. Prepare 1% cresol red in water (0.5 g/50 ml). To prepare loading dye, dissolve 17 g sucrose in a total volume of 19 ml of water. Add 1 ml 1% cresol red.

EB Buffer (Qiagen Recipe)

10 mM Tris-Cl pH 8.3

EDTA Stock

To prepare 1 liter, 0.5M EDTA pH 8.0: Add 186.1 g of disodium EDTA-2H 2O to 800 ml of H_2O. Stir vigorously on a magnetic stirrer. Adjust the pH to 8.0 with NaOH (approx. 20 g of NaOH pellets). Dispense into aliquots and sterilize by autoclaving. The disodium salt of EDTA will not go into solution until the pH of the solution is adjusted to approx. 8.0 by the addition of NaOH. For tetrasodium EDTA, use 226.1 g of EDTA and adjust pH with HCl.

6x gel loading buffer
0.25% Bromophenol blue
0.25%Xylene cyanol FF
15% Ficoll Type 4000
120 mM EDTA

IPTG

IPTG is isopropylthio-b-D-galactoside. Make a 20% (w/v, 0.8 M) solution of IPTG by dissolving 2 g of IPTG in 8 ml of distilled H_2O. Adjust the volume of the solution to 10 ml with H_2O and sterilize by passing it through a 0.22µm disposable filter. Dispense the solution into 1 ml aliquots and store them at -20°C.

LB Medium

To make 1 liter, use 10 g tryptone, 5 g yeast extract, 10 g NaCl. Adjust pH to 7.0. Sterilize by autoclaving.

LB Agar

Dispense 15 g per liter of agar directly into final vessel. Prepare LB medium as above and add to agar. NOTE: Agar will not go into solution until it is autoclaved (or boiled). If adding antibiotics, autoclave medium first and allow to cool until warm to the touch, then add the antibiotic. Dispense about 30 ml per plate. Allow plates to dry either at 37°C overnight or 20 minutes in a laminar flow hood (lids removed). Store in original Petri plate bags, inverted, at 4°C for up to 2 weeks.

NaCl

To prepare 1 liter of a 5 M solution: Dissolve 292 g of NaCl in 800 ml of H_2O. Adjust the volume to 1 liter with H_2O. Dispense into aliquots and sterilize by autoclaving. Store the NaCl solution at room temperature.

NaOH

The preparation of 10 N NaOH involves a highly exothermic reaction, which can cause breakage of glass containers. Prepare this solution with extreme care in plastic beakers. To 800 ml of H_2O, slowly add 400g of NaOH pellets, stirring continuously. As an added precaution, place the beaker on ice. When the pellets have dissolved completely, adjust the volume to 1 liter with H_2O. Store the solution in a plastic container at room temperature. Sterilization is not necessary.

20X SB (electrophoresis buffer)

(Buffer diluted to 1X should be 10 mM Sodium hydroxide and pH 8.5) for 1 liter, weigh out 8 g NaOH and ~40 g boric acid - add water, dissolve and add additional boric acid until pH = 8.0; bring final volume to 1 liter.

SDS stock

10% or 20% (w/v) SDS. Also called sodium lauryl (or dodecyl) sulfate. To prepare a 20% (w/v) solution, dissolve 200 g of electrophoresis-grade SDS in 900 ml of H_2O. Heat to 68°C and stir with a magnetic stirrer to assist dissolution. If necessary, adjust the pH to 7.2 by adding a few drops of concentrated HCl. Adjust the volume to 1 liter with H_2O. Store at room temperature. Sterilization is not necessary. Do not autoclave.

Use a mask when weighing this out.

20x SSC

0.3M Na(3) citrate

3M NaCl

SOB Medium: per liter

Bacto-tryptone 20 g

Yeast extract 5 g

NaCl 0.584 g

KCl 0.186 g Mix components and adjust pH to 7.0 with NaOH and autoclave.

2 M Mg ++ stock

MgCl 2-6H_2O 20.33 g

MgSO 4 -7H_2O 24.65 g

Distilled water to 100 ml. Autoclave or filter sterilize.

2 M Glucose

Glucose 36.04 g

Distilled water to 100 ml. Filter sterilize.

For SOB Medium + magnesium: Add 1 ml of 2 M Mg ++ stock to 99 ml SOB Medium.

For SOC Medium: Add 1 ml of 2 M Mg ++ stock and 1 ml of 2 M Glucose to 98 ml of SOB Medium.

3M Sodium Acetate - pH 5.2

To prepare a 3 M solution: Dissolve 408.3 g of sodium acetate-3H_2O in 800 ml of H_2O. Adjust the pH to 5.2 with glacial acetic acid. Adjust the volume to 1 liter with H_2O. Dispense into aliquots and sterilize by autoclaving.

Southern Solutions

Depurination solution (for Southern blotting) 250mM HCl

Denaturation solution (for Southern blotting) 1.5M NaCl and 0.5M NaOH

Neutralization solution (for Southern blotting)

1.5M NaCl

0.5M Tris-HCl, pH adjusted to 7.5

50x TAE Prepare a 50x stock solution in 1 liter of H_2O

242 g of Tris base

57.1 ml of glacial acetic acid

100 ml of 0.5 M EDTA (pH 8.0)

The 1x working solution is 40 mM Tris-acetate/1 mM EDTA.

5X (or 10X) TBE

Prepare a 5x stock solution in 1 liter of H_2O:

54 g of Tris base

27.5 g of boric acid

20 ml of 0.5 M EDTA (pH 8.0)

The pH of the concentrated stock buffer should be approx. 8.3. Some investigators prefer to use more concentrated stock solutions of TBE (10x as opposed to 5x). However, 5x stock solution is more stable because the solutes do not precipitate during storage. Passing the 5x or 10x buffer stocks through a 0.22ìm filter can prevent or delay formation of precipitates.

TE buffer

10 mM Tris-Cl (pH, usually 7.6 or 8.0)

1 mM EDTA (pH 8.0)

Use concentrated stock solutions to prepare. If sterile water and sterile stocks are used, there is no need to autoclave. Otherwise, sterilize solutions by autoclaving for 20 minutes. Store the buffer at room temperature.

1 M Tris-Cl used at various pHs

Using Tris base : To make 1 liter, dissolve 121 g Tris Base in 800 ml of water. Adjust pH to the desired value by adding approximately the following:

pH = 7.4 about 70 ml of concentrated HCl

pH = 7.6 about 60 ml of concentrated HCl

pH = 8.0 about 42 ml of concentrated HCl

Make sure solution is at room temperature before making final pH adjustments. Bring final volume to 1 liter. Sterilize by autoclaving.

Using Trizma tables: an alternate procedure for preparing Tris solutions is to combine the proper amount of Tris Base and Tris Hydrochloride to achieve the desired value using Sigma's Tris tables.

WB (10% redistilled glycerol, 90% distilled water, v/v)

In a 1-liter graduated cylinder, add 100 ml of glycerol and 900 ml of distilled water. Cover with parafilm and mix thoroughly. Sterilized by autoclaving, and chill to 4°C.

Western Blotting Solutions

1X Transfer buffer 1: 25 mM Tris, 192 mM Glycine, pH 8.3

Mix 3.03 g Tris, 14.4 g glycine; add dd water to 1 liter do not adjust pH.

1X Transfer buffer 2: 25 mM Tris, 192 mM Glycine, pH 8.3, 20 % methanol

Mix 3.03 g Tris, 14.4 g glycine; add 200 ml methanol; add dd water to 1 liter do not adjust pH. (NOTE: methanol is not needed for PVDF membranes)

10X Western Buffer: 200 mM Tris pH = 7.5; 1.5 M NaCl

To prepare 1X Western Buffer, dilute 10X buffer to 1X, adding Tween-20 to 0.1%. Remove 50 ml and set aside for the last two washes. To the remainder, add I-Block to 0.2%, heating gently with constant stirring until dissolved. Bring to room temperature before using.

X-gal 5-bromo-4-chloro-3-indolyl-b-D-galactoside (same recipe for X-phosphate)

Make a 2% (w/v) stock solution by dissolving X-gal in dimethylformamide at a concentration of 20 mg/ml solution. Use a glass or polypropylene tube. Wrap the tube containing the solution in aluminum foil to prevent damage by light and store at -20 ° C. It is not necessary to sterilize X-gal solutions.

□□□

References

Acquaah, G. (1992) Practical protein electrophoresis for genetic research. Dioscorides Press, Portland.

Aguadé, M. W. Meyers, A. D. Long, and C. H. Langley. (1994) Reduced DNA sequence polymorphism in the su(s) and su(wa) regions of the Drosophilia melanogaster as revealed by SSCP and stratified DNA sequencing.

Akey DT, Akey JM, Zhang K, Jin L. Assaying DNA methylation based on high-throughput melting curve approaches. *Genomics* 2002, 80:376-384

Alberts, B., Johnson, A., Lewis, J. Raff, M., Roberts, K., Walter, P. (2008). Molecular Biology of the Cell, pp 538-539, 5th ed. Garland Science, Taylor & Francis Group, NY.

Alwine JC, Kemp DJ, Stark GR (1977). Method for detection of specific RNAs in agarose gels by transfer to diazobenzyloxymethyl-paper and hybridization with DNA probes. Proc. Natl. Acad. Sci. U. S. A. 74 (12): 5350–4.

Ambroz K., (2006). Improving quantification accuracy for western blots Image Analysis 9/2006.

Anderson, M. D., T. K. Prasad and C. R. Steward (1995). Changes in isozyme profiles of catalase, peroxidase, and glutathione reductase during acclimation to chilling in mesocotyls of maize seedlings. *Plant Physiol.* 109: 1247–1257.

Andrews, A. T. (1993) Electrophoresis, theory, techniques and biochemical and clinical applications. Claredon Press. Oxford.

Ausubel, *et al.*, (eds). (1993) Current Protocols in Molecular Biology. Green Publishing and Wiley Interscience, New York.

B. Lyon (1999). DNA markers and the molecular breeding of cotton. The Australian Cottongrower 20(5) 80-83.

Baldwin, D., Crane, V., Rice, D. (1999) A comparison of gel-based, nylon filter and microarray techniques to detect differential RNA expression in plants. *Current Opinion in Plant Biol.* 2: 96-103.

Barrett, S. C. H. and B. C. Husband (1990) The genetics of plant migration and colonization. In: A. H. D. Brown, M. T. Clegg, A. L. Kahler, B. S. Weir (eds.), Plant population genetics, breeding, and geneticresources. Sinauer Associates, Inc, Sunderland, 254–278.

Barrett, S. C. H. and J. S. Shore (1989) Isozyme variation in colonizing plants. In: D. E. Soltis, P. S. Soltis [eds.], Isozymes in plant biology. Dioscorides Press, Portland, Oregon 106–125.

Behl, R. K., Chhabra, A. K. and Satyavir (2003) Frontiers in DNA Applications, CCSHAU, Hisar.

Belgrader, P., Smith, J. K., Weedn, V. W., and Northrup, M. A. (1998b). Rapid PCR for identity testing using a battery-powered miniature thermal cycler. *J Forensic Sci* 43, 315-9.

Bhattramakki Dinakar, A. K. Chhabra and Gary E. Hart (1998) Optimization of Nebulization conditions for shearing sorghum genomic DNA. *International Sorghum and Millet Newsletter* ICRISAT 39:92-94.

Blouin, M. S., Parsons, M., Lacaille, V. & Lotz, S. (1996). "Use of microsatellite loci to classify individuals by relatedness". *Molecular Ecology* 5: 393–401.

Bogdanov *et al.* (1996). A Phospholipid Acts as a Chaperone in Assembly of a Membrane Transport Protein. *Journal of Biological Chemistry* 271: 11615-11618.

Bor (2006), Northern Blot analysis of mRNA from mammalian polyribosomes, *Nature* Protocols, 10. 1038/nprot. 2006. 216

Botstein, D., White, R. L., Skolnick, M., and Davis, R. W. (1980). Construction of a genetic linkage map in man using restriction fragment length polymorphisms. *Am. J. Human Genet.*, 32, 314-331.

Botstein, D; White; R. L; Skolnick, M; Davis, R. W. (1980) *Am. J. Hum. Genet.* 32:314-337.

Brock TD (1994) Life at high temperatures. Bacteriology 303: *Procaryotic Microbiology*.

Brown SM & Kresovich S (1996). Molecular characterization for plant genetic resources conservation, pp. 85-93. *In: Paterson H (ed.), Genome Mapping of Plants. Academic Press, San Diego*

Browne, D. L. and Litt, M. (1992) Characterization of (CA)n microsatellites with degenerate sequencing primers. *Nucleic Acids Res.* 20, 141.

Bruford, M. W. and R. K. Wayne. (1993) Microsatellites and their application to population genetic studies. *Current Opinion in Genetics and Development.*

Bunyan DJ, Eccles DM, Sillibourne J, Wilkins E, Thomas NS, Shea-Simonds J, Duncan PJ, Curtis CE, Robinson DO, Harvey JF, Cross NC (2004). "Dosage analysis of cancer predisposition genes by Multiplex Ligation-Dependent Probe Amplification". *Br. J. Cancer* 916: 1155–1159.

Burr, B., Burr, F. A. (1991) Recombinant inbred lines for molecular mapping in maize. *Theor. Appl. Genet.* 85:55-60.

Buxbaum *et al.* (2002). "Cationic electrophoresis and electrotransfer of membrane glycoproteins". *Analytical Biochemistry* 314: 70-76.

Cardy, B. J., Stuber, C. W., Wendel, J. F. and Goodman, M. M. (1983). Techniques for starch gel electrophoresis of isozymes from maize (*Zea mays* L.). Institute of Statistics Mimeograph Series No. 1317, North Carolina State University, Raleigh, NC

Catherine J. Howarth and Rattan S. Yadav (2002) Successful marker assisted selection for drought tolerance and disease resistance in pearl millet, IGER Innovations. 18-21.

Chhabra, A. K. (1995). Molecular characterization of cytoplasmic nuclear male sterility sources and tall/dwarf near-isogenic lines in pearl millet, PhD thesis, CCSHAU, Hisar.

Chhabra, A. K. (2001). Role of Molecular Markers in Varietal Identification, *Manual on ICAR Short Course on Seed Technology.*

Chhabra, A. K., Bob Klein, Dinakar Bhattramakki and Gary E. Hart (1998). Mapping DNA simple-sequence-repeat loci in Sorghum bicolor (L.) Moench. Impact of Molecular Biology on Crop Production, August 30-September 1, 1998, Minneapolis, Minneasota (Abstract p17).

Chhabra, A. K., R. K. Behl and N. Albassam (2001) Role of isozymes in pearl millet improvement Landbauforschung Völkenrode 4 51:165-173.

Chhabra, A. K. and Somveer (2002). Molecular Markers: Concept and approaches for crop improvement. In: Manual on Advances in Vegetable Production *ICAR-NATP Short Course on Vegetable Improvement.*

Chhabra, A. K., I. S. Khariwal., K. N. Rai, C. T. Hash and A. K. Murthy (1997). Influence of cytoplasmic-nuclear male sterility systems on microsporogenesis in pearl millet (*Pennisetum glaucum* (L.) R. Br.) *Euphytica* 98: 1-10.

Chhabra, A. K., John Mullet, P. Klein and Gary E. Hart (1999). DNA markers for fertility restoration gene (Rf1) in sorghum. APS / CPS Joint Annual Meeting, August 7-11, 1999, Palais des Congrès de Montréal, Montréal, Quebec, Canada.

Chhabra, A. K., K. N. Rai, I. S. Khariwal., S. Sivaramakrishan and C. T. Hash (1998) Mitochondrial DNA- RFLP analysis distinguishes new CMS sources in pearl millet (*Pennisetum glaucum (L. R. Br.). Journal of Plant Biochemistry and Biotechnology* 7(2): 85-92.

Chhabra, A. K., K. N. Rai, I. S. Khariwal., S. Sivaramakrishan and C. T. Hash (2003) Effect of seed and DNA source on RFLP pattern of mtDNA in pearl millet. *International Sorghum and Millet Newsletter* 42: 16-21.

Chowdhury. V. K., Jain, R. K., and Patel, R. K. (2005) Applications of genomics, molecular markers and transformation tools for crop improvement, Department of Biotechnology and Molecular Biology, CCSHAU, Hisar.

Chowdhury. V. K., Santosh Dhillon and P. Kharb (2005) Manual on Introductory Plant Biotechnology, Deptt. of BMB, CCSHAU, Hisar

Clarke, B. C., L. B. Moran, and R. Appels. (1989) DNA analyses in wheat breeding. *Genome* 32:334-339.

Clinton, M. and R. K. Scougall. (1995) Detection and capture of 35S-Labeled gas released from reaction tubes during differential display PCR. *BioTechniques* 19(5):798-799.

Condit, R. and Hubbell, S. P. (1991). Abundance and DNA sequence of two-base repeat regions in tropical tree genomes. *Genome* 34, 66-71.

Crivello, J. F. (1997). In The PCR Technique: Quantitative PCR (J. W. Larrick, ed.), pp. 170-172. Eaton Publishing, Mountain View, CA.

D. B. Goldstein, A. R. Linares, L. L. Cavalli-Sforza, and M. W. Feldman (1995) "An Evaluation of Genetic Distances for Use With Microsatellite Loci". *Genetics* 139: 463–471.

D. Ishikawa and T. Taki (1998) Micro-scale analysis of lipids by far-eastern blot (TLC blot), *Nihon yukagaku kaishi* 47 (10): 963-970.

Dakin, E. E. & Avise, J. C. (2004) Microsatellite null alleles in parentage analysis. *Heredity* 93: 504 – 509.

Dean, M. and B. Gerrard. (199x) Helpful hints for the detection of single-stranded conformation polymorphisms. *BioTechniques*.

Dellaporta, S. L., Wood, J. and Hicks, J. B. (1983) A plant DNA minipreparation : Version II. *Plant Mol. Biol. Rep.* 1 : 19-21.

Dieffenbach, C. W, and Dveksler, G. S. (1995). Design, in PCR Primer, A Laboratory Manual, Ed., Cold Spring Harbor Laboratory Press, New York, 133-155.

Deutsches Hepatitis C Forum e. V. Homepage – Qualitativer HCV-RNA Nachweis

Di Rienzo, A., A. C. Peterson, J. C. Garza, A. M. Valdes, M. Slatkin, and N. B. Freimer. (1994) Mutational processes of simple-sequence repeat loci in human populations. *Proceedings of the National Academy of Sciences*, USA 91: 3166-3170.

Dieffenbach, C. W., Lowe, T. M. J., Dveksler, G. S., General Concepts for PCR Primer.

Dinakar Bhattramakki, Jianmin Dong, Ashok K. Chhabra, and Gary E. Hart (2000) An integrated SSR and RFLP linkage map of *Sorghum bicolor* (L.) Moench. *Genome* 43: 988-1002

DNA markers. *Nucleic Acids Res* 17(16): 6463-6471.

Dohner DE, Dehner MS, Gelb LD. (1995) Inhibition of PCR by mineral oil exposed to UV irradiation for prolonged periods. *Biotechniques*; 18(6): 964-7.

Dos Santos JB, Nienhuis J, Skroch P, Tivang J & Slocum MK (1994). Comparison of RAPD and RFLP genetic markers in determining genetic similarity among *Brassica oleracea* L. genotypes. *Theoretical and Applied Genetics*, 87: 909-915.

Dudley, J. W. (1993). Molecular markers in plant improvement: manipulation of genes affecting quantitative traits. *Crop Sci.*, 33: 660-668.

Durand, G. M., Zukin, R. S. (1993) Developmental Regulation of mRNAs Encoding Rat Brain Kainate/AMPA Receptors: A Northern Analysis Study. *J. Neurochem.* 61: 2239-2246.

E. Zietkiewicz, A. Rafalski, and D. Labuda (1994). Genome fingerprinting by simple sequence repeat (SSR)-anchored polymerase chain reaction amplification. *Genomics* 20 (2): 176-83.

Eads CA, Danenberg KD, Kawakami K, Saltz LB, Blake C, Shibata D, Danenberg PV, Laird PW (2000) MethyLight: a high-throughput assay to measure DNA methylation. *Nucleic Acids Res.*, 28:E32

Edwards, A., Civitello, A., Hammond, H. A., and Caskey, C. T. (1991). DNA typing and genetic mapping with trimeric and tetrameric tandem repeats. *Am. J. Hum. Genet.* 49, 746-756.

Edwards, M. D., Stuber, C. W. and Wendel, J. F. C. (1987). Molecular marker facilitated investigations of quantitative trait loci in maize. *Genetics*, 116, 113-125.

Engler-Blum, G., Meier, M., Frank, J., Muller, G. A. (1993) Reduction of Background Problems in Nonradioactive Northern and Southern Blot Analysis Enables Higher Sensitivity Than 32P-Based Hybridizations. *Anal. Biochem.* 210: 235-244.

Fang, G., Hammar, S. and Grumet. R. (1992) A quick and inexpensive method for removing polysaccharides from plant genomic DNA . *BioTechniques* 13 : 52-56.

Flavell AJ, Knox MR, Pearce SR & Ellis THN (1998). Retrotransposon-based insertion polymorphisms (RBIP) for high throughput marker analysis. *Plant Journal*, 16: 643-650.

Frary A, Nestbitt T. C., Frary A, Grandille S and van der Knapp (2000). Fw2. 2: A quatitative trait locus key to the evolution of tomato fruit size. *Science*: 289, 85-88.

Freeze, HH (1993). Preparation and analysis of glycoconjugates. *Current Protocols in Molecular Biology*: 17. 7. 1-17. 7. 8.

Fukuda *et al.* (2006). New staining methodology: Eastern blotting for glycosides in the field of Kampo medicines. *Journal of Natural Medicines* 60: 21-27.

G, Buiatti M, Maestri E, Malcevshi A, Marmiroli N, Aert R, Volckaert G, Rueda J, Linacero R, Vazquez A and Karp A (1997). Reproducibility testing of RAPD, AFLP and SSR markers in plants by a network of European laboratories. *Mol Breed* 3: 381-390.

Gaur, P. M. (1990). Inheritance and linkage of isozymes and morphological traits genes in chickpea saskatoon.

Garland, S; Lewin, L; Blakeney, A; Reinke, R; Henry, R. (2000) *Theor. Appl. Genet.* 101: 364.

George, M. L. C., Prasanna, B. M., Rathore, R. S., Setty, T. A. S., Kasim, F., Azrai, M., Vasal, S., Balla, O., Hautea, D., Canama, A., Regalado, E., Vargas, M., Khairallah, M., Jeffers, D. and Hoisington. D. (2003). Identification of QTLs conferring resistance to downy mildews of maize in Asia. *Theor. Appl. Genet.*, 107: 544-551.

Gerdes T, Kirchhoff M, Lind AM, Larsen GV, Schwartz M, Lundsteen C (2005). Computer-assisted prenatal aneuploidy screening for chromosome 13, 18, 21, X and Y based on multiplex ligation-dependent probe amplification (MLPA). *Eur. J. Hum. Genet.* 13 (2): 171–175.

Glenn, T. C. and S. J. Glenn. (1994). Rapid elution of DNA from agarose gels using polyester plug spin inserts (PEPSIs). *Trends in Genetics* 10(10):344.

Gortner, G., Pfenninger, M., Kahl, G., Weising, K. (1996) Northern blot analysis of simple repetitive sequence transcription in plants. *Electrophoresis*. 17: 1183-1189.

Griffiths, A. J. F., Miller, J. F., Suzuki, D. T., Lewontin, R. C. & Gelbart, W. M. (1996). Introduction to Genetic Analysis, 5th Edition. W. H. Freeman, New York.

Guldberg, P. and F. Guttler. (1993). 'Broad-range' DGGE for single -step mutation scanning of entire genes: application to human phenylalanine hydroxylase gene. *Nucleic Acids Research* 22(5): 880-881.

Gupta P. K. (2002). Molecular markers and QTL analysis in crop plants. *Curr Sci*ence: 83, 113-114.

Gupta PK and Varshney RK (2000) The development and use of microsatellite markers for genetic analysis and plant breeding with emphasis on bread wheat. *Euphytica* 113: 163-185.

Gupta, P. K. (2004). Biotechnology and Genomics. Rastogi Publications, Meerut, India.

Gupta, P. K; Roy, J, K; Prasad, M. (2001) *Curr. Sci* 80: 524.

Hamada, H., Petrino, M. G., and Kakunaga, T. (1982). A novel repeated element with Z-DNA-forming potential is widely found in evolutionarily diverse eukaryotic genomes. *Proc. Natl. Acad. Sci.* USA 79: 6465-6469.

Hamasaki H, Aoyagi M, Kasama T, Handa S, Hirakawa K, Taki T (January 1999). GT1b in human metastatic brain tumors: GT1b as a brain metastasis-associated ganglioside. *Biochim. Biophys. Acta* 1437 (1): 93–9.

Hands on Training course on DNA Markers: Development and applications, Feb 25-March 12, 2004, CCMB, Hyderabad.

Hansen M, Halldén C and Säll T (1998) Error rates and polymorphism frequencies for three RAPD protocols. *Plant Mol Biol Rep* 16: 139-146.

Hayashi K (1992). PCR-SSCP: A method for detection of mutations. *Genetic Analysis: Techniques and Applications*: 9: 73-79.

Hearne CM, Ghosh S & Todd JA (1992). Microsatellites for linkage analysis of genetic traits. *Trends in Genetics*: 8: 288-294.

Hempelmann E, Schirmer RH, Fritsch G, Hundt E, Gröschel-Stewart U (1987). "Studies on glutathione reductase and methemoglobin from human erythrocytes parasitized with Plasmodium falciparum. ". *Mol Biochem Parasitol.* 23 (1): 19–24. doi:10. 1016/ 0166-6851(87)90182-4.

Herman JG, Graff JR, Myohanen S, Nelkin BD, Baylin SB (1996) Methylation-specific PCR: a novel PCR assay for methylation status of CpG islands. *Proc Natl Acad Sci* USA, 93:9821-9826.

Higuchi R, Dollinger G, Walsh PS, Griffith R. (1992) Simultaneous amplification and detection of specific DNA sequences. *Biotechnology* 10: 413–417.

Higuchi R, Krummel B, Saiki R (1988). "A general method of *in vitro* preparation and specific mutagenesis of DNA fragments: study of protein and DNA interactions". *Nucleic Acids Res* 16 (15): 7351–67.

Hochstenbach R, Meijer J, van de Brug J, Vossebeld-Hoff I, Jansen R, van der Luijt RB, Sinke RJ, Page-Christiaens GC, Ploos van Amstel JK, de Pater JM (2005). Rapid detection of chromosomal aneuploidies in uncultured amniocytes by multiplex ligation-dependent probe amplification (MLPA). *Prenat. Diagn.* 25 (11): 1032–1039.

http://en. wikipedia. org/wiki/Variable_number_tandem_repeat

http://entomology. wisc. edu/~dshoemak/Publications/Pub. htm

http://gslc. genetics. utah/units/disorders/karyotype/image/FISH_technique. jpg

http://hepatitis-c. de/pcr1. htm

http://hepatitis-c. de/pcr1. htm

http://hepatitis-c. de/pcr1. htmhttp://www. bact. wisc. edu/Bact303/b27

http://images. google. com/imgres?imgurl= http://cotton. pi. csiro. au/Publicat/conf.

http://images. google. com/imgres?imgurl= http://cotton. pi. csiro. au/Publicat/conf/

http://images. google. com/imgres?imgurl= http://cotton. pi. csiro. au/Publicat/conf.

http://insects. ucr. edu/people/heraty. html

http://linkinghub. elsevier. com/retrieve/pii/S1388-1981(98)00003-1.

http://ncbi. nih. gov/books/bv. fcgi?call=bv. View. ShowSection&rid=handbook. glossary. 1237

http://pest. cabweb. org/PDF/BER/Ber88-6/Ber88577. pdf.

http://seqcore. brcf. med. umich. edu/doc/educ/dnapr/mbglossary/mbgloss. html

http://upload. wikimedia. org/wikipedia/en/b/b8/DNA-structure-and-bases. png

http://uwadmnweb. uwyo. edu/zoology/mcdonald/molmark/mcdgloss. html

http://www. answer. com/fluorescence. in. situ. hybridization

http://www. ars. usda. gov/research/projects/projects. htm?ACCN_NO= 405094 & fy = 2004

http://www. asicoaquaticmarkers. com/Glossary. htm

http://www. bact. wisc. edu/Bact303/b27

http://www. bact. wisc. edu/Bact303/b27

http://www. bact. wisc. edu/Bact303/b27

http://www. bergen. org/AAST/Projects/Gel/index. html

http://www. bergen. org/AAST/Projects/Gel/index. html

http://www. biowire. com/nucleus/nucleus_1_1. jsp

http://www. biowire. com/nucleus/nucleus_1_1. jsp

http://www. cgn. wageningn-ur. nl/pgr/research/molgen/verview. htm

http://www. cgn. wageningn-ur. nl/pgr/research/molgen/verview. htm

http://www. cgn. wageningn-ur. nl/pgr/research/molgen/verview. htm

http://www. cgn. wageningn-ur. nl/pgr/research/molgen/verview. htm

http://www. cgn. wageningn-ur. nl/pgr/research/molgen/verview. htm

http://www. cgn. wageningn-ur. nl/pgr/research/molgen/verview. htm

http://www. colostate. edu/Depts/Entomology/courses/en575/en575. html

http://www. everythingbio. com/glos/definition. php?word=Cleaved +amplified+ polymorphic+sequence

http://www. everythingbio. com/glos/definition. php?word=Cleaved+ amplified+ polymorphic+sequence

http://www. exactantigen. com/review/western-blot-antibody. html.

http://www. google. co. in/search?hl=en&lr=&oi=defmore&defl=en&q=define:DNA

http://www. greenfacts. org/glossary/mno/molecular-marker. htm

http://www. ias. ac. in/currsci/feb252005/541. pdf

http://www. intl-pag. org/5/abstracts/p-5c-159. html

http://www. intl-pag. org/5/abstracts/p-5c-159. html

http://www. ion. ucl. ac. uk/images/Fluorescence-Microscope. jpg

http://www. karymullis. com

http://www. le. ac. uk/biology/phh4/addprot. htm

http://www. md. ucl. ac. be/didac/sbim2520/pcr. html

http://www. md. ucl. ac. be/didac/sbim2520/pcr. html

http://www. mrcindia. org/mol-ent. htm

http://www. ncbi. nlm. nih. gov/entrez/query. fcgi?cmd= Retrieve&db= PubMed& list_uids=10331278&dopt=Abstract

http://www. ndsu. nodak. edu/instruct/mcclean/plsc731/mapping/mapping2. htm

http://www. ndsu. nodk. edu/instruct/mcclan/plsc731/mapping/mapping4. htm

http://www. ndsu. nodk. edu/instruct/mcclan/plsc731/mapping/mapping4. htm

http://www. ndsu. nodk. edu/instruct/mcclan/plsc731/mapping/mapping4. htm

http://www. ndsu. nodk. edu/instruct/mcclan/plsc731/mapping/mapping4. htm

http://www. ndsu. nodk. edu/instruct/mcclan/plsc731/mapping/mapping4. htm

http://www. ndsu. nodk. edu/instruct/mcclan/plsc731/mapping/mapping4. htm

http://www. nhm. ac. uk/hosted_sites/acarology/saas/saa/pdf07/003-014. pdf

http://www. nobel. se/chemistry/laureates/1993/mullis-lecture. html

http://www. nobel. se/chemistry/laureates/1993/mullis-lecture. html

http://www. nobel. se/chemistry/laureates/1993/mullis-lecture. html

http://www. nobel. se/chemistry/laureates/1993/mullis-lecture. html

http://www. oznet. ksu. edu/biotech/glossary. htm

http://www. phschool. com/science/biology_place/biocoach/dnarep/intro. html

http://www. pparc. ac. uk/frontiers/archive/update. asp?id=18U8&style=update

http://www. scipub. net/botany/molecular-markers-plant-genetics-biotechnology. html

http://www. scipub. net/entomology/index. html

http://www. vetscite. org/issue1/ tools/leute_4_080. htm

http://www. vetscite. org/issue1/ tools/leute_4_080. htm

http://www. vetscite. org/issue1/ tools/leute_4_080. htm

http://www.icrisat.org/manual on molecular marker

https://www. who. int/tdr/grants/workplans/entomol. htm

Hughes, C. R. and D. C. Queller. (1993) Detection of highly polymorphic microsatellite loci in a species with little allozyme polymorphism. *Molecular Ecology* 2: 131-137.

Innis, M. A., and Gelfand, D.H. (1994) Optimization of PCRs, in PCR protocols, A Guide to Methods and Applications, Innis, M. A., Gelfand, D. H., Sninsky, J. J., and White, T. J., Ed., CRC Press, London, 5-11.

Ishikawa & Taki (2000). *Methods in Enzymology* 312: 145.

Jacob, H. J., Lindpaintner, K., Lincoln, S. E., Kusumi, K., Bunker, R. K., Mao, Yi-Pei, Ganten, D. Dzau, V. J. and Lander, E.S. (1991) Genetic mapping of a gene causing hypertensive rat. *Cell* 67, 213-224.

Jansen, R. C. and Stam, P. (1994). High resolution of quantitative traits into multiple loci via interval mapping. *Genetics*, 136, 1447-1455.

Jarne, P. & Lagoda, P. J. L. (1996). Microsatellites, from molecules to populations and back. *Trends in Ecology and Evolution* 11: 424 – 429.

Jayasena (1999). Aptamers: An Emerging Class of Molecules That Rival Antibodies in Diagnostics. *Clinical Chemistry* 45: 1628-1650.

Jeffreys, A. J., Wilson, V., and Thein, S. L. (1985). Hypervariable minisatellite regions in human DNA. *Nature* 314, 67-73.

Jeffreys, A. J., Wilson, V., Newumann, R., and Keyte, J. (1988). Amplification of human minisatellites by the polymerase chain reaction: Towards DNA fingerprinting of single cells. *Nucleic Acids Res.* 16, 10953-10971.

Jones, A. S. 1953. The isolation of bacterial nucleic acids using cetyltrimethylammonium bromide (cetavlon). Biochem. Biophys. *Acta.* 10 : 607-12.

Jourdren C, Barret P, Horvais R, Foisset N, Delourme R, Renard M (1996) Identification of RAPD markers linked to the loci controlling erucic acid level in rapeseed. *Molecular Breeding* 2: 61-71.

Kaeppler, S. M., Philips, R. L., Kim, T. S. (1993) Use of near-isogenic lines derived by backcrossing or selfing to map quantitative traits. *Theor. Appl. Genet.* 87: 233-237.

Karen Sumire Kubo, R. M. Stuart, J. Freitas-Astúa1, R. Antonioli-Luizon, E. C. Locali-Fabris, H. D. Coletta-Filho1, M. A. Machado and E. W. Kitajima (2009) Evaluation of the genetic variability of orchid fleck virus by single-strand conformational polymorphism analysis and nucleotide sequencing of a fragment from the nucleocapsid gene.

Karl, S. A., B. W. Bowen and J. C. Avise. (1992) Global population genetic structure and male-mediated gene flow in the green turtle (*Chelonia mydas*): RFLP analyses of anonymous nuclear loci. *Genetics* 131: 163-173.

Karp A, Kresovich S, Bhat KV, Ayada WG and Hodgkin T (1997) Molecular tools in plant genetic resources conservation: a guide to the technologies. *IPGRI technical bulletin No 2*. International Plant Genetic Resources Institute, Rome, Italy.

Karp A, Seberg O & Buiatti M (1996). Molecular techniques in the assessment of botanical diversity. *Annals of Botany*, 78: 143-149.

Kearsey, M. J. (1998). The principles of QTL analysis (a minimal mathematics approach). *J. Exp. Bot.*, 49, 1619-1623.

Kearsey, M. J. and Pooni, H. S. (1996). The Genetical Analysis of Quantitative Traits. Chapman and Hall, London.

Kearsey, M. J. and Farquhar, A. G. L. (1998). QTL analysis in plants; where are we now? *Heredity*, 80, 137-142.

Kephart SR (1990). Starch gel electrophoresis of plant isozymes: a comparative analysis of techniques. *American Journal of Botany*, 77: 693-712.

Kevil, C. G., Walsh, L., Laroux, F. S., Kalogeris, T., Grisham, M. B., Alexander, J. S. (1997) An Improved, Rapid Northern Protocol. *Biochem. and Biophys. Research Comm.* 238:277-279.

Kimpton, C. P., P. Gill, A. Walton, A. Urquhart, E. S. Millican, and M. Adams. (1993) Automated DNA profiling employing multiplex amplification of short tandem repeat loci. PCR *Methods and Applications* 3: 13-22.

Kirkpatrick, B. W. and G. L. Hart. (1994) Conformation polymorphisms and targeted marker development. *Animal Genetics* 25: 77-82.

Kirkpatrick, B. W., B. M. Huff, and E. Casas-Carillo. (1993) Double-strand DNA conformation polymorphisms as a source of highly polymorphic genetic markers. *Animal Genetics* 24: 155-161.

Klien, R. R., P. E. Klien, Ashok K. Chhabra, J. Dong, S. Pammi, K. I. Childs, J. E. Mullet, W. I. Rooney, K. F. Schertz (2001). Molecular mapping of the *RF1* gene for pollen fertility restoration in sorghum (*Sorghum bicolor* L.) *Theoretical and Applied Genetics* 102: 1206-1212

Kochert, Gary (1985). Introduction to RFLP Mapping and Plant Breeding Applications, Department of Botany, University of Georgia, Athens, GA 30602.

Kresovich, S; Szewc-Mc Fadden, A. K; Block, S. M. and Mc-Terson, J. R. (1995) *Theor. Appl. Genet.* 91:206-211.

Kristensen LS, Mikeska T, Krypuy M, Dobrovic A: (2008) Sensitive Melting Analysis after Real Time- Methylation Specific PCR (SMART-MSP): high-throughput and probe-free quantitative DNA methylation detection. *Nucleic Acids Res.* 36:e42

Kukita, Y., *et al.* (1997) SSCP Analysis of Long DNA Fragments in Low pH Gel. *Human Mutation*; 1997, (10): 400-7.

Kurien & Scofield (2006). Western Blotting. *Methods* 38: 283-293.

Kwok, S., Kellog, D. E. McKinney, N., Spasic, D., Goda, L., Levenson, C., and Sninsky, J. J., Effects of primer-template mismatches on the polymerase chain reaction: Human Immunodeficiency Virus 1 model studies. *Nucleic Acids Res.* 18:999-1005, 1990.

Lander, E. S. and Botstein, D. (1989). Mapping Mendelian factors underlying quantitative traits using RFLP linkage maps. *Genetics,* 121, 185-199.

Langemeier, J. L., R. F. Cook, C. J. Issel, and R. C. Montelaro. (1994) Application of cycle dideoxy fingerprinting to screening heterogeneous populations of the equine infectious anemia virus. *BioTechniques* 17(3): 484-490.

Leca-Bouvier & Blum (2005). "Biosensors for protein detection: A review". *Analytical Letters* 38: 1491.

Lefebvre, V., Palloix, A., Caranata, C., Pochard, E. (1995) Construction of an intraspecific linkage map of pepper using molecular markers and doubled-haploid progenies. *Genome* 38:112-121.

Lessa, E. P. and G. Applebaum. (1993) Screening techniques for detecting allelic variation in DNA sequences. *Molecular Ecology* 2: 119-129.

Liang, P. Pardee, A. B. (1995) Recent advances in differential display. *Current Opinion Immunol.* 7: 274-280.

Lincoln, S., Daly, M. and Lander, E. (1992). Mapping genes controlling quantitative traits with MAPMAKER/QTL. *Whitehead Institute Technical Report,* 2nd edition, Whitehead Institute, MA.

Litt, M. and Luty, J. A. (1989). A hypervariable microsatellite revealed by in vitro amplification of a dinucleotide repeat within the cardiac muscle actin gene. *Am. J. Hum. Genet.* 44, 397-401.

Liu, B. H. (2002). Statistical Genomics: Linkage, Mapping and QTL Analysis. 2nd Edition. CRC Press, Boca Raton, Florida.

Livak, K. J., Marmaro, J., and Todd, J. A. (1995). Towards fully automated genome-wide polymorphism screening. *Nat Genet* 9, 341-2.

Longo, M. C. and Hartley, J. L. (1986) *Focus* 8:3, 3.

Longo, M. C., M. S. Berninger, and J. L. Hartley. (1990) Use of uracil DNA glycosylase to control carry-over contamination in polymerase chain reactions. *Gene* 93: 125-128.

Lyon, K. J. Fiona F. Ballard and Bruce R. Lyon. (2006) Utilisation of Disease Resistance Genes in Cotton . Australian Cotton Cooperative Research Centre, School of Biological Sciences A12, The University of Sydney, Sydney, NSW.

Maekawa, M., K. Sudo, and T. Kanno. (1993) Search for improved electrophoretic conditions for PCR-Single-Strand Conformation Polymorphism analysis: is an SDS buffer condition useful? *PCR Methods and Applications* 3:130-132.

Maniatis, T., Fritsch, E. F. and Sambrook, J. (1982) Molecular Cloning: A Laboratory Manual, Cold Spring Harbor Laboratory.

Mann M and Jensen ON (2003). Proteomic analysis of post-translational modifications. *Nature Biotechnology* 21, 255 - 261.

Mauricio R (2001). Mapping quantitative trait loci in plants: uses and caveats for evolutionary biology. *Nature Reviews/Genetics*, 2, 370-375.

Meyer W, Michell TG, Freedman EZ and Vilgalys R (1993) Hybridization probes for conventional DNA fingerprinting used as single primers in polymerase chain reaction to distinguish strain of Cryptococcus neoformans. *J Clin Biol* 31: 2274-2280.

Michael P. Weiner, Tim Gackstetter, Gina L. Costa, John C. Bauer, and Keith A. Kretz (1995) Molecular Biology: Current Innovations and Future Trends. Eds. A. M. Griffin and H. G. Griffin.

Michelmore RW, Paran I, Kesseli RV (1991) Identification of markers linked to disease-resistance genes by bulked segregant analysis: A rapid method to detect markers in specific genomic regions by using segregating populations. *Proc. Natl. Acad. Sci.* USA 88: 9828-9832.

Mohan, M; Nari, S; Bhagwat, A; Krishna, T. G; Yano, M; Bhatia, C. R; Sasaki, T. (1997) Mol. Breed. 3:87.

Morgante M & Olivieri AM (1993). PCR-amplified microsatellites as markers in plant genetics. *The Plant Journal*, 3: 175-182.

Mori, H., Takeda-Yoshikawa, Y., Hara-Nishimura, I., Nishimura, M. (1991) Pumpkin malate synthase Cloning and sequencing of the cDNA and Northern blot analysis. *Eur. J. Biochem.* 197: 331-336.

Mullis KB, Faloona FA: (1987) Specific synthesis of DNA in vitro via a polymerase-catalyzed chain reaction. *Methods Enzymol* 155: 335–350.

Mullis KB: (1993) The polymerase chain reaction. Nobel Lecture, December 8.

Mullis, K., Faloona, F., Scharf, S., Saiki, R., Horn, G., and Erilich, H. (1986). Specific enzymatic amplification of DNA in vitro: The polymerase chain reaction. Cold Spring Harbor Symp. *Quant. Biol.* 51:263-273.

Mullis, K. B., and Faloona, F. A. 1987. Specific synthesis of DNA *in vitro* via a polymerase-catalyzed chain reaction. *Methods in Enzymology* 155: 335-50.

Murray, M. G. and Thompson, W. F. (1980) Rapid isolation of high molecular weight plant DNA. *Nucl. Acids Res.* 8 : 4321-4325.

Muthukrishnan, S., Gill, B. S., and Mehta, S. L. (1991). Laboratory Manual on RFLP Mapping in Wheat Genome, Division of Biochemistry, IARI, New Delhi.

Nagaoka T and Ogihara Y (1997) Applicability of inter-simple sequence repeat polymorphisms in wheat for use as DNA markers in comparison to RFLP and RAPD markers. *Theor Appl Genet* 94: 597-602.

Nair, S. K., Prasanna, B. M., Garg, A., Rathore, R. S., Setty, T. A. S. and Singh, N. N. (2005). Identification and validation of QTLs conferring resistance to sorghum downy mildew (*Peronosclerospora sorghi*) and Rajasthan downy mildew (*P. heteropogoni*) in maize. *Theor. Appl. Genet.*, 110, 1384-1392.

Nakamura, Y., Leppert, M., O'Connell, P., Wolff, R., Holm, T., Culver, M., Martin, C., Fujimoto E., Hoff, M., Kumlin, E., and White, R. (1987). Variable number tandem repeat (VNTR) markers for human gene mapping. *Science* 235, 1616-1622.

O. Malley, D., Wheeler, N. C. and Guries, R. P. (1980). A manual for starch gel electrophoresis. Staff paper series #11. University of Wisconsin-Madison, Madison, WI.

Ochman H, Gerber AS, Hartl DL. (1988) *Genetics*. Nov; 120(3):621-3.

Orita, M., Iwahana, H., Kanazawa, H., Hayashi, K., and Sekiya, T. (1989) Detection of polymorphism of human DNA by gel electrophoresis as single-strand conformation polymorphisms. *Proc. Natl. Acad. Sci.* USA 86, 2766-2770.

Orita, M., Y. Suzuki, T. Sekiya, and K. Hayashi. (1989) Rapid and sensitive detection of point mutations and DNA polymorphisms using the polymerase chain reaction. *Genomics* 5: 874-879.

Pang, J., Modlin, J., and Yolken, R. (1992) Use of modified nucleotides and uracil-DNA glycosylase (UNG) for the control of contamination in the PCR-based amplification of RNA. *Mol Cel Probes* 6, 251-6.

Paterson, A. H., Tanskley, S. D. and Sorrells, M. E. (1991). DNA markers in plant improvement. *Adv. Agronomy*, 46, 39-90.

Paterson, A. H; Tanksley, S. D; Sorells, M. E. (1991) *Adv. Agro.* 46: 39.

Poulsen GB, Kahl G and Weising K (1994) Differential abundance of simple repetitive sequences in species of Brassica and related Brassicaceae. *Plant Syst Evol* 190: 21-30.

Procter M, Chou LS, Tang W, Jama M, Mao R (2006). "Molecular diagnosis of Prader-Willi and Angelman syndromes by methylation-specific melting analysis and methylation-specific multiplex ligation-dependent probe amplification". *Clin. Chem.* 52 (7): 1276–1283.

Queller, D. C., Strassman, J. E. & Hughes, C.R. (1993). Microsatellites and Kinship. *Trends in Ecology and Evolution* 8: 285–288.

Ramsay G (1998). DNA chips: State-of-the art. *Nature Biotechnology*, 16: 40-44.

Rand K, Qu W, Ho T, Clark SJ, Molloy P: Conversion-specific detection of DNA methylation using real-time polymerase chain reaction (ConLight-MSP) to avoid false positives. *Methods* 2002, 27:114-120.

Reidhaar-Olson JF, Hammer J: April 2001. The impact of genomics tools on target discovery. *Curr Drug Discovery*.

Renart J, Reiser J, Stark GR (1979). "Transfer of proteins from gels to diazobenzyloxymethyl-paper and detection with antisera: a method for studying antibody specificity and antigen structure". *Proc Natl Acad Sci* USA. 76 (7): 3116–3120.

Rickwood, D. and Hames, B. D. (1982) Gel Electrophoresis of Nucleic Acids: A Practical Approach, IRL Press, Oxford, England.

Rongwen, J; Akkaya, M. S; Bhagwat, A. A; Lavi, U. and Cregan, P. B. (1995). *Theor. Appl. Genet.* 90; 43-48.

S. Nimabal, A. K. Chhabra, R. K. Behl, K. P. Singh, S. K. Sharma and C. Ram (2002). Characterization of backcross inbred lines of bread wheat using protein and isozyme markers. *In: Proceedings (Ed. Li Dajue): Second International Conference on Sustainable Agriculture for Food, Energy and Industry (Vol. II)* pp: 1112-1116, held at Institute of Botany, Beijing, China, Sept. 8-13, 2002.

Saghai-Maroof, M. A., K. Soliman, R. A. Jorgensen, and R. W. Allard. (1984). Ribosomal DNA spacer-length polymorphisms in barley: Mendelian inheritance, chromosomal location, and population dynamics. *PNAS* 81:8014-8018.

Salvi S, Tuberosa R, Chiapparino E, Maccaferri M, Veillet S, van Beuningen L, Issac P, Edwards K and Phillips RL (2002). Towards positional cloning of Vgt1, a QTL controlling the transition from the vegetative to the reproductive phase in maize. *Plant Mol Biol*, 48, 601-613.

Sambrook, Fritsch and Maniatis (1987). Molecular Cloning: A Laboratory Manual.

Sanchez de la Hoz MP, Dàvila JA, Loarce Y and Ferrer E (1996) Simple sequence repeat primers used in polymerase chain reaction amplifications to study genetic diversity in barley. *Genome* 39: 112-117.

Sanghi-Maroof, M. A; Biyashav, R. M; Yang, G. P; Zhang, Q. and Allard, R. W. (1994) *Proc. Natl. Acad. Sci.* USA 91: 5466-5470.

SAP (or juice) extractor: MEKU Erich Pollähne G. m. b. H. - 3015 Wennigsen, Am Weingarten 14, Germany.

Sarkar, G., Paynton, C., and Sommer, S. S. (1991). Segments containing alternating purine and pyrimidine dinucleotides: patterns of polymorphism in humans and prevalence throughout phylogeny. *Nucleic Acids Res.* 19, 631-636.

Sax, K. (1923). The association of size differences with seed-coat pattern and pigmentation in Phaseolus vulgaris. *Genetics*: 8, 552-560.

Schlamp, K., Weinmann, A., Krupp, M., Maass, T., Galle, P. R., Teufel, A. (2008) BlotBase: A northern blot database. *Gene.* 427: 47-50.

Schouten JP, McElgunn CJ, Waaijer R, Zwijnenburg D, Diepvens F, Pals G (2002). Relative quantification of 40 nucleic acid sequences by multiplex ligation-dependent probe amplification. *Nucleic Acids Res.* 30 (12): e57.

Sharma PC, Huttel B, Winter P, Kahl G, Gardner RC and Weising K (1995) The potential of microsatellites for hybridization and polymerase chain reaction-based DNA fingerprinting of chickpea (*Cicer arietium* L.) and related species. *Electrophoresis* 16: 1755-1761.

Sheffield, V. C., J. S. Beck, A. E. Kwitek, D. W. Sandstrom, and E. M. Stone. 1993. The sensitivity of Single-Strand Conformation Polymorphism analysis for the dection of single base substitutions. *Genomics* 16: 325-332.

Sieben, V. J. ; C. S. Debes Marun, P. M. Pilarski, G. V. Kaigala, L. M. Pilarski, C. J. Backhouse (2007-06). "FISH and chips: chromosomal analysis on microfluidic platforms". IET Nanobiotechnology 1 (3): 27-35. doi:10. 1049/iet-nbt:20060021. *http://link. aip. org/link/?NBT/1/27/1. Retrieved on 2009-01-26.*

Sieben, V. J. ; C. S. Debes-Marun, L. M. Pilarski, C. J. Backhouse (2008-11). An integrated microfluidic chip for chromosome enumeration using fluorescence in situ hybridization. Lab on a Chip 8 (12): 2151-2156. doi:10. 1039/b812443d. *http:// www. rsc. org/publishing/journals/LC/article. asp?doi=b812443d. Retrieved on 2009-03-24.*

Simpson, S.P. (1989) Detection of linkage between quantitative trait loci and restriction fragment length polymorphisms using inbred lines. *Theor. Appl. Genet.* 77: 815-819.

Singh, B. D. (2005). Biotechnology : Expanding Horizons, Kalyani Publication, New Delhi.

Slade, R. W., C. Moritz, A. Heideman, and P. T. Hale. Rapid assessment of single-copy nuclear DNA variation in diverse species. *Molecular Ecology* 2: 359-373.

Soltis, D. E., Haufler, C. H., Darrow, D. C. and Gastony, G. J. (1983). Starch gel electrophoresis of ferns: A compilation of grinding buffers, gel and electrode buffers, and staining schedules. *Amer. Fern. J.*, 73(1):9-27.

Southern, E. M. (1975) *J. Mol. Biol.* 98: 503-515.

Spinardi, L., R. Mazars, and C. Theillet. 1991. Protocols for an improved detection of point mutations by SSCP. *Nucleic Acids Research* 19(14): 4009.

Stallings, R. L., Torney, D. C., Hildebrand, C. E., Longmire, J. L., Deaven, L. L., Jett, J. H., Doggett, N. A., and Moyzis, R. K. (1990). Physical mapping of human chromosomes by repetitive sequence fingerprinting. Proc. *Natl. Acad. Sci.* USA 87, 6218- 6222.

Stam, P. (1993). Construction of integrated genetic linkage map by means of a computer package: Joinmap. *The Plant J.*, 5, 739-744.

Staub JE, Serquen FC & Gupta M (1996). Genetic markers, map construction, and their application in plant breeding. *HortScience*, 31: 729-741.

Streit, S., Michalski, C. W., Erkan, M., Kleef, J., Friess, H. (2008) Northern blot analysis for detection of RNA in pancreatic cancer cells and tissues. *Nature Protocols*. 4: 37-43.

Stuber, C. W. (1995). Mapping and manipulating quantitative traits in maize. *Trends Genet.*, 11, 477-481.

Sunnucks, P., *et al.* (2000) SSCP Is Not So Difficult: The Application and Utility of Single-Stranded Conformation Polymorphism in Evolutionary Biology and Molecular Ecology. *Molecular Ecology*, (9): 1699-710.

Sweetman, W. A., R. Rash, B. Sykes, P. Beighton, J. T. Hecht, B. Zabel, J. T. Thomas, R. Boot-Handford, M. E. Grant, and G. A. Wallis. (1992). SSCP and segregation analysis of the human type X collagen gene (COL10A1) in heritable forms of chondrodysplasia. *American Journal of Human Genetics* 51:841-849.

Taki *et al.* (1994). Blotting of glycolipids and phospholipids from a high-performance thin-layer chromatogram to a polyvinylidene difluoride membrane. *Analytical Biochemistry* 221: 312-316.

Tamulonis, J. P., Luzzi, B. M., Hussey, R. S., Parrot, W. A., Boerma, H. R. (1997). DNA marker analysis of loci conferring resistance to root-knot nematode in soybean. *Theor. Appl. Genet.* 95: 664.

Taniguchi, M., Miura, K., Iwao, H., Yamanaka, S. (2001) Quantitative Assessment of DNA Microarrays – Comparison with Northern Blot Analysis. *Genomics*. 71:34-39.

Tanksley, S. D. (1993). Mapping polygenes. *Annu. Rev. Genet.*, 27, 205-233.

Tanksley, S. D. and McCouch, S. R. (1997). Seed banks and molecular maps: unlocking genetic potential from the wild. *Science*, 277, 1063-1066.

Tanner, MJ and Anstee, DJ (1976). A method for the direct demonstration of the lectin-binding components of the human erythrocyte membrane. *Biochemistry Journal* 153: 265-270.

Tansley, S. D. and Nelson, J. C. (1996). Advanced backcross QTL analysis: a method for simultaneous discovery and transfer of valuable QTLs from unadapted germplasm into elite breeding lines. *Theor. Appl. Genet.*, 92, 191-203.

Tautz, D. (1989) Hypervariability of simple sequences as a general source for polymorphic DNA markers. *Nucleic Acids Research* 17(16): 6463-6471.

Tautz, D., Trick, M., and Dover, G. A. (1996). Cryptic simplicity in DNA is a major source of genetic variation. *Nature* 322, 652-656.

Thomas *et al.* (2009). Antigenic protein modifications in Ehrlichia. *Parasite Immunology* 31. 296-303.

Thornton, C. G., J. L. Hartley, and A. Rashtchian. (1992) Utilizing uracil DNA glycosylase to control carryover contamination: characterization of residual UDG activity following thermal cycling. *BioTechniques* 13(2): 180-184.

Toroser D, Thormann CE, Osborn TC, Mithen R (1995) RFLP mapping of quantitative trait loci controlling seed aliphatic-glucosinolate content in oilseed rape (*Brassica napus* L.). *Theor. Appl. Genet.* 91: 802-808.

Towbin *et al.* (1979). Electrophoretic transfer of proteins from polyacrylamide gels to nitrocellulose sheets: procedure and some applications. *PNAS* 76: 4350.

Towbin *et al.* (1984). Glycosphingolipid-blotting: an immunological detection procedure after separation by thin layer chromatography. *Journal of Immunological Methods* 72: 471.

Towbin H, Staehelin T, Gordon J. (1979) Electrophoretic transfer of proteins from polyacrylamide gels to nitrocellulose sheets: procedure and some applications. *Proc Natl Acad Sci* USA. 76 (9): 4350–4354.

Trayhurn, P. (1996) Northern Blotting. *Pro. Nutrition Soc.* 55:583-589.

Tsumura Y, Ohba K and Strauss SH (1996) Diversity and inheritance of inter-simple sequence repeat polymorphisms in Douglas-fir (*Pseudotsuga menziesii*) and sugi (*Cryptomeria japonica*). *Theor Appl Genet* 92: 40-45.

Turnpenny, P. & Ellard, S. (2005). Emery's Elements of Medical Genetics, 12th. ed. Elsevier, London.

UNIDO-International Centre for Biotechnology and Genetic Engineering, Workshop Handbook (1991), RFLP in Plant Breeding, Rockefeller Foundation, USA.

Utans, U., Liang, P., Wyner, L. R., Karnovsky, M. J., Russel, M. E. (1994) Chronic cardiac rejection: Identification of five upregulated genes in transplanted hearts by differential mRNA display. *Proc. Natl. Acad. Sci.* USA. 91:6463-6467.

Valoczi, A., Hornyik, C., Varga, N., Burgyan, J., Kauppinen, S., Havelda, Z. (2004) Sensitive and specific detection of microRNAs by northern blot analysis using LNA-modified oligonucleotide probes. *Nuc. Acids Research.* 32: e175.

Vidal-Puig, A., and D. E. Moller. (1994). Comparative sensitivity of alternative single-strand conformation polymorphism (SSCP) methods. *BioTechniques* 17(3): 490-496.

Virk PS, Zhu J, Newbury HJ, Bryan GJ, Jackson MT and Ford-Lloyd BV (2000) Effectiveness of different classes of molecular markers for classifying and revealing variations in rice (Oryza sativa) germplasm. *Euphytica* 112: 275-284.

Volikos E, Robinson J, Aittomaki K, Mecklin JP, Jarvinen H, Westerman AM, de Rooji FW, Vogel T, Moeslein G, Launonen V, Tomlinson IP, Silver AR, Aaltonen LA (2006). LKB1 exonic and whole gene deletions are a common cause of Peutz-Jeghers syndrome. *J. Med. Genet.* 43 (5): e18.

Vos P, Hogers R, Bleeker M, Reijans M, van der Lee T, Hornes M, Frijters A, Pot J, Peleman J, Kuiper M, Zabeau M (1995) AFLP: a new technique for DNA fingerprinting. *Nucleic Acid Research* 23 (21): 4407-4414.

W. Neal Burnette (April 1981). "'Western blotting': electrophoretic transfer of proteins from sodium dodecyl sulfate — polyacrylamide gels to unmodified nitrocellulose and radiographic detection with antibody and radioiodinated protein A". Analytical Biochemistry (United States: Academic Press) 112 (2): 195–203. doi:10. 1016/0003-2697(81)90281-5. ISSN 0003-2697.

Wagner, John. Screening Methods for Detection of Unknown Point Mutations. *<http:// wwwusers. med. cornell. edu/~jawagne/screening_for_mutations. html#Single-Strand. Conformational. Polymorphism>*. Accessed 2003 February 17.

Walsh CT, Garneau-Tsodikova S, and Gatto GJ Jr. (2005). Protein posttranslational modifications: The chemistry of proteome diversifications. Angewandte Chemie International Edition in English 44, 7342-7372.

Walsh, S. P., D. A. Metzger, and R. Higuchi. (1991) Chelex 100 as a medium for simple extraction of DNA for PCR-based typing from forensic material. *Biotechniques* 10(4): 506-513.

Walton M (1993). Molecular markers: Which ones to use? *Seed World*, 131: 22-29.

Wang, A. M., Doyle, M. V., and Mark, D. F. 1989. Quantitation of mRNA by the polymerase chain reaction. *Proc Natl Acad Sci USA* 86, 9717-21.

Watson, J. D. and Crick, F. H. . (1953). Genetical implications of the struture of deoxyribonucleic acid. *Nature*. 171, 964-967.

Weber, J. L. (1990). Informativeness of human (dC-dA)n. (dG-dT)n polymorphisms. *Genomics* 7: 524-530.

Western blot antibody. exactantigen. com.

Weber, J. L. and May, P. E. (1989). Abundant class of human DNA polymorhisms which can be typed using the polymerare chain reaction. *Am. J. Hum. Genet.* 44, 388-396.

Weghorst, C. M. and G. S. Buzard. 1993. Enhanced single-strand conformation polymorphism (SSCP) detection of point mutations utilizing methylmercury hydroxide. *BioTechniques* 15(3): 397-400.

Wehner M, Mangold E, Sengteller M, Friedrichs N, Aretz S, Friedl W, Propping P, Pagenstecher C (2005). "Hereditary nonpolyposis colorectal cancer: pitfalls in deletion screening in MSH2 and MLH1 genes". *Eur. J. Hum. Genet.* 13 (8): 983–986.

Wendel, J. F. (1980). Enzyme extraction from a tannin-rich plant. *Isozyme Bull.* 13:116.

Wendel, J. F. and Weeden, N. F. (1989). Visualization and interpretation of plant isozymes. *In*: D. E. Soltis and P. S. Soltis (eds.). *Isozymes in Plant Biology*. pp. 5-45. Dioscorides Press, Portland, Oregon, USA.

Whitkus R, Doebley J & Wendel JF (1994). Nuclear DNA markers in systematics and evolution, pp. 116-141. *In: Phillips RL & Vasil IK (eds.), DNA-Based Markers in Plants (Advances in Cellular and Molecular Biology of Plants, Vol. 1)*. Kluwer Academic Publishers, Dordrecht.

Wilfingseder D, Stoiber H: Quantifizierung von PCR Produktmengen durch real-time PCRVerfahren. Antibiotika Monitor, Heft 1/2/2002.

Wilkins, M.H.F., Stokes, A.R. and Wilson, H. R. (1953). Molecular structure of deoxypentose nucleic acids. *Nature*. 171, 738-740.

Williams JGK, Kubelik AR, Livak KJ, Rafalski JA & Tingey SV (1990). DNA polymorphisms amplified by arbitrary primers are useful as genetic markers. *Nucleic Acid Research*, 18: 6531-6535.

Williams, J. G. K., Kubelik, A. R., Livak, K. J., Rafalski, J. A., and Tingey, S. V. (1990). DNA polymorphisms amplified by arbitrary primers are useful as genetic markers. *Nucleic Acids Res.* 18, 6531-6535.

Wilson AJ and Chourey PS (1984) A rapid inexpensive method for the isolation of restrictable mitochondrial DNA from various plant sources. *Plant Cell Rep* 3: 237–239.

Wilson, Keith (2000) Principles and Techniques of Practical Biochemistry: 4th ed. pp. 425-429. Cambridge University Press.

Wilting SM, Snijders PJ, Meijer GA, Ylstra B, van den Ijssel PR, Snijders AM, Albertson DG, Coffa J, Schouten JP, van de Wiel MA, Meijer CJ, Steenbergen RD (2006). "Increased gene copy numbers at chromosome 20q are frequent in both squamous cell carcinomas and adenocarcinomas of the cervix. *J. Pathol.* 209 (2): 220–230.

Wyman, A. R. and White, R. (1980). A highly polymorphic locus in human DNA. Proc. *Nat. Acad. Sci.* USA 77, 6754-6758.

www. olympusmicro. com/. . . / bx51fluorescence. html

Yamanaka, S., Poksay, K. S., Arnold, K. S., Innerarity, T. L. (1997) A novel translational repressor mRNA is edited extensively in livers containing tumors caused by the transgene expression of the apoB mRNA-editing enzyme. *Genes Dev.* 11: 321-333.

Yang GP, Saghai Maroof MA, Xu CG, Zhang Q & Biyashev RM (1994). Comparative analysis of microsatellite DNA polymorphism in landraces and cultivars of rice. *Molecular and General Genetics*, 245: 187-194.

Yang W, de Oliveira AC, Godwin I, Schertz K & Bennetzen JL (1996). Comparison of DNA marker technologies in characterizing plant genome diversity: variability in Chinese sorghums. *Crop Science*, 36: 1669-1676.

Yang, H., McLeese, J., Weisbart, M., Dionne, J. -L., Lemaire, I., Aubin, R. A. (1993) Simplified high throughput protocol for Northern hybridization. *Nucleic Acids Research*. 21: 3337-3338.

Yano, M., Katayose, Y., Ashikari, M., Yamanouchi, U., Monna, I., Fuse, T., Baba, T., Yamamoto, K., Umihara, Y., Magamura, Y. and Sasaki, T. (2000). Hd1, a major photoperiod sensitivity quantitative trait locus in rice, is closely related to the Arabidopsis flowering time gene CONSTANS. *Plant Cell* 12, 2473-2484.

Yap, E. P. H. and J. O'D. McGee. (1993). Nonisotopic discontinuous phase single strand conformation polymorphism (DP-SSCP): genetic profiling of D-loop of human mitochondrial (mt) DNA. Nucleic Acids Research 21(17): 4155. Akagi, H; Yokozeki, Y; Inagaki, A and Fujimura, T. (1996) *Theor. Appl. Genet.* 93: 1071-1077.

Yau SC, Bobrow M, Mathew CG, Abbs SJ (1996). "Accurate diagnosis of carriers of deletions and duplications in Duchenne/Becker muscular dystrophy by fluorescent dosage analysis". *J. Med. Genet.* 33 (7): 550–558.

Zeng, Z. B. (1994). Precision mapping of quantitative trait loci. *Genetics*, 136, 1457-1468.

Zeven AC, Dehmer KJ, Gladis T, Hammer K & Lux H (1998). Are the duplicates of perennial kale (*Brassica oleracea* L. var. *ramosa* DC.) true duplicates as determined by RAPD analysis?. *Genetic Resources and Crop Evolution*, 45: 105-111.

Zhao, X. and Kochert, G. (1993) *Plant Mol. Biol.* 21: 607-614.

Zheng, K. L., Subudi, P. K, Domingo, J, Magpantay, G. and Huang, N. 1995. Rapid DNA isolation for marker-assisted selection in rice breeding. *Rice Genetics Newsletter*. 12 : 14.

Zietkiewicz E, Rafalski A & Labuda D (1994). Genome fingerprinting by simple sequence repeat (SSR)-anchored polymerase chain reaction amplification. *Genomics*, 20: 176-183.

Zietkiewicz, E; Rafalski, A. and Labuda, D. (1994) *Genomics* 20;176.

Index

D

E

F

G

H

I

K

L

M

N

P

Q

R